10th International LISA Symposium 2014

Journal of Physics: Conference Series
Volume 610

Gainesville, Florida, USA
18 - 23 May 2014

ISBN: 978-1-5108-0795-2
ISSN: 1742-6588

Printed from e-media with permission by:

Curran Associates, Inc.
57 Morehouse Lane
Red Hook, NY 12571

Some format issues inherent in the e-media version may also appear in this print version.

Printed by Curran Associates, Inc. (2015)

For permission requests, please contact the Institute of Physics
at the address below.

Institute of Physics
Dirac House, Temple Back
Bristol BS1 6BE UK

Phone: 44 1 17 929 7481
Fax: 44 1 17 920 0979

techtracking@iop.org

Additional copies of this publication are available from:

Curran Associates, Inc.
57 Morehouse Lane
Red Hook, NY 12571 USA
Phone: 845-758-0400
Fax: 845-758-2634
Email: curran@proceedings.com
Web: www.proceedings.com

Table of contents

Volume 610

10th International LISA Symposium
18–23 May 2014, Florida, USA

Accepted papers received: 2 April 2015
Published online: 11 May 2015

Preface

Papers

LISA Science Overview

LTP Overview

Other space experiments (overview)

Ground based (overview)

Ground based (detail)

Technologies for LISA-like missions

Sources

Journal of Physics: Conference Series **610** (2015) 011001 doi:10.1088/1742-6596/610/1/011001

Proceedings of the 10th International LISA Symposium
University of Florida, FL, USA
May 18-23, 2014

Preface:
The LISA Symposia have become a mainstay of the gravitational wave community. Held every two years, they are the prime opportunity for our community to discuss the exciting science, technology, mission designs, and progress of the Laser Interferometer Space Antenna. The 8th LISA symposium, held at Stanford University in the summer of 2010 was the largest symposium so far and was dominated by progress and hopes that the LISA mission will soon excel following the expected launch of the LISA pathfinder (LPF), no later than 2012, and the expected prioritization by the Decadal survey which was released 6 weeks later. The following years were challenging. Although the Decadal survey ranked LISA very high, NASA's budget issues, mostly due to the cost increase of the James Webb Space Telescope, and continued delays in LPF put too much stress on the LISA project and it officially ended in 2011. The LISA International Science Team (LIST), the core group of LISA scientists and technologists, was dissolved and the community in the U.S. was struggling to maintain cohesion.

In the wake of these events, ESA started a new selection process for their next three large missions, L1, L2, and L3, and the European LISA team developed the New Gravitational wave Observatory (NGO), an evolved LISA concept, as an ESA only L1 candidate. A few weeks before the 9th LISA Symposium, held in Paris in May 2012, ESA announced its decision to select JUICE, a planetary mission to Jupiter and its moons, as its next large science mission (L1). Despite having the highest ranked science case, NGO was not selected due to further delays in LPF and the general feeling outside the GW community that the technology is perhaps too challenging to be pulled off in time for the L1 launch in 2022. Many U.S. members of the LISA community cancelled their travel plans and the mood at that symposium ranged from resignation to defiance. Hope for a somewhat timely launch of a LISA-like mission rested upon L2, the next large mission in Europe, and a potential comprehensive technology development program followed by a number one selection in the 2020 Decadal Survey in the U.S.

The selection of L2 was combined with the selection of L3 and the newly formed eLISA consortium submitted an updated NGO concept under the name eLISA, or Evolved LISA, to the competition. It was widely believed that the launch date of 2028 for L2, would be seen by the selection committee as providing sufficient time to retire any remaining technological risks for LISA. However, the committee selected the 'Hot and Energetic Universe', an X-ray mission, as the science theme for L2 and the 'Gravitational Universe', the eLISA science theme, for L3. Although very disappointed, it was not a surprising decision. LPF did experience further delays just prior to and during the selection process, which may have influenced the decision.

The strong technology program in the U.S. never materialized because WFIRST, the highest priority large mission in the 2010 Decadal following JWST, not only moved ahead but was also up-scoped significantly. The L3 selection, the WFIRST schedule, and the missing comprehensive technology development in the U.S. will make a launch of a GW mission in the 2020s very difficult. Although many in the LISA community, including ourselves, did not want to accept this harsh reality, this was the situation just prior to the 10th LISA symposium. However, despite all of this, the LISA team is now hopeful!

In May of 2014 the LISA community gathered at the University of Florida in Gainesville to discuss progress in both the science and technology of LISA. The most notable plenary and contributed sessions included updates on the progress of LISA Pathfinder, which remains on track for launch in the second half of 2015(!), the science of LISA which ranges from super-massive black hole mergers and cosmology to the study of compact binaries within our own galaxy, and updates from other programs that share some of LISA's science or technology. Plenary talks from the pulsar timing and ground-based laser interferometer groups told of the reasonable expectation of gravitational wave detection within the next 4 to 8 years. We also heard about the GRACE Follow-on mission, which will demonstrate a precision laser ranging system in space in 2017, using technology that is somewhat similar to that of LISA. Presentations on the Large Synoptic Survey Telescope, Athena, the Cherenkov Telescope Array, and WFIRST provided data on the landscape in which LISA will live in the 2030s.

Beyond the 10th symposium there is much to look forward to. There is high-expectation that LISA Pathfinder will launch in 2015, prior to the 11th symposium in Zürich, which, for the first time, will be dedicated to the results of LPF and not its preparation. Ground-based gravitational wave observatories, especially Advanced LIGO, are rapidly approaching their required sensitivities and could make the first direct detection before the 12th LISA symposium. Advanced VIRGO and KAGRA, and the pulsar timing community are also hopeful that they will reach the required sensitivity within this decade or shortly thereafter. These events will dramatically improve the perception of gravitational wave science by the broader astronomy and astrophysics communities. The U.S. LISA team is also embolden by the announcement that NASA is now planning to join ESA in the gravitational wave L3 mission as a junior partner and will begin funding a technology development program to support this partnership. A space-based gravitational wave mission is inevitable.

At the time of the 10th LISA Symposium, it was not clear if the gradient of LISA's trajectory was perceived as positive or negative. But in hindsight, 2014 will hopefully be seen as a time when LISA regained some of the ground recently lost and began accelerating towards launch.

Acknowledgements:

No conference of the size of a LISA Symposium is organized by a single person. It always requires many helping hands, thinking heads, and the always welcome and badly needed financial support. We gratefully acknowledge the financial support from the Alachua County Tourist Development Council; although the associated paperwork looked initially like an insurmountable obstacle, it turned out to be fairly straightforward and simple thanks to the help of Nancy Fisher and John Pricher in the county office. We also appreciate the generous support from the Florida Space Grant Consortium. There probably has never been a LISA Symposium which did not receive financial support from the Albert Einstein Institute in Germany or a large gravitational wave meeting in the US without support from the LIGO Lab. We highly appreciate and acknowledge their support. The entire University of Florida shared the honor (and some of the financial burden) to host the symposium; financial support was provided by the Institute of High Energy Physics and Astrophysics (IHEPA), the Departments of Physics and of Mechanical and Aerospace Engineering, the College of Liberal Arts and Sciences, and the Office of Sponsored Research. Still, the cost per participant including travel and hotel is always high and we appreciate the support all participants received from their respective funding agencies and the tax payers who continue to invest money into our dreams.

We also like to thank the international organizing committee who provided us with invaluable advice during the organization of the scientific program of the symposium. Our local organizing committee, Steve Eikenberry, David Tanner, Cliff Will and Bernard Whiting, helped in more ways we can list here and this includes the food and wine selection during the banquet. Representative of all graduate students, we like to thank in particular Anh Nguyen, Taiwo Olatunde, Andrew Chilton and Ryan Shelley for their help. We know that many other students provided much needed additional help but we can't list you all; in some cases a student from another group might even have jumped in when needed and we don't even know. We are very grateful to all of you. You are the future.

Last but not least, this would have been impossible without the help of the two VIPs in the Department of Physics: Janet Germany and Kristin Nichola. There is no way we would have taken this on without you.

Giacomo Ciani, John W. Conklin, Guido Mueller

Journal of Physics: Conference Series **610** (2015) 011002 doi:10.1088/1742-6596/610/1/011002

Peer review statement

All papers published in this volume of *Journal of Physics: Conference Series* have been peer reviewed through processes administered by the proceedings Editors. Reviews were conducted by expert referees to the professional and scientific standards expected of a proceedings journal published by IOP Publishing.

Journal of Physics: Conference Series **610** (2015) 012001 doi:10.1088/1742-6596/610/1/012001

Massive Black Hole Science with eLISA

Enrico Barausse[1,2], **Jillian Bellovary**[3,4], **Emanuele Berti**[5], **Kelly Holley-Bockelmann**[3,4], **Brian Farris**[6,7], **Bangalore Sathyaprakash**[8], **Alberto Sesana**[9]

[1]CNRS, UMR 7095, Institut d'Astrophysique de Paris, 98bis Bd Arago, 75014 Paris, France
[2]Sorbonne Universités, UPMC Univ Paris 06, UMR 7095, 98bis Bd Arago, 75014 Paris, France
[3]Department of Physics and Astronomy, Vanderbilt University, Nashville, TN 37235, USA
[4]Department of Physics, Fisk University, Nashville, TN 37208, USA
[5]Department of Physics and Astronomy, The University of Mississippi, University, MS 38677, USA
[6]Center for Cosmology and Particle Physics, Physics Department, New York University, New York, NY 10003, USA
[7]Department of Astronomy, Columbia University, 550 West 120th Street, New York, NY 10027, USA
[8]School of Physics and Astronomy, Cardiff University, 5, The Parade, Cardiff, CF24 3AA, United Kingdom[1]
[9]Max-Planck-Institut für Gravitationsphysik, Albert Einstein Institut, D-14476, Golm, Germany

Abstract. The evolving Laser Interferometer Space Antenna (eLISA) will revolutionize our understanding of the formation and evolution of massive black holes (MBHs) along cosmic history, by probing massive black hole binaries (MBHBs) in the $10^3 - 10^7 M_\odot$ range out to redshift $z \gtrsim 10$. High signal-to-noise ratio detections of $\sim 10 - 100$ MBHB coalescences per year will allow accurate measurements of the parameters of individual MBHBs (such as their masses, spins and luminosity distance), and a deep understanding of the underlying cosmic MBH parent population. This wealth of unprecedented information can lead to breakthroughs in many areas of physics, including astrophysics, cosmology and fundamental physics. We review the current status of the field, recent progress and future challenges.

1. Introduction

The evolving Laser Interferometer Space Antenna (eLISA, [1]) is designed to be sensitive to gravitational waves (GWs) at mHz frequencies. One of the strongest sources in this frequency window are MBHBs merging throughout the Universe [2]. According to our current understanding of structure formation in a ΛCDM Universe, MBHBs frequently form along cosmic history following galaxy mergers. MBHs we see in today's galaxies are expected to be the natural end-product of a complex evolutionary path, in which black holes (BHs) seeded in proto-galaxies at high redshift grow through cosmic history via a sequence of MBHB mergers and accretion episodes [3, 4]. However, our current observational knowledge of the MBH population is limited to a small fraction of these objects: either those that are active (see e.g. [5]), or those in our neighborhood, where stellar- and gas-dynamical measurements are possible (see [6] for a review).

[1] Currently on sabbatical leave at LIGO Laboratory, California Institute of Technology, MS 100-36, Pasadena, CA 91125

eLISA will revolutionize this picture by probing MBHBs in the $10^3 - 10^7 M_\odot$ range out to redshift $z \gtrsim 10$ [2]. In the current design [1], eLISA will be capable of detecting $\sim 10 - 100$ MBHB coalescences per year and of accurately measuring the parameters of individual MBHBs, such as their masses, spins and luminosity distance. This wealth of unprecedented measurements has the potential to revolutionize many areas of physics, ranging from astrophysics to cosmology and fundamental physics. This paper summarizes contributions to the "LISA Symposium X" that were devoted to this subject.

We start in Section 2 by describing current models for MBH formation and evolution, highlighting present uncertainties due either to the lack of observations or to poor theoretical constraints. eLISA will dramatically change this situation by probing the very first coalescences of seed BHs at high redshift. The interaction of MBHs with the environment has a key role in bringing MBHBs close enough that GW emission becomes efficient. Section 3 describes recent developments on this front, focusing on observational signatures of merging MBHBs in the electromagnetic (EM) domain. Coincident detection of MBHB mergers as both GW *and* EM sources will pave the way to multimessenger astronomy, promising extraordinary advances in the understanding of accretion physics. In Section 4 we show how eLISA will open a new era of precision measurements of MBH spins. Individual spin measurements will allow us to test gravity in the strong-field regime with unprecedented accuracy; the *collective* properties of spin and mass distributions of the whole MBH population carry information on the physics of accretion flows, and in general of the intimate link between MBHs and galaxy evolution across cosmic time. Section 5 explores eLISA's potential for cosmology. GW observations can yield a direct measurement of the luminosity distance to the source. If an EM counterpart is detected, the coincident measurement of the source redshift will provide an appealing opportunity for calibration-free cosmography. Last but not least, Section 6 touches on the invaluable insights that eLISA will provide in terms of fundamental physics. Dynamical measurements of the behavior of gravity in the strong-field regime will constrain the geometry of the spacetime around BHs, telling us whether the Kerr solution actually describes these objects and potentially even yielding smoking guns of new physics beyond General Relativity (GR).

2. Birth and growth of massive black holes

There is mounting observational evidence that supermassive BHs with masses between 10^6 and $10^{10} M_\odot$ reside at the heart of nearly every galaxy [7, 8, 9]. Though MBHs are an observational certainty, nearly every aspect of their evolution – from their birth, to their fuel source, to their basic dynamics – is a matter of lively debate.

The existence of bright quasars at high redshift implies that billion solar mass BHs must be in place within their host galaxies less than a billion years after the Big Bang. This remarkable observational fact suggests that the seeds of these most massive MBHs are sown during, or even before, the formation of protogalaxies. However, the precise MBH seed formation mechanism is not known, nor is it clear that there is only one seed formation channel at play over the entire MBH mass spectrum.

Currently, the two most widely accepted MBH birth scenarios are a) the remnants of the first generation of stars (Population III: see e.g. [10, 11, 12, 13, 14]), and b) the direct collapse of pristine gas within massive dark matter halos [15, 16, 17, 18, 19, 20]. In both these models, overdensities within a gas cloud collapse as the gas radiates away energy; if the gas can cool efficiently, then it gravitationally fragments into ever smaller clouds. The mass of the final collapsed object, be it a protostellar cloud or a seed black hole, is determined by the Jeans mass of the gas that can no longer cool efficiently enough to fragment further.

In the Population III seed scenario, metal-free gas is cooled by H_2 and HD, but since these molecules are much less effective than metals in radiating away energy, the Jeans mass is likely to be larger than that of stars formed in the present era. However, one of the difficulties in this scenario is that the actual initial mass function of this first generation of stars is hotly debated, and the issue will not easily resolve itself with the current state-of-the-art simulations. Early work placed the stellar mass at $100 - 1000\,M_{\odot}$ [21, 22], but later work suggested that fragmentation is far more common [23, 24, 25, 26, 27]. A recent study examining the role of turbulence in Population III star formation suggested a flat initial mass function with a characteristic mass less than $100\,M_{\odot}$ (e.g. [28]). The newest work tends to indicate that, even though the protostellar clouds do fragment, they tend to quickly merge within a dynamical time, essentially generating the very massive Population III stars once again [29, 30]. While these theoretical advances are promising, a more thorough treatment involving radiative transfer and 3D magnetohydrodynamics is required before we will truly understand the nature of the first stars and whether they can spawn MBH seeds.

In the fiducial direct collapse model, the only coolant is atomic Hydrogen, and this requires that the gas sit in dark matter halos with virial temperatures higher than 10^4 K. To inhibit fragmentation and to suppress H_2 formation, there is an additional photon bath that heats the gas; one of the most likely candidates for this radiation field, the Lyman-Werner background, is thought to be generated by nearby Population III stars [31, 32, 33] at redshifts $z \sim 10 - 20$. Preventing efficient cooling and fragmentation in this way results in a huge Jeans mass, and the eventual seed BHs are of order $10^4 - 10^6\,M_{\odot}$. These direct collapse BHs are the most obvious candidates to grow into the MBHs powering high-redshift quasars, because they require no stringent assumptions on the accretion or merger history to grow to $10^9\,M_{\odot}$ by $z \sim 6$.

As with the Population III scenario, the main epoch of direct collapse seed formation is brief, ending as metals pollute the halo gas and the radiation field drops below a critical heating threshold as the Universe expands [34, 35]. Still, pockets of pristine and irradiated gas could remain at low redshift that could collapse into MBH seeds [36, 37]. One potential issue with this direct collapse model is that the intensity of radiation needed to prevent fragmentation is under debate, and the actual Lyman-Warner background generated at redshift $z \sim 20$ is not well-constrained, as it depends in part on the occupation fraction and initial mass function of Population III stars. Therefore, direct collapse BH formation may arguably be too rare to account for the quasar population, much less the entire MBH mass spectrum (see however [38]).

One of the great promises of a future space-based GW observatory like eLISA is that it may be able to pin down the relative efficiency of light versus MBH seed formation channels [39, 40, 41, 42]. These pathways affect the occupation fraction of MBHs in local dwarf galaxies, the existence of intermediate-mass BHs, and the scaling relations between MBHs and their host galaxies (i.e. the M–σ relation) at low masses. Studies have attempted to predict MBH-galaxy occupation fractions [43, 37] and the low-mass M–σ relation [44] for a range of seed models, but observations in this regime are very difficult, and thus constraints are difficult to obtain. Direct GW detections will elucidate much regarding the hidden MBH population and their seeding mechanism.

Turning now to MBH growth, it is tempting to think that a clearer picture emerges. The famous Soltan argument has often been invoked to claim that MBHs are fueled nearly entirely by gas during the brief quasar epoch. The logic goes like this. If we assume that quasars are powered by gas accretion onto MBHs, then we can turn the observed energy and number density of optically-bright quasars into an estimate of the gas mass accreted by MBHs during the quasar era. Happily, if we compare the mass density of accreted gas during the quasar phase (assuming a radiative efficiency of 10%) to the mass density locked up in the local MBH population, the numbers agree. This implies that 90% of MBH masses are built from gas that is accreted at the Eddington limit before redshift 2; it also implies that MBHs do not seem to accrete in a low-efficiency, or "quiet" mode, nor are there many obscured or undetected quasars within the MBH mass budget. Though this argument seems pat and iron-clad, there are many uncertainties involved. For example, converting the optically observed quasar luminosity function to a bolometric energy density is fraught with difficulty. Furthermore, it is very clear that optical quasar surveys do miss a large fraction of the real quasar population, those that are radio-loud or X-ray bright (see e.g. [45, 46]). In addition, many theoretical efforts call into question the assumption of Eddington-limited accretion, some advocating super-Eddington accretion along filaments to feed ultramassive BHs [47], and others promoting a "radio" mode of quiescent gas accretion [48, 49]. Along this line, simulations indicate that the high-redshift universe may be rife with hidden MBHs that are accreting at less than the Eddington rate [50, 51].

Despite these uncertainties, it is widely accepted that gas is the primary fuel for MBH growth. It is thought that galaxy mergers help to restock the gas reservoir around the BH as the gas is shocked and then falls toward the galactic center. Indeed one of the best explanations of the M–σ relation invokes galaxy mergers that drive gas toward the BH, which fuels it and subsequently generates a prodigious radiative "feedback" that pushes the remaining gas toward the galaxy outskirts and cuts off the MBH fuel source (see e.g. [52]). Once the gas is in the galaxy bulge potential, it slowly cools and forms new stars with high velocity dispersion [53, 54]. Though early simulations of equal-mass galaxy mergers were very promising in supporting this view, later unequal-mass mergers fail to grow the primary MBHs enough to fall on the M-σ relation [55, 51]. The problem is that most low-mass MBHs, like the one in our Milky Way, simply don't undergo equal-mass mergers at the rate needed to build the MBH and to couple this growth to the bulge in this framework.

Fear not: mergers are not the only process which may trigger MBH growth. In fact, galaxies which lack a bulge (often thought to have extremely quiescent merger histories) may host MBHs as well, and even actively growing ones. One thought is that MBHs may grow efficiently through the accretion of cold, unshocked gas (i.e. "cold flows") which enter the galaxy through filamentary accretion. Some galaxies acquire most of their gas through this process [56, 57], and MBHs may efficiently accrete this gas as well [58, 50]. Another method which may be prevalent consists of mergers of MBHs with other MBHs [59]. In the simulations presented in [50], the central MBHs in massive galaxies at $z = 4$ have built up over half of their mass through MBHB mergers. This is a regime where GW observations can verify the accuracy of these predictions, which depend on many properties of MBH seed formation (initial mass, formation redshift, and efficiency of formation). By directly collecting GW data from the growth of high-redshift MBHs, we can determine if this phenomenon is dominated by MBH mergers, or whether it is correlated with galaxy merger rates, or some combination of the two.

Overall, it is likely that there are several mechanisms to grow a MBH, depending on the assembly history of the host galaxy. The most massive MBHs are likely fueled via gas accretion from major mergers at high redshift [60], which explains the relative tightness of the M–σ relation in this regime. Lower-mass galaxies, which experience fewer major mergers, exhibit more scatter within the local scaling relations, likely because their MBHs grow through more stochastic processes such as cold flow accretion, or through secular processes like bar-driven gas inflow. The combination of these processes plus the unknown contribution of MBHB mergers leads to a myriad of possibilities for MBH growth, which will be possible to disentangle by means of low-frequency GW observations.

3. Gas accretion onto massive black hole binaries and their electromagnetic signature

According to the evolution picture emerging from Section 2, MBHB coalescences must be common events at all redshifts. These systems do not live in isolation, but they are embedded in dense galactic nuclei, surrounded by gas and stars. Therefore, the interaction of MBHBs with their environment may provide a unique opportunity to observe EM signatures as well as GWs, opening new avenues in multimessenger astronomy. Information from a simultaneous detection of EM and GWs may be useful for studying fundamental aspects of gravitational physics. For

Journal of Physics: Conference Series **610** (2015) 012001 doi:10.1088/1742-6596/610/1/012001

example, in some modified gravity scenarios, the propagation velocity for gravitons may differ from that of photons [61, 62]. Additionally, the measurement of the luminosity distance from the GW signal at an accuracy of $1 - 10\%$, coupled with the redshift information from the EM detection, could serve as a cosmological "standard siren" of unprecedented accuracy (better than $\sim 1\%$) [63]. Such detections may also combine accurate measurements of MBH spins and masses obtained from GW signals with EM observations to probe MBH accretion physics in great detail [64]. Since most eLISA sources will be relatively low-mass systems at high redshift, where gas-rich environments are common, we focus here on the interaction between a MBH and a putative massive circumbinary disk.

The standard picture of circumbinary accretion disks can be described as follows. Tidal torques from the binary tend to drive gas outward, clearing an evacuated cavity in the innermost region of the disk. Meanwhile, viscous torques transport angular momentum outward in the disk, allowing gas to flow inward and refill this cavity. The balance of tidal and viscous torques determines the location of the inner edge of the circumbinary disk at $r \approx 2a$, where a is the binary separation. This balance can be maintained, provided the timescale $t_{\rm gw}$ for inspiral of the binary due to GW emission is much longer than the viscous timescale of the disk, $t_{\rm visc}$. This is known as the "pre-decoupling" epoch.

To date, a number of "dual" systems in which two MBHs occupy the same galaxy but are too widely separated to be gravitationally bound have been observed [65, 66, 67, 68], as well as several candidate binary systems (see e.g. [69, 70] and references therein). Proposed EM signatures of such binaries include spatially resolving two AGN-like point sources, identifying double-peaked broad emission lines, spatial structures in radio jets, characteristic time variability in quasar emission, and characteristic features in quasar spectra.

Theoretical aspects of this problem have been studied analytically [71, 72, 73, 74, 75, 76, 77, 78, 79], often using approximate angle-averaged tidal torque formulae. While these techniques have proven very useful in highlighting qualitative features of the accretion, they tend to overestimate the barrier to accretion imposed by binary torques by imposing symmetry in the accretion flow. As EM counterparts to MBH mergers depend sensitively on the amount of gas available for accretion, it is important to understand non-axisymmetric effects which may provide a mechanism for delivering more gas to the MBHs. Indeed, 2D and 3D simulations have demonstrated that gas streams may be stripped from the inner cavity, accreting directly onto the binary. Prior numerical simulations have been performed in 2D [80, 81] and in 3D [82, 83, 84, 85, 86, 87, 88]. While 3D codes have been useful in probing the gas dynamics during the final orbits prior to MBH merger, long viscous timescales render them prohibitively costly for simulating the quasi-steady-state flow during the "pre-decoupling" epoch. For this epoch, 2D simulations such as those performed using the *DISCO* code [89] have proven quite useful. These simulations include the inner cavity in the computational domain, and use shock-capturing Godunov-type methods to evolve thin ($h/r \sim 0.03$) disks over the viscous timescales necessary to accurately capture the steady-state accretion at high resolution.

The *DISCO* code is a moving-mesh code which allows one the freedom to specify the motion of the computational cells [90]. For binary accretion, the chosen rotation profile matches the nearly Keplerian fluid motion outside the cavity, while transitioning to uniform rotation at the binary orbital frequency inside the cavity. Because grid cells move azimuthally with the fluid, advection errors are minimized, allowing for the accurate capture of the dynamics of accretion streams which penetrate into the cavity.

The quasi-steady-state solutions which the simulations relax to after several viscous timescales can be interpreted as generalizations of the Shakura-Sunyaev disk solutions [91], with the central gravitating object replaced by a binary. The initial disk configurations consist of the "middle region" Shakura-Sunyaev solution for a steady-state, geometrically thin, optically thick accretion disk, assuming a gas-pressure dominated fluid with electron scattering as the dominant opacity.

The fluid evolves according the the 2D viscous Navier-Stokes equations, assuming an α-law viscosity prescription. A Γ-law equation of state of the form $P = (\Gamma - 1)\epsilon$ is chosen, where ϵ is the internal energy density, and the adiabatic index is set to $\Gamma = 5/3$. Appropriate radiative cooling and viscous heating terms are added to the energy equation. The cooling rate for an optically thick, geometrically thin disk is $q_{cool} = 4\sigma/3\tau T^4$, where T is the mid-plane temperature, and τ is the optical depth for electron scattering ($\tau = \Sigma\sigma_T/m_p$). It is assumed that the fluid is gas-pressure dominated everywhere, ignoring radiation pressure. In each simulation the binary is chosen to have zero eccentricity.

The important findings of these simulations are the following:

(i) As shown in Fig. 1, gas enters the circumbinary cavity along accretion streams, and the interaction of these streams with the cavity wall causes the cavity to become lopsided (in agreement with e.g. [80, 83, 88, 87, 86, 92]). See [92] and [87] for a description of the mechanism driving the growth of this lopsidedness.

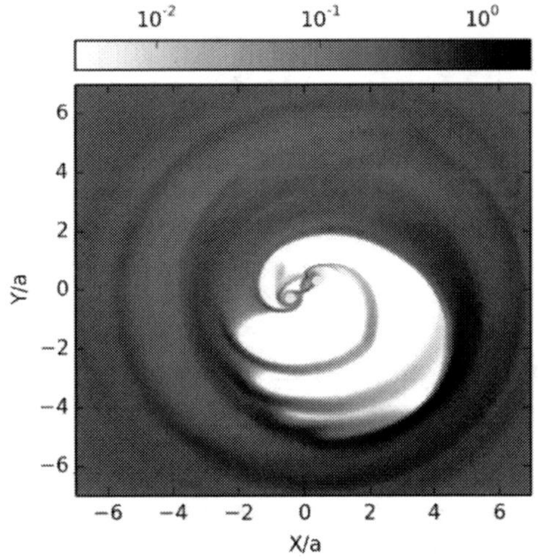

Figure 1. Snapshot of surface density Σ during quasi-steady state after $t \gtrsim t_{vis}$. Surface density is normalized by the maximum value at $t = 0$ and plotted on a logarithmic scale in the inner $\pm 6a$. Orbital motion is in the counter-clockwise direction.

(ii) These simulations support the growing consensus that the accretion rate onto the MBHs is not significantly reduced by the presence of a binary, when compared to the accretion rate onto a single MBH of the same mass [81, 87, 86, 88]. This is the case in spite of the fact that much of the inner cavity is cleared of gas by the action of the binary torques, and it is due to the effectiveness of the narrow accretion stream in delivering gas from the circumbinary disk inner edge to the individual black holes.

(iii) For each mass ratio considered, "mini-disks" surrounding each MBH are formed. In each case, the mini-disks are persistent, as their accretion timescale greatly exceeds the binary orbital timescale. For the binary mass ratios $q = 0.11$ and $q = 0.43$, the size of these mini-disks is in rough agreement with the semi-analytic predictions of [72].

(iv) Significant periodicity in the accretion rates emerges for $q \gtrsim 0.1$. At these mass ratios, the binary torques are strong enough to excite eccentricity in the inner cavity and create an overdense lump, whose interaction with the passing MBHs leads to periodicity in the accretion rate. The strongest peak in the periodograms for these cases corresponds to the orbital frequency of the lump, with many associated harmonics for the $q \gtrsim 0.43$ cases. This periodicity may constitute a unique observational signature of MBHBs.

(v) For each case considered, the accretion rate onto the secondary is sufficiently large relative to that of the primary, so that the mass ratio q is increasing. Similar results have been found previously in SPH calculations [82, 83, 93, 88]. As MBHs are expected to gain a significant fraction of their mass through gas accretion [94, 95, 96], this suggests a mechanism which may bias the distribution of binaries near merger toward higher mass ratios.

(vi) The emission associated with the shock heating in the accretion streams is sufficient to bring the emission from within the cavity above that of a disk around a single MBH. The peak in dL/dr which appears at $r/a \approx 5$ corresponds to the shock heating of gas in the stream which is not directly accreted, but rather impacts the cavity wall as seen along the lower-right edge of the cavity in Fig. 1, leading to the thin strip of bright emission. Scaled to a $10^8 M_\odot$ binary with a separation near decoupling at $a/M = 100$, this enhancement is significant in soft and hard X-rays (see Fig. 2). This X-ray enhancement has been predicted by [97], who estimated the characteristic frequency of mini-disk "hot spot" emission by estimating the amount of energy released due to shock heating when accretion streams impact the minidisks. Instruments sensitive to X-ray emission from AGN such as XMM-Newton[2], NuSTAR[3], the upcoming eROSITA[4] all-sky survey, and the proposed ATHENA[5] X-ray observatory may be sensitive to these signatures.

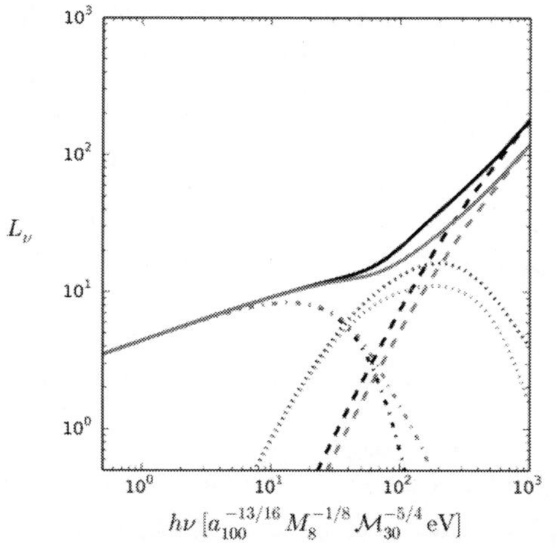

Figure 2. Thermal spectra computed from simulation snapshot at $t \gtrsim t_{vis}$. Red curves are calculated from disk data $\approx 3t_{bin}$ after that of black curves. The full spectrum is represented by solid lines, the component arising only from minidisk regions within distance $d < 0.5a$ of either MBH is represented by dashed lines, the "cavity" emission is represented by dotted lines, and the emission from the "outer region" is represented by dashed-dotted lines. We have introduced the scaling parameters $a_{100} \equiv a/100M$, $M_8 \equiv M/10^8 M_\odot$, and $\mathcal{M}_{30} \equiv \mathcal{M}_a/30$.

While there remain no confirmed observations of EM counterparts for MBHBs, simulations have already provided tantalizing evidence that significant gas can accrete onto such a binary, even at separations in the eLISA band. A number of distinguishing observational signatures of such systems have been proposed, including directly resolving two AGN-like point sources, searching for offset emission lines, spatial structures in radio jets, characteristic time variability in quasar emission, and characteristic features in quasar spectra. As we enter the age of GW astronomy the possibility of a simultaneous measurement of both gravitational and EM radiation from a merging MBHB appears increasingly likely.

[2] http://sci.esa.int/xmm-newton/
[3] http://www.nustar.caltech.edu/
[4] http://www.mpe.mpg.de/erosita/
[5] http://www.the-athena-x-ray-observatory.eu/

4. Massive black hole spins as gravitational and cosmological probes

In GR, MBHs are described by the Kerr metric and are completely characterized by three "charges" (or "hairs"), the mass M, the spin S and an electric charge Q, with the spin satisfying the Kerr bound $a = cS/(GM^2) \leq 1$ (if the spin is larger than this bound, the Kerr metric does not describe a BH but a naked singularity). Also, the charge Q is usually expected to be negligible for astrophysical BHs, due to the presence of plasma that quickly neutralizes it, and also because of quantum effects such as Schwinger pair production or vacuum breakdown mechanisms producing cascades of electron-positron pairs near the BH horizon. BHs in gravity theories different from GR are still characterized by the mass and spin, but may also present additional charges that might provide a way to test gravity in strong-field regimes (cf. Section 6 for a related discussion).

The effect of the mass in GR is qualitatively similar to Newtonian gravity, i.e. it sources an attractive force decaying as GM/r^2 for $r \gg GM/c^2$ (r being the distance from the BH). Because of the long-range character of the force it sources, the mass can be estimated rather easily with EM observations, e.g. (in the case of MBHs) by observations of the nuclear dynamics of stars and gas, fitting of the spectral energy distribution of galaxies, spectroscopic single epoch measurements and reverberation mapping. EM measurements of MBH masses are typically accurate only within $20 - 50\%$ (e.g. even for our own SgrA*, the mass can only be estimated to within $\sim 10\%$), but still allowed discovering correlations with galactic properties, hinting at a symbiotic co-evolution of MBHs with their galactic hosts [98, 99, 100, 101, 102].

Measuring BH spins is more complicated, because they do not enter the dynamics at Newtonian order, but only at 1.5 post-Newtonian order, i.e. they change the Newtonian equations of motion only by corrections $\sim \mathcal{O}(v/c)^3$, where v is the characteristic velocity of the gas and/or stars surrounding the MBH. This results in significant effects only very close to the event horizon. The most promising EM technique to estimate the spins of MBHs is through the spectra of relativistically broadened Kα iron lines (cf. [103, 104] for recent reviews on this topic). Indeed, X-ray observatories such as XMM-Newton and Suzaku have by now measured significant samples of spins, and more constraints are becoming available from NuSTAR's hard X-ray data [105, 106, 107]. The masses and spins of MBHs in binary systems will be measured very accurately by eLISA, with errors $\delta M \lesssim 0.1\%$ and $\delta a \lesssim 0.01$ [108, 2]. These GW measurements are very clean compared to EM ones, because they are not affected by the systematics usually present in EM data (which are due to poor understanding of the dissipative gas physics). This is because GWs are emitted in the latest stages of a binary system's evolution, when the effect of the gas on the orbital dynamics is typically negligible [109, 110].

More specifically, eLISA will detect GWs from MBHBs both during the long post-Newtonian inspiral, which radiates in the eLISA band for separations less than about 200 gravitational radii for binaries with total mass $\sim 10^5 - 10^6 M_\odot$ (i.e. those to which eLISA is most sensitive), and during the final plunge, merger and ringdown of the system. This will allow testing a plethora of strong-field general-relativistic effects, namely:

(i) The spin-orbit coupling (also known as "frame dragging") [111, 112], which causes the BH spins to precess during the inspiral, thus producing amplitude modulations in the emitted gravitational waveforms [112, 113, 114, 115, 116], which increase the signal-to-noise ratio and improve the estimation of the source parameters (and in particular the sky localization). The precessional dynamics can be very rich: for example, to a first approximation (and on short timescales) the spins precess around the binary's total angular momentum, but in certain configurations they can undergo a more complicated "transitional precession" [112] (whereby spins almost anti-aligned with the orbital angular momentum change their orientation on a very short timescale at some point during the inspiral), or get locked in secularly stable resonant configurations that tend to align or anti-align the spins [117, 118, 119, 120];

(ii) The spin-spin coupling [121, 122], which appears at higher post-Newtonian order than the spin-orbit coupling and thus modifies the spin precession in the later inspiral, causing additional modulations in the gravitational waveforms and improving parameter estimation;

(iii) The non-linear relations between the final mass, spin and recoil velocity of the BH remnant forming from the merger, and the masses and spin vectors of the binary's components at large separations. These relations have been studied in detail with fully general-relativistic numerical simulations (see e.g. [123, 124, 125] for reviews on this topic), whose results can be extrapolated to generic binary configurations by exploiting knowledge of the test-particle, self-force and post-Newtonian dynamics [126, 127, 128, 129, 130, 131, 132, 133];

(iv) The quasi-normal mode ringing of the BH remnant (see [134] for a review), whose frequencies and decay times are functions of the remnant's mass and spin alone if GR is correct. Measuring these frequencies therefore constitutes a genuine strong-field test of the gravity theory (cf. Section 6).

Besides testing the strong-field general-relativistic dynamics through these effects, eLISA measurements of the spins will also provide useful information on the cosmological evolution of MBHs. In fact, although the impact of gas on the dynamics of MBHBs is negligible once they enter the eLISA band [109, 110], radiatively efficient accretion of gas is known to be the main driver of the evolution of the masses and spins on cosmological timescales [94, 135, 136, 137]. Also, because eLISA will detect GWs from MBHBs at redshift $z \gtrsim 10$, mass and spin measurements will give precious information about the high-redshift formation of the first generation of BH seeds, from which present-day MBHs are believed to descend.

Early studies [41] in this direction highlighted for instance that eLISA mass measurements will be able to discriminate with high confidence between a scenario in which MBHs evolve from "heavy" ($\sim 10^4 - 10^5 M_\odot$) seeds forming at $z = 10 - 15$ from the collapse of protogalactic disks, and a "light-seed" scenario in which the seeds form at $z \sim 20$ from the collapse of Population III stars into BHs with masses $\sim 50 - 300 M_\odot$, shedding light on the nature of the first seed BHs forming in the young universe (cf. Section 2). As for the spin evolution, [138] (see also [139, 140, 141, 142]) showed that eLISA should be able to tell a coherent accretion scenario (where gas accretion always happens on prograde orbits on the equatorial plane) from a chaotic scenario (where the MBH captures clouds that are isotropically distributed around it with randomly oriented angular momenta). Clearly, the first scenario predicts almost extremal spins (because accretion is always prograde), while the second predicts small spins $a \lesssim 0.3$, because the angular momentum transferred by the clouds tends to cancel out on long timescales.

Two ingredients were missing from these early attempts, namely *(a)* the strong-field spin-orbit coupling reviewed above, and *(b)* a more realistic connection between the properties of the accretion flow and those of the host galaxy. Regarding *(a)*, [143, 144, 145, 146] showed that the interaction between the spin-orbit coupling and the viscous stresses active inside an off-equatorial, geometrically thin accretion disk results in a quick alignment between the MBH spin $\boldsymbol{S}$ and the disk's orbital angular momentum $\boldsymbol{L}$ (a phenomenon known as "Bardeen-Petterson effect"), provided that $|\boldsymbol{L}| > 2|\boldsymbol{S}|$. Because this alignment takes place on a timescale much shorter than the accretion timescale, accretion will be essentially coherent if a MBH is hosted in a gas-rich galactic nucleus (where it is more likely that the condition $|\boldsymbol{L}| > 2|\boldsymbol{S}|$ will be satisfied), while in gas-poor nuclei accretion will be more likely to resemble the chaotic accretion scenario outlined above. The Bardeen-Petterson effect is also expected to be important for the orientation of the spins of the merging binaries detectable with eLISA. Indeed, binaries forming in gas-rich nuclei will likely have almost aligned spins, and thus produce lower recoil velocities for the merger remnant, which is therefore unlikely to escape from the host galaxy [147].

An investigation of the impact of effects *(a)* on the cosmological evolution of the MBH spins was first performed by [148], which also accounted for *(b)* by simulating the co-evolution between the MBHs and their host galaxies with a semi-analytical galaxy formation model, including both the evolution of dark-matter (via merger trees) and the evolution of baryonic structures (intergalactic and interstellar media, galactic disks, star-forming spheroids, as well as MBHs with their accretion disks). Ref. [148] also obtained predictions (in principle testable with eLISA) for the fraction of MBH mergers in gas-rich environments (and thus with aligned spins) as opposed to ones in gas-poor environments (and thus with misaligned spins). However, in spite of its sophistication, the model of [148] still assumed that the accretion flow onto the MBH had vanishing average angular momentum in gas-poor regimes (i.e. whenever the condition $|\boldsymbol{L}| > 2|\boldsymbol{S}|$ is not satisfied). This assumption is an idealization, as galaxies do have a non-zero angular momentum, and one would therefore expect the average angular momentum $\bar{L}$ of the clouds accreting onto the MBH to be non-zero. Allowing for $\bar{L} \neq 0$ can indeed have important consequences for the spin evolution of MBHs, as shown in [149].

Ref. [149] left $\bar{L}$ as a free parameter, because a first-principle calculation would be extremely challenging: galaxy formation simulations are not yet capable of resolving the MBH's sphere of influence and the accretion disk, and in any case it is far from clear whether such simulations have full control of the subgrid/dissipative physics at small scales (see however [150] for a recent attempt to extract the angular momentum of the clouds from a hydrodynamical simulation, with resolution up to 10 pc.) Alternatively, a bold attempt can be made at trying to connect $\bar{L}$ to measurements of the velocity dispersion v/σ of the gas and stars in galaxies (e.g. if v/σ were zero, it is clear that accretion would be perfectly isotropic, i.e. $\bar{L} = 0$). Ref. [151] adopted this approach and used the semi-analytical galaxy formation model of [148] with the spin-evolution model of [149], connecting $\bar{L}$ to measurements of v/σ of the gas and stellar components in various galaxy morphologies. While still debatable because these measurements are currently only available at distances of $\gtrsim 100$ pc from the MBH (and thus far from the accretion disk and the MBH's sphere of influence), this approach allowed the authors of [151] to produce testable predictions for the spin distribution in competing models for the "isotropy" $\bar{L}$ of the accretion flow. In particular, three models were considered: (A) one connecting $\bar{L}$ to the velocity dispersion of the gaseous component of galaxies; (B) one connecting $\bar{L}$ to the velocity dispersion of the stellar component; and (C) a hybrid model. A comparison with existing iron-Kα measurements of MBH spins shows that model (A) is ruled out quite convincingly, while both models (B) and (C) are in agreement with observations (with marginally significant evidence in favor of the hybrid model (C)). Ref. [151] also showed that the idealized accretion prescriptions discussed above (namely purely coherent and purely chaotic accretion, as well as the original model of [148]) are disfavored by existing iron-Kα measurements. Clearly, eLISA's measurements of the spins of merging binaries will allow discriminating between these competing models with much higher significance, thus providing a way to test the properties of the accretion flow onto MBHs with unprecedented accuracy.

5. Cosmography with space-based detectors

The geometry, large-scale structure and dynamics of the Universe can be inferred with precision if we can accurately measure the distance and redshift to sources distributed throughout the Universe. This is because the luminosity distance $D_{\rm L}$ to a source as a function of its redshift z depends on the geometry of the Universe and a number of cosmological parameters, such as the Hubble parameter, relative fractions of density in dark energy, dark matter, baryonic matter, etcetera [152]. Redshift is very well measured by spectroscopic methods, except for sources at low redshifts ($z \ll 1$), where peculiar velocities due to the gradient of the local gravitational potential could be comparable to cosmological expansion. The real challenge, however, is to accurately measure distances to cosmological sources.

Precision cosmography is enabled by sources whose intrinsic luminosity L can be deduced by some observed property of a source, e.g. its time variability, so that one can infer the luminosity distance $D_{\rm L}$ from its apparent luminosity F, namely $D_{\rm L} = \sqrt{L/4\pi F}$. In 1986 Schutz pointed out how GW observations of inspiralling compact binaries could provide an astronomer's ideal standard candle [153]. Since then there has been quite a lot of work in trying to understand how observations of mergers involving MBHs by LISA (and eLISA) could be used to measure cosmological parameters [154, 155, 156, 157]. Here we will briefly review the basic idea, challenges posed by observations and some solutions. The current perspective on the problem is that eLISA's application for cosmography will be limited by weak gravitational lensing, and precision measurements of cosmological parameters are only possible if eLISA observes several tens of sources during its lifetime. In this regard, ground-based detectors are likely to be more useful for cosmology as they are expected to observe a very large number of sources, which helps to mitigate problems posed by weak lensing [158].

5.1. Self-calibrating standard sirens

Inspiralling compact binaries are often referred to as self-calibrating standard sirens [63]. The word "siren" is used to indicate that GWs are more akin to sound waves than EM waves, and eLISA is a detector that "listens" to its sources. They are "self-calibrating" as no other calibration process is required to measure the distance to an inspiralling binary other than the description of its dynamics by GR [153]. This favorable situation should be contrasted with the astrophysical modelling of sources and construction of a cosmic distance ladder, that are required to calibrate the distance to cosmological sources [152].

The response of eLISA to GWs from the inspiral phase of a coalescing binary depends on its distance $D_{\rm L}$ from the Earth, sky position (θ, φ), component masses (m_1, m_2) and their spins, orbital eccentricity and orbital angular momentum $(\vec{S}_1, \vec{S}_2, e, \vec{L})$ (all at some fiducial time) [63]. In order to measure the luminosity distance to the source it is necessary to disentangle all the other parameters. The shape of a signal that spends a sufficiently long time (say several months) in the eLISA band depends quite sensitively on its intrinsic parameters (masses, spins and eccentricity), and the motion of eLISA with respect to the source induces amplitude and phase modulations in the signal that depend on the position of the source and the orientation of its orbital angular momentum. By fitting the data with precomputed templates that depend on the different parameters of the signal one can resolve all the source parameters. Significant correlations between the luminosity distance and other parameters (most notably the orbital inclination) corrupt the accuracy with which we can estimate the distance. Even so, due to the large signal-to-noise ratios expected from supermassive BH binaries, eLISA should be able to measure distances at $z \sim 1$ to better than 0.1%-1% accuracy [63, 156, 108].

Schutz also pointed out that although GW observations can measure distances, they are not able to measure the source redshift. Recent work, however, has shown that at least in the case of binary neutron stars, and quite possibly also for neutron star-BH binaries, it should be possible to also measure the source's redshift from the effect of tides on the waveform phasing [159, 160]. The intrinsic mass $M_{\rm int}$ of the neutron star will be imprinted in the tidal effects, and once the intrinsic mass is known the redshift can be inferred from the fact that the observed mass $M_{\rm obs}$ is given by $M_{\rm obs} = (1+z)M_{\rm int}$.

For MBHBs, however, there is no way to infer the source's redshift from GW observations alone. Since eLISA will not be sensitive to merging neutron star binaries (at least not at cosmological distances), we have to rely on EM follow-up of GW events to infer the source's redshift. Thus, it was argued that there is great synergy in multi-messenger observations of MBH binaries (cf. Section 3). GW observations would provide distance measurements, while EM observations would provide redshift, and the two observations together would be a new tool for cosmology.

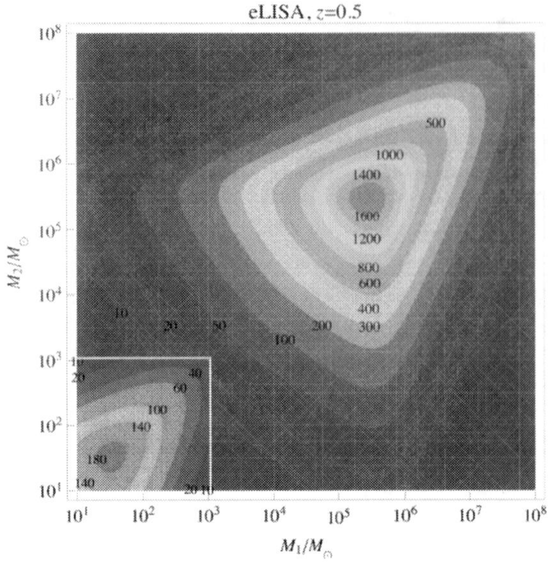

Figure 3. Signal-to-noise ratio (SNR) of the inspiral phase of coalescing BH binaries as a function of the component masses of binary sources at $z = 0.5$ Signals are assumed to last for a year before they merge in the eLISA band. The large SNR helps in measuring the distance with great accuracy in spite of correlations between other signal parameters. The inset shows the SNR for Einstein Telescope (ET)—a third generation underground GW detector. Although eLISA and ET are sensitive to different range of masses, there are some sources that could be observed by both eLISA and ET.

5.2. Challenges in using eLISA for cosmography

At first it was thought that eLISA will be a powerful new tool for cosmography. However, it soon became clear that significant challenges are posed by different aspects of the measurement, making the problem quite hard [63, 161]. In the following we will discuss the most important challenges and what solutions, if any, have been found in confronting them.

Localizing sources. GW interferometers like eLISA are quadrupole antennas with very good sky coverage [162]. They are able to observe better than 2π steradians of the sky at any one time, but they are generally not very good at resolving the sky position of the source. The angular resolution of eLISA for sources that use only the dominant quadrupole harmonic of the signal at twice the orbital frequency is $\sim$ 10's of square degrees [63]. At a redshift of $z \sim 0.5$ (the relevant redshift where one can expect to detect one or two merging MBHBs each year) a sky patch of that size contains thousands of galaxies, and therefore there seems to be no hope of identifying the host galaxy to measure the source's redshift. Moore and Hellings [163, 164] first realized that if one includes in the search templates higher-order signal harmonics, in addition to the dominant harmonic at twice the orbital frequency, the angular resolution can get sensibly smaller. More complete analyses including higher harmonics [156] and/or spin-induced precessional effects [114] showed that sources are resolved to $\sim$ 1 square degree. This angular resolution is much smaller than before – not small enough to contain just the host galaxy, but perhaps small enough to contain the (not yet fully virialized) galaxy cluster to which the host belongs. Moreover, higher harmonics also help break the degeneracy between the inclination angle and the distance, reducing the fractional error on distance measurement by a factor of 2 to 5 [156].

The Task Force that was appointed to systematically explore how well LISA might determine source parameters and constrain dark energy concluded that each year LISA could observe a few events for which distance can be measured to within 1% *and* the source can be localized to within one square degree – accuracies that are sufficient to pin down the dark energy equation-of-state parameter w at the level of one percent [108]. Although the improvement in sky resolution brought through higher-order harmonics and spin effects is sufficient to measure the host redshift, it would be far better if sources could be identified by the EM counterpart that the merger event might produce, which we will discuss next.

Electromagnetic counterparts. Quite a lot of work has gone on in understanding the EM counterparts produced by a merger event. A MBHB merger by itself does not produce any EM radiation. The environment in which a MBHB merges consists of hot diffuse gas, circumbinary accretion disks and stars. As highlighted in Section 3, cold gas forming a putative circumbinary disk can stream toward the two MBHs and form a dense, small disk surviving at the moment of the MBHB coalescence. Numerical simulations of MBHBs seem to suggest that the merger event could shock and heat the surrounding gas particles to high temperatures, leading to bright EM counterparts in coincidence to the coalescence. If the immediate vicinity of the binary is devoided of gas at coalescence, this sort of radiation can be expected within months to years of merger, hence monitoring the sky position of merger could indeed be interesting from an astrophysical point of view as well as for cosmography (see [165] for a comprehensive review of EM counterparts to coalescing MBHBs).

Weak gravitational lensing. The space between a GW source and eLISA is not empty. Small-scale inhomogeneities that are ever present in the path of a signal from a source to eLISA render the signal brighter or dimmer, depending on the integrated effect of the inhomogeneities along a given path. This is called weak gravitational lensing, and all cosmological sources are subject to this phenomenon. Weak lensing cannot be detected accurately, and there is no way to fully correct for its effect on particular sources. Although weak lensing can significantly affect individual detections, the effect averages out when considering a large population of sources. At redshift $z \sim 1$, systematic errors in the measurement of distance due to weak lensing are typically $\sim$ 2-5%, an order of magnitude larger than statistical errors for an average signal at this distance [161].

It is possible to correct for the systematics to some extent, but the best possible technique (in fact a combination of different observations) can decrease the error by a factor of 2 [166, 167]. Hence weak lensing will severely limit how well distances can be inferred, and it is not possible to make a meaningful measurement of dark energy with a single source. If eLISA manages to accumulate $\gtrsim 30$ events at $z < 2$ during its lifetime (which implies MBHB merger rates at the upper end of what is currently estimated and/or a mission lifetime $\gtrsim 5$ years) then it should be possible to measure w to an accuracy of $\sim 4\%$ [157].

5.3. Outlook

The foregoing summary is largely based on studies that were conducted in the context of LISA. Future studies should repeat these investigations in the context of eLISA to have a concrete evaluation of its science potential. It would be useful to study what optimizations can be made within the existing mission framework to maximize the science potential. Estimates of event rates in the context of eLISA, which has significantly better sensitivity than LISA at higher frequencies (at the expense of poorer sensitivity at lower frequencies) would be useful, as higher rates are obviously the cleanest way to defeat the effect of weak lensing. Studies on how weak lensing biases can be further reduced would be useful, as also more reliable estimates of the EM radiation that would be produced after merger. In particular, it is important to study the extent to which gas and disks could affect the dynamics of the binary, and whether this could bias the estimation of signal parameters.

Two concepts beyond eLISA have been studied: Big Bang Observer [168] in the US and Deci-Hertz Gravitational Wave Observatory (DECIGO) [169] in Japan. These ambitious projects with more than an order of magnitude improvement in strain sensitivity will obviously not have the limitations of LISA or eLISA for cosmography (see e.g. [170, 171]). The problem here might be one of data analysis. The large number of sources they might detect could pose problems with unambiguous estimation of signal and source parameters. It is important to study whether our knowledge of the waveforms predicted by GR is good enough to cleanly disentangle the signals.

Journal of Physics: Conference Series **610** (2015) 012001 doi:10.1088/1742-6596/610/1/012001

6. Black hole mergers as probes of strong-field gravity

Einstein's GR is certainly one of the most elegant and successful physical theories, and so far it has passed all experimental tests with flying colors [172]. However there are theoretical and experimental reasons suggesting that the theory must be modified at some level. From a theoretical point of view, GR is a purely classical theory, and power counting arguments indicate that it is not renormalizable in the standard field-theory sense; however, one can build a renormalizable theory by adding quadratic curvature terms – i.e., high-energy/high-curvature corrections – to the Einstein-Hilbert action [173]. Furthermore, high-energy (ultraviolet) corrections seem necessary to avoid singularities that are inevitable in classical GR, as shown by the Hawking-Penrose singularity theorems [174]. From an observational point of view, cosmological measurements are usually interpreted as providing evidence for dark matter and a nonzero cosmological constant ("dark energy"). This interpretation poses serious conceptual issues, including the cosmological constant problem ("why is the observed value of the cosmological constant so small in Planck units?") and the coincidence problem ("why is the energy density of the cosmological constant so close to the present matter density?"). No dynamical solution of the cosmological constant problem is possible within GR [175]. It seems reasonable that ultraviolet corrections to GR would inevitably "leak" down to cosmological scales, showing up as low-energy (infrared) corrections.

The arguments summarized above suggest that Einstein's theory of gravity should be modified at both low and high energies, but it is not easy to introduce modifications to GR that respect these requirements without facing additional problems. Einstein's theory is the unique interacting theory of a Lorentz-invariant massless helicity-2 particle [176], and therefore new physics in the gravitational sector must introduce additional degrees of freedom. Any additional degrees of freedom must modify the theory at both low and high energies *while being consistent with GR in the intermediate-energy regime*, i.e. at length scales between $\sim 1\,\mu$m and about one Astronomical Unit, where the theory is extremely well tested. Intermediate-energy constraints include laboratory experiments, Solar System experiments (that verify the Einstein Equivalence Principle to remarkable accuracy, and force parametrized post-Newtonian parameters such as β and γ to be extremely close to their GR values) and binary pulsar experiments (that place stringent bounds on popular extensions of GR such as scalar-tensor theories, Lorentz-violating theories and TeVeS): see [172, 177] for reviews.

eLISA is arguably the best strong-gravity laboratory one could hope for, as it will probe the strong-field dynamics of GR out to cosmological distances to levels unachievable by binary pulsars[6], Earth-based interferometers or Pulsar Timing Arrays (see [179, 180] for reviews). As we will argue below, space-based GW interferometers could provide crucial hints about the nature of the compact objects at galactic centers and about the nature of gravity itself.

It is useful to classify "tests of strong field gravity" as belonging to two – qualitatively very different – categories:

(1) External tests: can laboratory experiments, astrophysical observations or future GW measurements determine whether GR is the correct theory of gravity? To frame this question in terms of hypothesis testing, one would like to have a valid opponent to GR. What constitutes a "valid opponent" is a matter of taste. For our purpose (i.e., tests of strong-field gravity with BH mergers) it should be a cosmologically viable fundamental theory with a well-posed initial value formulation, and field equations that follow from an action principle. Furthermore, the theory should be simple enough to allow calculations of (say) BH solutions and GW emission. There are countless attempts to modify GR [181], but all modifications must introduce some sort of screening mechanism in order to be viable at intermediate energies. Screening mechanisms

[6] One of the most extraordinary laboratories for strong-gravity tests is PSR J0348+0432 [178]. Even this binary system, which is highly relativistic for binary-pulsar standards, has an orbital velocity $v \simeq 2\times10^{-3}c$, much smaller than the orbital velocities $v \approx c/3$ of an astrophysical BH binary near merger.

include chameleons, symmetrons, dilatons, MOND-like dynamics, the Vainshtein mechanism, etcetera [182]. Since we don't have a full theory of quantum gravity, an effective field-theory approach is often invoked when constructing phenomenological alternatives to GR [183, 184]. For example, one can start with the most generic four-dimensional theories of gravity including quadratic curvature invariants generically coupled to a single scalar field ϕ:

$$S = \int \frac{d^4x\,\sqrt{-g}}{16\pi}\left[R - 2\nabla_a\phi\nabla^a\phi - V(\phi) + f_1(\phi)R^2 + f_2(\phi)R_{ab}R^{ab} + f_3(\phi)R_{abcd}R^{abcd} + f_4(\phi)^*RR\right] + S_{\rm mat}\left[\gamma(\phi)g_{\mu\nu},\, \Psi_{\rm mat}\right]\,, \tag{1}$$

where $\Psi_{\rm mat}$ collectively denotes matter fields, $V(\phi)$ is the scalar self-potential, $f_i(\phi)$ are generic coupling functions, the Chern-Simons term $^*RR \equiv \frac{1}{2}R_{abcd}\epsilon^{baef}R^{cd}{}_{ef}$, ϵ^{abcd} is the Levi-Civita tensor, and in the matter action $S_{\rm mat}$ we allow for a nonminimal coupling that violates the (weak) equivalence principle. Not all of these theories are acceptable: for example, to avoid higher-order derivatives in the equations of motion one must generally assume the couplings to be small and treat the theory as an effective field theory (the equations are second-order in the strong-coupling limit only if the quadratic invariants enter in the special "Gauss-Bonnet" combination). The action above may seem complicated, but it actually represents a very restricted class of theories, and calculations of gravitational radiation have been performed only in very specific subcases, such as scalar-tensor theories with specific choices of the potential and couplings [185, 186, 187, 188] and some forms of quadratic gravity [189, 190]. The bottom line is that there are very few "serious" alternatives to GR (in the sense that they are well posed, follow from a Lagrangian, make sensible predictions...) and even fewer for which GW calculations have been carried out. For these theories, GW observations usually yield constraints that are comparable to, and often better than, binary pulsar and Solar System bounds [179, 180]. *(2) Internal tests: is GR "internally" consistent with astrophysical observations?* One of the most striking predictions of GR is the existence of BHs. Astronomers commonly believe that the compact objects that harbor galactic centers are the BHs of GR, but this "BH paradigm" rests on somewhat shaky foundations. Evidence that these objects possess event horizons (or more correctly, apparent horizons) rather than solid surfaces usually rests on plausibility arguments based on accretion physics [191, 192], that leave room for some skepticism[7] [193]. It is also important to stress that, strictly speaking, any tests that probe the Kerr *metric* alone (such as tests based on matter accretion or ray-tracing of photon trajectories) are of little value as internal tests of GR. The reason is that most alternative theories (including generic scalar-tensor theories [194] and a large class of higher-curvature theories [195]) admit the Kerr metric as a solution, and the theories that don't (e.g. Einstein-dilaton Gauss-Bonnet [196], Dynamical Chern-Simons [197], and Lorentz-violating gravity [198, 199, 200, 201]) predict BH solutions that differ from GR by amounts that should be astrophysically unmeasurable (see, however, Refs. [202, 203] for a family of BH solutions whose deviations from the Kerr metric may be important). Many "quasi-Kerr metrics" that have been proposed in this context should be viewed as unnatural strawmen: they often have serious pathologies [204], and they are therefore unacceptable even for the limited scope of parametrizing deviations from the Kerr metric [205].

These considerations imply that the only way to unambiguously verify that the compact objects in galactic centers are actually Kerr BHs is **via their GW dynamics**, especially in the strong-field merger/ringdown phase [206, 207, 208]. This is why space-based GW

[7] From a theorist's point of view, one of the most convincing arguments in favor of the BH paradigm is that the alternatives are either unstable (as in the case of dense star clusters, fermion stars or naked singularities), unnatural (e.g. "exotic" matter violating some of the energy conditions), contrived (such as gravastars), implausible as the end-point of collapse in astrophysical settings (boson stars) or nearly indistinguishable from Kerr (this is the case for BH solutions in alternative theories with coupling parameters that are reasonable from a fundamental physics point of view).

observations hold great promise to constrain strong-field gravity [179]. Their qualitative advantage over Earth-based detectors is simple to understand. The fundamental oscillation mode of a nonrotating BH has frequency $f \simeq 1.2 \times 10^{-2}(10^6 M_\odot/M)$ Hz. This frequency lies exactly in the "bucket" of eLISA's noise power spectral density for "light" BHs of mass $M \sim 10^6 M_\odot$, that were presumably the building blocks of the large BHs we see at galactic centers today. Advanced LIGO, by contrast, has maximum sensitivity at $f \sim 10^2$ Hz, i.e. for intermediate-mass BHs of mass $M \sim 10^2 M_\odot$, whose very existence is still highly uncertain [209]. These oscillation modes are called "quasinormal" modes, because they are damped by GW emission. The no-hair theorem stated in Section 4 implies that, in GR, the frequencies and damping times of all QNMs depend only on the BH mass M and spin a. A measurement of the dominant mode's frequency and damping time yields both M and a; the measurement of *any* other frequency and/or damping time can then be used to verify that the BH formed as a result of the merger is indeed a Kerr BH, as predicted by GR. The feasibility of this (internal) test of GR depends on the measurability of QNM frequencies/damping times and on our ability to resolve modes. Both measurability and resolvability scale like $1/\rho$, where $\rho \sim h/S_n$ is the signal-to-noise ratio of the merger event [206, 207]. The maximum sensitivity S_n of Earth-based and space-based detectors is comparable in order of magnitude, but (as we saw above) space-based detectors target sources that are $\sim 10^4$ times more massive. The GW amplitude $h \sim \sqrt{\epsilon_{\rm rd} M}$, where $\epsilon_{\rm rd} \sim 10^{-2}(4\eta)^2$ [210] is a "ringdown efficiency" and $\eta \equiv m_1 m_2/(m_1+m_2)^2 \in [0, 0.25]$ is the so-called symmetric mass ratio. The punchline of this argument is that $\rho \sim \sqrt{M}$, so *no-hair theorem tests with space-based observations of BH binary mergers are typically* $\sim 10^2$ *stronger than Earth-based GW tests.*

In summary, BH mergers are extraordinary (local) probes of the no-hair theorem that are potentially detectable by eLISA throughout the entire Universe. Each merger event allows us to do much more than this: it can give us tests of consistency of the post-merger Kerr remnant with the pre-merger binary dynamics [211, 212, 213], allow us to test the area theorem [214], and perhaps even allow us to peer into the nonlinear dynamics responsible for the coupling of different quasinormal modes [215]. Even more interestingly, each BH binary merger can be thought of as a local probe of whether *strong-field* GR is valid *at the redshift z at which the merger occurred.* This is an incredible opportunity for tests of GR, because it would verify that Einstein's gravity accurately describes BHs at least out to redshift z. This would place even more stringent constraints on the viable modifications of GR. This idea has one drawback: it requires (ideally) the determination of the merging binary's redshift z, or at the very least the determination of a *lower bound* on the source redshift. As described in Section 5, source localization and distance determination are intimately related in a LISA-like mission, and they get significantly better for a three-arm mission [108]. In conclusion, a three-arm mission would make a big difference to test the no-hair theorem at cosmological redshift - and thus to place tight constraints on the strong-field behavior of the theory at distances where cosmological observations are relevant.

Another aspect worth emphasizing is that, unlike Solar System tests (that only probe the "static", quasi-Newtonian behavior of gravitational fields) and binary pulsars (that essentially measure the energy flux predicted by a given theory of gravity, and not much else in terms of the dynamics of the gravitational field) the *direct* observation of GWs can probe the number of polarization states as well as the propagation properties of GWs, as encoded in their dispersion relation. Tests of the dispersion relation place constraints on a putative nonzero graviton mass, and they get better, as expected, for sources at cosmological distance. In fact it has been shown that eLISA can constrain the mass of the graviton to a level that is ~ 4 orders of magnitude better than current Solar System constraints [216, 217]. Quite naturally, constraints on any given alternative theory of gravity get better with multiple observations because of the improved statistics. Calculations in the case of a hypothetical massive graviton show that the improvement is better than the naive Poisson-statistic expectation of $\sqrt{N}$, where N is the number of events, because the louder events "carry more weight" in determining the combined bound [218].

Last but not least, the discovery space of eLISA is potentially enormous. The potential of eLISA for cosmology was discussed in Section 5; here we will focus on the discovery space related to MBH observations. Binary pulsars are already constraining phenomena like "spontaneous scalarization" to levels that are comparable to what eLISA could do by observing neutron stars spiraling into MBHs [217]. However, one can imagine scenarios where modified gravity would produce smoking-gun signatures of deviations from GR that would be observable by eLISA, and completely invisible to binary-pulsar tests. The simplest case study are extreme mass-ratio inspirals in scalar-tensor theory. If the scalar has a mass, Kerr BHs become vulnerable to the so-called "black-hole bomb" instability: superradiance can amplify incident waves, the mass of the scalar acts like a mirror that reflects amplified waves back onto the BH, and therefore the superresonant amplification of incident waves can grow without bound [219]. This instability could have striking effects, such as the existence of "floating orbits" [220] at which the inspiralling body could stall, emitting essentially monochromatic radiation – a "GW laser"! Similar effects could occur if the energy conditions are violated [221, 222], so eLISA could hint at exotic physics responsible for possible violations of the energy conditions. Finally, if the objects at galactic centers were not BHs but (say) gravastars or boson stars, the oscillation modes of these exotic objects would inevitably be excited by orbiting bodies, leaving characteristic signatures in the energy flux that are potentially detectable by eLISA [223, 224].

Opening new observational windows on the Universe inevitably reveals more than we anticipated. We should be ready for surprises.

7. Conclusions

We have reviewed our current understanding of MBH formation and evolution. The mysteries that surround the birth and growth of these cosmic monsters are enormous, and they are intimately related to the growth of structure in our Universe. Observations of gravitational radiation from MBH mergers – especially in combination with EM counterparts – have the potential to measure spins to levels unachievable by other means, clarify the role of accretion in MBH growth, and constrain fundamental physics in unprecedented ways. An eLISA-like mission will be a spectacular, unrivalled laboratory for fundamental physics and astrophysics.

Acknowledgments

Enrico Barausse and Alberto Sesana thank their co-authors Massimo Dotti and Elena Maria Rossi for insightful conversations. Enrico Barausse acknowledges support from the European Union's Seventh Framework Programme (FP7/PEOPLE-2011-CIG) through the Marie Curie Career Integration Grant GALFORMBHS PCIG11-GA-2012-321608. Emanuele Berti is supported by NSF CAREER Grant No. PHY-1055103. Bangalore S. Sathyaprakash acknowledges the support of the LIGO Visitor Program through the National Science Foundation award PHY-0757058 and STFC grant ST/J000345/1.

References

[1] Amaro-Seoane P A *et al.* 2013 *ArXiv e-prints* (*Preprint* `1305.5720`)
[2] Amaro-Seoane P, Aoudia S, Babak S, Binétruy P, Berti E, Bohé A, Caprini C, Colpi M, Cornish N J, Danzmann K, Dufaux J F, Gair J, Hinder I, Jennrich O, Jetzer P, Klein A, Lang R N, Lobo A, Littenberg T, McWilliams S T, Nelemans G, Petiteau A, Porter E K, Schutz B F, Sesana A, Stebbins R, Sumner T, Vallisneri M, Vitale S, Volonteri M, Ward H and Wardell B 2013 *GW Notes, Vol. 6, p. 4-110* **6** 4–110 (*Preprint* `1201.3621`)
[3] Kauffmann G and Haehnelt M 2000 MNRAS **311** 576–588 (*Preprint* `astro-ph/9906493`)
[4] Volonteri M, Haardt F and Madau P 2003 ApJ **582** 559–573 (*Preprint* `astro-ph/0207276`)

[5] Véron-Cetty M P and Véron P 2010 A&A **518** A10
[6] Kormendy J and Ho L C 2013 ARA&A **51** 511–653 (*Preprint* 1304.7762)
[7] Gehren T, Fried J, Wehinger P A and Wyckoff S 1984 ApJ **278** 11–27
[8] Kormendy J and Ho L C 2013 ARA&A **51** 511–653 (*Preprint* 1304.7762)
[9] Ferrarese L and Ford H 2005 Space Sci. Rev. **116** 523–624 (*Preprint* arXiv:astro-ph/0411247)
[10] Bond J R, Arnett W D and Carr B J 1984 ApJ **280** 825–847
[11] Couchman H M P and Rees M J 1986 MNRAS **221** 53–62
[12] Madau P and Rees M J 2001 ApJ **551** L27–L30 (*Preprint* astro-ph/0101223)
[13] Abel T, Bryan G L and Norman M L 2002 *Science* **295** 93–98 (*Preprint* arXiv:astro-ph/0112088)
[14] Bromm V and Larson R B 2004 ARA&A **42** 79–118 (*Preprint* arXiv:astro-ph/0311019)
[15] Loeb A and Rasio F A 1994 ApJ **432** 52–61 (*Preprint* arXiv:astro-ph/9401026)
[16] Oh S P and Haiman Z 2002 ApJ **569** 558–572 (*Preprint* astro-ph/0108071)
[17] Begelman M C, Volonteri M and Rees M J 2006 MNRAS **370** 289–298 (*Preprint* astro-ph/0602363)
[18] Lodato G and Natarajan P 2006 MNRAS **371** 1813–1823 (*Preprint* arXiv:astro-ph/0606159)
[19] Wise J H, Turk M J and Abel T 2008 ApJ **682** 745–757 (*Preprint* 0710.1678)
[20] Choi J H, Shlosman I and Begelman M C 2013 ApJ **774** 149 (*Preprint* 1304.1369)
[21] Abel T, Bryan G L and Norman M L 2002 *Science* **295** 93–98 (*Preprint* astro-ph/0112088)
[22] Bromm V, Coppi P S and Larson R B 2002 ApJ **564** 23–51 (*Preprint* astro-ph/0102503)
[23] Stacy A, Greif T H and Bromm V 2010 MNRAS **403** 45–60 (*Preprint* 0908.0712)
[24] Jappsen A K, Klessen R S, Glover S C O and Mac Low M M 2009 ApJ **696** 1065–1074 (*Preprint* 0709.3530)
[25] Machida M N, Matsumoto T and Inutsuka S i 2008 ApJ **685** 690–704 (*Preprint* 0803.1224)
[26] Greif T H, Bromm V, Clark P C, Glover S C O, Smith R J, Klessen R S, Yoshida N and Springel V 2012 MNRAS **424** 399–415
[27] Susa H, Hasegawa K and Tominaga N 2014 ApJ **792** 32 (*Preprint* 1407.1374)
[28] Clark P C, Glover S C O, Smith R J, Greif T H, Klessen R S and Bromm V 2011 *Science* **331** 1040– (*Preprint* 1101.5284)
[29] Inayoshi K and Haiman Z 2014 *ArXiv e-prints* (*Preprint* 1406.5058)
[30] Inayoshi K, Omukai K and Tasker E J 2014 *ArXiv e-prints* (*Preprint* 1404.4630)
[31] Dijkstra M, Haiman Z, Mesinger A and Wyithe J S B 2008 MNRAS **391** 1961–1972 (*Preprint* 0810.0014)
[32] Regan J A and Haehnelt M G 2009 MNRAS **396** 343–353 (*Preprint* 0810.2802)
[33] Shang C, Bryan G L and Haiman Z 2010 MNRAS **402** 1249–1262 (*Preprint* 0906.4773)
[34] Yue B, Ferrara A, Salvaterra R, Xu Y and Chen X 2014 MNRAS **440** 1263–1273 (*Preprint* 1402.5675)
[35] Ferrara A, Salvadori S, Yue B and Schleicher D 2014 MNRAS **443** 2410–2425 (*Preprint* 1406.6685)
[36] Jimenez R and Haiman Z 2006 Nature **440** 501–504 (*Preprint* arXiv:astro-ph/0602450)
[37] Bellovary J, Volonteri M, Governato F, Shen S, Quinn T and Wadsley J 2011 ApJ **742** 13 (*Preprint* 1104.3858)
[38] Visbal E, Haiman Z and Bryan G L 2014 *ArXiv e-prints* (*Preprint* 1406.7020)
[39] Schneider R, Ferrara A, Ciardi B, Ferrari V and Matarrese S 2000 MNRAS **317** 385–390 (*Preprint* astro-ph/9909419)
[40] Sesana A, Volonteri M and Haardt F 2007 MNRAS **377** 1711–1716 (*Preprint* astro-ph/0701556)
[41] Sesana A, Gair J, Berti E and Volonteri M 2011 *Phys. Rev. D* **83** 044036 (*Preprint* 1011.5893)
[42] Plowman J E, Hellings R W and Tsuruta S 2011 MNRAS **415** 333–352 (*Preprint* 1009.0765)
[43] Tanaka T and Haiman Z 2009 ApJ **696** 1798–1822 (*Preprint* 0807.4702)
[44] Volonteri M and Natarajan P 2009 MNRAS **400** 1911–1918 (*Preprint* 0903.2262)
[45] Brusa M, Civano F, Comastri A, Miyaji T, Salvato M, Zamorani G, Cappelluti N, Fiore F, Hasinger G, Mainieri V, Merloni A, Bongiorno A, Capak P, Elvis M, Gilli R, Hao H, Jahnke K, Koekemoer A M, Ilbert O, Le Floc'h E, Lusso E, Mignoli M, Schinnerer E, Silverman J D, Treister E, Trump J D, Vignali C, Zamojski M, Aldcroft T, Aussel H, Bardelli S, Bolzonella M, Cappi A, Caputi K, Contini T, Finoguenov A, Fruscione A, Garilli B, Impey C D, Iovino A, Iwasawa K, Kampczyk P, Kartaltepe J, Kneib J P, Knobel C, Kovac K, Lamareille F, Leborgne J F, Le Brun V, Le Fevre O, Lilly S J, Maier C, McCracken H J, Pello R, Peng Y J, Perez-Montero E, de Ravel L, Sanders D, Scodeggio M, Scoville N Z, Tanaka M, Taniguchi Y, Tasca L, de la Torre S, Tresse L, Vergani D and Zucca E 2010 ApJ **716** 348–369 (*Preprint* 1004.2790)
[46] Eckart M E, McGreer I D, Stern D, Harrison F A and Helfand D J 2010 ApJ **708** 584–597 (*Preprint* 0911.2448)
[47] Natarajan P and Treister E 2009 MNRAS **393** 838–845 (*Preprint* 0808.2813)
[48] Hardcastle M J, Evans D A and Croston J H 2007 MNRAS **376** 1849–1856 (*Preprint* astro-ph/0701857)
[49] Merloni A and Heinz S 2008 MNRAS **388** 1011–1030 (*Preprint* 0805.2499)
[50] Bellovary J, Brooks A, Volonteri M, Governato F, Quinn T and Wadsley J 2013 ApJ **779** 136 (*Preprint*

1307.0856)
[51] Micic M, Holley-Bockelmann K and Sigurdsson S 2011 MNRAS **414** 1127–1144 (*Preprint* 1102.0327)
[52] Hopkins P F, Hernquist L, Cox T J, Di Matteo T, Robertson B and Springel V 2006 ApJS **163** 1–49 (*Preprint* arXiv:astro-ph/0506398)
[53] Di Matteo T, Springel V and Hernquist L 2005 Nature **433** 604–607 (*Preprint* arXiv:astro-ph/0502199)
[54] Mayer L, Kazantzidis S, Madau P, Colpi M, Quinn T and Wadsley J 2007 *Science* **316** 1874– (*Preprint* 0706.1562)
[55] Van Wassenhove S, Volonteri M, Mayer L, Dotti M, Bellovary J and Callegari S 2012 ApJ **748** L7 (*Preprint* 1111.0223)
[56] Kereš D, Katz N, Weinberg D H and Davé R 2005 MNRAS **363** 2–28 (*Preprint* arXiv:astro-ph/0407095)
[57] Brooks A M, Governato F, Quinn T, Brook C B and Wadsley J 2009 ApJ **694** 396–410 (*Preprint* 0812.0007)
[58] Di Matteo T, Khandai N, DeGraf C, Feng Y, Croft R A C, Lopez J and Springel V 2012 ApJ **745** L29 (*Preprint* 1107.1253)
[59] Holley-Bockelmann K, Micic M, Sigurdsson S and Rubbo L J 2010 ApJ **713** 1016–1025 (*Preprint* 1002.3378)
[60] Treister E, Schawinski K, Urry C M and Simmons B D 2012 ApJ **758** L39 (*Preprint* 1209.5393)
[61] Kocsis B, Haiman Z and Menou K 2008 ApJ **684** 870–887 (*Preprint* 0712.1144)
[62] Deffayet C and Menou K 2007 ApJ **668** L143–L146 (*Preprint* 0709.0003)
[63] Holz D E and Hughes S A 2005 ApJ **629** 15–22 (*Preprint* astro-ph/0504616)
[64] Kocsis B, Frei Z, Haiman Z and Menou K 2006 ApJ **637** 27–37 (*Preprint* astro-ph/0505394)
[65] Komossa S, Burwitz V, Hasinger G, Predehl P, Kaastra J S and Ikebe Y 2003 ApJ **582** L15–L19 (*Preprint* astro-ph/0212099)
[66] Comerford J M, Schluns K, Greene J E and Cool R J 2013 ApJ **777** 64 (*Preprint* 1309.2284)
[67] Liu X, Civano F, Shen Y, Green P, Greene J E and Strauss M A 2013 ApJ **762** 110 (*Preprint* 1209.5418)
[68] Woo J H, Cho H, Husemann B, Komossa S, Park D and Bennert V N 2014 MNRAS **437** 32–37 (*Preprint* 1401.3042)
[69] Dotti M, Sesana A and Decarli R 2012 *Advances in Astronomy* **2012** 940568 (*Preprint* 1111.0664)
[70] Liu F K, Li S and Komossa S 2014 ApJ **786** 103 (*Preprint* 1404.4933)
[71] Goldreich P and Tremaine S 1980 ApJ **241** 425–441
[72] Artymowicz P and Lubow S H 1994 ApJ **421** 651–667
[73] Armitage P J and Natarajan P 2002 ApJ **567** L9–L12 (*Preprint* arXiv:astro-ph/0201318)
[74] Milosavljević M and Phinney E S 2005 ApJ **622** L93–L96 (*Preprint* arXiv:astro-ph/0410343)
[75] Chang P, Strubbe L E, Menou K and Quataert E 2010 MNRAS **407** 2007–2016 (*Preprint* 0906.0825)
[76] Haiman Z, Kocsis B, Menou K, Lippai Z and Frei Z 2009 *Classical and Quantum Gravity* **26** 094032 (*Preprint* 0811.1920)
[77] Shapiro S L 2010 Phys. Rev. D **81** 024019 (*Preprint* 0912.2345)
[78] Kocsis B, Haiman Z and Loeb A 2012 MNRAS **427** 2660–2679 (*Preprint* 1205.4714)
[79] Tanaka T L 2013 *ArXiv e-prints* (*Preprint* 1303.6279)
[80] MacFadyen A I and Milosavljević M 2008 ApJ **672** 83–93 (*Preprint* arXiv:astro-ph/0607467)
[81] D'Orazio D J, Haiman Z and MacFadyen A 2012 *ArXiv e-prints* (*Preprint* 1210.0536)
[82] Hayasaki K, Mineshige S and Sudou H 2007 PASJ **59** 427–441 (*Preprint* arXiv:astro-ph/0609144)
[83] Cuadra J, Armitage P J, Alexander R D and Begelman M C 2009 MNRAS **393** 1423–1432 (*Preprint* 0809.0311)
[84] Farris B D, Liu Y T and Shapiro S L 2011 Phys. Rev. D **84** 024024 (*Preprint* 1105.2821)
[85] Farris B D, Gold R, Paschalidis V, Etienne Z B and Shapiro S L 2012 *Physical Review Letters* **109** 221102 (*Preprint* 1207.3354)
[86] Noble S C, Mundim B C, Nakano H, Krolik J H, Campanelli M, Zlochower Y and Yunes N 2012 ApJ **755** 51 (*Preprint* 1204.1073)
[87] Shi J M, Krolik J H, Lubow S H and Hawley J F 2012 ApJ **749** 118 (*Preprint* 1110.4866)
[88] Roedig C, Sesana A, Dotti M, Cuadra J, Amaro-Seoane P and Haardt F 2012 A&A **545** A127 (*Preprint* 1202.6063)
[89] Duffell P and MacFadyen A I 2014 *in prep.*
[90] Duffell P C and MacFadyen A I 2013 ApJ **769** 41 (*Preprint* 1302.1934)
[91] Shakura N I and Sunyaev R A 1973 A&A **24** 337–355
[92] Farris B D, Duffell P, MacFadyen A I and Haiman Z 2014 ApJ **783** 134 (*Preprint* 1310.0492)
[93] Roedig C, Dotti M, Sesana A, Cuadra J and Colpi M 2011 MNRAS **415** 3033–3041 (*Preprint* 1104.3868)
[94] Soltan A 1982 *Mon. Not. R. Astron. Soc.* **200** 115–122
[95] Yu Q and Tremaine S 2002 MNRAS **335** 965–976 (*Preprint* arXiv:astro-ph/0203082)
[96] Elvis M, Risaliti G and Zamorani G 2002 ApJ **565** L75–L77 (*Preprint* arXiv:astro-ph/0112413)
[97] Roedig C, Krolik J H and Miller M C 2014 ApJ **785** 115 (*Preprint* 1402.7098)

[98] Gebhardt K, Bender R, Bower G, Dressler A, Faber S M, Filippenko A V, Green R, Grillmair C, Ho L C, Kormendy J, Lauer T R, Magorrian J, Pinkney J, Richstone D and Tremaine S 2000 *Astrophys. J. Lett.* **539** L13–L16 (*Preprint* astro-ph/0006289)
[99] Ferrarese L and Merritt D 2000 *Astrophys. J. Lett.* **539** L9–L12 (*Preprint* astro-ph/0006053)
[100] Marconi A and Hunt L K 2003 *Astrophys. J. Lett.* **589** L21–L24 (*Preprint* astro-ph/0304274)
[101] Häring N and Rix H W 2004 *Astrophys. J. Lett.* **604** L89–L92 (*Preprint* astro-ph/0402376)
[102] Gültekin K, Richstone D O, Gebhardt K, Lauer T R, Tremaine S, Aller M C, Bender R, Dressler A, Faber S M, Filippenko A V, Green R, Ho L C, Kormendy J, Magorrian J, Pinkney J and Siopis C 2009 *Astrophys. J.* **698** 198–221 (*Preprint* 0903.4897)
[103] Reynolds C S 2013 *ArXiv e-prints* (*Preprint* 1307.3246)
[104] Brenneman L 2013 *Measuring the Angular Momentum of Supermassive Black Holes*
[105] Risaliti G, Harrison F A, Madsen K K, Walton D J, Boggs S E, Christensen F E, Craig W W, Grefenstette B W, Hailey C J, Nardini E, Stern D and Zhang W W 2013 *Nature* **494** 449–451 (*Preprint* 1302.7002)
[106] Marinucci A, Matt G, Kara E, Miniutti G, Elvis M, Arevalo P, Ballantyne D R, Baloković M, Bauer F, Brenneman L, Boggs S E, Cappi M, Christensen F E, Craig W W, Fabian A C, Fuerst F, Hailey C J, Harrison F A, Risaliti G, Reynolds C S, Stern D K, Walton D J and Zhang W 2014 *Mon. Not. R. Astron. Soc.* **440** 2347–2356 (*Preprint* 1402.7245)
[107] Marinucci A, Matt G, Miniutti G, Guainazzi M, Parker M L, Brenneman L, Fabian A C, Kara E, Arevalo P, Ballantyne D R, Boggs S E, Cappi M, Christensen F E, Craig W W, Elvis M, Hailey C J, Harrison F A, Reynolds C S, Risaliti G, Stern D K, Walton D J and Zhang W 2014 *ArXiv e-prints* (*Preprint* 1404.3561)
[108] Arun K G, Babak S, Berti E, Cornish N, Cutler C, Gair J, Hughes S A, Iyer B R, Lang R N, Mandel I, Porter E K, Sathyaprakash B S, Sinha S, Sintes A M, Trias M, Van Den Broeck C and Volonteri M 2009 *Classical and Quantum Gravity* **26** 094027 (*Preprint* 0811.1011)
[109] Barausse E, Cardoso V and Pani P 2014 *Phys. Rev. D* **89** 104059 (*Preprint* 1404.7149)
[110] Barausse E, Cardoso V and Pani P 2014 *ArXiv e-prints* (*Preprint* 1404.7140)
[111] Barker B M and O'Connell R F 1975 *Phys. Rev. D* **12** 329–335
[112] Apostolatos T A, Cutler C, Sussman G J and Thorne K S 1994 *Phys. Rev. D* **49** 6274–6297
[113] Vecchio A 2004 Phys. Rev. D **70** 042001 (*Preprint* astro-ph/0304051)
[114] Lang R N and Hughes S A 2006 *Phys. Rev. D* **74** 122001 (*Preprint* gr-qc/0608062)
[115] Lang R N, Hughes S A and Cornish N J 2011 *Phys. Rev. D* **84** 022002 (*Preprint* 1101.3591)
[116] Lang R N and Hughes S A 2008 *Astrophys. J.* **677** 1184–1200 (*Preprint* 0710.3795)
[117] Schnittman J D 2004 *Phys. Rev. D* **70** 124020 (*Preprint* astro-ph/0409174)
[118] Kesden M, Sperhake U and Berti E 2010 *Phys. Rev. D* **81** 084054 (*Preprint* 1002.2643)
[119] Kesden M, Sperhake U and Berti E 2010 ApJ **715** 1006–1011 (*Preprint* 1003.4993)
[120] Berti E, Kesden M and Sperhake U 2012 Phys. Rev. D **85** 124049 (*Preprint* 1203.2920)
[121] Kidder L E 1995 *Phys. Rev. D* **52** 821–847 (*Preprint* gr-qc/9506022)
[122] Kidder L E, Will C M and Wiseman A G 1993 *Phys. Rev. D* **47** 4183 (*Preprint* gr-qc/9211025)
[123] Centrella J, Baker J G, Kelly B J and van Meter J R 2010 *Reviews of Modern Physics* **82** 3069–3119 (*Preprint* 1010.5260)
[124] Sperhake U, Berti E and Cardoso V 2011 *ArXiv e-prints* (*Preprint* 1107.2819)
[125] Lehner L and Pretorius F 2014 *ArXiv e-prints* (*Preprint* 1405.4840)
[126] Buonanno A, Kidder L E and Lehner L 2008 *Phys. Rev. D* **77** 026004 (*Preprint* 0709.3839)
[127] Kesden M 2008 *Phys. Rev. D* **78** 084030 (*Preprint* 0807.3043)
[128] Rezzolla L, Barausse E, Dorband E N, Pollney D, Reisswig C, Seiler J and Husa S 2008 *Phys. Rev. D* **78** 044002 (*Preprint* 0712.3541)
[129] Barausse E and Rezzolla L 2009 *Astrophys. J.l* **704** L40–L44 (*Preprint* 0904.2577)
[130] Tichy W and Marronetti P 2008 *Phys. Rev. D* **78** 081501 (*Preprint* 0807.2985)
[131] Healy J, Lousto C O and Zlochower Y 2014 *ArXiv e-prints* (*Preprint* 1406.7295)
[132] Barausse E, Morozova V and Rezzolla L 2012 *Astrophys. J.* **758** 63 (*Preprint* 1206.3803)
[133] van Meter J R, Miller M C, Baker J G, Boggs W D and Kelly B J 2010 *Astrophys. J.* **719** 1427–1432 (*Preprint* 1003.3865)
[134] Berti E, Cardoso V and Starinets A O 2009 *Classical and Quantum Gravity* **26** 163001 (*Preprint* 0905.2975)
[135] Merloni A, Rudnick G and Di Matteo T 2004 *Mon. Not. R. Astron. Soc.* **354** L37–L42 (*Preprint* astro-ph/0409187)
[136] Shankar F, Salucci P, Granato G L, De Zotti G and Danese L 2004 *Mon. Not. R. Astron. Soc.* **354** 1020–1030 (*Preprint* astro-ph/0405585)
[137] Shankar F, Weinberg D H and Miralda-Escudé J 2013 *Mon. Not. R. Astron. Soc.* **428** 421–446 (*Preprint* 1111.3574)

[138] Berti E and Volonteri M 2008 *Astrophys. J.* **684** 822–828 (*Preprint* 0802.0025)
[139] Volonteri M, Madau P, Quataert E and Rees M J 2005 *Astrophys. J.* **620** 69–77 (*Preprint* astro-ph/0410342)
[140] Lagos C D P, Padilla N D and Cora S A 2009 *Mon. Not. R. Astron. Soc.* **395** 625–636 (*Preprint* 0901.0547)
[141] Fanidakis N, Baugh C M, Benson A J, Bower R G, Cole S, Done C and Frenk C S 2011 *Mon. Not. R. Astron. Soc.* **410** 53–74 (*Preprint* 0911.1128)
[142] Fanidakis N, Baugh C M, Benson A J, Bower R G, Cole S, Done C, Frenk C S, Hickox R C, Lacey C and Del P Lagos C 2012 *Mon. Not. R. Astron. Soc.* **419** 2797–2820 (*Preprint* 1011.5222)
[143] Bardeen J M and Petterson J A 1975 *Astrophys. J. Lett.* **195** L65
[144] King A R, Lubow S H, Ogilvie G I and Pringle J E 2005 *Mon. Not. R. Astron. Soc.* **363** 49–56 (*Preprint* astro-ph/0507098)
[145] King A R and Pringle J E 2006 *Mon. Not. R. Astron. Soc.* **373** L90–L92 (*Preprint* astro-ph/0609598)
[146] Perego A, Dotti M, Colpi M and Volonteri M 2009 *Mon. Not. R. Astron. Soc.* **399** 2249–2263 (*Preprint* 0907.3742)
[147] Bogdanović T, Reynolds C S and Miller M C 2007 *Astrophys. J. Lett.* **661** L147–L150 (*Preprint* astro-ph/0703054)
[148] Barausse E 2012 *Mon. Not. R. Astron. Soc.* **423** 2533–2557 (*Preprint* 1201.5888)
[149] Dotti M, Colpi M, Pallini S, Perego A and Volonteri M 2013 *Astrophys. J.* **762** 68 (*Preprint* 1211.4871)
[150] Dubois Y, Volonteri M and Silk J 2013 *ArXiv e-prints* (*Preprint* 1304.4583)
[151] Sesana A, Barausse E, Dotti M and Rossi E M 2014 *ArXiv e-prints* (*Preprint* 1402.7088)
[152] Liddle A 2003 *An Introduction to Modern Cosmology, Second Edition*
[153] Schutz B F 1986 *Nature* **323** 310
[154] Holz D E and Hughes S A 2005 *Astrophys.J.* **629** 15–22 (*Preprint* astro-ph/0504616)
[155] Dalal N, Holz D E, Hughes S A and Jain B 2006 *Phys. Rev. D* **74** 063006 (*Preprint* astro-ph/0601275)
[156] Arun K G, Iyer B R, Sathyaprakash B S, Sinha S and Van Den Broeck C 2007 Phys. Rev. D **76** 104016
[157] Petiteau A, Babak S and Sesana A 2011 *Astrophys.J.* **732** 82 (*Preprint* 1102.0769)
[158] Sathyaprakash B, Schutz B and Van Den Broeck C 2010 *Class. Quant. Grav.* **27** 215006 (*Preprint* 0906.4151)
[159] Messenger C and Read J 2012 *Phys. Rev. Lett* **108** 091101 (*Preprint* 1107.5725)
[160] Messenger C, Takami K, Gossan S, Rezzolla L and Sathyaprakash B S 2014 *Phys. Rev. X* **4** 040001 (*Preprint* 1312.1862)
[161] Van Den Broeck C, Trias M, Sathyaprakash B and Sintes A 2010 *Phys.Rev.* **D81** 124031 (*Preprint* 1001.3099)
[162] Sathyaprakash B and Schutz B 2009 *Living Rev.Rel.* **12** 2 (*Preprint* 0903.0338)
[163] Moore T A and Hellings R W 2002 Phys. Rev. D **65** 062001 (*Preprint* gr-qc/9910116)
[164] Hellings R W and Moore T A 2003 *Classical and Quantum Gravity* **20** 181 (*Preprint* gr-qc/0207102)
[165] Schnittman J D 2011 *Class.Quant.Grav.* **28** 094021 (*Preprint* 1010.3250)
[166] Shapiro C, Bacon D, Hendry M and Hoyle B 2010 *Mon.Not.Roy.Astron.Soc.* **404** 858–866 (*Preprint* 0907.3635)
[167] Hirata C M, Holz D E and Cutler C 2010 *Phys.Rev.* **D81** 124046 (*Preprint* 1004.3988)
[168] Harry G, Fritschel P, Shaddock D, Folkner W and Phinney E 2006 *Class.Quant.Grav.* **23** 4887–4894
[169] Kawamura S, Ando M, Seto N, Sato S, Nakamura T *et al.* 2011 *Class.Quant.Grav.* **28** 094011
[170] Cutler C and Holz D E 2009 Phys. Rev. D **80** 104009 (*Preprint* 0906.3752)
[171] Hirata C M, Holz D E and Cutler C 2010 Phys. Rev. D **81** 124046 (*Preprint* 1004.3988)
[172] Will C M 2014 *Living Reviews in Relativity* **17** 4 (*Preprint* 1403.7377)
[173] Stelle K S 1977 Phys. Rev. D **16** 953–969
[174] Hawking S W and Penrose R 1970 *Royal Society of London Proceedings Series A* **314** 529–548
[175] Weinberg S 1989 *Reviews of Modern Physics* **61** 1–23
[176] Deser S 1970 *General Relativity and Gravitation* **1** 9–18 (*Preprint* gr-qc/0411023)
[177] Wex N 2014 *ArXiv e-prints* (*Preprint* 1402.5594)
[178] Antoniadis J, Freire P C C, Wex N, Tauris T M, Lynch R S, van Kerkwijk M H, Kramer M, Bassa C, Dhillon V S, Driebe T, Hessels J W T, Kaspi V M, Kondratiev V I, Langer N, Marsh T R, McLaughlin M A, Pennucci T T, Ransom S M, Stairs I H, van Leeuwen J, Verbiest J P W and Whelan D G 2013 *Science* **340** 448 (*Preprint* 1304.6875)
[179] Gair J R, Vallisneri M, Larson S L and Baker J G 2013 *Living Reviews in Relativity* **16** 7 (*Preprint* 1212.5575)
[180] Yunes N and Siemens X 2013 *Living Reviews in Relativity* **16** 9 (*Preprint* 1304.3473)
[181] Clifton T, Ferreira P G, Padilla A and Skordis C 2012 Phys. Rep. **513** 1–189 (*Preprint* 1106.2476)
[182] Joyce A, Jain B, Khoury J and Trodden M 2014 *ArXiv e-prints* (*Preprint* 1407.0059)

[183] Burgess C P 2004 *Living Reviews in Relativity* **7** 5 (*Preprint* gr-qc/0311082)
[184] Burgess C P 2007 *Annual Review of Nuclear and Particle Science* **57** 329–362 (*Preprint* hep-th/0701053)
[185] Damour T and Esposito-Farèse G 1996 Phys. Rev. D **54** 1474–1491 (*Preprint* gr-qc/9602056)
[186] Alsing J, Berti E, Will C M and Zaglauer H 2012 Phys. Rev. D **85** 064041 (*Preprint* 1112.4903)
[187] Mirshekari S and Will C M 2013 Phys. Rev. D **87** 084070 (*Preprint* 1301.4680)
[188] Lang R N 2014 Phys. Rev. D **89** 084014 (*Preprint* 1310.3320)
[189] Yagi K, Stein L C, Yunes N and Tanaka T 2012 Phys. Rev. D **85** 064022 (*Preprint* 1110.5950)
[190] Yagi K, Stein L C, Yunes N and Tanaka T 2013 Phys. Rev. D **87** 084058 (*Preprint* 1302.1918)
[191] Narayan R 2005 *New Journal of Physics* **7** 199 (*Preprint* gr-qc/0506078)
[192] Narayan R and McClintock J E 2013 *ArXiv e-prints* (*Preprint* 1312.6698)
[193] Abramowicz M A, Kluźniak W and Lasota J P 2002 A&A **396** L31–L34 (*Preprint* astro-ph/0207270)
[194] Sotiriou T P and Faraoni V 2012 *Physical Review Letters* **108** 081103 (*Preprint* 1109.6324)
[195] Psaltis D, Perrodin D, Dienes K R and Mocioiu I 2008 *Physical Review Letters* **100** 091101
[196] Kleihaus B, Kunz J and Radu E 2011 *Physical Review Letters* **106** 151104 (*Preprint* 1101.2868)
[197] Ayzenberg D and Yunes N 2014 *ArXiv e-prints* (*Preprint* 1405.2133)
[198] Barausse E, Jacobson T and Sotiriou T P 2011 Phys. Rev. D **83** 124043 (*Preprint* 1104.2889)
[199] Barausse E and Sotiriou T P 2012 *Physical Review Letters* **109** 181101; erratum 2103 ibidem **110** 039902 (*Preprint* 1207.6370)
[200] Barausse E and Sotiriou T P 2013 Phys. Rev. D **87** 087504 (*Preprint* 1212.1334)
[201] Barausse E and Sotiriou T P 2013 *Classical and Quantum Gravity* **30** 244010 (*Preprint* 1307.3359)
[202] Herdeiro C A R and Radu E 2014 *Physical Review Letters* **112** 221101 (*Preprint* 1403.2757)
[203] Degollado J C and Herdeiro C A R 2014 Phys. Rev. D **90** 065019 (*Preprint* 1408.2589)
[204] Johannsen T 2013 Phys. Rev. D **87** 124017 (*Preprint* 1304.7786)
[205] Cardoso V, Pani P and Rico J 2014 Phys. Rev. D **89** 064007 (*Preprint* 1401.0528)
[206] Berti E, Cardoso V and Will C M 2006 Phys. Rev. D **73** 064030 (*Preprint* gr-qc/0512160)
[207] Berti E, Cardoso J, Cardoso V and Cavaglià M 2007 Phys. Rev. D **76** 104044 (*Preprint* 0707.1202)
[208] Barausse E and Sotiriou T P 2008 *Physical Review Letters* **101** 099001 (*Preprint* 0803.3433)
[209] Miller M C and Colbert E J M 2004 *International Journal of Modern Physics D* **13** 1–64
[210] Berti E, Cardoso V, Gonzalez J A, Sperhake U, Hannam M, Husa S and Brügmann B 2007 Phys. Rev. D **76** 064034 (*Preprint* gr-qc/0703053)
[211] Kamaretsos I, Hannam M, Husa S and Sathyaprakash B S 2012 Phys. Rev. D **85** 024018 (*Preprint* 1107.0854)
[212] Gossan S, Veitch J and Sathyaprakash B S 2012 Phys. Rev. D **85** 124056 (*Preprint* 1111.5819)
[213] Kamaretsos I, Hannam M and Sathyaprakash B S 2012 *Physical Review Letters* **109** 141102 (*Preprint* 1207.0399)
[214] Hughes S A and Menou K 2005 ApJ **623** 689–699 (*Preprint* astro-ph/0410148)
[215] London L, Healy J and Shoemaker D 2014 *ArXiv e-prints* (*Preprint* 1404.3197)
[216] Will C M 1998 Phys. Rev. D **57** 2061–2068 (*Preprint* gr-qc/9709011)
[217] Berti E, Buonanno A and Will C M 2005 Phys. Rev. D **71** 084025 (*Preprint* gr-qc/0411129)
[218] Berti E, Gair J and Sesana A 2011 Phys. Rev. D **84** 101501 (*Preprint* 1107.3528)
[219] Press W H and Teukolsky S A 1972 Nature **238** 211–212
[220] Cardoso V, Chakrabarti S, Pani P, Berti E and Gualtieri L 2011 *Physical Review Letters* **107** 241101 (*Preprint* 1109.6021)
[221] Cardoso V, Carucci I P, Pani P and Sotiriou T P 2013 *Physical Review Letters* **111** 111101 (*Preprint* 1308.6587)
[222] Cardoso V, Carucci I P, Pani P and Sotiriou T P 2013 Phys. Rev. D **88** 044056 (*Preprint* 1305.6936)
[223] Pani P, Berti E, Cardoso V, Chen Y and Norte R 2010 Phys. Rev. D **81** 084011 (*Preprint* 1001.3031)
[224] Macedo C F B, Pani P, Cardoso V and Crispino L C B 2013 Phys. Rev. D **88** 064046 (*Preprint* 1307.4812)

Research Update on Extreme-Mass-Ratio Inspirals

Pau Amaro-Seoane[1], Jonathan R. Gair[2], Adam Pound[3], Scott A. Hughes[4] and Carlos F. Sopuerta[5]

[1] Max Planck Institut für Gravitationsphysik (Albert-Einstein-Institut), D-14476 Potsdam, Germany
[2] Institute of Astronomy, University of Cambridge, Madingley Road, Cambridge, CB3 0HA, UK
[3] Mathematical Sciences, University of Southampton, Southampton, SO17 1BJ, UK
[4] Department of Physics and MIT Kavli Institute, Massachusetts Institute of Technology, Cambridge, MA 02139, USA
[5] Institut de Ciències de l'Espai (CSIC-IEEC), Campus UAB, Facultat de Ciències, 08193 Bellaterra, Spain

E-mail: sopuerta@ieec.uab.es

Abstract. The inspirals of stellar-mass mass compact objects into massive black holes in the centres of galaxies are one of the most important sources of gravitational radiation for space-based detectors like LISA or eLISA. These extreme-mass-ratio inspirals (EMRIs) will enable an ambitious research program with implications for astrophysics, cosmology, and fundamental physics. This article is a summary of the talks delivered at the plenary session on EMRIs at the 10th International LISA Symposium. It contains research updates on the following topics: astrophysics of EMRIs; EMRI science potential; and EMRI modeling.

1. Introduction

One of the most important sources of gravitational waves (GWs) for a space-based detector like eLISA [1, 2, 3] are extreme-mass-ratio binaries in the stage where the dynamics are driven by GW emission. These systems, called EMRIs, are composed of a stellar-mass compact object (SCO) (a white dwarf, a neutron star, or a stellar-mass black hole (BH)) that inspirals into a massive BH (MBH) located at a galactic centre. The masses of interest for the SCO are in the range $m = 1 - 10^2\ M_\odot$, and for the MBH in the range $M = 10^5 - 10^7\ M_\odot$. Then, the mass-ratio for these systems is in the interval $\mu = m/M \sim 10^{-7} - 10^{-3}$. The GW signals emitted by EMRIs are long lasting (months to years) and contain many GW cycles (of the order of 10^5 during the last year before the SCO plunges into the MBH). Many of these cycles are spent in the neighbourhood of the MBH horizon, meaning EMRI GWs encode a map of the strong-field region of the MBH. These extraordinary features of EMRIs allow for a revolutionary research program, which could lead to understanding different aspects of stellar dynamics in galactic centres, tests of the geometry of BHs, and tests of General Relativity (GR) and alternative theories of gravity. However, to implement such an ambitious program, several preliminary studies need to be done. In particular, we need to know how many events we expect to observe, which requires understanding which kind of astrophysical mechanisms can produce ERMI events. Even more importantly, given the complexity of EMRI signals (which are essentially composed of a number of harmonics of three independent frequencies), we need a very precise theoretical description of EMRI waveforms and also efficient data analysis search algorithms. Finally, it is important to carry out theoretical studies in astrophysics, cosmology, and fundamental physics that will enable us to exploit all the scientific potential of EMRI GW observations.

This writeup offers a broad discussion of these themes, summarizing the three talks given in the plenary session on EMRIs [4] at the 10th International LISA Symposium [5], which was organized and chaired by Scott A. Hughes and Carlos F. Sopuerta. The topics and organization of these talks are the following: the astrophysics of EMRIs (Pau Amaro-Seoane), in Sec. 2; an overview of EMRI science (Jonathan R. Gair), in Sec. 3; and the status of EMRI modeling (Adam Pound), in Sec. 4.

2. Astrophysics of EMRIs

2.1. Introduction

One of the most exciting results of modern astronomy is the discovery, mostly through high-resolution observations of the kinematics of stars and gas, that most, if not all, nearby bright galaxies harbor a dark, massive, compact object at their centers. The most spectacular case is our own galaxy, the Milky Way. By tracking and interpreting the stellar dynamics at the centre of our galaxy, we have the most well-established evidence for the existence of a MBH, with a mass of $\sim 4 \times 10^6 \, M_\odot$ [6, 7, 8, 9].

These observations are difficult, especially for low-mass MBHs (ranging between 10^5 and $10^7 \, M_\odot$) and even more so for the even less-massive intermediate MBHs. Nowadays, using adaptive optics we could optimistically hope to get a handful of measurements of stellar velocities of such targets about ~ 5 kpc away in some ten years. This is what makes an experiment such as eLISA [1, 2, 3] so appealing. Whilst main-sequence stars are tidally disrupted when approaching the central MBH, SCOs slowly spiral into the MBH and are swallowed after some $\sim 10^{3-5}$ orbits in the eLISA band. At the closest approach to the MBH, the system emits a burst of GWs which contains information about the spacetime geometry, the masses of the system, and the spins of the MBH. We can regard each such burst as a snapshot of the system. By extracting the information encoded in the GWs of this scenario, we can determine the redshifted mass of the central MBH with an error of $\sim 1 - 0.1\,\%$ and for the spin of the MBH with $0.01\,\%$.

2.2. The role of the MBH spin

For a binary of an MBH and an SCO to be in the eLISA band, it has to have a frequency of between roughly 10^{-4} and 1 Hz. The emission of GWs is more efficient as the SCO approaches its last stable orbit (LSO), so that eLISA will detect the sources when they are close to the LSO line. The total mass required to observe systems with frequencies between 10^{-4} and 1 Hz is $10^4 - 10^7 \, M_\odot$. For masses larger than $10^7 \, M_\odot$, the frequencies close to the LSO will be too low, so that their detection will be very difficult. On the other hand, for a total mass of less than $10^3 \, M_\odot$ we could in principal detect the EMRI at an early stage, but then the amplitude of the GWs would be rather low.

Naively one could assume that the inspiral time is dominated by GW emission and that if this is shorter than a Hubble time, the SCO will enter into an EMRI with the MBH. This is wrong, because one has to take into account the relaxation of the stellar system. Whilst it certainly can increase the eccentricity of the SCO's orbit, it can also perturb the orbit and circularize it, so that the inspiral time, $t_{\rm GW}$, becomes larger than a Hubble time. The condition for the SCO to be in an EMRI is that it is on an orbit for which $t_{\rm GW} \ll (1-e)\, t_{\rm r}$ [10], with $t_{\rm r}$ being the *local* relaxation time and e the orbital eccentricity. When the orbit has a semi-major axis for which the condition is not fulfilled, the SCO will have to be already on a so-called "plunging orbit", with $e \geq e_{\rm plunge} \equiv 1 - 4\, R_{\rm Schw}/a_{\rm plunge}$, where $R_{\rm Schw}$ is the Schwarzschild radius of the MBH (i.e., $R_{\rm Schw} = 2GM/c^2$, with M being the MBH mass), and $a_{\rm plunge}$ is the semi-major axis.

It has been claimed a number of times by different authors that these plunging orbits would result in too short a burst of gravitational radiation, which could only be detected if it originated in our own Galactic Center [11] because one needs a coherent integration of some few thousand repeated passages through the periapsis in the eLISA bandwidth. Therefore, such plunging objects would be lost from the GW signal, since they would be plunging directly through the horizon of the MBH and only a final burst of GWs would be emitted, and (i) such a burst would be very difficult to recover, since the very short signal would be buried in a sea of instrumental and confusion noise, and (ii) the information contained in the signal would be practically nil. There has been some work on the detectability of such bursts [12, 11, 13], but they would only be detectable in our galaxy or in the close neighborhood, and the event rates are rather low, even in the most optimistic scenarios.

However, this picture is based on orbits around a non-spinning, Schwarzschild MBH, and it is significantly altered when the spin of the MBH is accounted for. Reference [14] estimated the number of cycles that certain EMRI orbital configurations that were thought to be plunging orbits (or orbits with insufficiently many cycles) in the case of non-spinning MBHs can spend in a frequency regime of $f \in [10^{-4}, 1]$ Hz during their last year(s) of inspiral before plunging into a spinning MBH. It is important to assess how many of these EMRIs will have sufficient Signal-to-Noise Ratio (SNR) to be detectable. Reference [14] found that (prograde) EMRIs in a very radial orbit (that would lead to a plunge in the case of a Schwarzschild MBH) actually spend a significant number of cycles near the LSO in the Kerr case, more than sufficient to be detectable with good SNR. The number of cycles has previously been associated with N_φ (the number of times that the azimuthal angle φ advances 2π), which is usual for binary systems. However, as we mentioned above, the structure of the waveforms from EMRIs is quite rich since they contain harmonics of three different frequencies: f_φ, f_r (associated with the time to go from apoapsis to periapsis and back), and f_θ (associated with the time to complete an oscillation around the orbital plane in the case of inclined orbits). Therefore the waveforms have cycles associated with these three frequencies, $(f_r, f_\theta, f_\varphi)$, which makes them quite complex, and in principle this is good for detectability (assuming we have the correct waveform templates). Moreover, these cycles happen just before plunge and take place in the strong field region very near the MBH horizon. So these cycles should contribute more to the SNR than cycles taking place farther away from the MBH horizon.

More interestingly, very highly eccentric EMRIs have an impact on event rates. One can prove that

$$\frac{a_{\rm EMRI}^{\rm Kerr}}{a_{\rm EMRI}^{\rm Schw}} = \mathcal{W}^{\frac{-5}{6-2\gamma}}(\iota,\, s)\,, \qquad (1)$$

$$\frac{\dot{N}_{\rm EMRI}^{\rm Kerr}}{\dot{N}_{\rm EMRI}^{\rm Schw}} = \mathcal{W}^{\frac{20\gamma-45}{12-4\gamma}}(\iota,\, s)\,, \qquad (2)$$

where $a_{\rm EMRI}$ is the maximum radius within which we estimate the event rate $\dot{N}_{\rm EMRI}$, and the "Schw"/"Kerr" labels refer to results for a non-spinning (Schwarzschild)/spinning (Kerr) MBH. In these expressions, $\mathcal{W}$ is a function that depends on ι, the inclination of the EMRI orbit, and on s, the MBH dimensionless spin parameter[1]. We also have assumed that the stellar BHs distribute around the central MBH following a power-law cusp of exponent γ; i.e., that the density profile follows $\rho \propto r^{-\gamma}$ within the region where the gravity of the MBH dominates over the gravity of the stars, with γ ranging between 1.75 and 2 for the heavy stellar components

[1] For the derivation and some examples of values for $\mathcal{W}$, we refer the reader to the work of [14].

[15, 16, 17, 18, 19, 20, 21, 22]. See [23] for an interesting first appearance of this concept[2].

For instance, for a spin of $s = 0.999$ and an inclination of $\iota = 0.4\,\text{rad}$, it is estimated that $\mathcal{W} \sim 0.26$ and, thus, $N_{\rm EMRI}^{\rm Kerr} \sim 30$. That is, *the event rates are boosted by a factor of 30* in comparison to a non-rotating MBH.

Very eccentric EMRIs spend enough cycles inside the band of eLISA to be detectable. We note here that whilst it is true that EMRIs very near the new separatrix shifted by the spin effect will probably not contribute enough cycles to be detected, it is equally true for the old separatrix (Schwarzschild, without spin). In this sense, it is found that the spin generically increases the number of cycles inside the band for prograde EMRIs in such a way that EMRIs very near to the non-spin separatrix, which contributed few cycles, become detectable EMRIs. In summary, spin increases the area, in configuration space, of detectable EMRIs. One can thus predict that EMRIs will be highly dominated by prograde orbits.

Moreover, because spin allows for stable orbits very near the horizon in the prograde case, the contribution of each cycle to the SNR is significantly bigger than each cycle of an EMRI around a non-spinning MBH.

It has also been shown that vectorial resonant relaxation will not be efficient enough to change prograde orbits into retrograde orbits once GW evolution dominates (which would make the EMRIs plunge instantaneously, as they would be in a non-allowed region of phase space).

More remarkably, the very eccentric EMRIs we have described here are solely produced by two-body relaxation, a chaotic process, and as such, they are ignorant of the Schwarschild barrier [24, 25], a phenomenon that drastically suppresses EMRI rates close to the MBH. While low-eccentricity EMRIs run into the problem of having to find a way to traverse this barrier, very eccentric EMRIs do not. One can therefore predict that EMRI rates will be dominated by high-eccentricity binaries, with the proviso that the central MBH is described by the Kerr solution of GR.

2.3. Conclusions

The event rate of very eccentric EMRIs is much larger than that of low-eccentricity EMRIs, as a number of different studies by different authors using different methods have found (although in earlier works very eccentric EMRIs were enviasaged to be direct plunges since they only considered Schwarzschild MBHs). Up to now, spin effects of the central MBH have been neglected. Hence, the question arises, whether a plunge is really a plunge when the central MBH is spinning. This consideration has previously been ignored.

To estimate EMRI event rates, one needs to know whether the orbital configuration of the SCO is stable or not, because this is the kernel of the difference between an EMRI and a plunge. Reference [14] takes into account the fact that the spin brings the LSO much closer to the horizon in the case of prograde orbits but it pushes it away for retrograde orbits. Since the modifications introduced by the spin are not symmetric with respect to the non-spinning case, and they are more dramatic for prograde orbits, one can prove that the inclusion of spin increases the number of EMRI events by a significant factor. The exact factor of this enhancement depends on the spin, but the effect is already quite important for spins around $s \sim 0.7$.

Reference [14] proves that these very eccentric EMRIs do spend enough cycles inside the band of eLISA to be detectable. Although EMRIs very near the new separatrix shifted by the spin effect will probably not contribute enough cycles to be detected, it is equally true for the old separatrix, without spin. The conclusion is that the presence of spin increases the area, in configuration space, of detectable EMRIs. As a consequence, EMRIs will be highly dominated by prograde orbits.

[2] The authors obtained a similar solution for how electrons distribute around a positively charged Coulomb centre.

It has also been shown that these new kind of EMRIs originate in a region of phase-space such that they will be ignorant of the Schwarschild barrier [24, 25], which drastically reduces the rates for low eccentricity EMRIs. The reason for this is that they are driven by two-body relaxation and not resonant relaxation. While the boost in EMRI rates due to resonant relaxation is affected by the Schwarzschild barrier, so that low-eccentricity EMRIs run into the problem of having to find a way to cross it, high-eccentricity EMRIs are already in the right place and led by two-body relaxation, a chaotic process ignorant of secular effects. The barrier affects the production of EMRIs via torques, but not two-body relaxation, which is the mechanism producing the high-eccentricity EMRIs. Moreover, because the supression in the rates is severe for those EMRIs with semi-major axes with values approximately $a \gtrapprox 0.03$ pc, it has been predicted that the rates will be dominated by the kind of EMRIs described in Ref. [14].

In short, the existence of these highly eccentric EMRIs is good news: They sample regions closer to the MBHs' event horizons, offer richer signal sources, are not suppressed by secular effects (unlike lower-eccentricity EMRIs), enhance the event rates, and are louder and, hence, probe larger horizon distances.

3. The scientific potential of EMRI observations

It is expected that a space-based gravitational wave detector similar to eLISA [1] will observe a few tens of EMRI events per year [3], although this number is subject to large astrophysical uncertainties. These events will be seen out to redshifts $z \sim 0.7$ and they will primarily be the inspirals of stellar BHs with mass $m \sim 10M_\odot$ into MBHs with mass $M \sim 10^5$–$10^6 M_\odot$ [26]. For every EMRI observed, eLISA will measure the system parameters to high precision, with typical fractional errors on masses of $\sim 10^{-4}$, typical errors on the spin of the central black hole of a few times 10^{-4} and typical precision on the sky location and distance of a source of a few square degrees and $\sim 10\%$ respectively. These precise measurements come from accurately tracking the phase of the inspiral over the $\sim 10^5$ waveform cycles observed for a typical system and are therefore not very dependent on the final design of the detector. The precise measurements for tens of systems that eLISA EMRIs will provide have tremendous potential to inform us about astrophysics, cosmology, and fundamental physics.

3.1. Astrophysics

Massive black holes in the low-mass range that will be observed by eLISA are not well constrained by electromagnetic observations, and there are therefore big uncertainties in the demographics of this population. EMRI observations will measure the mass and spin distributions of these MBHs in the local Universe for the first time. Using a simple power law model for the BH mass function, $dn/d\log M = AM^\alpha$, it was shown that with N eLISA observations of EMRIs, constraints would be placed on the BH mass function at the level [27]

$$\Delta(\ln A) \approx 1.1\sqrt{10/N}, \qquad \Delta(\alpha) \approx 0.35\sqrt{10/N}. \tag{3}$$

The current uncertainty in the slope of the mass function is $\sim \pm 0.3$, and so eLISA will improve on this with as few as 10 EMRI observations. Predicted event rates are a factor of several larger than this. eLISA observations of EMRIs by themselves will not be able to probe evolution in the mass function with redshift [27]. The analysis in [27] assumed that the scaling of the astrophysical EMRI rate with the mass of the central MBH was known, so these constraints are really on the slope of the EMRI-rate-weighted mass function. While analytic and numerical calculations might improve our understanding of the rate scaling before eLISA is launched, the best chance to break this degeneracy is probably to use both eLISA observations of EMRIs and those of MBH mergers together to probe the mass function and its evolution.

EMRI observations have the potential to inform us about several other aspects of astrophysics, although these have not yet been investigated in detail. The properties of EMRI orbits measured by eLISA encode details of EMRI formation mechanisms. In the standard capture scenario, EMRIs would be expected to be on eccentric orbits inclined to the equatorial plane of the MBH, but other formation scenarios make different predictions: EMRI formation by the splitting of a binary would tend to create circular but inclined EMRIs; massive star formation in a disc would lead to circular and equatorial EMRIs; and formation of EMRIs via the tidal stripping of massive stars would lead to circular, inclined EMRIs involving low mass compact objects (white dwarfs) [10]. eLISA observations will indicate the relative importance of these different mechanisms. In addition, the rates, orbital properties and compact object masses of EMRI events probe the dynamical processes, such as relaxation and mass segregation, that drive them and the properties of stellar populations (e.g., mass functions) in galactic centres.

3.2. Cosmology

The amplitude of a gravitational wave source scales as $(1+z)M/D_L(z)$, where M is the mass, z the redshift and $D_L(z)$ the luminosity of the source. The redshifted mass, $(1+z)M$, can be determined very precisely from the observed phase evolution of the source, so provided that the mass-redshift degeneracy can be broken, gravitational wave sources can be used as "standard sirens" to measure $D_L(z)$ and probe the cosmological expansion history of the Universe [28]. Redshifts can be measured directly if an electromagnetic counterpart to a GW source is observed. For EMRIs, the only possible counterpart would be the tidal disruption of a white dwarf by a low-mass, highly-spinning BH [29]. However, it is extremely unlikely that any such EMRIs will be observed [26]. Constraints can also be derived statistically using many EMRI events and this was examined in [30]. For each event, every observed galaxy consistent with the error bars from the GW measurement can be used to estimate the redshift of the source and hence a value for the luminosity distance or cosmological parameters. Averaging over multiple EMRI events provides final constraints on the cosmology. It was found that LISA observations of ~ 20 events at redshift $z \lesssim 0.5$ would be sufficient to measure the Hubble constant, H_0, to $\sim 1\%$. eLISA will have slightly larger error bars for each EMRI, and therefore more events will be required to reach the same precision. However, we expect eLISA to observe several tens of EMRIs, all in this redshift range, and so we should be able to place constraints on H_0 at the level of $1-2\%$. Although electromagnetic constraints are likely to be more precise than this before eLISA is launched, the GW derived constraints will be completely independent, providing an important verification of existing results and a way to identify previously unknown systematics.

3.3. Fundamental physics

GW observations will probe sources in a regime of strong-field, non-linear and dynamical gravity that has never been explored, thus providing new tests of our understanding of the theory of gravity. All GW sources and detectors can be used for tests of fundamental physics, but EMRIs will provide particularly powerful probes as they will generate many tens of thousands of observable cycles in the strong-field regime, they are "clean" systems, as we expect most EMRIs to be BH binaries, and the orbital dynamics are rich—we expect most EMRIs to be both eccentric and inclined, so the emitted waveforms comprise a complex superposition of three fundamental frequencies—and the inspiralling object explores a large fraction of the space-time before it plunges. For a thorough review of the tests of fundamental physics that will be possible using EMRIs and other eLISA sources, we refer the interested reader to [31], but we will summarise some of the main ideas here.

In general relativity, with some additional physicality assumptions, it is known that the Kerr BH is the unique end state of gravitational collapse [32]. This is often referred to as the "no-hair" theorem, since the Kerr metric depends on only two parameters, mass M and spin a, and all higher mass (M_l) and spin (S_l) multipole moments of the space-time are related to these as $M_l + iS_l = M(ia)^l$. The GWs emitted by an EMRI encode the orbital dynamics of the inspiralling object, which are in turn determined by the space-time structure. EMRI GWs thus encode a map of the space-time structure, which can be used to verify the no-hair property of the central black hole. This was first discussed in [33], in which a space-time with arbitrary multipole moments was used to show that the multipoles were separately encoded in gravitational wave observables, specifically the orbital precession rate and inspiral rate. The multipole expansion is not particularly practical, as an infinite number of multipoles are needed to represent Kerr. Recent work has instead focussed on "bumpy" black holes [34], in which the metric is Kerr plus a small perturbation. There have been many different studies of EMRI constraints on bumpy black holes [34, 35, 36, 37], including BH solutions motivated by alternative theories of gravity [38, 39]. These studies conclude that EMRI observations will be able to simultaneously measure the MBH mass and spin to $\sim 0.01\%$ and a deviation in the quadrupole moment of the black hole of $\sim 0.1\%$. In addition, EMRI observations could place a bound on the deviation parameter characterising dynamical Chern Simons modified gravity (a parity-violating alternative to GR inspired by string theory) of $\xi^{\frac{1}{4}} < 10^4$km [39].

GWs from EMRIs also encode qualitative information about the surface of the central object, whether or not the external space-time differs from Kerr. The presence/absence of a horizon would be indicated by a cut-off/continuation of GW emission after plunge (e.g., persistent emission for a massive boson star [40]); the properties of the surface of the central object can be inferred by comparing the observed inspiral rate and observed energy flux to deduce the energy lost to tidal interaction between the two objects [41], with $O(100\%)$ changes in the tidal energy loss being detectable with eLISA; and if the quasi-normal mode (QNM) spectrum of the central object is sufficiently different from that of Kerr, an EMRI could tidally excite QNMs as its orbital frequency passes through the QNM frequencies during an inspiral, leading to a resonant signature in the emitted GWs—e.g., for orbits around "gravastars", which comprise a thin shell of material with a de Sitter interior and Schwarzschild black hole exterior [42].

Measurable departures of an EMRI signal from the predictions of GR could also arise from astrophysical perturbations. The gravitational effect of a sufficiently dense and massive accretion torus could measurably perturb an EMRI [43], although this imprint can be mimicked by an error in the MBH mass and spin. In addition, hydrodynamic drag from material that the EMRI is moving through will qualitatively change the inspiral, by driving the orbit to become more prograde, while a GW driven inspiral should become more retrograde [44]. However, in both cases the imprints will only be measurable if the amount of matter close to the central BH is unphysically high. If the inspiralling object in an EMRI is migrating through a massive disc a ~ 1 radian dephasing could accumulate over a typical observation [45]. Finally, the presence of a second MBH within ~ 0.1pc would leave a detectable imprint [46], while the presence of a second small object near the EMRI could lead to chaotic motion for $\sim 1\%$ of EMRIs [47]. For each of these mechanisms, the number of systems in which a detectable signature could accumulate is small and, overall, we expect that these effects should leave a measurable imprint in only a small fraction ($\lesssim 10\%$) of observed EMRIs. These imprints will be qualitatively different from fundamental physics effects and so will not be misinterpreted as the latter, but instead will provide constraints on the underlying astrophysical process.

EMRI observations can also be used in several other ways to test fundamental physics. Kerr orbits have a complete set of constants of motion, allowing the orbital motion to be separated, but this will not be true in generic space times, leading to qualitative differences in the dynamics that are potentially observable, such as the appearance of chaos late in the inspiral [37] or the appearance of persistent resonances during the inspiral [48]. EMRI observations can be used to test specific alternative theories of gravity (e.g., scalar-tensor gravity [49, 50], dynamical Chern Simons modified gravity [39] and scalar Gauss-Bonnet gravity [51]) or to place constraints on generic deviations from GR using phenomenological models [52, 53]. In most cases, the constraints possible with EMRIs are one or more orders of magnitude better than those that are available currently or likely to be obtained in the near future. EMRIs can be used to test that energy losses to gravitational waves are consistent with the quadrupole formula by comparing the inspiral rate to the quadrupole prediction. Indeed, the constraints on scalar-tensor gravity come primarily from detecting the excess energy lost to dipole radiation [49]. Finally, EMRI observations can be used to test that GW polarisation is consistent with the two transverse-tensor modes expected in GR. Up to four alternative polarisation states exist in alternative theories of gravity, and eLISA's sensitivity to longitudinal modes is ten times greater than that to transverse modes at high frequencies [54]. Whether or not such modes are detectable will of course depend on the relative amplitude of excitement of the different modes in astrophysical sources. For a detailed discussion of all of these possibilities, see [31] and references therein.

4. Status of EMRI modeling

4.1. Basic features and challenges

Realizing the scientific potential of EMRI observations will require an accurate model of their waveforms. This poses many distinct challenges. The orbital motion is highly relativistic and the fields are strong, meaning post-Newtonian approximations are inapplicable. The system contains disparate lengthscales, the SCO's size $\sim m$ and the orbital scale $\sim M$, meaning full numerical relativity is unsuitable. Furthermore, the system contains disparate timescales: in addition to the orbital period $\sim M$, there is the inspiral time $t_{GW} \sim M/\mu$. For an astrophysically relevant mass ratio $\mu \sim 10^{-5}$, the inspiral time corresponds to an enormous $\sim 10^5$ wavecycles. Hence, to extract system parameters from an observed signal, a model must accurately match the true waveform's phase over this same, extraordinary number of orbits.

Despite these obstacles, substantial progress has been made in modeling EMRIs in recent years. Inspirals can now be simulated with an accuracy that may suffice for signal detection, if not for parameter estimation [55, 56]. Numerous highly accurate calculations on the shorter, orbital timescale have been performed for generic orbits about a nonrotating BH [57, 58, 59, 60, 61], as well as some for circular, equatorial orbits about a rotating one [62, 63]. And the fundamental ingredients are now in place to simulate a complete inspiral accurately enough for parameter estimation, assuming the system is "clean" and described by classical GR [64, 65, 66, 67]. This progress has largely stemmed from advances in the self-force program [68, 69], in which the SCO is treated as a source of perturbation of the background geometry of the MBH. Before detailing the various calculations that have been performed, we will briefly review the formalism.

4.2. The self-force formalism

The gravitational self-force program, initiated nearly twenty years ago [70, 71], is now mature and on firm mathematical footing [72, 65, 73]. Its essential idea is to write the full metric of the system, $\mathsf{g}_{\mu\nu}$, as an expansion

$$\mathsf{g}_{\mu\nu} = g_{\mu\nu} + h^1_{\mu\nu} + h^2_{\mu\nu} + O(\mu^3), \tag{4}$$

where $g_{\mu\nu}$ is the metric of the MBH, and $h^n_{\mu\nu} \sim \mu^n$ is the nth-order perturbation due to the SCO. The perturbations travel to infinity, where they can be observed as gravitational waves, but they also distort the geometry around the small object, influencing its motion; this effect on the motion is termed the self-force.

To cope with the challenges of EMRI modeling, this general picture must be tailored to accommodate both the multiple lengthscales and the multiple timescales in the problem.

4.2.1. Matched asymptotic expansions The multiple lengthscales are handled using the method of matched asymptotic expansions [70, 72, 65]. One first defines some appropriate worldline

z^μ to represent the SCO's bulk motion in the background spacetime. Near the SCO, in a region of radius $r \sim m$ around z^μ, the gravity of m dominates over that of M, and the 'outer expansion' (4) fails to be accurate. In this region, a complementary 'inner expansion' is assumed to exist, one that effectively zooms in on the SCO by letting $\mu \to 0$ while holding r/m fixed. In this inner expansion, the background metric is that of the SCO itself. In a 'buffer region', where $m \ll r \ll M$, both the inner and outer expansions are valid, and information about the SCO can be fed from the inner expansion to the outer. It turns out that in this procedure, only minimal information about the SCO is required—specifically, the SCO's multipole moments. An algorithm now exists for calculating the metric in the buffer region to arbitrary order in perturbation theory [74], and the end result is a local expression for each $h^n_{\mu\nu}$ in terms of the SCO's multipole moments.

4.2.2. A generalized equivalence principle More precisely, the result of the local analysis in the buffer region is a split of $h^n_{\mu\nu}$ into two pieces:

$$h^n_{\mu\nu} = h^{Sn}_{\mu\nu} + h^{Rn}_{\mu\nu}. \tag{5}$$

The first piece, $h^{Sn}_{\mu\nu}$, loosely represents the SCO's bound field, being constructed locally from the SCO's multipole moments. The second piece, $h^{Rn}_{\mu\nu}$, propagates independently of the SCO and cannot be determined locally in the buffer region; it is determined only from the global boundary conditions imposed on the perturbations $h^n_{\mu\nu}$.

The fields $h^{Rn}_{\mu\nu}$ together form a smooth vacuum metric $\hat{\mathrm{g}}_{\mu\nu} = g_{\mu\nu} + \sum_n h^{Rn}_{\mu\nu}$, an effective metric in the neighbourhood of the SCO. This effective metric has proven to be extremely important: at least through second order in μ, a sufficiently slowly spinning and sufficiently spherical[3] SCO is known to move on a geodesic of $\hat{\mathrm{g}}_{\mu\nu}$ [66, 75]. In terms of the fields $h^{Rn}_{\mu\nu}$ on the background $g_{\mu\nu}$, the geodesic equation in $\hat{\mathrm{g}}_{\mu\nu}$ can be written as

$$\begin{aligned}\frac{D^2 z^\mu}{d\tau^2} &= \frac{1}{2}\left(g^{\mu\nu} + u^\mu u^\nu\right)\left(g_\nu{}^\rho - h^{R\rho}_\nu\right)\left(h^R_{\sigma\lambda;\rho} - 2h^R_{\rho\sigma;\lambda}\right)u^\sigma u^\lambda + O(\mu^3) \\ &\equiv F^\mu_1 + F^\mu_2 + O(\mu^3),\end{aligned} \tag{6}$$

where F^μ_n is the nth-order self-force, $h^R_{\mu\nu} = h^{R1}_{\mu\nu} + h^{R2}_{\mu\nu}$, τ is proper time on z^μ as measured in $g_{\mu\nu}$, $u^\mu = \frac{dz^\mu}{d\tau}$, and a ';' denotes covariant differentiation compatible with $g_{\mu\nu}$. Equation (6) is a statement of a *generalized equivalence principle*: the SCO, regardless of its internal composition, falls freely in a certain geometry $\hat{\mathrm{g}}_{\mu\nu}$, which in some sense it feels to be the ambient vacuum around it.

This generalized equivalence principle was first formulated at linear order in μ by Detweiler and Whiting [76]. A variant of it has been found to hold true even in the fully nonlinear, non-perturbative context of a material body with generic multipole structure [73].

4.2.3. Point particles and punctures To utilize the above results in a practical model, one must replace the inner expansion with something more useful in the region $r \sim m$. One does so by simply extending the result (5) from the buffer region down to $r = 0$. After this extension is performed, $h^{Sn}_{\mu\nu}$ gets called the singular field, because it diverges on z^μ, while $h^{Rn}_{\mu\nu}$ gets called the regular field, because it is smooth at z^μ. Crucially, this replacement of the physical metric in the SCO's interior with the extension from the buffer region does not alter the motion of the SCO or the metric outside the SCO.

[3] For a generic SCO, leading-order spin and quadrupole moment contributions to Eq. (6) can be obtained straightforwardly [72, 65, 74, 73], but order-$(m/M)^2$ subleading spin effects are unknown in the perturbative context [74].

Journal of Physics: Conference Series **610** (2015) 012002 doi:10.1088/1742-6596/610/1/012002

At first order, the extended field $h^1_{\mu\nu}$ is identical to the perturbation produced by a point mass in linearized gravity [72, 65], meaning it can be calculated numerically from the linearized Einstein equation with a point-mass source. Beyond first order, the point particle picture fails, but one can still use this extension procedure to effectively replace the complicated physics inside the SCO with the simple field $h^{Sn}_{\mu\nu}$, which appears as a 'puncture' (rather than a partcle) on the manifold. A globally well-defined field equation for the extended $h^n_{\mu\nu}$ is not known in this nonlinear case, but the field equations for $h^{Rn}_{\mu\nu}$ [4] in a region covering the SCO can easily be combined with equations for $h^n_{\mu\nu}$ outside the SCO to obtain a global solution [77, 66, 74, 67].

4.2.4. Representations of motion In Sec. 4.2.1, we stated that "One first defines some appropriate worldline z^μ." That simple statement elides several complexities involved in the choice of worldline [65]. In brief, on the orbital timescale $\sim M$, the orbit can be approximated as a geodesic of $g_{\mu\nu}$, and the perturbation can be calculated with the particle/puncture moving on that geodesic. Self-force effects then appear as order-μ corrections to the geodesic, and these corrections appear as terms in $h^2_{\mu\nu}$ rather than altering the position of the particle/puncture. But on the inspiral timescale t_{GW}, the orbit deviates by a large amount from any reference geodesic, and to maintain accuracy, the effect of the self-force must be incorporated into the motion of the particle/puncture.

There are two ways to incorporate this long-term backreaction of the perturbation on the worldline. First, one might solve the equation of motion (6) coupled to the field equations for $h^n_{\mu\nu}$ and $h^{Rn}_{\mu\nu}$, with the particle/puncture moving on z^μ; this is called the 'self-consistent' approach [72, 65]. Alternatively, on the right-hand side of Eq. (6), but not on the left, one might replace z^μ with an expansion $z^\mu = z^\mu_{0(\tau)} + z^\mu_{1(\tau)} + O(\mu^2)$, where $z^\mu_{0(\tau)}$ is the geodesic of $g_{\mu\nu}$ that is instantaneously tangential to z^μ at time τ. At each instant τ on z^μ, F^μ_1 is then calculated from the solution to the linearized Einstein equation with the particle moving on $z^\mu_{0(\tau)}$, while F^μ_2 picks up the effect of the correction $z^\mu_{1(\tau)}$. Call this the 'geodesic source + osculation' approach, named after the fact that the source for $h^1_{\mu\nu}$ is at each instant on z^μ constructed from the osculating geodesic $z^\mu_{0(\tau)}$ [78, 56].

4.3. Self-forced motion in Kerr spacetime: A hierarchy of models

The self-force formalism can, in principle, be carried to any order in its perturbative expansion. But how high must we go to build useful EMRI models?

Geodesics in Kerr are characterized by their energy E, angular momentum L_z, and Carter constant Q, or equivalently, by their three orbital frequencies f_r, f_θ, and f_ϕ. When the self-force is accounted for, its dissipative piece causes slow changes in these 'constants' of motion. However, the conservative piece of the self-force also causes long-term changes that cannot be neglected [78]. The relative importance of the various pieces of the self-force, through second order, was first systematically studied by Hinderer and Flanagan [64]. They used a two-timescale expansion of accelerated motion in Kerr to study generic behavior on both the orbital and inspiral timescales, leading to the following expansion of the orbital phase:

$$\phi(t,\mu) = \frac{1}{\mu}\left[\phi_0(\tilde{t}) + \mu\phi_1(\tilde{t}) + O(\mu^2)\right] \tag{7}$$

where $\tilde{t} \equiv \mu t$ is $\sim M$ when $t \sim t_{GW}$. In this expansion, the leading term, ϕ_0, called the adiabatic order, is determined by the averaged dissipative piece of F^μ_1; at this order, only the

[4] More accurately, numerically practical field equations are formulated for 'residual fields' $h^{\mathcal{R}n}_{\mu\nu} \approx h^{Rn}_{\mu\nu}$, which can be chosen such that their values on z^μ, and the values of any finite number of their derivatives, agree with those of $h^{Rn}_{\mu\nu}$ and its derivatives on z^μ.

Journal of Physics: Conference Series **610** (2015) 012002 doi:10.1088/1742-6596/610/1/012002

leading averaged rates of change of E, L_z, and Q are required. The subleading term, ϕ_1, called the post-1 adiabatic order, is determined by the averaged dissipative piece of F_2^μ, the conservative piece of F_1^μ, and the oscillatory dissipative piece of F_1^μ. A model that captures only ϕ_0 is likely to suffice for signal detection but not for parameter estimation. A model that captures both ϕ_0 and ϕ_1 ensures all errors are small, $\sim \mu$, over an inspiral; therefore, it is expected to be sufficient for parameter estimation.

This description assumes the orbital frequencies f_r and f_θ are incommensurate. When the frequencies are related by a rational ratio, orbital resonances occur and substantially complicate the picture. Because the orbital frequencies continually evolve, the resonance will last only a relatively short time $\sim M/\sqrt{\mu} \ll t_{GW}$.[5] But during this time, the orbital elements change rapidly, leading to large, $\sim \sqrt{1/\mu}$ shifts in the waveform's phase [80]. These transient resonances occur generically in most inspirals [81]. They have interesting mathematical properties [82] and observational relevance even outside of gravitational wave astronomy [83], but for the purpose of EMRI modeling, they are of great importance for one reason: even for signal detection, their effects might have to be included, but a leading-order adiabatic evolution does not capture those effects. Correctly modeling resonances is thought to have the same requirements as a post-1 adiabatic approximation [80]—the complete F_1^μ and the averaged dissipative effect of F_2^μ. However, for detection, it might instead suffice to use the adiabatic approximation in the sections of waveform between resonances.

We now detail the calculations that have been performed, moving in order of increasing completeness, from adiabatic evolution to the full F_1^μ to F_2^μ.

4.3.1. Adiabatic evolution Building on seminal work by Mino [84], several authors developed practical adiabatic approximation schemes that express the averaged rates of change $\langle \dot{E} \rangle$, $\langle \dot{L}_z \rangle$, and $\langle \dot{Q} \rangle$ in terms of the solution to the Teukolsky equation [85, 86, 87, 88]. Compared to obtaining the complete F_1^μ, these schemes are extremely simple, and they have been implemented for generic (inclined and eccentric) orbits in Kerr [55]. However, these calculations do not account for orbital resonances, possibly limiting their usefulness in signal detection. As a first step to overcoming this limitation, the effects of resonances on $\langle \dot{E} \rangle$, $\langle \dot{L}_z \rangle$, and $\langle \dot{Q} \rangle$ have been calculated [89, 90], and "post-adiabatic" methods for evolving through resonances are under development.

4.3.2. First-order self-force There are now several active codes capable of calculating first-order effects in both the time domain [57, 91] and the frequency domain [62, 92, 93]. One of two routes is typically used in these calculations. Either one calculates $h^1_{\mu\nu}$ using a point particle source and then subtracts the contribution of $h^{S1}_{\mu\nu}$ to find $h^{R1}_{\mu\nu}$ and F_1^μ, as in mode-sum schemes [94, 95, 96, 97]; or one calculates the regular quantities $h^{R1}_{\mu\nu}$ and F_1^μ directly, as in puncture/effective-source schemes [98, 99, 100].

Numerous calculations have been performed, and they are best divided into two categories: those for orbits about a Schwarzschild BH, and those for orbits about a Kerr BH. In all cases, calculations have used a geodesic source orbit, meaning they are accurate only on short timescales but can be incorporated into two-timescale or osculating-geodesic long-term evolution schemes.

For orbits in Schwarzschild, inspirals of moderate eccentricity have been simulated [56] using the 'geodesic source + osculation' approach. Furthermore, many physical effects have been calculated from the conservative sector of the dynamics on the orbital timescale: orbital precession [61], the shift in frequency of the innermost stable circular orbit (ISCO) in Schwarzschild [59, 101, 102], Detweiler's redshift variable (the ratio of proper time measured

[5] Although it is possible for an orbit to become trapped near a resonance, such an event is extremely unlikely to be observed [79].

along the geodesic in the regular metric to the time measured by an inertial observer at infinity) [58, 61], spin precession [103], and tidal effects [104].

For orbits in Kerr (the astrophysically relevant case), no inspirals have yet been simulated. However, calculations have been performed for circular orbits [62, 63], and from those results, Detweiler's redshift variable and the ISCO shift have been found [105]. Also, recent work has established how to obtain self-force effects by reconstructing $h^1_{\mu\nu}$ from Teukolsky variables, rather than via the more technically demanding route of directly solving the nonseparable linearized Einstein equation [62, 106]; this should spark a new round of calculations for more generic orbits.

Although short-term conservative effects in Schwarzschild and Kerr are not directly relevant to inspiral simulations, they have engendered a fruitful cross-cultural pollination between self-force models and other binary models [58, 107, 108, 109, 101, 102, 110, 111], setting benchmarks for numerical relativity, fixing high-order PN parameters and calibrating effective-one-body theory (EOB) [112, 108, 101, 111, 113, 114, 115, 116]. More strikingly, the comparisons with numerical relativity have shown that self-force results are surprisingly accurate even in the limit of comparable masses [109, 102, 110]. See Ref. [117] for a recent review of this synergy between models.

All of the results mentioned in this section have been for the astrophysically relevant case of a gravitational perturbation. Significantly more progress has been made in the case of a scalar field, which has proven to be an important testbed for numerical methods in the self-force program [118, 119, 98, 99, 120, 121]. Fully self-consistent inspirals have been evolved in Schwarzschild in this case [122]. And in the geodesic-source approximation, many calculations have been performed for orbits in Kerr [123, 124, 125], leading to successful calculations of scalar self-force effects for both inclined [126] and highly eccentric orbits in Kerr [127]. Much of the numerical infrastructure developed for this work, and the physical insight gained from it, should carry over to the gravitational case.

4.3.3. Second-order self-force After several preliminary studies of the second-order problem [128, 129, 65, 130], in recent years complete second-order formalisms have been derived [67, 66, 74, 75]. Because the point-particle description fails beyond linear order, mode-sum regularization can no longer be used. However, puncture schemes arise naturally. As of yet, no numerical implementations of these puncture schemes have been performed, but practical numerical ingredients and plans have been steadily developing [131, 132].

4.4. Summary of progress and current status

The self-force formalism is now well developed and robust, extendable to any order in perturbation theory and to spinning objects of any multipolar structure.

To detect EMRI signals, the adiabatic approximation to the self-forced evolution is the most promising avenue. Excluding resonance effects, adiabatic evolutions have been simulated for generic orbits in Kerr [55], and a method has been proposed to evolve through resonances without requiring a complete self-force calculation [90]. However, while they are relatively cheap compared to a full self-force calculation, these adiabatic schemes may be too slow to be used directly to populate a template bank. For the purposes of testing data analysis methods and gleaning broad features of EMRIs, simple 'kludge' models can be used instead [133, 134, 135, 136]. For building a template bank for real data analysis, one possible tool is the use of adiabatic approximations or full self-force data to calibrate the extreme-mass-ratio limit of EOB, which can then be used in EMRI source modeling [137, 138, 139, 140].

To accurately extract system parameters, we must improve upon these adiabatic models by including both the full first-order self-force and the time-averaged, dissipative effects of the second-order force. Progress toward this end has been steady, but is far from complete. Long-term evolutions incorporating the full first-order self-force have been performed for orbits of moderate eccentricity in a Schwarzschild background [56]. Short-term effects of the first-order self-force have been calculated for generic orbits in Schwarzschild [60] and for circular, equatorial orbits in Kerr [62, 63]. And work has commenced to extend these results to generic orbits in Kerr and to incorporate them into orbital evolutions, either self-consistently or using a 'geodesic source + osculation' approach.

The most daunting challenge now remaining is the incorporation of second-order self-force effects into inspiral models. This remains a distant prospect, and as of this writing, no numerical results have yet been obtained. However, the formalism is now at hand, and numerical calculations are underway [132]. Furthermore, there is some evidence that for some types of orbits, particularly quasicircular ones, accurate models could be constructed using post-Newtonian approximations of second-order effects, potentially sidestepping the need for complete second-order calculations for those classes of orbits [141].

5. Conclusions

We have summarized recent progress in three fundamental aspects of EMRI science: (i) the astrophysical mechanisms and event rates for EMRIs, (ii) the scientific potential of EMRI GW observations for astrophysics, cosmology, and fundamental physics, and (iii) the modeling of the dynamics of EMRIs and their GW signals.

Although there has been a lot of progress in these research topics, there is still a lot of work to be done in order to be ready for a fruitful exploitation of EMRI GW observations with a space-based detector like eLISA. These three areas of research involve great theoretical challenges, which can be seen as part of the price to be paid for the extraordinary scientific results that can be obtained with EMRIs. The case of EMRI modeling involves the solution of the general-relativistic two-body problem in the extreme-mass ratio regime, an interesting problem by itself. As we have described, there is still much work to be done to solve the self-force problem for spinning MBHs, and even more for introducing second-order perturbative effects that are needed for an accurate description of EMRI gravitational waveforms. In the case of the study of the astrophysical mechanisms and event rates for EMRIs, the astrophysical problem involves a huge range of scales that cannot be handled with direct N-body simulations, and clever approximations have to be made. In the case of the scientific potential of EMRIs, there are a lot of possibilities, as we have described. An example to illustrate the amount and complexity of the work that can be required to make progress in this direction is the question of testing alternative theories of gravity with EMRIs. To that end, we would need a good description of EMRIs in those alternative theories of gravity. Taking into account that the basic features of the GW emission mechanism and the properties of BHs may be quite different from those in GR, where EMRI dynamics is already quite challenging, making predictions for EMRI GW observations in alternative theories of gravity can be in general a very difficult problem. However, by trying to solve this kind of problem we may learn new aspects of these alternative theories, especially in the strong-field gravity regime where not much is known.

In conclusion, the challenges originated by these questions about EMRIs and their intrinsic scientific interest require a great deal of human and computational resources. Despite the fact that the L3 ESA mission devoted to the science of the Gravitational Universe white paper [1] is expected to be launched around 2034, the problems we face are of sufficient magnitude to deserve increased attention in both the short and long term.

Acknowledgments

PAS' work has been supported by the Transregio 7 "Gravitational Wave Astronomy" financed by the Deutsche Forschungsgemeinschaft DFG (German Research Foundation). AP acknowledges support from the European Research Council under the European Union's Seventh Framework Programme (FP7/2007-2013)/ERC grant agreement no. 304978. CFS acknowledges support from contracts 2009-SGR-935 (Catalan Agency for Research Funding, AGAUR) and AYA-2010-15709 (Spanish Ministry of Science and Innovation, MICINN). SAH's work on EMRIs has most recently been supported in part by NSF grant PHY-1403261, by the John Simon Guggenheim Memorial Foundation, and with sabbatical support from the Canadian Institute for Theoretical Astrophysics and the Perimeter Institute for Theoretical Physics.

References

[1] Amaro-Seoane P *et al.* (eLISA Collaboration) 2013 (*Preprint* 1305.5720)
[2] Amaro-Seoane P, Aoudia S, Babak S, Binetruy P, Berti E *et al.* 2012 *Class. Quant. Grav.* **29** 124016 (*Preprint* 1202.0839)
[3] Amaro-Seoane P, Aoudia S, Babak S, Binetruy P, Berti E *et al.* 2013 *GW Notes* **6** 4–110 (*Preprint* 1201.3621)
[4] EMRI plenary session: http://www.phys.ufl.edu/lisasymposiumx/program_tu1.html
[5] 10th International LISA Symposium: http://www.phys.ufl.edu/lisasymposiumx/index.html
[6] Eisenhauer F, Genzel R, Alexander T, Abuter R, Paumard T, Ott T, Gilbert A, Gillessen S, Horrobin M, Trippe S, Bonnet H, Dumas C, Hubin N, Kaufer A, Kissler-Patig M, Monnet G, Ströbele S, Szeifert T, Eckart A, Schödel R and Zucker S 2005 *Astrophys. J.* **628** 246–259 (*Preprint* arXiv:astro-ph/0502129)
[7] Ghez A M, Salim S, Hornstein S D, Tanner A, Lu J R, Morris M, Becklin E E and Duchêne G 2005 *Astrophys. J.* **620** 744–757 (*Preprint* astro-ph/0306130)
[8] Ghez A M, Salim S, Weinberg N N, Lu J R, Do T, Dunn J K, Matthews K, Morris M R, Yelda S, Becklin E E, Kremenek T, Milosavljevic M and Naiman J 2008 *Astrophys. J.* **689** 1044–1062 (*Preprint* 0808.2870)
[9] Gillessen S, Eisenhauer F, Trippe S, Alexander T, Genzel R, Martins F and Ott T 2009 *Astrophys. J.* **692** 1075–1109 (*Preprint* 0810.4674)
[10] Amaro-Seoane P, Gair J R, Freitag M, Miller M C, Mandel I, Cutler C J and Babak S 2007 *Class. Quant. Grav.* **24** 113 (*Preprint* arXiv:astro-ph/0703495)
[11] Hopman C, Freitag M and Larson S L 2007 *Mon. Not. Roy. Astron. Soc.* **378** 129–136 (*Preprint* astro-ph/0612337)
[12] Rubbo L J, Holley-Bockelmann K and Finn L S 2006 *Astrophys. J. Lett.* **649** L25–L28
[13] Yunes N, Sopuerta C F, Rubbo L J and Holley-Bockelmann K 2008 *Astrophys. J.* **675** 604–613 (*Preprint* 0704.2612)
[14] Amaro-Seoane P, Sopuerta C F and Freitag M D 2013 *Mon. Not. Roy. Astron. Soc.* **429** 3155–3165 (*Preprint* 1205.4713)
[15] Peebles P J E 1972 *Astrophys. J.* **178** 371–376
[16] Bahcall J N and Wolf R A 1976 *Astrophys. J.* **209** 214–232
[17] Bahcall J N and Wolf R A 1977 *Astrophys. J.* **216** 883–907
[18] Amaro-Seoane P, Freitag M and Spurzem R 2004 *Mon. Not. Roy. Astron. Soc.* (*Preprint* astro-ph/0401163)
[19] Preto M, Merritt D and Spurzem R 2004 *Astrophys. J. Lett.* **613** L109–L112 (*Preprint* astro-ph/0406324)
[20] Alexander T and Hopman C 2009 *Astrophys. J.* **697** 1861–1869 (*Preprint* 0808.3150)
[21] Preto M and Amaro-Seoane P 2010 *Astrophys. J. Lett.* **708** L42–L46 (*Preprint* 0910.3206)
[22] Amaro-Seoane P and Preto M 2011 *Class. Quant. Grav.* **28** 094017 (*Preprint* 1010.5781)
[23] Gurevich A 1964 *Geomag. Aeronom.* **4** 247–255
[24] Merritt D, Alexander T, Mikkola S and Will C M 2011 *Phys.Rev.* **D84** 044024 (*Preprint* 1102.3180)
[25] Brem P, Amaro-Seoane P and Sopuerta C F 2014 *Mon. Not. Roy. Astron. Soc.* **437** 1259–1267 (*Preprint* 1211.5601)
[26] Gair J R 2009 *Class. Quant. Grav.* **26** 094034 (*Preprint* 0811.0188)
[27] Gair J R, Tang C and Volonteri M 2010 *Phys. Rev. D* **81** 104014 (*Preprint* 1004.1921)
[28] Schutz B F 1986 *Nature* **323** 310–311
[29] Sesana A, Vecchio A, Eracleous M and Sigurdsson S 2008 *Mon. Not. Roy. Astron. Soc.* **391** 718–726 (*Preprint* 0806.0624)
[30] MacLeod C L and Hogan C J 2008 *Phys. Rev. D* **77** 043512 (*Preprint* 0712.0618)
[31] Gair J R, Vallisneri M, Larson S L and Baker J G 2013 *Living Rev. Rel.* **16** 7 (*Preprint* 1212.5575)
[32] Hawking S and Ellis G 1973 *The Large scale structure of space-time* (Cambridge: Cambridge University Press)
[33] Ryan F D 1995 *Phys. Rev. D* **52** 5707–5718
[34] Collins N A and Hughes S A 2004 *Phys. Rev. D* **69** 124022 (*Preprint* gr-qc/0402063)
[35] Glampedakis K and Babak S 2006 *Class. Quant. Grav.* **23** 4167–4188 (*Preprint* gr-qc/0510057)
[36] Barack L and Cutler C 2007 *Phys. Rev. D* **75** 042003 (*Preprint* gr-qc/0612029)
[37] Gair J R, Li C and Mandel I 2008 *Phys. Rev. D* **77** 024035 (*Preprint* 0708.0628)

[38] Sopuerta C F and Yunes N 2009 *Phys. Rev. D* **80** 064006 (*Preprint* 0904.4501)
[39] Canizares P, Gair J R and Sopuerta C F 2012 *Phys. Rev. D* **86** 044010 (*Preprint* 1205.1253)
[40] Kesden M, Gair J R and Kamionkowski M 2005 *Phys. Rev. D* **71** 044015 (*Preprint* astro-ph/0411478)
[41] Li C and Lovelace G 2008 *Phys. Rev. D* **77** 064022 (*Preprint* gr-qc/0702146)
[42] Pani P, Berti E, Cardoso V, Chen Y and Norte R 2009 *Phys. Rev. D* **80** 124047 (*Preprint* 0909.0287)
[43] Barausse E, Rezzolla L, Petroff D and Ansorg M 2007 *Phys. Rev. D* **75** 064026 (*Preprint* gr-qc/0612123)
[44] Barausse E and Rezzolla L 2008 *Phys. Rev. D* **77** 104027 (*Preprint* 0711.4558)
[45] Yunes N, Kocsis B, Loeb A and Haiman Z 2011 *Phys. Rev. Lett.* **107** 171103 (*Preprint* 1103.4609)
[46] Yunes N, Coleman Miller M and Thornburg J 2011 *Phys. Rev. D* **83** 044030 (*Preprint* 1010.1721)
[47] Amaro-Seoane P, Brem P, Cuadra J and Armitage P J 2012 *Astrophys. J. Lett.* **744** L20 (*Preprint* 1108.5174)
[48] Apostolatos T A, Lukes-Gerakopoulos G and Contopoulos G 2009 *Phys. Rev. Lett.* **103** 111101 (*Preprint* 0906.0093)
[49] Scharre P D and Will C M 2002 *Phys. Rev. D* **65** 042002 (*Preprint* gr-qc/0109044)
[50] Berti E, Buonanno A and Will C M 2005 *Phys. Rev. D* **71** 084025 (*Preprint* gr-qc/0411129)
[51] Yagi K 2012 *Phys. Rev. D* **86** 081504 (*Preprint* 1204.4524)
[52] Yunes N and Pretorius F 2009 *Phys. Rev. D* **80** 122003 (*Preprint* 0909.3328)
[53] Gair J R and Yunes N 2011 *Phys. Rev. D* **84** 064016 (*Preprint* 1106.6313)
[54] Tinto M and da Silva Alves M E 2010 *Phys. Rev. D* **82** 122003 (*Preprint* 1010.1302)
[55] Fujita R, Hikida W and Tagoshi H 2009 *Prog. Theor. Phys.* **121** 843–874 (*Preprint* 0904.3810)
[56] Warburton N, Akcay S, Barack L, Gair J R and Sago N 2012 *Phys. Rev. D* **85** 061501 (*Preprint* 1111.6908)
[57] Barack L and Sago N 2007 *Phys. Rev. D* **75** 064021 (*Preprint* gr-qc/0701069)
[58] Detweiler S L 2008 *Phys. Rev. D* **77** 124026 (*Preprint* 0804.3529)
[59] Barack L and Sago N 2009 *Phys. Rev. Lett.* **102** 191101 (*Preprint* 0902.0573)
[60] Barack L and Sago N 2010 *Phys. Rev. D* **81** 084021 (*Preprint* 1002.2386)
[61] Barack L and Sago N 2011 *Phys. Rev. D* **83** 084023 (*Preprint* 1101.3331)
[62] Shah A G, Friedman J L and Keidl T S 2012 *Phys. Rev. D* **86** 084059 (*Preprint* 1207.5595)
[63] Dolan S 2013 Approaches to self-force calculations on Kerr presented at 16th Capra Meeting on Radiation Reaction in General Relativity, University College Dublin URL **http://maths.ucd.ie/capra16/talks/Dolan.pdf**
[64] Hinderer T and Flanagan E E 2008 *Phys. Rev. D* **78** 064028 (*Preprint* 0805.3337)
[65] Pound A 2010 *Phys. Rev. D* **81** 024023 (*Preprint* 0907.5197)
[66] Pound A 2012 *Phys. Rev. Lett.* **109** 051101 (*Preprint* 1201.5089)
[67] Gralla S E 2012 *Phys. Rev. D* **85** 124011 (*Preprint* 1203.3189)
[68] Barack L 2009 *Class. Quant. Grav.* **26** 213001 (*Preprint* 0908.1664)
[69] Poisson E, Pound A and Vega I 2011 *Living Rev. Relativity* **14** 7 (*Preprint* 1102.0529) URL **http://www.livingreviews.org/lrr-2011-7**
[70] Mino Y, Sasaki M and Tanaka T 1997 *Phys. Rev. D* **55** 3457–3476 (*Preprint* gr-qc/9606018)
[71] Quinn T C and Wald R M 1997 *Phys. Rev. D* **56** 3381–3394 (*Preprint* gr-qc/9610053)
[72] Gralla S E and Wald R M 2008 *Class. Quant. Grav.* **25** 205009 (*Preprint* 0806.3293)
[73] Harte A I 2012 *Class. Quant. Grav.* **29** 055012 (*Preprint* 1103.0543)
[74] Pound A 2012 *Phys. Rev. D* **86** 084019 (*Preprint* 1206.6538)
[75] Pound A 2014 Nonlinear gravitational self-force: Second-order equations of motion for a monopolar body in preparation
[76] Detweiler S L and Whiting B F 2003 *Phys. Rev. D* **67** 024025 (*Preprint* gr-qc/0202086)
[77] Detweiler S L 2012 *Phys. Rev. D* **85** 044048 (*Preprint* 1107.2098)
[78] Pound A and Poisson E 2008 *Phys. Rev. D* **77** 044013 (*Preprint* 0708.3033)
[79] van de Meent M 2014 *Phys. Rev. D* **89** 084033 (*Preprint* 1311.4457)
[80] Flanagan E E and Hinderer T 2012 *Phys. Rev. Lett.* **109** 071102 (*Preprint* 1009.4923)
[81] Ruangsri U and Hughes S A 2014 *Phys. Rev. D* **89** 084036 (*Preprint* 1307.6483)
[82] Gair J R, Yunes N and Bender C M 2012 *J. Math. Phys.* **53** 032503 (*Preprint* 1111.3605)
[83] Brink J, Geyer M and Hinderer T 2013 (*Preprint* 1304.0330)
[84] Mino Y 2003 *Phys. Rev. D* **67** 084027 (*Preprint* gr-qc/0302075)
[85] Drasco S, Flanagan E E and Hughes S A 2005 *Class. Quant. Grav.* **22** S801–S846 (*Preprint* gr-qc/0505075)
[86] Sago N, Tanaka T, Hikida W and Nakano H 2005 *Prog. Theor. Phys.* **114** 509–514 (*Preprint* gr-qc/0506092)
[87] Sago N, Tanaka T, Hikida W, Ganz K and Nakano H 2006 *Prog. Theor. Phys.* **115** 873–907 (*Preprint* gr-qc/0511151)
[88] Ganz K, Hikida W, Nakano H, Sago N and Tanaka T 2007 *Prog. Theor. Phys.* **117** 1041–1066 (*Preprint* gr-qc/0702054)

[89] Flanagan E E, Hughes S A and Ruangsri U 2014 *Phys. Rev. D* **89** 084028 (*Preprint* 1208.3906)
[90] Isoyama S, Fujita R, Nakano H, Sago N and Tanaka T 2013 *PTEP* **2013** 063E01 (*Preprint* 1302.4035)
[91] Dolan S R and Barack L 2013 *Phys. Rev. D* **87** 084066 (*Preprint* 1211.4586)
[92] Akcay S, Warburton N and Barack L 2013 *Phys. Rev. D* **88** 104009 (*Preprint* 1308.5223)
[93] Osburn T, Forseth E, Evans C and Hopper S 2014 (*Preprint* 1409.4419)
[94] Barack L and Ori A 2000 *Phys. Rev. D* **61** 061502 (*Preprint* gr-qc/9912010)
[95] Barack L, Mino Y, Nakano H, Ori A and Sasaki M 2002 *Phys. Rev. Lett.* **88** 091101 (*Preprint* gr-qc/0111001)
[96] Mino Y, Nakano H and Sasaki M 2002 *Prog. Theor. Phys.* **108** 1039–1064 (*Preprint* gr-qc/0111074)
[97] Barack L and Ori A 2003 *Phys. Rev. Lett.* **90** 111101 (*Preprint* gr-qc/0212103)
[98] Barack L and Golbourn D A 2007 *Phys. Rev. D* **76** 044020 (*Preprint* 0705.3620)
[99] Vega I and Detweiler S L 2008 *Phys. Rev. D* **77** 084008 (*Preprint* 0712.4405)
[100] Wardell B, Vega I, Thornburg J and Diener P 2012 *Phys. Rev. D* **85** 104044 (*Preprint* 1112.6355)
[101] Favata M 2011 *Phys. Rev. D* **83** 024028 (*Preprint* 1010.2553)
[102] Le Tiec A, Barausse E and Buonanno A 2012 *Phys. Rev. Lett.* **108** 131103 (*Preprint* 1111.5609)
[103] Dolan S R, Warburton N, Harte A I, Le Tiec A, Wardell B *et al.* 2014 *Phys. Rev. D* **89** 064011 (*Preprint* 1312.0775)
[104] Dolan S R, Nolan P, Ottewill A C, Warburton N and Wardell B 2014 (*Preprint* 1406.4890)
[105] Isoyama S, Barack L, Dolan S R, Le Tiec A, Nakano H *et al.* 2014 (*Preprint* 1404.6133)
[106] Pound A, Merlin C and Barack L 2014 *Phys. Rev. D* **89** 024009 (*Preprint* 1310.1513)
[107] Blanchet L, Detweiler S L, Le Tiec A and Whiting B F 2010 *Phys. Rev. D* **81** 064004 (*Preprint* 0910.0207)
[108] Blanchet L, Detweiler S L, Le Tiec A and Whiting B F 2010 *Phys. Rev. D* **81** 084033 (*Preprint* 1002.0726)
[109] Le Tiec A, Mroue A H, Barack L, Buonanno A, Pfeiffer H P *et al.* 2011 *Phys. Rev. Lett.* **107** 141101 (*Preprint* 1106.3278)
[110] Le Tiec A, Buonanno A, Mrou A H, Pfeiffer H P, Hemberger D A *et al.* 2013 *Phys. Rev. D* **88** 124027 (*Preprint* 1309.0541)
[111] Shah A G, Friedman J L and Whiting B F 2013 (*Preprint* 1312.1952)
[112] Damour T 2010 *Phys. Rev. D* **81** 024017 (*Preprint* 0910.5533)
[113] Barack L, Damour T and Sago N 2010 *Phys. Rev. D* **82** 084036 (*Preprint* 1008.0935)
[114] Barausse E, Buonanno A and Le Tiec A 2012 *Phys. Rev. D* **85** 064010 (*Preprint* 1111.5610)
[115] Akcay S, Barack L, Damour T and Sago N 2012 *Phys. Rev. D* **86** 104041 (*Preprint* 1209.0964)
[116] Bini D and Damour T 2013 (*Preprint* 1312.2503)
[117] Le Tiec A 2014 (*Preprint* 1408.5505)
[118] Detweiler S L, Messaritaki E and Whiting B F 2003 *Phys. Rev. D* **67** 104016 (*Preprint* gr-qc/0205079)
[119] Haas R 2007 *Phys. Rev. D* **75** 124011 (*Preprint* 0704.0797)
[120] Canizares P, Sopuerta C F and Jaramillo J L 2010 *Phys. Rev. D* **82** 044023 (*Preprint* 1006.3201)
[121] Wardell B, Galley C R, Zenginoglu A, Casals M, Dolan S R *et al.* 2014 *Phys. Rev. D* **89** 084021 (*Preprint* 1401.1506)
[122] Diener P, Vega I, Wardell B and Detweiler S L 2012 *Phys. Rev. Lett.* **108** 191102 (*Preprint* 1112.4821)
[123] Warburton N and Barack L 2010 *Phys. Rev. D* **81** 084039 (*Preprint* 1003.1860)
[124] Dolan S R, Barack L and Wardell B 2011 *Phys. Rev. D* **84** 084001 (*Preprint* 1107.0012)
[125] Warburton N and Barack L 2011 *Phys. Rev. D* **83** 124038 (*Preprint* 1103.0287)
[126] Warburton N 2014 (*Preprint* 1408.2885)
[127] Thornburg J and Wardell B 2014 Scalar self-force for highly eccentric orbits in Kerr spacetime presented at 17th Capra Meeting on Radiation Reaction in General Relativity, California Institute of Technology URL http://www.tapir.caltech.edu/ capra17/cgi-bin/abstract.cgi?id=a425ec3f30f7f652d27107b1ec5538c8
[128] Rosenthal E 2006 *Phys. Rev. D* **73** 044034 (*Preprint* gr-qc/0602066)
[129] Rosenthal E 2006 *Phys. Rev. D* **74** 084018 (*Preprint* gr-qc/0609069)
[130] Detweiler S L 2012 *Phys. Rev. D* **85** 044048 (*Preprint* 1107.2098)
[131] Pound A and Miller J 2014 *Phys. Rev. D* **89** 104020 (*Preprint* 1403.1843)
[132] Pound A 2014 (*Preprint* 1404.1543)
[133] Barack L and Cutler C 2004 *Phys. Rev. D* **69** 082005 (*Preprint* gr-qc/0310125)
[134] Gair J R and Glampedakis K 2006 *Phys. Rev. D* **73** 064037 (*Preprint* gr-qc/0510129)
[135] Babak S, Fang H, Gair J R, Glampedakis K and Hughes S A 2007 *Phys. Rev. D* **75** 024005 (*Preprint* gr-qc/0607007)
[136] Sopuerta C F and Yunes N 2011 *Phys. Rev. D* **84** 124060 (*Preprint* 1109.0572)
[137] Damour T and Nagar A 2007 *Phys. Rev. D* **76** 064028 (*Preprint* 0705.2519)
[138] Yunes N, Buonanno A, Hughes S A, Coleman Miller M and Pan Y 2010 *Phys. Rev. Lett.* **104** 091102 (*Preprint* 0909.4263)

[139] Yunes N, Buonanno A, Hughes S A, Pan Y, Barausse E *et al.* 2011 *Phys. Rev. D* **83** 044044 (*Preprint* 1009.6013)
[140] Harms E, Bernuzzi S, Nagar A and Zenginoglu A 2014 (*Preprint* 1406.5983)
[141] Isoyama S, Fujita R, Sago N, Tagoshi H and Tanaka T 2013 *Phys. Rev. D* **87** 024010 (*Preprint* 1210.2569)

Ultra-compact binaries as gravitational wave sources

Sweta Shah[1], **Shane L Larson**[2,3] **and Warren Brown**[4]

[1] Department of Astrophysics/ IMAPP, Radboud University Nijmegen, P.O. Box 9010, 6500 GL Nijmegen, The Netherlands

[2] CIERA, 2145 Sheridan Road, Northwestern University, Evanston, IL 60208, USA

[3] Department of Astronomy, Adler Planetarium, Chicago, IL 60605, USA

[4] Smithsonian Astrophysical Observatory, 60 Garden St, Cambridge, MA 02138, USA

E-mail: `s.shah@astro.ru.nl`

E-mail: `s.larson@northwestern.edu`

E-mail: `wbrown@cfa.harvard.edu`

Abstract. Ultra-compact binaries are among the most numerous sources in the millihertz gravitational wave band, and as such represent one of the primary sources for LISA-like detectors (gravitational wave interferometers in space). Already there are almost 60 ultra-compact binaries that have been detected by electromagnetic means, and more are being discovered in dedicated searches as time goes on. Prominent in this population is the doubly-degenerate white dwarf system J0651, whose orbital decay has been measured and shown to track accurately with the predicted evolution due to the emission of gravitational wave emission. This paper reviews the current understanding of the ultra-compact binary population, recent progress in electromagnetic studies, and prospects for multi-messenger astronomy of these systems once a LISA-like detector is operational.

1. Introduction

Like the electromagnetic (EM) spectrum, the gravitational wave (GW) spectrum covers an enormous range of frequencies, from the extremely low frequencies of cosmological gravitational waves (gravitational wavelengths on the order of the horizon scale), to ultra-high frequencies for tiny compact sources such as cosmic string cusps or primordial black holes (with appropriately short gravitational wavelengths). While all of the frequencies are allowable, whether or not they are astrophysically relevant is a matter of extrapolation from known sources, as well as intense speculation and debate.

Restricting attention to massive astrophysical binaries from stellar masses to supermassive black holes, the relevant parts of the gravitational wave spectrum are generally related to the astrophysical timescales of binary orbits. Thus, the most extensively studied part of the spectrum ranges from the ultra-low-frequency band at $f_{gw} \sim 10^{-9}$Hz (SMBH inspirals, detectable by pulsar timing arrays), to the high frequency band at $f_{gw} \sim 1000$ Hz (merging stellar mass black holes and neutron stars, detectable by ground-based interferometers like LIGO and Virgo).

There is no doubt that many of these prospective sources exist; they have all been detected by other observational means using traditional telescopes. But GW observations offer an altogether new and unique way to probe the universe, observing high-energy astrophysical systems not on the interacting scales of atoms (where most electromagnetic ration originates from), but on the large system scales dominated by the bulk motion of matter. GW observatories will provide data that is wholly complementary to traditional EM observations, enabling new scientific investigations into the structure and evolution of systems that are currently poorly understood or poorly constrained.

As with all astronomical observations, it is exceedingly useful to have a large catalog of sources. In addition to providing an opportunity to conduct *population studies*, large catalogs also alleviate uncertainty in what is "normal" in a particular physical context by identifying outliers and odd events with respect to the entire catalog. This is often a particular worry in GW astronomy because current model predictions often forecast event rates as low as a handful per year.

In the millihertz band covered by LISA ($10^{-5}\text{Hz} \lesssim f_{gw} \lesssim 0.1\text{Hz}$) the sources are expected to be more numerous because a wide variety of systems are that are long-lived spend long periods of their evolutionary time in this band, changing slowly under the steady emission of gravitational waves. Foremost among these are the *ultra-compact binaries* (UCBs): binary systems where the compact components are degenerate stellar remnants (white dwarfs, neutron stars, and stellar mass black holes).

The galaxy is replete with UCBs, with even conservative estimates suggesting there should be $\sim 10^6$ or more radiating in the band covered by LISA; most of these will be close white dwarf binaries.

1.1. *Confusion Limit*

The UCBs are persistent and long-lived sources in the millihertz band. If evolving purely under GW emission, their orbital frequency changes according to

$$\dot{f}_{orb} = \frac{48}{5}\frac{c^3}{G}\frac{f_{orb}}{\mathcal{M}_c}\left(\frac{G}{c^3}2\pi f_{orb}\mathcal{M}_c\right)^{8/3}, \tag{1}$$

where the mass parameter $\mathcal{M}_c = (m_1 m_2)^{3/5}/(m_1+m_2)^{1/5}$ is called the *chirp mass.* The ability to detect frequency evolution for a LISA-like interferometer is fixed by the total observing time, T_{obs}, giving resolving power of $\delta f \sim 1/T_{obs}$. For a 1 year observation, this is $\delta f \sim 3.17 \times 10^{-8}$ Hz. Integrating Eq. 1 one can find the time t that a system takes to cross a frequency δf, equivalent to the time LISA needs to observe a system to see it evolve in frequency. For typical UCB masses of interest, with equal masses and total mass around the Chandresekhar mass, sources are stationary (they are *monochromatic*) in the LISA data below $f_{orb} \sim 2.5 \times 10^{-3}$ Hz.

Population synthesis models [1], suggest that below $f_{orb} \sim 1-2\times 10^{-3}$ Hz, there will be more than one UCB in the galaxy with orbital periods in any given frequency interval f to $f+\delta f$. This means that in the LISA data there will be an accumulated signal at those frequencies that is the incoherent sum of the GW from multiple UCB sources located in different parts of the sky; the sources are "confused" and cannot be individually resolved. In the context of analyzing data, there will be some bright binaries whose signals stand out because they are intrinsically strong GW emitters or because they are close to Earth; data analysis studies suggest on the order of $\sim 10^4$ of these "resolvable binaries" will be detectable [2]. Below the bright binaries, the remaining confused binaries will merge to form an irreducible astrophysical noise, a *confusion limited foreground.* For most millihertz band, LISA-style interferometers, this irreducible foreground is the limiting source of noise in source analysis below $f_{gw} \sim 3\times 10^{-3}$ Hz.

2. Verification binaries

Among the resolvable sources, there will be a subset of *verification binaries.* These are known UCBs that are detectable in both gravitational waves and using electromagnetic telescopes.

Journal of Physics: Conference Series **610** (2015) 012003 doi:10.1088/1742-6596/610/1/012003

Many of these have already been discovered and are the subject of ongoing electromagnetic observing campaigns.

2.1. Evolutionary paradigm
The classically understood evolutionary pathway for isolated stars with masses $m \lesssim 8M_\odot$ is that when the core hydrogen fuel is expended, the core collapses to smaller size, increasing the core temperature and causing the star to expand into a red giant phase. Helium fusion begins in the core, leaving a core of carbon and oxygen. White dwarfs are the degenerate, Earth-sized remnants of that core. They are born at temperatures of $\sim 100,000$K, and are comprised largely of carbon and oxygen with thin atmosphere layers of helium and hydrogen (the atmosphere comprising only 2% of the total mass).

This picture is modified in binary systems. If the companion star is close enough during the red giant phase, the giant star can fill its Roche lobe and transfer mass to the companion. If the mass transfer is stable, then the result is a common envelope that contains both stars. The details of this process is complicated and poorly constrained, but simple prescriptions have been developed that conserve the energy or the angular momentum of the common envelope. Experimentally, common envelope binaries are common [3]. Observations suggest a new system forms every few years in the Milky Way. The process is brief, but can cause a star to lose most of its mass so that its remnant ends up as a low mass white dwarf. Extremely low mass white dwarfs require two phases of common envelope evolution and so are rare, but are the unambiguous signposts of compact binaries. Exactly the type of systems that make good gravitational wave sources.

2.2. Finding low mass white dwarfs
Recent surveys to find low-mass white dwarf binaries [4, 5] have been successful using optical spectroscopy to identify targets and measure radial velocities. Figure 1 shows typical white dwarf spectra, with absorption lines from hydrogen in the atmospheres. The width of the lines directly measures the surface gravity of these stars; broad lines are a signature of a white dwarf, which has a high surface gravity.

Another important feature in these spectra is the Balmer decrement, the ratio of line strengths between different lines in the Balmer sequence. Imaging passbands spanning this spectral feature will produce a measure of an object's color, enabling the identification of low-mass white dwarf binaries.

Figure 2 shows a color-color diagram of stars in the Sloan Digital Sky Survey (SDSS). In this region, the g-r color reflects the effective temperature of these stars (brighter, more negative g-r value, are hotter) and u-g reflects their surface gravity (brighter, smaller u-g values have higher surface gravity). The entire plane is populated by sources (black dots); on the lower half of this diagram, at low surface gravity, the sources are normal A-type stars; at higher surface gravities the sources are normal WDs in the same temperature range.

The Extremely Low Mass (ELM) white dwarf survey targets the middle region of this color-color diagram, where low-mass white dwarfs are located due to their lower surface gravities. The first group of white dwarf binaries in this region were fortuitously discovered in other surveys, but targeted work over the last couple of years has more than tripled the number of known white dwarf binaries.

The net outcome of the survey is dozens of white dwarf binaries with inferred merger times less than a Hubble time, and many with orbital periods that put them firmly in the millihertz gravitational wave band observable by LISA. Figure 3 shows the current detected population.

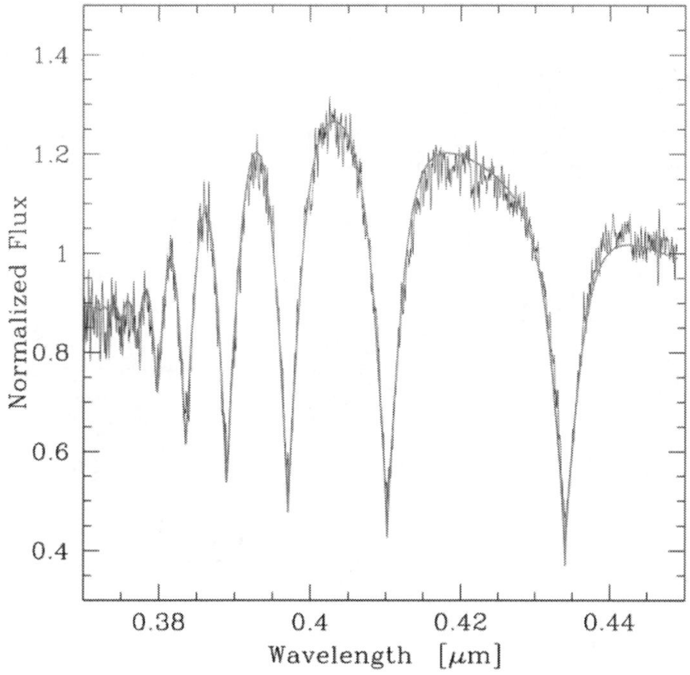

Figure 1. Typical spectra of a white dwarf, showing the deep hydrogen absorption lines from the atmosphere.

2.3. New LISA Verification Binaries

Two important systems have been discovered using these low-mass white dwarf surveys, both of them with short periods and expected to be strong sources in the LISA band: J0651, an eclipsing white dwarf system [6], and WD 0931, a face-on binary system [7]. Figure 4 shows these sources and several other ELM binaries plotted against an average LISA strain curve.

WD 0931 has a 20 minute orbital period, and shows only modest radial velocity variations, suggesting the system is close to face on, making it a strong gravitational wave source. Estimates suggest a strain of $h \sim 10^{-22}$, making it a strong verification source in the millihertz band.

J0651 is a particularly important system because the eclipses enable a precise measurement of the orbital ephemeris. Over just three years of observations, the time of eclipse in this system has shifted by 1 minute, about the same amount that the Hulse-Taylor pulsar has shifted in 40 years [8]. The shift is due to gravitational wave emission; this system is currently emitting more energy in gravitational wave radiation – 2.6 solar luminosities – than in light.

3. UCB Multi-Messenger Astronomy

The ultra-compact binary population of the galaxy is a preserved fossil record of binary evolution in the Milky Way [9, 10]. As such, detecting large numbers of UCBs in gravitational waves will provide important constraints on common envelope evolution, composition and distributions of SN Ia progenitor systems, probes of white dwarf structure, studies of tides between close stellar companions, and new precision tests of gravity. Some of this science can be accomplished using GW observations alone, but the synergy between GW and EM observations promises to enable

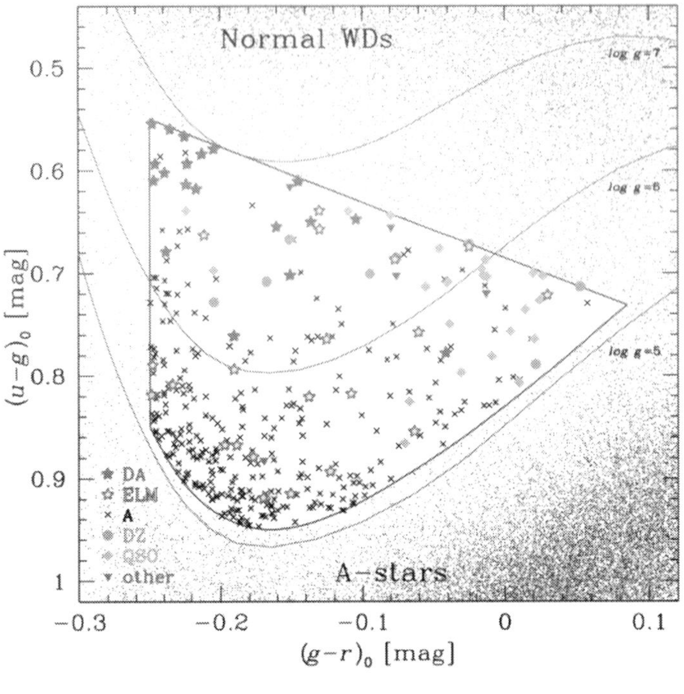

Figure 2. Typical color-color plot showing $(u-g)$ colors vs. $(g-r)$ colors. Note the clustering of A stars near the bottom and white dwarfs near the top; low-mass white dwarfs are being found in the central regions (red stars). Symbols indicate the spectroscopic identification of ELM (Extreme Low Mass) survey targets: red stars are the DA (hydrogen dominated) white dwarfs, red stars with yellow centers are the ELM white dwarfs, black x's are normal A-type stars, green circles are DZ (helium dominated) WDs, cyan diamonds are quasars, and magenta triangles are other miscellaneous objects.

high-fidelity probes of the structure and interaction of close binary stars.

3.1. Multi-messenger population

Significant numbers of UCBs have been detected already in the EM spectrum, and more are being found in surveys that are now targeting them (a running collection of prospective sources that have been detected by EM means is maintained at [11]). These observations provide useful constraints on expectations for the GW emission, but GW data collection is still somewhat in the future. This begs an important question: given what we already know, how many systems will be simultaneously detectable in *both* gravitational waves and electromagnetic astronomy? Simultaneous observations in both GW and EM that will enable a maximal science return.

Despite being observable with telescopes, not all of the close white dwarf binaries found in surveys like ELM will be detectable by a LISA like mission. Many of them will radiate at too low a frequency, or will simply be lost in the confusion noise of the 10 million other binaries in the galaxy. Currently we are discovering systems with EM telescopes and predicting if they will be observable with LISA. Once LISA flies, we will experience a dramatic role-reversal where we will detect many with LISA first and then have to do an EM search to determine if they can

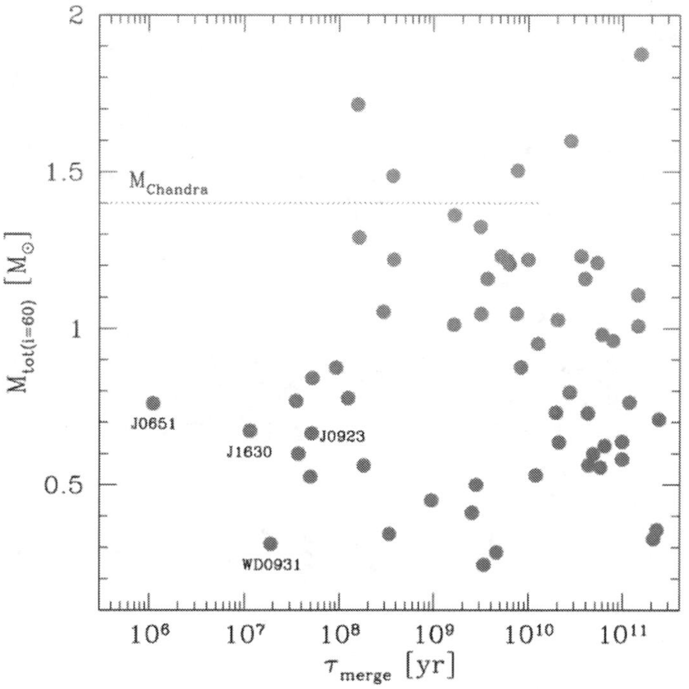

Figure 3. Gravitational wave merger time vs. total system mass for the low mass white dwarf binaries discovered by the ELM Survey (Brown et al 2013). They estimate system mass assuming $\iota = 60°$ when orbital inclination is unknown.

be seen in light. Until a GW observatory in space flies, we will not know precisely how many multi-messenger sources will exist, but good estimates can be made using population synthesis galaxies.

Population synthesis endeavours to simulate the entire galactic population of doubly degenerate stars by simulating the lives of millions of star systems at once, incorporating as much stellar evolution physics as possible. There are certain assumptions that have to be made, but existing synthesized galaxies have extremely high fidelity, producing not only distributions of masses and orbital periods, but galactic positions, EM brightnesses, and stellar temperatures. Figure 5 shows populations of multi-messenger binaries extracted from a population synthesis galaxy, plotted against possible future configurations of a LISA like detector [12]. From this one sees that for *any* plausible GW observatory in space, (1) the UCBs are an important and relevant source, and (2) there will be a sizable population of multi-messenger verification binaries.

Overall, the multi-messenger population will be dominated by white dwarfs (mostly detached, with a handful of interacting binaries), as well as a few accreting sources involving neutron stars and stellar mass black holes (though these will be rarer).

3.2. Multi-messenger binary data

The advantage of working with multi-messenger UCBs is that the EM and GW data provide highly complementary probes of the physical parameters that characterize a given binary. The precise parameters of a UCB define the evolutionary track that the binary is on, so measuring their values is of the utmost importance in mapping out the sequence and influence of different

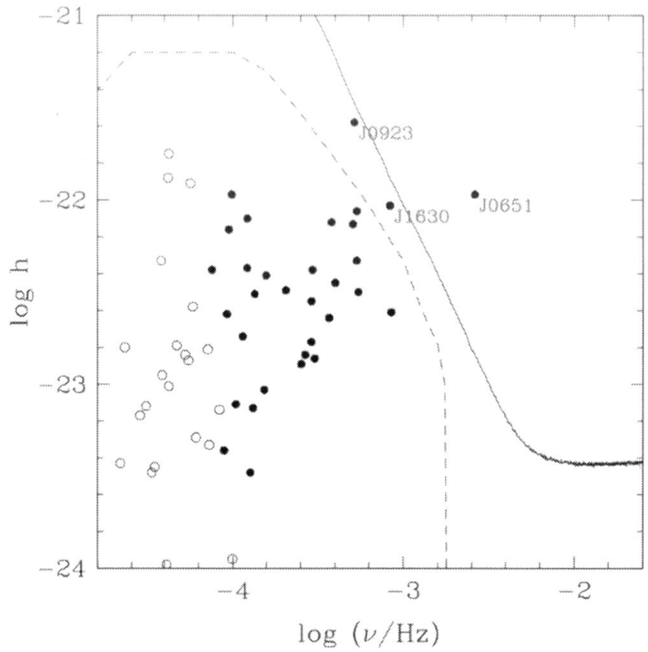

Figure 4. Known low mass white dwarf binaries from the ELM Survey, plotted with expected GW strengths against eLISA sensitivity configurations (solid) and the expected confusion limit (dashed line). Filled circles represent systems that will merge in less than a Hubble time, and open circles correspond to systems that will merge in more than a Hubble time.

phases of binary evolution. Many processes are known to play major roles in binary evolution but are poorly understood because the phases are either short-lived, or difficult to observe in action. As a case in point, consider common envelope evolution. This is a short lived phase in the binary evolution believed to be responsible for bringing the components from a wide orbit to a very tight orbit due to the orbital angular momentum loss due to the friction that this star suffers as it orbits. The efficiency with which this happens and the exact outcome of this is not understood. It could, however, be constrained if we had more accurate masses for the binary systems, something GW observations could provide.

There are many parameters which are difficult to measure with EM or GW observations alone, either because they do not have a clear signature in the data, or because they are strongly correlated with other parameters. Consider the GW scaling amplitude, $\mathcal{A}$, which can

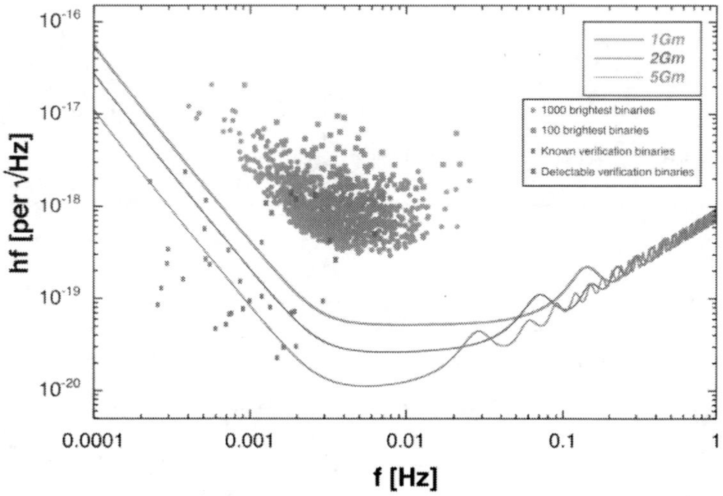

Figure 5. Multi-messenger binaries extracted from a population synthesis galaxy, plotted against future LISA-like detector sensitivities.

be expressed as a function of the system parameters as

$$\mathcal{A} = \frac{4(G\mathcal{M}_c)^{5/3}}{c^4 d}(\pi f_{gw})^{2/3} , \tag{2}$$

where d is the distance to the source. This is the amplitude for an optimally oriented system, but for unknown inclinations, ι, the measured value of $\mathcal{A}$ is uncertain by large factors, as shown by the long, elongated error ellipses in Figure 6. Parameter determination is further complicated by degeneracy in Eq. 2 between the distance d and chirp mass $\mathcal{M}_c$, the effect of which cannot be isolated by measuring the amplitude and frequency of the gravitational waves alone. As a completely independent detection method, adding EM data can (for instance) greatly constrain the inclinations, which narrows the acceptable values of $\mathcal{A}$.

The amplitude could be further constrained if distances are known. Once the GAIA survey is complete, it is expected that the known white dwarf population will increase by more than an order of magnitude, to more than 2×10^5 sources [13]. With a magnitude limit in the range of $m_V \simeq 20 - 25$, estimates from population synthesis galaxies suggest as many as 10^3 multi-messenger binaries may be detectable [12], meaning GAIA could provide a significant catalog of distances from astrometry.

The synergy between EM and GW observations of verification binaries extends to other parameters as well, most importantly to the sky position, which is obviously known for all verification binaries discovered in the pre-LISA era. Knowledge of the sky position greatly enhances our ability to model the GW waveforms in terms of their amplitude and modulation.

4. UCB Science Enabled by gravitational wave observations

4.1. Population studies

There have been many directed studies to examine how specific astrophysical effects can be characterized using GW data. For instance, the large number of of UCBs that LISA will detect provides a method for measuring the shape parameters of the galaxy. UCBs will be detectable across the entire Milky Way, allowing the radial and vertical scales heights of the disk to be mapped out to a few percent accuracy [14].

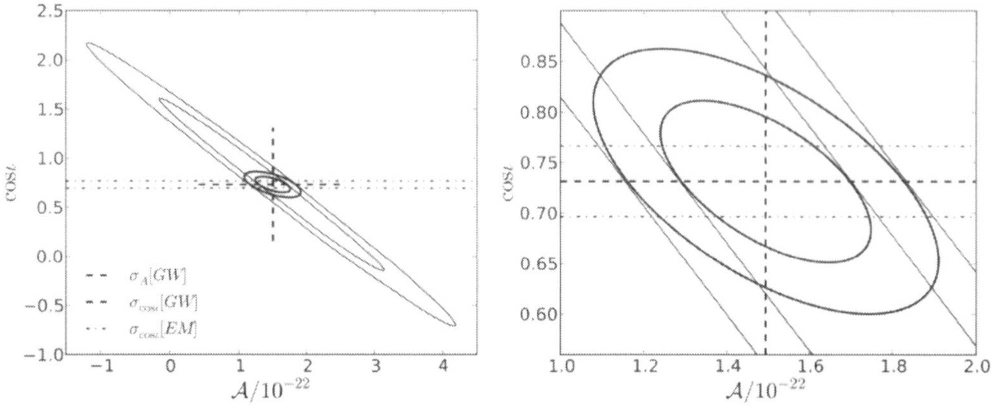

Figure 6. Left: Two-dimensional GW error ellipses of $\mathcal{A}$ and $\cos\iota$ for AM CVn. The two thin black lines represent 1-σ and 2-σ ellipses and the black dashed lines represent the 1-σ GW errors in $\mathcal{A}$ and $\cos\iota$ with the cross at the true value of the parameters. The unphysical values (i.e. negative numbers along the amplitude axis) are caused by the Gaussian tails of the parameter uncertainty about their true values. The right panel is a zoom of the area that is constrained by the 1-σ EM error of 4° shown in dash-dotted lines. The corresponding error ellipses for the reduced PDF are shown as thick black ellipses for 1-σ and 2-σ.

LISA will also be an effective tool for identifying the galactic population of supernova Ia progenitors, as shown in Figure 7. Many of these systems are chirping (evolving in frequency), placing them in a part of the band where they are individually resolvable from the confused population. Furthermore, measuring the chirp breaks many of the parameter degeneracies that plague GW system characterization. LISA will have the ability to identify doubly degenerate systems, Type Ia progenitors, throughout the galaxy [15].

4.2. Tidal interactions

At higher frequencies, where the chirp $\dot{f}$ and chirp evolution, $\ddot{f}$ can be measured, the UCBs are extremely compact, raising the possibility that tidal effects could be measured. In system J0651, tidal distortions of the low-mass companion can already be detected in the EM data. How can the synergy of GW and EM data help understand tidal effects in this and other UCBs?

Most of the energy in these systems is dissipated in gravitational waves, but the tidal distortion also bleeds additional energy out of the orbit, owing to the misalignment of the white dwarfs. Looking at the overall contributions to the frequency evolution, there are both gravitational wave and tidal terms that contribute to $\dot{f}$. The question we would like to answer with multi-messenger data is what is the relative size of the contributions to $\dot{f}$? The existence of non-GW contributions to the frequency evolution of a UCB can be determined from a well known observable constructed from the frequency evolution [16]

$$\left(\frac{\ddot{f}f}{\dot{f}^2}\right)_{GW} = \frac{11}{3}\,, \tag{3}$$

though it is expected to be difficult to measure for most systems [17].

The combination of EM and GW observations, especially for well characterized systems like J0651, can allow for greatly improved measurements of the system parameters that are

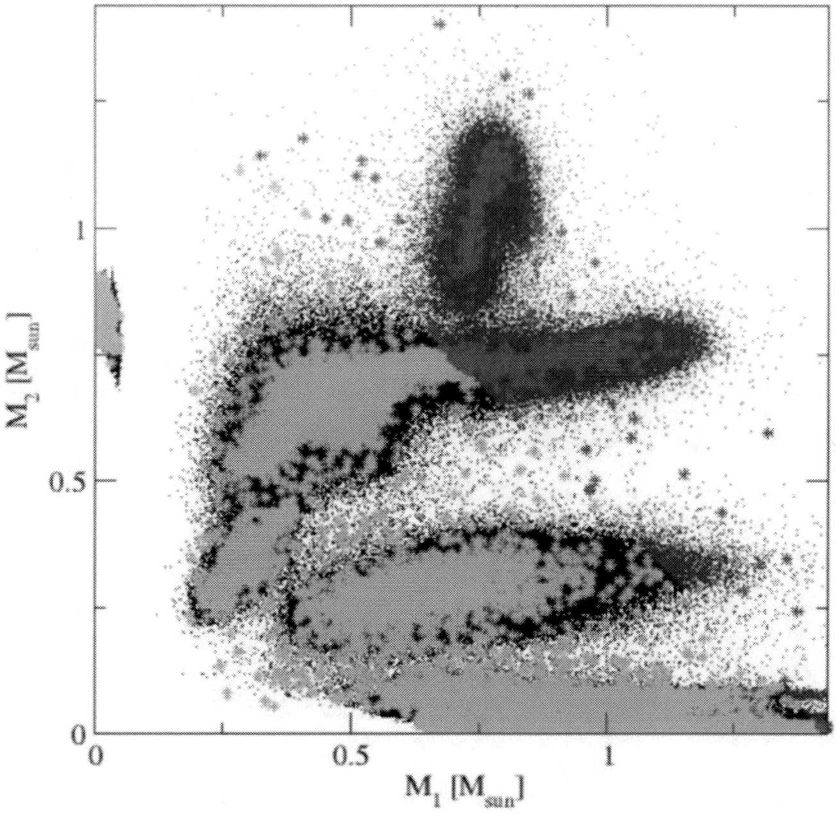

Figure 7. A population synthesis realization shows clear populations in the $\{m_1, m_2\}$ plane. All systems that have a total mass $m_1 + m_2$ greater than the Chandrasekhar mass (nominally, Type Ia progenitors) are colored purple and located in the upper right. LISA resolvable binaries are colored green (sub-Chandresekar mass) and blue (Chandrasekhar mass or greater).

degenerate when explaining tidal energy losses in UCBs [18], though for most UCBs the expected parameter errors suggest the constraints may not be useful even in joint observations; an extended EM observing campaign (e.g. $\sim$ 10 year) could mitigate this concern. For gravitational wave observations to be highly effective in constraining $\dot{f}$ and $\ddot{f}$, sytems will need to be extremely massive and at high frequencies ($f \gtrsim 6$ mHz), where LISA-like detectors are most sensitive and the confusion foreground has died away [19].

4.3. Structure probes

Another role that multi-messenger observations can play in the observation of UCBs is in probing the internal structure and dynamics, such as characterizing the accretion disks in cataclysmic variable systems.

Compact interacting binaries spill mass through the inner Lagrange point, and the resulting stream impacts on the accretion disk creating a *bright spot.* This spot emits $\sim 30\%$ of the total light in the system, and manifests in the light curve as *orbital humps.* The phase of the light curve is related to the profile of the bright spot with respect to the observer. By contrast, the phase fronts of the gravitational wave signal arrive depending on the orientation of the binary axis with respect to the observer, allowing a geometrical orientation of the components of the binary to be known. Knowing the location of the individual components allows the measurement of the bright spot impact location angle, α, which is a proxy for the accretion disk radius as shown in Figure 8 [20]. The expected errors are comparable to the current existing errors from systems where the accretion disk is eclipsed, but the method will not be restricted to eclipsing systems.

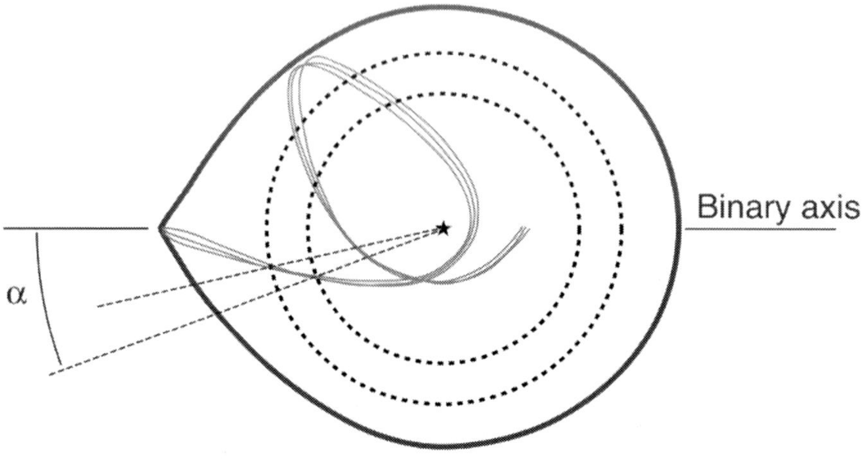

Figure 8. Matter streaming through the inner Lagrange point impacts on the accretion disk, making a bright spot that shows as a feature in the EM light curve. GW observations locate the binary axis, giving a reference point to measure the angle α.

5. Summary

In much of the LISA literature, the confusion background is characterized as a source of noise which must be accounted for when considering observations of other sources of gravitational waves. It is a sad fact that the UCBs are most often thought of *as noise*, a nuisance which interferes with seeing more interesting sources of gravitational waves (like supermassive black hole binaries or extreme mass ratio inspirals). In reality, however, the galactic binary foreground is an important astrophysical signal which will be easily visible to LISA and have profound implications for our understanding of the galaxy.

As tracers of galactic astrophysics, binary populations will provide observational access to many different components of the galaxy, including the disk and bulge [21], the dark galactic halo [22, 23], and even globular cluster systems [24]. Observations of the galactic binary foreground in gravitational waves provide a key way to measure the properties of an assumed but unseen component of the galactic stellar population. The total population of highly evolved compact binaries provides a unique record of galactic stellar evolution which should be easily accessible to an instrument like LISA. Near the frequencies where LISA is most sensitive, it is also plausible that galactic binary signals from the Milky Way's nearest companions might also be detected, and that there will be a low level confusion background of all the galactic binaries in the local Universe [25], which would form an ultimate noise floor in the low frequency gravitational wave band.

As a prospective source population, ultra-compact galactic binaries promise to be one of the richest areas for discovery as gravitational wave astronomy becomes a more prominent part of our observing toolbox.

SLL acknowledges support from NASA award NNX13AM10G.

References

[1] Nelemans G, Yungelson L R and Portegies Zwart S F 2001 *Astron. Astrophys.* **375** 890–898 (*Preprint* astro-ph/0105221)
[2] Belczynski K, Benacquista M and Bulik T 2010 *Astrophys. J.* **725** 816–823 (*Preprint* 0811.1602)
[3] Kochanek C S, Adams S M and Belczynski K 2014 *MNRAS* **443** 1319–1328 (*Preprint* 1405.1042)
[4] Brown W R, Kilic M, Allende Prieto C, Gianninas A and Kenyon S J 2013 *Astrophys. J.* **769** 66 (*Preprint* 1304.4248)
[5] Gianninas A, Dufour P, Kilic M, Brown W R, Bergeron P and Hermes J J 2014 *Astrophys. J.* **794** 35 (*Preprint* 1408.3118)
[6] Brown W R, Kilic M, Hermes J J, Allende Prieto C, Kenyon S J and Winget D E 2011 *Astrophys. J.* **737** L23 (*Preprint* 1107.2389)
[7] Kilic M, Brown W R, Gianninas A, Hermes J J, Allende Prieto C and Kenyon S J 2014 *MNRAS* **444** L1–L5 (*Preprint* 1406.3346)
[8] Hermes J J, Kilic M, Brown W R, Winget D E, Allende Prieto C, Gianninas A, Mukadam A S, Cabrera-Lavers A and Kenyon S J 2012 *Astrophys. J.* **757** L21 (*Preprint* 1208.5051)
[9] Yu S and Jeffery C S 2013 *MNRAS* **429** 1602–1608 (*Preprint* 1211.5091)
[10] Yu S and Jeffery C S 2011 *MNRAS* **417** 1392–1401 (*Preprint* 1107.0611)
[11] Nelemans G 2013 G. Nelemans maintains a wiki with the most up to date list of verification binary parameters http://www.astro.ru.nl/~nelemans/dokuwiki/doku.php?id=lisa_wiki
[12] Littenberg T B, Larson S L, Nelemans G and Cornish N J 2013 *MNRAS* **429** 2361–2365 (*Preprint* 1207.4848)
[13] Carrasco J M, Catalán S, Jordi C, Tremblay P E, Napiwotzki R, Luri X, Robin A C and Kowalski P M 2014 *Astron. Astroph.* **565** A11 (*Preprint* 1403.6045)
[14] Adams M R, Cornish N J and Littenberg T B 2012 *Phys. Rev. D* **86** 124032 (*Preprint* 1209.6286)
[15] Stroeer A, Benacquista M and Ceballos F 2013 Detecting Double Degenerate Progenitors of SNe Ia with LISA *IAU Symposium* (*IAU Symposium* vol 281) ed Di Stefano R, Orio M and Moe M pp 217–220
[16] Webbink R F and Han Z 1998 Gravitational radiation from close double white dwarfs *Laser Interferometer Space Antenna, Second International LISA Symposium on the Detection and Observation of Gravitational Waves in Space* (*American Institute of Physics Conference Series* vol 456) ed Folkner W M pp 61–67
[17] Shah S and Nelemans G 2014 *Astrophys. J.* **791** 76 (*Preprint* 1406.3603)
[18] Benacquista M J 2011 *Astrophys. J.* **740** L54 (*Preprint* 1109.2744)
[19] Shah S and Nelemans G 2014 *Astrophys. J.* **790** 161 (*Preprint* 1406.3599)
[20] Addison E, Breivik K and Larson S 2014 *submitted to Astrophys. J.*
[21] Seto N and Cooray A 2004 *Phys. Rev. D* **70** 123005 (*Preprint* astro-ph/0403259)
[22] Hiscock W A, Larson S L, Routzahn J R and Kulick B 2000 *Astrophys. J.* **540** L5–L8 (*Preprint* gr-qc/0005134)
[23] Hiscock W A 1998 *Astrophys. J.* **509** L101–L104 (*Preprint* gr-qc/9809018)
[24] Benacquista M J, Portegies Zwart S and Rasio F A 2001 *Classical and Quantum Gravity* **18** 4025–4031 (*Preprint* gr-qc/0010020)
[25] Farmer A J and Phinney E S 2003 *MNRAS* **346** 1197–1214 (*Preprint* astro-ph/0304393)

Stochastic background of gravitational waves from cosmological sources

Chiara Caprini

CNRS, URA 2306 and CEA, IPhT, F-91191 Gif-sur-Yvette, France

E-mail: chiara.caprini@cea.fr

Abstract. Gravitational waves (GW) can constitute a unique probe of the primordial universe. In many cases, the characteristic frequency of the emitted GW is directly related to the energy scale at which the GW source is operating in the early universe. Consequently, different GW detectors can probe different energy scales in the evolution of the universe. After a general introduction on the properties of a GW stochastic background of primordial origin, some examples of cosmological sources are presented, which may lead to observable GW signals.

1. Introduction

Several processes operating in the very early universe can act as sources of a stochastic background of GW. GW of cosmological origin form a fossil radiation: expansion prevents them from coming in thermal equilibrium with the other components of the universe, because of the weakness of the gravitational interaction. Therefore, a stochastic background of GW can contain in its amplitude, spectral shape and frequency range much information on the nature of the source that generated it. The detection of such a fossil background would thus have a profound impact on our knowledge of the early universe, possibly analogous to the impact of the measurement of the Cosmic Microwave Background (CMB) and of its temperature fluctuations. One difference is that GW are out of thermal equilibrium since the Planck scale, while photons decoupled at a temperature of around 0.3 eV. Therefore, relic GW are a potential source of information on the state of the universe at high energies, even higher than those that can be reached in terrestrial laboratories.

2. Overview of the GW signal from primordial sources

In a cosmological context, GW may be represented by a tensor perturbation h_{ij} $(i,j=1,2,3)$ of the FRW metric $ds^2=-dt^2+a^2(t)\,(\delta_{ij}+h_{ij})\,dx^idx^j$ which is transverse and traceless $\partial_i h_{ij}=h_{ii}=0$ (we assume flat spatial sections, $a(t)$ is the scale factor and repeated indices are summed). The transverse-traceless condition leaves only two independent physical degrees of freedom. In Fourier space, their linearized equation of motion is $\ddot{h}_{ij}(\mathbf{k},t)+3H\,\dot{h}_{ij}(\mathbf{k},t)+k^2\,h_{ij}(\mathbf{k},t)=16\pi G\,\Pi^{(TT)}_{ij}(\mathbf{k},t)$, where G is the Newton constant, H is the Hubble rate, a dot denotes derivative with respect to t, k is the physical wavenumber and $\Pi^{(TT)}_{ij}$ is the transverse-traceless part of the anisotropic stress Π_{ij}. The latter is given by $a^2\,\Pi_{ij}=T_{ij}-p\,a^2\,(\delta_{ij}+h_{ij})$, where T_{ij} is the energy-momentum tensor and p the background pressure. Processes that give rise to a non-zero tensor anisotropic stress in the early universe can directly source GW: as we will see,

Journal of Physics: Conference Series **610** (2015) 012004 doi:10.1088/1742-6596/610/1/012004

common examples of sources are electromagnetic fields, a scalar field with spatial gradients in its distribution, or the presence of velocity perturbations in the early universe fluid.

Furthermore, GW can be generated also in the absence of tensor anisotropic stresses, by the amplification of vacuum fluctuations during inflation [1]. In this case, the equation of motion for the canonically normalized free field $v_\pm = M_{Pl}\, a\, h_\pm/\sqrt{2}$ is a homogeneous wave equation, and reads $v''_\pm + a^2(k^2 - 2H^2)v_\pm = 0$ (where a prime denotes derivative with respect to conformal time, $\pm$ are the two polarizations of the tensor modes, M_{Pl} is the reduced Planck mass, and we have set $a''/a \simeq 2a^2H^2$ at lowest order in slow roll). From this equation it appears that particle creation occurs for super-horizon modes with $k \ll H$, because of the fast expansion of the background. The tensor modes thus generated outside the horizon remain constant, until they re-enter the horizon during the radiation or the matter dominated eras: afterwards, they propagate as GW, oscillating and being redshifted by the expansion of the universe.

The energy density of GW today can be written as $\rho_{gw} = \langle \dot{h}_{ij}\, \dot{h}_{ij} \rangle/(32\pi G) = \int \frac{df}{f} \frac{d\rho_{gw}}{d\log f}$, where $f = (k/2\pi)(a/a_0)$ is the present-day GW frequency (a_0 denoting the scale factor today) and $\langle\rangle$ denotes ensemble average. A cosmological stochastic background is (at least in a first approximation) statistically isotropic, stationary and Gaussian: its main properties are then described by its power spectrum. One defines the spectrum of energy density per logarithmic frequency interval divided by the critical density ρ_c today, $h^2\,\Omega_{\rm GW}(f) = \frac{h^2}{\rho_c}\frac{d\rho_{gw}}{d\log f}$, where h parametrizes the Hubble constant $H_0 = 100\, h\,{\rm km/s/Mpc}$.

A causal GW source that operates at sub-Hubble scales at some time t_* after inflation emits GW with a characteristic wave-number k_* that is larger than the Hubble rate H_* at that time: $k_* = H_*/\epsilon_*$ with $\epsilon_* \leq 1$ [2]. The characteristic GW frequency today for a causal process is then given by $f_c = H_*/(2\pi\epsilon_*)(a_*/a_0)$. For GW generated in the radiation era when the plasma temperature is T_*, the characteristic frequency today can thus be written as $f_c \simeq 2.6 \times 10^{-5}\,{\rm Hz}\;\; \epsilon_*^{-1}\,(T_*/1\,{\rm TeV})\,(g_*/100)^{1/6}$, where we have assumed a standard thermal history for the evolution of the universe after GW production, and g_* is the number of relativistic degrees of freedom at temperature T_*. The parameter $\epsilon_* \leq 1$ depends on the dynamics of the specific GW source under consideration. Modulo the value of this parameter, it is therefore possible to establish a correspondence between the characteristic frequency today f_c of the GW emitted by a causal source, and the epoch at which the source was operating, identified by T_* (c.f. Fig. 1). Note that this correspondence only holds for GW emitted by causal sources, and not for the GW spectrum possibly generated by inflation, which leads to the amplification of super-horizon tensor modes and gives rise to a spectrum which is almost scale invariant over a wide frequency range, see Fig. 1.

As a result of the correspondence between f_c and T_*, different GW detectors operating at different frequency ranges can probe different energy scales in the history of the universe. For a first-order phase transition (PT), one may have for instance $\epsilon_* \sim 10^{-3} - 1$ (see e.g. [3]). In this case, the above equation for f_c shows that GW produced around the EW scale $T_* \simeq 100$ GeV are potentially interesting for detection with eLISA, which operates in the frequency range $10^{-5}\,{\rm Hz} < f < 1\,{\rm Hz}$ [4]. On the other hand, GW production from the QCD phase transition (QCDPT) at $T_* \simeq 100$ MeV can fall into the frequency range of detection with pulsar timing array (PTA) [5]. Earth-based interferometers such as LIGO/Virgo or the future Einstein Telescope (ET) [6], operating at a higher frequency range $1\,{\rm Hz} < f < 10^3$ Hz, can correspondingly probe higher energy scales, up to about $T_* < 10^{10}$ GeV. Currently operating and future GW detectors are therefore potentially capable of probing, via the measurement of relic GW, the state of the universe at energy scales which are currently inaccessible by any other mean: those comprised between the scale of inflation and the TeV scale.

Note that processes operating in the early universe are not the only possible sources of a stochastic background of GW: a contribution is expected also from the incoherent superposition of radiation due to astrophysical sources which cannot be individually resolved. For example, supermassive black hole binaries generate a stochastic background of GW in the PTA frequency band (see e.g. [7]), white dwarf binaries in the eLISA frequency band (see e.g. [8]), neutron star binaries in the frequency band of Earth-based detectors (see e.g. [9]). Building accurate models of the astrophysical background is very important for two reasons [10, 11]. First of all, this background is interesting in itself since it carries information about the properties of the populations of compact objects that generate it. Secondly, it can be a foreground for the cosmological background (and vice versa): therefore, it has to be modeled accurately in order either to be subtracted, or to define the best frequency windows for searching for the cosmological background.

3. Status of the observations of a GW stochastic background

Current observational upper bounds on the amplitude of the spectrum $h^2\Omega_{\rm GW}(f)$ are shown in Fig. 1. Nucleosynthesis (BBN) and the CMB monopole can be used to establish indirect upper bounds on $h^2\Omega_{\rm GW}(f)$, since they provide a measurement of the total relativistic energy density of the universe respectively at the time of BBN (about 1 MeV), and at photon decoupling (about 0.3 eV). These bounds extend on frequencies higher than the frequency corresponding to the horizon size at BBN and at photon decoupling, and read respectively $h^2\Omega_{\rm GW}^{\rm BBN}(f) \leq 7.8 \times 10^{-6}$ for $f_{\rm BBN} > 10^{-10}$ Hz [12], and $h^2\Omega_{\rm GW}^{\rm CMB}(f) \leq 3.8 \times 10^{-6}$ for $f_{\rm CMB} > 10^{-16}$ Hz [12, 13]. They are represented in Fig. 1 respectively by the purple and red horizontal lines. CMB temperature anisotropies have also been used to set an upper bound on $h^2\Omega_{\rm GW}(f)$, since the presence of tensor perturbations in the FRW metric causes the CMB photons to move on perturbed geodesics and changes their temperature (Sachs Wolfe effect) [14]. The bound applies to frequencies corresponding to tensor modes that entered the horizon after the epoch of equality between radiation and matter, and are sub-horizon today: 10^{-18} Hz $< f < 10^{-16}$ Hz. If the GW spectrum was initially almost scale invariant, as is the case for a tensor spectrum generated during slow roll inflation, the transfer function for tensor modes implies that $\Omega_{\rm GW}(f)$ today decreases as f^{-2} for modes that entered the horizon in the matter era [14]. The strongest bound on $h^2\Omega_{\rm GW}(f)$ is therefore obtained at $f \sim 10^{-16}$ Hz: $h^2\Omega_{\rm GW} <$ few $\times 10^{-14}$ [2]. The bound, at the level given in Ref. [2], is represented by the green line in Fig. 1.

Other bounds have been established directly by observations with LIGO/Virgo and with PTA: the most recent analyses give respectively, for LIGO/Virgo $\Omega_{\rm GW}(f) \leq 5.6 \times 10^{-6}$ at 41.5 Hz $< f <$ 169.25 Hz [15], and for the Parkes PTA $\Omega_{\rm GW}(f) \leq 1.3 \times 10^{-9}$ at $f = 2.8$ nHz [16]. These are represented respectively by the blue and magenta lines in Fig. 1 (the value $h = 0.67$ has been chosen here [17]).

Recently, the BICEP2 experiment detected the presence of B-polarization in the CMB spectrum [18]. A B-polarization component of primordial origin is generated at photon decoupling from Thomson scattering, if the quadrupole temperature anisotropy of the photons, caused by metric perturbations, has a tensor component. A few months later, new results from the Planck satellite showed that the BICEP2 signal is strongly contaminated by the contribution of dust from the galaxy [19]. This does not exclude a priori the presence of a contribution from primordial tensor modes of inflationary origin, but it shows that foreground cleaning is absolutely essential. Note that the detection of a stochastic background of GW from inflation would constitute a major discovery: it would imply that gravity can be quantized at the linear level analogously to any canonical quantum field, and would also set the value of the energy scale of inflation in the simplest slow-roll inflationary scenario: $V^{1/4} \simeq 10^{16}$ GeV. The corresponding GW spectrum today in terms of $h^2\Omega_{\rm GW}(f)$ is shown in Fig. 1 (black line): here, the tensor to scalar ratio is set to $r = 0.2$ and the tensor spectral index to $n_T = -r/8$, we have assumed a standard thermal history after inflation (therefore omitting scenarios as the one described in [20]), and we have neglected changes in the number of relativistic degrees of freedom. As explained above, the region $h^2\Omega_{\rm GW} \propto f^{-2}$, for 10^{-18} Hz $< f < 10^{-16}$ Hz, corresponds to

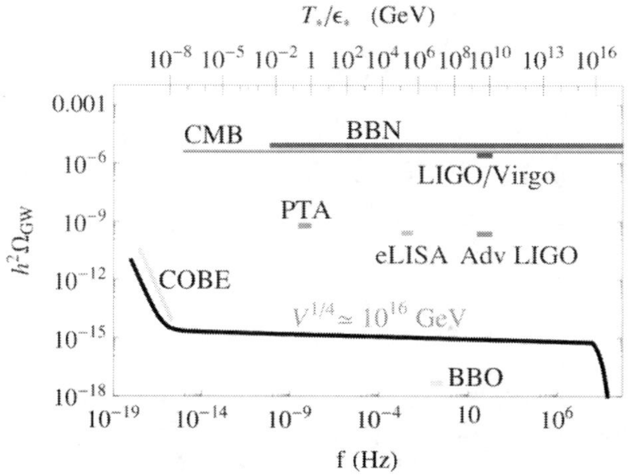

Figure 1. Observational upper bounds on $h^2\Omega_{GW}(f)$, together with the expected sensitivity of some GW detectors (eLISA, Advanced LIGO, BBO - see main text, Sec. 3). The upper horizontal axis shows the temperature in the radiation dominated universe at which a causal source can operate in order to emit GW with the characteristic frequency f shown in the lower horizontal axis (modulo the parameter ϵ_*). The black line shows the GW spectrum generated by slow roll inflation at the energy scale $V^{1/4} \simeq 10^{16}$ GeV.

the signal detectable both through temperature anisotropies and B-polarization in the CMB spectrum. On the other hand, the amplitude of the GW spectrum is too low to be directly observable both by currently operating and by forthcoming GW experiments, a part from the conceived space-based observatories BBO and DECIGO [21], which are especially designed to reach the inflationary background. The cyan line in Fig. 1 represents the maximal sensitivity of the BBO interferometer, taken from Ref. [22] to be at the level of $\Omega_{GW} \simeq 10^{-17}$ at $f \simeq 0.2$ Hz, after one year of observation time.

Figure 1 also shows the expected maximal sensitivity of the eLISA interferometer (orange line), taken from Ref. [23] at the level of $h^2\Omega_{GW} \simeq 2.5 \times 10^{-10}$ at $f \simeq 3$ mHz, and the expected maximal sensitivity for the correlation of the Hanford-Livingstone pair of Advanced LIGO detectors, taken from Ref. [22] at the level of $\Omega_{GW} \simeq 5 \times 10^{-10}$ at $f \simeq 30$ Hz.

4. Examples of proposed primordial GW sources

Besides inflation, there are other possible sources of GW in the early universe which are more promising for detection with future interferometers or PTA. These are all mechanisms that lead to non-zero tensor anisotropic stresses and thereby act as sources of GW. We present in the following some proposals that may lead to observable GW signals (note that this list is non-exhaustive).

• **Particle production during inflation**: in addition to the amplification of quantum vacuum fluctuations, GW can also be emitted classically during inflation if there are anisotropic stresses. Anisotropic stresses can be generated by particles: as the inflaton rolls down its potential, it provides a time-dependent effective mass to fields coupled to it. If such a field becomes effectively massless during inflation, particles of this field can be produced efficiently in a non-perturbative way [24]. The scenario which is the most favorable for detection arises when the inflaton ϕ couples to a gauge field through an interaction term of the form $\phi F_{\mu\nu}\tilde{F}_{\mu\nu}$, where $F_{\mu\nu}$ is the field strength and $\tilde{F}_{\mu\nu}$ its dual: this interaction is natural in models where the inflaton field is an axion [25]. The gauge field remains massless throughout inflation and its particles can be efficiently produced in a continuous way at horizon crossing, significantly enhancing and polarizing the inflationary GW spectrum up to a level observable by ground-based interferometers [25, 26, 27], see Fig. 2. Note however that, even though this enhancement operates far away from the scales probed by the CMB and by Large Scale Structure, it has been recently put forward that these models can be severely constrained by the current bounds on non-gaussianities [28].

• **Preheating**: GW production can occur also at the end of inflation, when the potential energy density driving inflation is converted into radiation energy density in the course of reheating. In many inflationary models, reheating starts with preheating, an explosive and non-perturbative decay of the inflaton condensate into fluctuations of itself and other fields coupled to it [30]. The system then evolves towards thermal equilibrium, but in a highly non-linear and turbulent way. The large field fluctuations lead to non-zero anisotropic stresses and therefore source GW, see e.g. [31]. Preheating is a causal process and therefore emits GW with a characteristic wavenumber $k_* = H_*/\epsilon_*$ at the time of production, where the Hubble rate during preheating H_* is directly related to the energy scale of inflation. One finds that the inflationary energy scale must be smaller than about 10^{11} GeV for the peak of the spectrum to fall into the frequency range of ground-based interferometers, and smaller than 10^7 GeV for the peak to fall into the eLISA frequency band. Thus GW from preheating could be observable if inflation occurs at sufficiently low energy: depending on the inflationary scenario, this could be problematic if a component of the BICEP2 signal turns out to be of primordial origin after the complete removal of dust contamination. The peak frequency and the shape of the GW spectrum depend on the inflationary model and on the nature and the couplings of the fields that are produced during preheating. In figure 3 we show, as an example, the GW spectrum from tachyonic preheating after hybrid inflation obtained in [32].

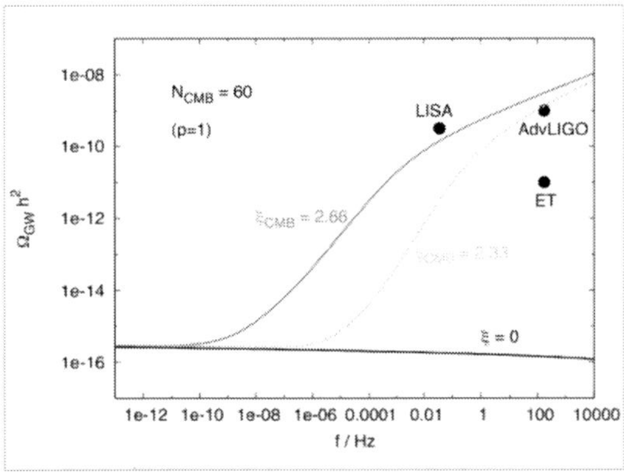

Figure 2. Taken from Ref. [25]. The GW spectra (green and red curves) sourced by particle production in a model of pseudoscalar inflation with linear potential ($p = 1$, see Ref. [25]) and 60 e-folds of CMB-observable inflation. The spectrum is observable by advanced LIGO if the coupling $\xi_{\rm CMB} = 2.33$ (taken at CMB scales). Higher values of the coupling that could possibly lead to a signal observable in LISA are forbidden by CMB non-Gaussianity constraints [29]. $\xi = 0$ corresponds to the standard inflationary spectrum.

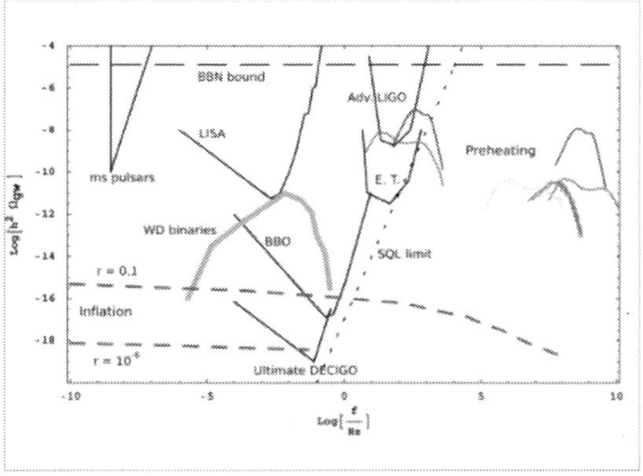

Figure 3. Taken from Ref. [32]. Examples of GW spectra from tachyonic preheating after hybrid inflation (blue, red and green curves) are shown together with the sensitivity of some GW detectors, the background from extragalactic white dwarf binaries, the inflationary background for two values of the tensor to scalar ratio r, and the spectrum from preheating in a model of chaotic inflation (thin grey curve). The amplitude and peak frequency of the spectra depend on the coupling between the field and the inflaton, its vev, and its initial velocity (c.f. [32]).

• **Cosmic strings**: cosmic strings can be left over by a PT occurring either at the end of inflation or during the thermal evolution of the universe. They are predicted by several high-energy physics models, both in field theory (grand unification, supersymmetry) and in string theory. Once produced, a network of stable cosmic strings evolves towards a self-similar regime characterized by a continuous energy loss: when long string segments cross each other and reconnect, they form smaller cosmic string loops, which oscillate relativistically and decay away by emitting GW (see e.g. [33]). Thus the cosmic string energy density is continually converted into GW via the production of loops: the resulting GW background covers a wide frequency range and can be looked for by different experiments, even simultaneously. Furthermore, cosmic strings also produce GW bursts, emitted by cusps and kinks. The bursts can be looked for either individually, or considering that the incoherent superposition of the burst signals from cusps [34] and kinks [35] also contributes to the stochastic GW background.

Two fundamental parameters enter in the GW spectrum, that depend on the underlying high-energy physics model: the string tension and the reconnection probability. The dimensionless parameter $G\mu$ representing the tension must be sufficiently small to satisfy observational constraints imposed by the CMB [36]. The reconnection probability is usually equal to one for cosmic strings in field theories, but it can be smaller for cosmic strings in string theory. The GW background from cosmic strings depends also on the typical size of the loops when they are produced: recently, different groups performing simulations reached an agreement on the loop size distribution, concluding that loops of all sizes are present, but the distribution is dominated by those produced within a few orders of magnitude of the horizon size [37, 38]. Figure 4, taken from [3], shows the GW background for large loops for different values of $G\mu$ and of the reconnection probability.

• **First order phase transitions**: in the course of its adiabatic expansion, the universe might have undergone several PTs driven by the temperature decrease. Their nature depends on the particle theory model, but if they are first order they proceed through the nucleation and collision of broken phase bubbles, which lead to anisotropic stresses and therefore to the generation of GW (for a review, see [3] and references therein). The EWPT in the standard model is a crossover, and it is not expected to lead to any appreciable cosmological signal. However, theories beyond the standard model can lead to a first order EWPT, giving a very rich phenomenology: besides a background of GW, a first order EWPT could provide Dark Matter candidates and baryogenesis. The recent discovery of the Higgs boson at the LHC confirms the paradigm of a scalar field-driven symmetry breaking in the early universe. There is yet no indication of new physics near the EW scale, but the order of the EWPT is not directly constrained by LHC data: several models leading to a first order EWPT remain viable, mainly based on expanding the Higgs sector with additional scalar states invisible at the energies probed by the LHC up to now (see e.g. [39]). If the EWPT is sufficiently strongly first order, it could lead to a GW signal detectable by the space-based interferometer eLISA. Similarly, the QCDPT is also predicted to be a crossover by lattice simulations but it can become first order if the neutrino chemical potential is sufficiently large [40]. GW detection would help to probe the nature of these PTs, and provide interesting information on the underlying particle theory. Towards the end of a first order PT, the true vacuum bubbles collide and convert the entire universe to the broken phase. The collisions break the spherical symmetry of the bubble walls, generating a non-zero anisotropic stress [41]. Moreover, they give rise to compression waves [42, 43] and MHD turbulence (see e.g. [44]) in the surrounding fluid: their anisotropic stresses can act as a source of GW even after the merging of the bubbles is completed. The characteristic frequency of the GW spectrum is determined by the average size of the bubble towards the end of the PT, and its amplitude by the amount of tensor-type stress energy that is available: both these factors are strongly related to the strength of the first order PT. Figure 5, taken from [3], shows the GW signal for one example of first order PT occurring around the EW scale [45].

5. Conclusions

We have little information about the physics and the processes operating in the very early universe, in particular for energies comprised between the scale of inflation (accessible through

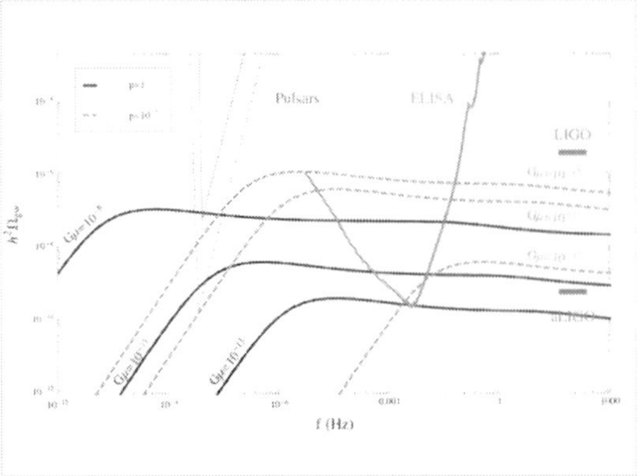

Figure 4. Taken from Ref. [3]. GW spectra from cosmic strings for different values of the string tension and of the reconnection probability p, for large loops, compared to the observational sensitivities of some GW detectors.

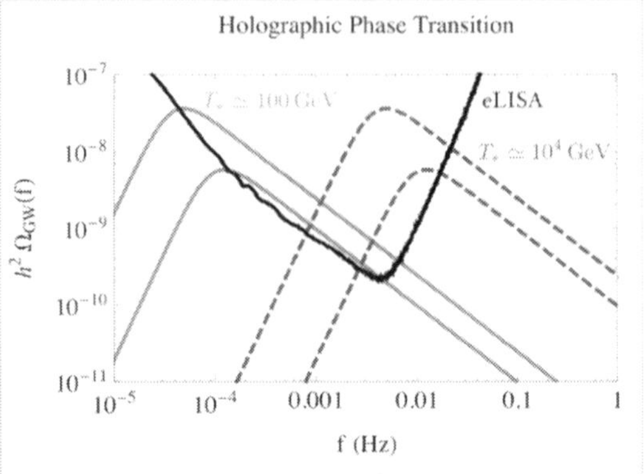

Figure 5. Taken from Ref. [3]. GW spectra in a model of first order PT connected to the radion stabilization in the Randal Sundrum model [45], for different values of the PT temperature and duration (for details, see [3]). The signal amplitude is high since in this particular model the phase transition is strongly first order, and the bubble walls move at the speed of light.

the observation of CMB anisotropies and polarization) and the EW scale (accessible at the LHC). Many interesting processes could take place in this energy range, such as reheating, baryogenesis, dark matter production, phase transitions and their remnants. Their occurrence and their characteristics ultimately depend on the high energy physics model that describes the very early universe. As discussed above, some of these processes can be powerful sources of GW, and due to their small interaction rate, GW would afterwards propagate freely until today. The detection of a GW signal from the early universe could therefore provide us with very significant information on the status of the universe at high energies. The characteristic frequency of the GW maps the temperature/energy scale of their generation process. GW by vacuum fluctuations during slow roll inflation are not visible by the next generation interferometers or by PTA. However, we have presented some examples of processes which generate a stochastic background of GW directly from anisotropic stresses, and which are more promising for detection. Therefore, GW are a powerful mean to learn about the early universe and high energy physics: the detection is extremely difficult, but it leads to a great payoff.

References

[1] L.P. Grishchuk, Zh. Ehksp. Teor. Fiz. **67** (1974) 825; A.A. Starobinsky, JETPL **30** (1979) 682
[2] B. Allen, In *Les Houches 1995, Relativistic gravitation and gravitational radiation* 373-417 [gr-qc/9604033].
[3] P. Binetruy *et al*, JCAP **1206** 027 (2012) [arXiv:1201.0983 [astro-ph]].
[4] P. A. Seoane *et al* [eLISA Collaboration], arXiv:1305.5720 [astro-ph.CO].
[5] C. Caprini *et al*, Phys. Rev. D **82** (2010) 063511 [arXiv:1007.1218 [astro-ph.CO]].
[6] http://www.ligo.org; http://www.ego-gw.it; http://www.et-gw.eu
[7] A. Sesana, A. Vecchio and C. N. Colacino, Mon. Not. Roy. Astron. Soc. **390** (2008) 192 [arXiv:0804.4476 [astro-ph]].
[8] A. J. Farmer and E. S. Phinney, Mon. Not. Roy. Astron. Soc. **346** (2003) 1197 [astro-ph/0304393].
[9] T. Regimbau and B. Chauvineau, Class. Quant. Grav. **24** (2007) S627 [arXiv:0707.4327 [gr-qc]].
[10] T. Regimbau and V. Mandic, Class. Quant. Grav. **25** (2008) 184018 [arXiv:0806.2794 [astro-ph]].
[11] D. Meacher, E. Thrane and T. Regimbau, Phys. Rev. D **89** (2014) 8, 084063 [arXiv:1402.6231 [astro-ph.CO]].

[12] T. L. Smith *et al*, Phys. Rev. Lett. **97** (2006) 021301 [astro-ph/0603144].
[13] S. Henrot-Versill *et al.*, arXiv:1408.5299 [astro-ph.CO].
[14] L. M. Krauss and M. J. White, Phys. Rev. Lett. **69** (1992) 869 [hep-ph/9205212]; B. Allen and S. Koranda, Phys. Rev. D **50** (1994) 3713 [astro-ph/9404068].
[15] J. Aasi *et al.* [LIGO Scientific and VIRGO Collaborations], arXiv:1406.4556 [gr-qc].
[16] R. M. Shannon *et al.*, Science **342** (2013) 6156, 334 [arXiv:1310.4569 [astro-ph.CO]].
[17] P. A. R. Ade *et al.* [Planck Collaboration], Astron. Astrophys. **571** (2014) A16 [arXiv:1303.5076 [astro-ph.CO]].
[18] P. A. R. Ade *et al.* [BICEP2 Collaboration], Phys. Rev. Lett. **112** (2014) 241101 [arXiv:1403.3985 [astro-ph.CO]].
[19] R. Adam *et al.* [Planck Collaboration], arXiv:1409.5738 [astro-ph.CO].
[20] L. A. Boyle and A. Buonanno, Phys. Rev. D **78** (2008) 043531 [arXiv:0708.2279 [astro-ph]].
[21] J. Crowder and N. J. Cornish, Phys. Rev. D **72** (2005) 083005 [gr-qc/0506015]; S. Kawamura *et al.*, Class. Quant. Grav. **23** (2006) S125.
[22] E. Thrane and J. D. Romano, Phys. Rev. D **88** (2013) 124032 arXiv:1310.5300 [astro-ph.IM].
[23] P. Amaro-Seoane, S. Aoudia, S. Babak, P. Binetruy, E. Berti, A. Bohe, C. Caprini and M. Colpi *et al.*, GW Notes **6** (2013) 4 [arXiv:1201.3621 [astro-ph.CO]].
[24] D. J. H. Chung *et al*, Phys. Rev. D **62** (2000) 043508 [hep-ph/9910437].
[25] J. L. Cook and L. Sorbo, Phys. Rev. D **85** (2012) 023534 [Erratum-ibid. D 86 (2012) 069901] [arXiv:1109.0022 [astro-ph.CO]].
[26] N. Barnaby *et al*, Phys. Rev. **D** 85 (2012) 023525 [arXiv:1110.3327 [astro-ph.CO]].
[27] S. G. Crowder, R. Namba, V. Mandic, S. Mukohyama and M. Peloso, Phys. Lett. B **726** (2013) 66 [arXiv:1212.4165 [astro-ph.CO]].
[28] R. Z. Ferreira and M. S. Sloth, arXiv:1409.5799 [hep-ph].
[29] P. D. Meerburg and E. Pajer, JCAP **1302** (2013) 017 [arXiv:1203.6076 [astro-ph.CO]].
[30] L. Kofman *et al*, Phys. Rev. Lett. **73**, 3195 (1994). L. Kofman *et al*, Phys. Rev. D **56**, 3258 (1997).
[31] S. Y. Khlebnikov and I. I. Tkachev, Phys. Rev. D **56**, 653 (1997); R. Easther *et al*, Phys. Rev. Lett. **99**, 221301 (2007) [arXiv:astro-ph/0612294]; J. García-Bellido and D. G. Figueroa, Phys. Rev. Lett. **98**, 061302 (2007) [arXiv:astro-ph/0701014]
[32] J. F. Dufaux *et al*, JCAP **0903**, 001 (2009) [arXiv:0812.2917 [astro-ph]].
[33] A. Vilenkin, Phys. Lett. B **107** 47 (1981); C. Hogan and M. Rees, Nature **311** 109 (1984); T. Vachaspati and A. Vilenkin, Phys.Rev. **D** 31, 3052 (1985); R. Caldwell and B. Allen, Phys.Rev. **D** 45, 3447 (1992); M.R. DePies and C.J. Hogan, Phys.Rev. **D** 75, 125006 (2007) [arXiv:astro-ph/0702335].
[34] X. Siemens, V. Mandic and J. Creighton, Phys. Rev. Lett. **98** (2007) 111101 [astro-ph/0610920].
[35] S. Olmez *et al*, Phys.Rev. **D** 81, 104028 (2010)
[36] P. A. R. Ade *et al.* [Planck Collaboration], Astron. Astrophys. **571** (2014) A25 [arXiv:1303.5085 [astro-ph.CO]].
[37] JCAP **1010**, 003 (2010) [arXiv:1006.0931 [astro-ph]]
[38] J.J. Blanco-Pillado *et al*, Phys. Rev. D **83**, 083514 (2011) [arXiv:1101.5173 [astro-ph]].
[39] M. Fairbairn and R. Hogan, JHEP **1309** (2013) 022 [arXiv:1305.3452 [hep-ph]]; G. C. Dorsch *et al*, JHEP **1310** (2013) 029 [arXiv:1305.6610 [hep-ph]]; W. Huang *et al*, JHEP **1303** (2013) 164 [arXiv:1210.0906 [hep-ph]]; M. Laine *et al*, JCAP **1301** (2013) 011 [arXiv:1211.7344 [hep-ph]]; H. H. Patel and M. J. Ramsey-Musolf, Phys. Rev. D **88** (2013) 035013 [arXiv:1212.5652 [hep-ph]]; W. Huang, Z. Kang, J. Shu, P. Wu and J. M. Yang, Phys. Rev. D **91** (2015) 2, 025006 [arXiv:1405.1152 [hep-ph]].
[40] D. J. Schwarz and M. Stuke, JCAP **0911** (2009) 025 [Erratum-ibid. **1010** (2010) E01] [arXiv:0906.3434 [hep-ph]].
[41] A. Kosowsky and M. S. Turner, Phys. Rev. D **47** (1993) 4372 [astro-ph/9211004].
[42] M. Hindmarsh, S. J. Huber, K. Rummukainen and D. J. Weir, Phys. Rev. Lett. **112** (2014) 041301 [arXiv:1304.2433 [hep-ph]].
[43] J. T. Giblin and J. B. Mertens, Phys. Rev. D **90** (2014) 2, 023532 [arXiv:1405.4005 [astro-ph.CO]].
[44] C. Caprini *et al*, JCAP **0912** (2009) 024 [arXiv:0909.0622 [astro-ph.CO]].
[45] L. Randall and G. Servant, JHEP **0705** (2007) 054 [hep-ph/0607158]; T. Konstandin and G. Servant, arXiv:1104.4791 [hep-ph].

Journal of Physics: Conference Series 610 (2015) 012005 doi:10.1088/1742-6596/610/1/012005

The LISA Pathfinder Mission

M Armano[a], H Audley[b], G Auger[c], J Baird[n], P Binetruy[c], M Born[b], D Bortoluzzi[d], N Brandt[e], A Bursi[t], M Caleno[f], A Cavalleri[g], A Cesarini[g], M Cruise[h], K Danzmann[b], I Diepholz[b], R Dolesi[g], N Dunbar[i], L Ferraioli[j], V Ferroni[g], E Fitzsimons[e], M Freschi[a], J Gallegos[a], C García Marirrodriga[f], R Gerndt[e], LI Gesa[k], F Gibert[k], D Giardini[j], R Giusteri[g], C Grimani[l], I Harrison[m], G Heinzel[b], M Hewitson[b], D Hollington[n], M Hueller[g], J Huesler[f], H Inchauspé[c], O Jennrich[f], P Jetzer[o], B Johlander[f], N Karnesis[k], B Kaune[b], N Korsakova[b], C Killow[p], I Lloro[k], R Maarschalkerweerd[m], S Madden[f], D Mance[j], V Martín[k], F Martin-Porqueras[a], I Mateos[k], P McNamara[f], J Mendes[m], L Mendes[a], A Moroni[t], M Nofrarias[k], S Paczkowski[b], M Perreur-Lloyd[p], A Petiteau[c], P Pivato[g], E Plagnol[c], P Prat[c], U Ragnit[f], J Ramos-Castro[qr], J Reiche[b], J A Romera Perez[f], D Robertson[p], H Rozemeijer[f], G Russano[g], P Sarra[t], A Schleicher[e], J Slutsky[s], C F Sopuerta[k], T Sumner[n], D Texier[a], J Thorpe[s], C Trenkel[i], H B Tu[g], D Vetrugno[g], S Vitale[g], G Wanner[b], H Ward[p], S Waschke[n], P Wass[n], D Wealthy[i], S Wen[g], W Weber[g], A Wittchen[b], C Zanoni[d], T Ziegler[e], P Zweifel[j]

[a] European Space Astronomy Centre, European Space Agency, Villanueva de la Cañada, 28692 Madrid, Spain
[b] Albert-Einstein-Institut, Max-Planck-Institut für Gravitationsphysik und Universität Hannover, 30167 Hannover, Germany
[c] APC UMR7164, Université Paris Diderot, Paris, France
[d] Department of Industrial Engineering, University of Trento, via Sommarive 9, 38123 Trento, and Trento Institute for Fundamental Physics and Application / INFN
[e] Airbus Defence and Space, Claude-Dornier-Strasse, 88090 Immenstaad, Germany
[f] European Space Technology Centre, European Space Agency, Keplerlaan 1, 2200 AG Noordwijk, The Netherlands
[g] Dipartimento di Fisica, Università di Trento and Trento Institute for Fundamental Physics and Application / INFN, 38123 Povo, Trento, Italy
[h] Department of Physics and Astronomy, University of Birmingham, Birmingham, UK
[i] Airbus Defence and Space, Gunnels Wood Road, Stevenage, Hertfordshire, SG1 2AS, UK
[j] Institut für Geophysik, ETH Zürich, Sonneggstrasse 5, CH-8092, Zürich, Switzerland
[k] Institut de Ciències de l'Espai (CSIC-IEEC), Campus UAB, Facultat de Ciències, 08193 Bellaterra, Spain
[l] Istituto di Fisica, Università degli Studi di Urbino/ INFN Urbino (PU), Italy
[m] European Space Operations Centre, European Space Agency, 64293 Darmstadt, Germany
[n] The Blackett Laboratory, Imperial College London, UK
[o] Physik Institut, Universität Zürich, Winterthurerstrasse 190, CH-8057 Zürich, Switzerland
[p] SUPA, Institute for Gravitational Research, School of Physics and Astronomy, University of Glasgow, Glasgow, G12 8QQ, UK
[q] Department d'Enginyeria Electrònica, Universitat Politècnica de Catalunya, 08034 Barcelona, Spain

[r] Institut d'Estudis Espacials de Catalunya (IEEC), C/ Gran Capità 2-4, 08034 Barcelona, Spain
[s] NASA Goddard Space Flight Center, 8800 Greenbelt Road, Greenbelt, MD 20771, USA
[t] CGS S.p.A, Compagnia Generale per lo Spazio, Via Gallarate, 150 - 20151 Milano, Italy

E-mail: paul.mcnamara@esa.int

Abstract. LISA Pathfinder (LPF), the second of the European Space Agency's Small Missions for Advanced Research in Technology (SMART), is a dedicated technology validation mission for future spaceborne gravitational wave detectors, such as the proposed eLISA mission. LISA Pathfinder, and its scientific payload - the LISA Technology Package - will test, in flight, the critical technologies required for low frequency gravitational wave detection: it will put two test masses in a near-perfect gravitational free-fall and control and measure their motion with unprecedented accuracy. This is achieved through technology comprising inertial sensors, high precision laser metrology, drag-free control and an ultra-precise micro-Newton propulsion system. LISA Pathfinder is due to be launched in mid-2015, with first results on the performance of the system being available 6 months thereafter.

The paper introduces the LISA Pathfinder mission, followed by an explanation of the physical principles of measurement concept and associated hardware. We then provide a detailed discussion of the LISA Technology Package, including both the inertial sensor and interferometric readout. As we approach the launch of the LISA Pathfinder, the focus of the development is shifting towards the science operations and data analysis - this is described in the final section of the paper

1. Introduction

LISA Pathfinder, the second of the European Space Agency's Small Missions for Advanced Research in Technology (SMART), is a technology validation mission for future low frequency, spaceborne, laser interferometric gravitational wave detectors. When adopted by the European Space Agency (ESA) in 2004, LISA Pathfinder was a precursor for the joint ESA/NASA mission, LISA (Laser Interferometer Space Antenna) [1]. However, due to budget constraints at NASA, the partnership was dissolved in 2011, and LISA was reformulated as an ESA-only mission called the New Gravitational-wave Observatory (NGO) [2]. NGO was proposed to ESA as the first Large-class mission (L1) in its Cosmic Visions (CV) programme. However, it was not selected.

The NGO concept was further refined and renamed eLISA (evolved-LISA). This concept was included as the strawman mission design included in the science white-paper, *The Gravitational Universe* [3], submitted to the ESA call for science themes for the 2nd and 3rd Large-class missions in the CV programme. After review, *The Gravitational Universe* was selected as the science theme for the 3rd Large-class mission.

LISA, and variants of the LISA mission concept, have continually been ranked as one of the most scientifically important missions under study [4]. However the very concept of low frequency gravitational wave detection, *i.e.* that a particle falling under the influence of gravity alone can be isolated to follow a geodesic in spacetime, has never been demonstrated to the required precision. This is the most basic assumption of Einstein's General Relativity: LISA Pathfinder was selected by the European Space Agency to experimentally demonstrate this hypothesis with unprecedented accuracy.

LISA Pathfinder essentially mimics one arm of the eLISA constellation by shrinking the 1 million kilometre armlength down to a few tens of centimetres, giving up the sensitivity to gravitational waves, but keeping the measurement technology. The scientific objective of the LISA Pathfinder mission consists then of the first in-flight test of low frequency gravitational wave detection metrology.

LISA Pathfinder was first proposed in 1998 as ELITE (European LIsa Technology

Experiment) [5]. This mission consisted of a single spacecraft in geostationary orbit with a differential acceleration goal of $10^{-14}\,\mathrm{ms}^{-2}/\sqrt{\mathrm{Hz}}$ over a frequency range of 1-100 mHz. This original proposal was refined and proposed to ESA in 2000 in response to the SMART-2 announcement of opportunity. At the time, the proposal called for a joint LISA and Darwin pathfinder mission, consisting of two free-flying spacecraft, with three payloads (LISA Technology Package, Darwin Technology Package, and a NASA provided LISA Technology Package). The goal of the mission was to demonstrate drag-free control (for LISA) and formation flying (for Darwin). The mission was approved by the Science Programme Committee (SPC) in November 2000. After an initial industrial study, the mission was descoped to a single spacecraft (the Darwin Pathfinder was cancelled) and renamed LISA Pathfinder (LPF). At the time, LPF carried two payloads, the European built LISA Technology Package (LTP), and the NASA provided Disturbance Reduction System (DRS). Both payloads consisted of two inertial sensors, a laser metrology system, micro-Newton thrusters and drag free control software. However, the DRS was descoped and now consists of micro-Newton thrusters and a dedicated processor running the drag-free and attitude control software, and will rely on the LTP for its inertial sensing capability.

LISA Pathfinder is due to be launched in 2015 on-board a dedicated small launch vehicle. The launcher selected is the European VEGA launcher, which will launch LPF from the European spaceport of Kourou (French Guyana) into a parking orbit with perigee at 200 km, apogee at ≈ 1600 km, and an inclination to the equator of 5.3°. After a series of apogee raising manoeuvres using an expendable propulsion module, LISA Pathfinder will enter a transfer orbit towards the first Sun-Earth Lagrange point (L1). After separation from the propulsion module, the LPF spacecraft will be stabilised using the micro-Newton thrusters, entering a 500,000 km by 800,000 km Lissajous orbit around L1.

Following the initial on-orbit check-out and instrument calibration, the in-flight demonstration of the LISA technology will then take place. The nominal lifetime of the science operations is 180 days; this includes the LTP and DRS operations.

2. The physics of LISA Pathfinder

eLISA will detect gravitational waves in the tidal acceleration they cause between distant free-falling test particles. This relative acceleration is observable in the variation of the Doppler frequency shift detected between a light beam received from the distant emitting test particle (e) and a local light beam generated with an equivalent laser at the receiver (r). The time derivative of this interference "Doppler beat note" can be expressed, for an optimal projection of gravitational wave strain h onto the axis defined by the test particles separation L,

$$\frac{\dot{\nu}_r - \dot{\nu}_e}{\nu} = \frac{1}{2}\left[\dot{h}\left(t\right) - \dot{h}\left(t - \frac{L}{c}\right)\right] + \frac{1}{c}\left[a_r\left(t\right) - a_e\left(t - \frac{L}{c}\right)\right] \ , \tag{1}$$

This highlights the competition of the gravitational wave signal with the difference in the acceleration between the two particles – or test masses, TM – due to any stray forces, measured relative to their locally free-falling reference frames at the time of emission and detection, $\Delta a \equiv \left[a_r\left(t\right) - a_e\left(t - \frac{L}{c}\right)\right]$, projected onto the axis separating the TM [6, 7]. This simple picture omits two complications relevant to eLISA: the Doppler shifts from laser frequency noise, handled by using two interferometry arms and time delay interferometry (TDI), and the composite nature of the TM-TM measurement, with a Doppler measurement between distant spacecraft (SC-SC) summed with two local measurements of relative TM-SC acceleration. At low frequencies, the eLISA noise budget is dominated by a relative stray TM acceleration Δa with spectral density $\sqrt{2} \times 3 \times 10^{-15}\ \mathrm{m/s^2/Hz^{1/2}}$.

LISA Pathfinder will demonstrate the possibility to perform a relative acceleration measurement between two TM in the same SC. The short, 38 cm, baseline, reduces the GW strain term related to the change in $\dot{h}$ to an unresolvable level. However, it will measure the stray differential accleration noise from stray forces acting on two eLISA TM inside their respective gravitational reference sensors (GRS) – Δa between two nominally free-falling test particles – and demonstrate a local interferometric measurement of TM acceleration, at levels that can guarantee most of the eLISA science return.

The most sensitive differential acceleration measurements to date in the mHz frequency band come from the geodesy mission GOCE, of order 10^{-10} m/s^2/Hz$^{1/2}$ at mHz frequencies[8]. The LPF measurement configuration is essentially that of a gravity gradiometer, measuring one component – $\frac{\partial g_x}{\partial x}$ – of the 3×3 tensor measured by GOCE and, like GOCE, features rectangular parallelepiped TM of several cm dimensions inside conducting housings that provide capacitive position sensing and electrostatic force actuation. However, because of the big proposed leap forward in the differential acceleration measurement precision, to a specification of 3×10^{-14} m/s^2/Hz$^{1/2}$ at 1 mHz, LPF represents a truly new class of measurement. This improvement is made possible first by a change in location, from low earth orbit (LEO) to L1. This reduces the drag forces acting on the satellite, but also, more fundamentally, drastically reduces the gravity difference felt by the two TM, from the μm/s^2 level caused by Terrestrial gravity gradients in LEO to order nm/s^2, limited by the imperfect mass balancing of the LPF spacecraft itself, thus reducing the force noise introduced by actively compensating a large differential acceleration between the TM that limits a geodesy gradiometer. The residual difference in local g remains an important limitation for LPF and will be addressed shortly. To take advantage of this reduction in the required forcing, LPF and eLISA must correspondingly reduce other sources of stray forces, arising in large part in the interaction between the TM and the surrounding conducting shield that doubles as an electrostatic sensor and actuator, referred to collectively as the gravitational reference sensor (GRS). Compared to the hardware used in GOCE and other space accelerometry missions to date, the LPF GRS, presented in detail in Sec. 3.1, features much larger gaps between the TM and surrounding electrostatic sensor, removal of the conducting wire used to discharge the TM, and an overall mechanical and electronic design aimed at minimizing the force noise on the TM. The price of this design is the need for a UV TM discharge system, and, due to the large gaps, increased difficulty in caging the TM during launch vibration and a reduced capacitive position sensitivity, which requires the use of an interferometric readout of the TM position.

The central experiment of LISA Pathfinder is that of measuring the noise in the relative stray TM acceleration, the differential stray force per unit mass or, equivalently, the "open loop" differential acceleration that they would experience in the absence of any known or calibrated control or elastic coupling forces [9]. Considering the control of the actuation force on the SC along the critical x axis, in the main science mode, the thruster system uses a stiff, roughly 1 Hz bandwidth, controller to keep the SC centered on TM1 according to the o_1 readout of the TM1 displacement relative to the optical bench, $(x_1 - x_{SC})$ (Figure 1). An electrostatic control force, F_{ES}, is applied to the second TM, using a slow, roughly 1 mHz bandwidth, to maintain a fixed displacement of the two TM $(x_2 - x_1)$, according to the relative displacement interferometer readout o_{12} (Figure 1). The equations of motion for this system along the measurement x axis simplify to

$$\begin{aligned} \ddot{x}_1 &= a_1 - \omega_1^2 (x_1 - x_{SC}) \\ \ddot{x}_2 &= a_2 - \omega_2^2 (x_2 - x_{SC}) + \frac{F_{ES}}{m} \end{aligned} \tag{2}$$

where ω_1^2 and ω_2^2 represent springlike coupling or "stiffness" to the satellite motion, and a_1 and a_2 are the stray forces per unit mass acting on the two TM. Combining these two equations and

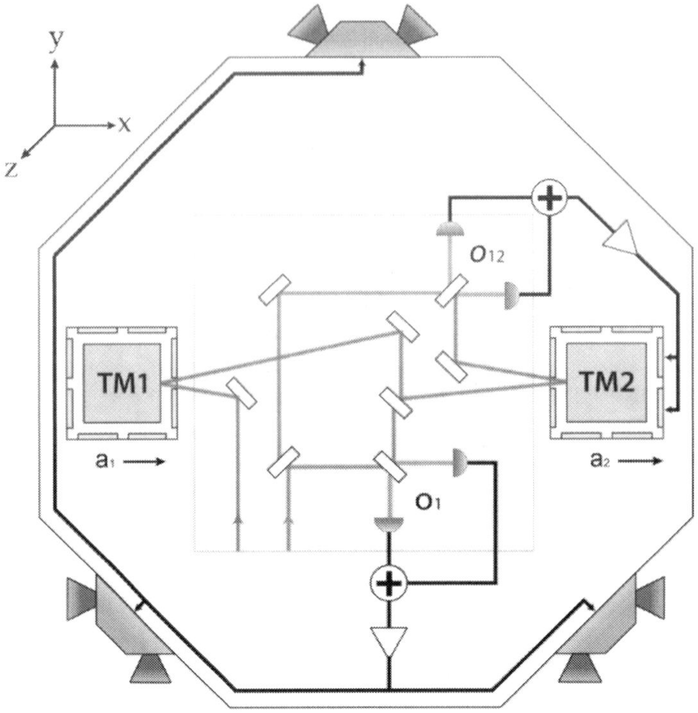

Figure 1. Cartoon of the LPF measurement configuration, showing the two TM inside their respective GRS housings, with an outline of the optical metrology for measuring the relative acceleration between the two TM (o_{12}) and that of TM1 relative to the spacecraft (o_1).

considering interferometric readout o_1 and o_{12}, allows generating $\Delta\hat{a}$, a time domain estimator of the differential stray force per unit mass,

$$\begin{aligned}\Delta\hat{a} &\equiv \ddot{o}_{12} - \frac{F_{ES}}{m} + \left(\omega_2^2 - \omega_1^2\right) o_1 + \omega_2^2 o_{12} \\ &= \Delta a + \ddot{n}_{12} + \left(\omega_2^2 - \omega_1^2\right) n_1 + \omega_2^2 n_{12}\end{aligned} \quad (3)$$

Thus $\Delta\hat{a}$ is an estimator of the desired differential stray acceleration, $\Delta a \equiv (a_2 - a_1)$. This is contaminated by noise in the interferometric displacement measurements, n_1 and n_{12}, directly in the differential acceleration $\ddot{n}_{12}$ but also, less significantly, in the calculation of the two stiffness coupling terms. This differential measurement of $\Delta\hat{a}$ also includes noise in the applied actuation force F_{ES}, which, as will be discussed later, is an important limiting factor in LPF but not in eLISA.

We note that the satellite coupling contributions are essentially subtracted off, along with the applied electrostatic forces, such that it is not the noisy satellite motion, but rather the *imprecision* in its measurement, that contributes noise through this elastic coupling. This requires that the parasitic stiffness ω_1, ω_2, and the scale factor for the commanded electrostatic actuators responsible for F_{ES}, be calibrated in a dedicated "system identification" experiment, in which large motion signals in both o_1 and o_{12} as well as large applied forces F_{ES} are excited, by applying modulated "guidance signals" that coherently move the drag-free and electrostatic

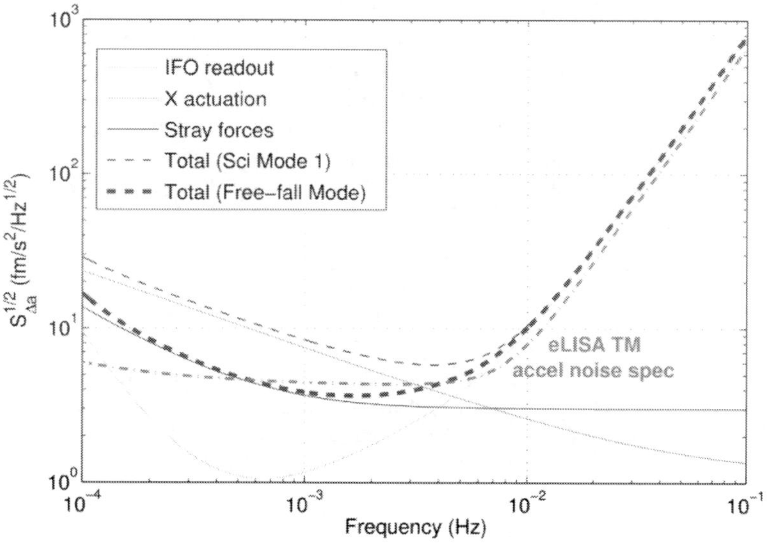

Figure 2. Current best estimates for the LPF differential acceleration noise measurement, in blue, with and without x axis force actuation, compared against the eLISA TM acceleration noise requirements in red (based mainly on the analysis in Ref.[10]).

controller setpoints. This measurement, and the data analysis strategy, will be discussed in Sec. 4.

Analysis of this main acceleration noise measurement will thus focus on an accurate estimation of the noise spectral density in $\Delta\hat{a}\,(t)$, with noise contributed by stray forces and interferometry noise. Our current best estimate of this stray acceleration noise measured in this configuration (known as "SCI Mode 1"), based on extensive ground testing with flight hardware and flight hardware prototypes in both torsion pendulum small force measurements and optical metrology tests, is shown in Fig. 2. The leading sources of force noise and interferometer measurement noise are discussed in Secs. 3.1 and 3.2.

As seen in the figure, the total low frequency acceleration noise in SCI Mode 1 is predicted to be dominated by the noise in the actuation along the x axis, which is needed to compensate the static spacecraft self-gravity imbalance between the two TM. Our estimate of this noise term is based on the target DC self-gravity imbalance limit of 0.65 nm/s^2, limited by modelling and integration precision of the spacecraft and payload mass, and an overall electrostatic actuator force stability, measured with the flight GRS FEE electronics to be slightly more than 10 ppm/Hz$^{1/2}$ at 1 mHz. Importantly, this force noise source is not present in eLISA, where two interferometry arms joining at a vertex ensure that the satellite control does not require TM forcing along the sensitive measurement axes: while LPF has two TM along a single sensitive axis in a single SC, requiring that one TM must be forced to follow the other, eLISA employs only a single TM per sensitive axis per spacecraft, and only spacecraft forcing is needed on these interferometry axes.

In order to measure the residual acceleration noise in the absence of this actuation noise which will not be present in eLISA, the LPF team is working on a "free-fall" mode experiment, in which the second TM is left free for periods up to several hundred seconds, in free-falling arcs of nearly parabolic motion – in nearly uniform sub-nano-g field relative to the inertial reference of TM1 – terminated by brief impulsive forces in order to limit the TM displacement and to compensate the differential gravity imbalance *on average*. This leaves a gapped-data series of actuation-free acceleration noise to analyze [11]. As shown in Fig. 2, successful implementation of this free-fall mode would allow a significant improvement in the overall acceleration noise sensitivity at low frequencies, to the level of the eLISA TM acceleration noise requirement around 1 mHz. Testing the measurement sequence and analysis algorithms for this gapped and high-dynamic range data series in order to accurately estimate the underlying low-frequency acceleration noise is an important part of the ground preparation for LPF as we approach launch.

Performing a differential acceleration measurement approaching the eLISA requirements is one side of the LPF charter; creating a physical model that quantitatively describes the residual noise in this measurement, to apply to eLISA and other future ambitious experimental gravitational measurements, is the other. As such, the challenge in preparing for the LPF mission is that of defining a measurement program that, in additional to the main acceleration noise measurements, allows dedicated measurements to isolate, measure, and, in many cases, mitigate specific sources of both force noise and interferometry measurement noise. With the ability to both measure and controlledly modify the thermal, electrostatic, magnetic, and radiation environments seen by the TM and surrounding apparatus[12], as well as modify the dynamical control schemes and operating points, LPF will thus represent an orbiting laboratory for realizing high precision references of geodesic motion for fundamental physics experiments in space.

3. The LISA Technology Package

Unlike traditional observatory or planetary missions, the payload in LISA Pathfinder cannot be considered as a discrete piece of hardware carried by the spacecraft. Instead, during science operations, the payload and the spacecraft act as a single unit: the attitude control of the spacecraft is driven by the payload. LISA Pathfinder will carry two payloads; the LISA Technology Package (LTP), and the Disturbance Reduction System (DRS). The LISA Technology Package is provided by a consortium of European national space agencies (France, Germany, Italy, Spain, Switzerland, The Netherlands, and the United Kingdom) and ESA, while the DRS is provided by NASA. Only the LTP will be described here.

Figure 3 shows an artists impression of the LTP. The LTP consists of two major subsystems; the Inertial Sensor Subsystem, and the Optical Metrology Subsystem. Both subsystems are described in further detail in the following sections.

3.1. The Inertial Sensor Subsystem

Each of the two test masses is hosted in what is referred to as the 'Inertial Sensor Subsystem' (ISS), whose integration has been completed and is currently undergoing the final stages of testing prior to being integrated with the optical bench.

A schematic section of the ISS is shown in Fig. 4 and includes a dedicated vacuum chamber, a capacitive sensor electrode housing for TM actuation and position measurement, a caging system to secure the test mass during launch and to release it in flight, the UV vacuum fibre feedthrough for test mass charge management and the diagnostic items.

The test mass with the surrounding capacitive sensor constitute the Gravitational Reference Sensor (GRS), considered a key element for achieving the demanding test mass free-fall purity required by space-based gravitational wave detectors. The GRS for LISA Pathfinder has been designed to be suitable for use on board LISA-like missions [13] [14], with the overall force noise performance relaxed only to accommodate the x-axis actuation that is needed for LPF and for corresponding relaxations in the spacecraft electromagnetic and thermal environment.

With the electrode housing being the closest part to the test mass, it is potentially the most relevant disturbance source: careful design, implementation and performance verification strategy have been critical in achieving what was considered a very challenging goal.

Figure 3. Artists impression of the LISA Technology Package (LTP). Credit: ESA

The electrode configuration design was driven by the need to have large gaps that reduce modelled and unmodelled surface forces, particularly from stray electrostatic potential fluctuations and Brownian motion from residual gas impacts, and by the aim of having comparable performance for all degrees of freedom in order to suppress possible cross-talk effects.

The configuration that has been chosen (see Fig. 5) is a rather symmetric, gap-sensing electrode design.

The main features include gaps of approximately 4 mm between the electrode housing and test mass, and injection electrodes on two axes instead of one to reduce off-axis stiffness. The choice of the test mass size was also important: in addition to the decrease in acceleration per given force with the total mass, increasing the TM size allows the surface area to achieve the 1.8 $\mathrm{nm/Hz}^{1/2}$ displacement sensitivity needed for SC control on the yz axes, while limiting the electrostatic stiffness associated with the needed AC sensing bias to a negligible contribution.

The implementation of the electrode housing has been particularly challenging because of the numerous acceleration noise contributions related to employed materials, construction techniques, surface finishing and geometrical tolerances.

We mention here briefly just the effects that represented the main drivers. Temperature gradient fluctuations across the electrode housing cause a fluctuation of the stray acceleration due to the radiometric, radiation pressure and asymmetric outgassing effects. To limit these disturbances, it is important to use material with high thermal conductance and low outgassing rate. In addition, the level of outgassing rate sets the base pressure of the residual gas around the test mass that give rise to Brownian noise, one of the potential main entries in the acceleration noise budget around 1 mHz. A proper choice of material is not the only precaution to be taken in order to control the outgassing: finishing and preparation of the surfaces, their handling during the entire process of integration, and avoiding constructive solutions that may give rise to subtle virtual leaks are all aspects to be monitored very carefully.

In order to reduce the effect of stray voltages due to inhomogeneous contact potential, all TM and facing electrode housing surfaces should be coated with the same conductive material and unavoidable non-coated dielectric paths around the electrodes should be minimized and electrostatically shielded. The control of the GRS internal surface properties has been considered

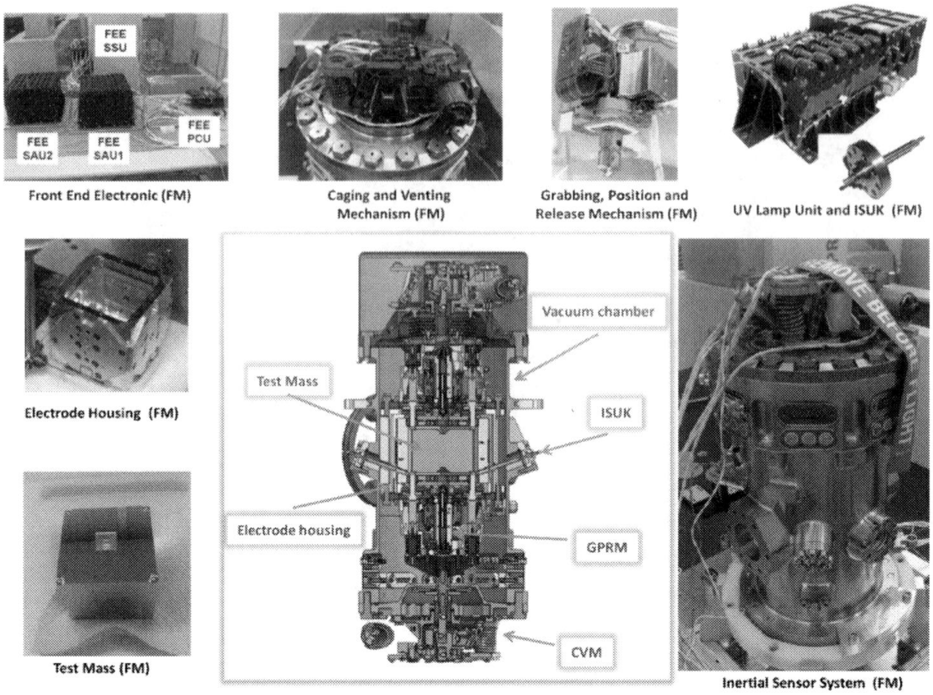

Figure 4. Schematic drawing and photographs of the Inertial Sensor Subsystem flight models (FM, courtesy of CGS). At the centre a section of the full model of the ISS. From bottom left: test mass and electrode housing (courtesy of CGS) ; GRS Front-End Electronics, Caging Mechanism (CM) composed by the Caging and Venting to space Mechanism (CVM) and the Grabbing Positioning and Release Mechanism (GPRM, courtesy of RUAG), UV Lamp Unit and a UV vacuum fibre feedthrough (Courtesy of ICL), Inertial Sensor Subsystem completely integrated (courtesy of CGS).

carefully also for their role in defining the performance of the test mass discharging system based on the UV photoelectric effect. It required extensive investigation of the overall process both through detailed simulation, experimental investigations on samples, and experiments performed in conditions as representative as possible of the hardware and of the operating conditions in flight, using the torsion pendulum facilities in Trento.

Finally, a challenging requirement arises in the need to reduce cross-talk and to reduce the GRS signal levels – and thus noise coming from sensing gain fluctuations – once the TM is aligned to the interferometry system, which requires geometrical tolerances in the range of 10-50 μm. Machining precision and clever design and assembly strategies have allowed this goal to be achieved without over-complicating the manufacturing process.

Following a demanding trade-off phase supported by an extensive and in-depth breadboarding activity, the industrial contractor, Compagnia Generale per lo Spazio (CGS), developed and qualified the two flight model electrode housings for LISA Pathfinder delivering them in October 2013. The electrode housing was realised as a molybdenum-sapphire composite structure where a molybdenum electrode cubic frame accommodates six faces with the electrodes made of gold-

Journal of Physics: Conference Series **610** (2015) 012005 doi:10.1088/1742-6596/610/1/012005

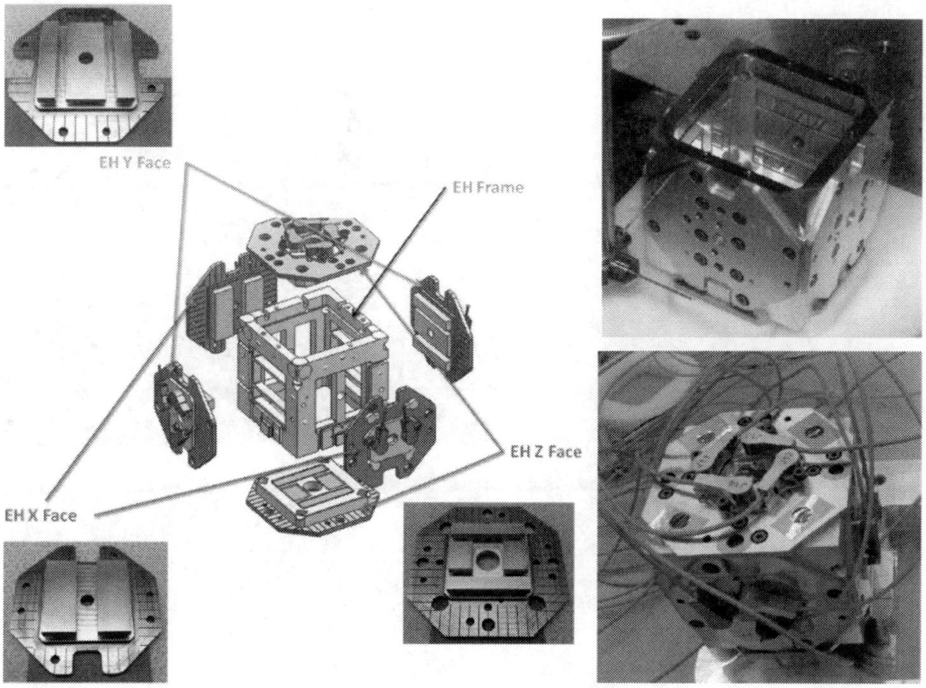

Figure 5. Flight Model electrode housing implementation: the exploded view shows the main parts of the electrode housing, including the electrode configurations on each face with their picture; on the top right, a photograph of the assembled electrode housing without one of the Z faces, showing how the electrode housing appears inside; bottom right, electrode housing completely assembled with connectors and cables (courtesy of CGS).

coated metal plates electrically isolated by means of sapphire spacers. All the inner surfaces facing the test mass are gold coated as well. An exploded view of this device is shown in Fig. 5, together with pictures of the flight model electrode housing at different integration stages and the inner side of the faces with the electrodes.

The GRS integration required analysis and design of gravitational compensation masses, with the final verification of the gravitational requirement based on a model calculated from measured values of the masses and positions of all the satelliteś elements. Before this final integration, a calibration test campaign was devoted to measure the sensitivity of the capacitive displacement sensor of the flight model electrode housing coupled to its front-end electronics (FEE) flight model. A white noise floor at the level of $\mathrm{aF}/\sqrt{\mathrm{Hz}}$ was also measured (Fig. 6), demonstrating the fulfilment of the required position sensitivity of $1.8\,\mathrm{nm}/\sqrt{\mathrm{Hz}}$.

Functional tests of the caging and vent mechanism (CVM), the TM launch lock device that provides a roughly 1200 N holding force, preventing damages due to vibrations at launch and opens the ISS vacuum chamber vent valve to space, and of the grabbing, positioning and release mechanism (GPRM), a second stage actuator that breaks the launch lock device adhesion to the TM and positions it before the final TM release phase, were carried out successfully.

The sensors are currently completing the final acceptance tests to pass to the next stage of

Journal of Physics: Conference Series **610** (2015) 012005 doi:10.1088/1742-6596/610/1/012005

integration with the optical metrology: vibration and thermal vacuum tests on both ISS were successfully performed and now the final bakeout procedure and discharging management system test are ongoing.

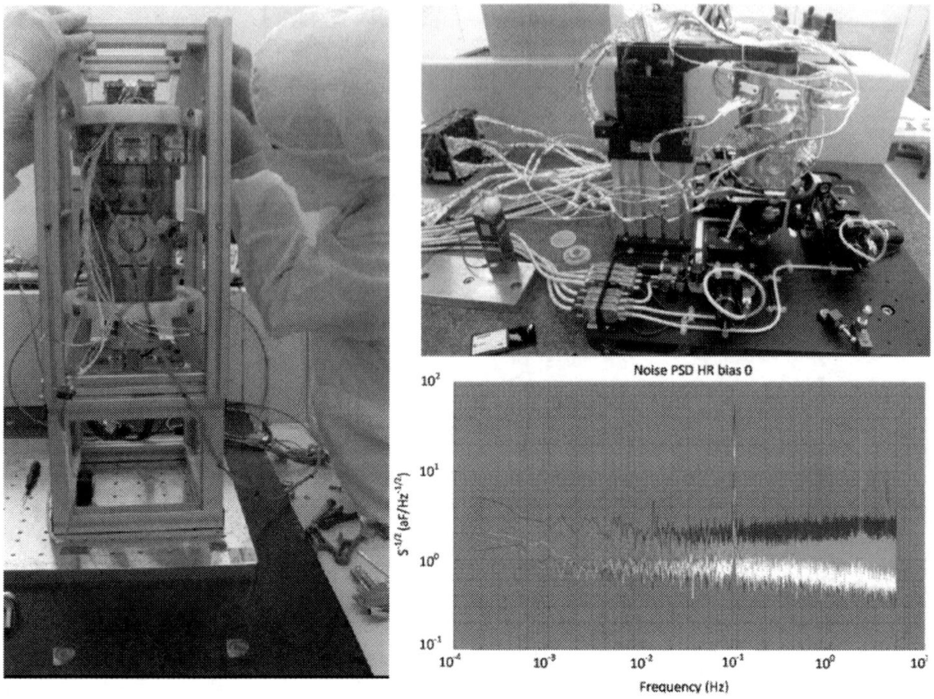

Figure 6. One of the Inertial Sensor Subsystems during the integration is shown in the picture on the left; the top right picture shows the calibration test set-up at the testing facility of Compagnia Generale per lo Spazio (CGS): it allowed the whole vacuum chamber, hosting the electrode housing connected to its flight model FEE to move in six degrees of freedom, with respect to the TM, while a CMM was used to verify the actual displacement within 1 μm accuracy ; bottom right, a plot of the white noise floor of the six capacitive readout channels of one of the ISS, that are at the level of $\mathrm{aF}/\sqrt{\mathrm{Hz}}$ in the measurement frequency bandwidth, in compliance with the required position sensitivity of $1.8\,\mathrm{nm}/\sqrt{\mathrm{Hz}}$ (courtesy of CGS).

3.2. The LISA Pathfinder Optical Metrology System

The aim of the Optical Metrology System (OMS) is to measure the relative displacement between the two TMs along the axis between them [15]. A secondary aim is to measure the two relative angular displacements of the test masses about this axis. The overall requirements are that the OMS should be able to measure the relative displacemets of the test masses with a noise level of below $6.5\mathrm{pm}/\sqrt{\mathrm{Hz}}$ in the measurement band of $3 - 10\,\mathrm{mHz}$. The angular measurement is achieved by a technique known as differential wavefront sensing (DWS) and has a target sensitivity of $2 \times 10^{-8}\mathrm{rad}/\sqrt{\mathrm{Hz}}$. The OMS is a complicated system with a number of different subsystems interlinked by various feedback control systems. An outline of the system and its interlinks can be seen in Figure 7

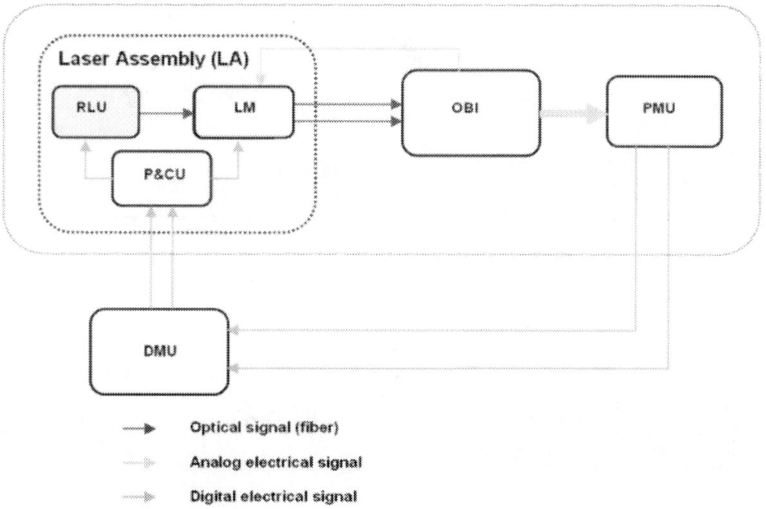

Figure 7. Diagram of the LPF OMS system showing the subsystems and their interlinks. The acronyms are explained in the text apart from the Power and Control Unit (P&CU).

- **Reference Laser Unit (RLU)** The RLU is the laser source for the OMS. The laser is an Nd:YAG NPRO operating at $\lambda = 1064\,\text{nm}$, and producing 35 mW of single mode, single frequency light. The light is coupled into a polarisation preserving single mode optical fibre.
- **Laser Modulator (LM)** The LM takes light from the RLU fibre, splits it into two beams and frequency shifts each beam by slightly different frequencies, both around 80 MHz so that the resulting two beams have a controlled frequency separation of around 1kHz. These two beams are then coupled back into two optical fibres.
- **Optical Bench Interferometer (OBI)** The OBI takes light from the two feed fibres from the LM, collimates the output beams, and uses them to form 4 interferometers on the optical bench. The four interferometers are:
  - **Reference** interferometer that provides a reference phase signal that can also be used to stabilise the optical path onto the optical bench;
  - **x12** interferometer measures the relative displacement and angles between the two test masses, these are the main science signals;
  - **x1** interferometer measures the relative displacement and angles between test mass 1 and the optical bench;
  - **Frequency Noise** interferometer to measure laser frequency noise.

  To perform the measurement with the required accuracy the OBI must direct the beam onto the nominal test masses positions with an accuracy of $\pm 25\mu$m. The OBI must also be extremely stable and should add less than $1\text{pm}/\sqrt{\text{Hz}}$ of noise in the measurement bandwidth. The optical bench is constructed from Zerodur low expansion material with the critical components hydroxy-catalysis bonded to the Zerodur[16]. The layout of the OBI is shown in Figure 8.
- **Phasemeter Unit (PMU)** Each interferometer is read out by two quadrant photodiodes for redundancy. The PMU takes the electrical signals from all of the quadrant photodiodes.

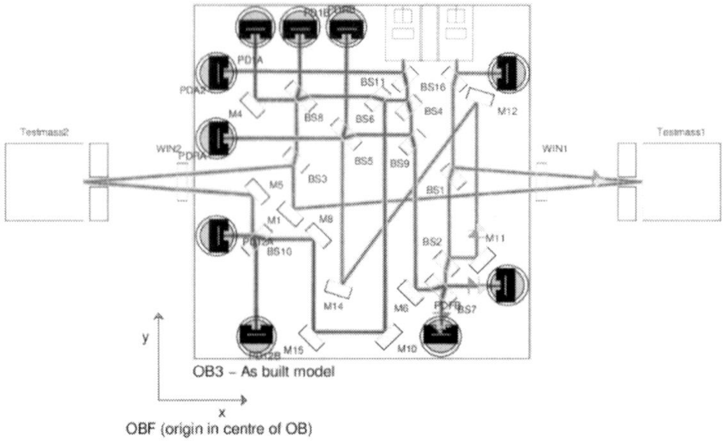

Figure 8. The optical model of the OBI. This is the as-built model. The two different laser frequencies are shown in different colours.

It calculates the phase of the heterodyne signal in each quadrant using a single bin discrete Fourier transform, and passes this data, plus the average light level to the DMU.

- **Data Management Unit (DMU)** The DMU is responsible for data handling for many of the LTP subsystems. One of its tasks is to take the signals from the PMU, process them to produce the required optical longitudinal and DWS signals. It also communicates relevant optical signals to other systems such as the Drag Free and Attitude Control System.

3.3. On station thermal test campaign
Throughout the Manufacture of the flight system, individual units were thoroughly tested, however it was only in October and November of 2011 that all of the subsystems were assembled together and their integrated performance tested as part of the On Station Thermal Test campaign (OSTT)[17]. All of the OMS systems used in the test were flight or flight-equivalent items, however the test masses were replaced by rigidly mounted dummy mirrors. The test were carried out in a vacuum and thermal environment representative of that endured by the spacecraft. The OMS performance met its requirements for both longitudinal and angular sensing. A graph of the longitudinal sensing performance is shown in Figure 9.

4. Operations and Data Analysis
The science operations phase of the LISA Pathfinder mission that focusses on the optimisation and characterisation of the LTP instrument, and the LPF satellite as a whole, takes place immediately after the industrial commissioning phase, and will last for around 90 days.

Each of these days will be packed with investigations and measurements with the aim of meeting the two primary scientific goals of the mission:

- Demonstrate the quietest possible test-mass free-fall by minimising the relative residual acceleration of the test-masses;

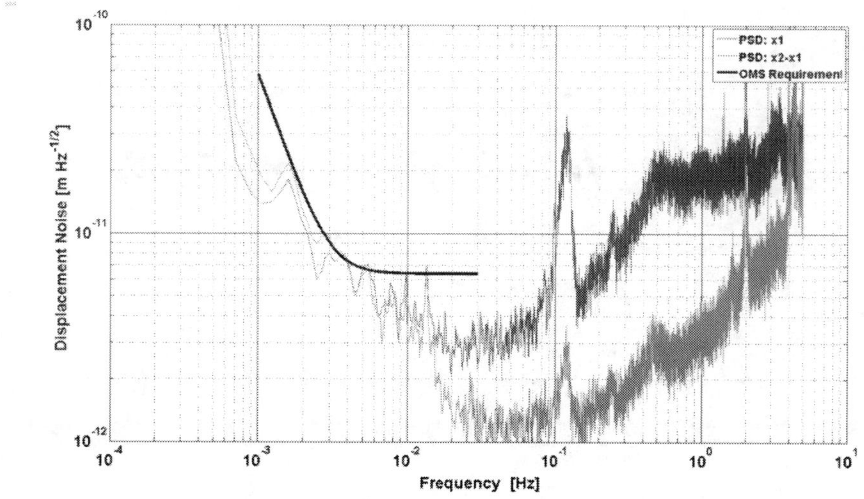

Figure 9. Longitudinal sensing noise of the OMS during the OSTT campaign. The noise is comfortably below requirement which is shown as the bold black line.

- Develop a detailed physical noise model of the entire system, showing the contribution of all the main disturbances to the obtained free-fall.

4.1. Science Investigations

Achieving these goals requires a careful planning and scheduling of the different experiments which can be carried out on-board the satellite. Table 1 lists examples of the experiments which will be performed at some stage of the operations phase, grouped into broad categories. The exact ordering of the investigations will, of course, depend somewhat on the results obtained, and as such can only be tentatively planned in advance. Nonetheless, due to the constraints on the mission planning, a consolidated time-line of investigations needs to be available two weeks in advance at all stages of the mission. As such, the initial weeks of the science operations phase have already been planned.

An example of such an investigation is the system identification investigation designed to probe the primary parameters which influence the two main Drag-Free and Attitude Control (DFACS) loops: the drag-free loop which maintains the SC centred on the reference test-mass, and the x-axis suspension loop which maintains the second test-mass at a fixed distance from the reference test-mass. This investigation involves injecting a sequence of sinusoidal signals into the set-points of the two loops. Figure 10 illustrates the first part of this investigation, the injection into the drag-free loop, and shows the signals from a simulation.

Planning and scheduling an investigation requires a number of activities to take place. Firstly, the investigation must be carefully designed to allow extraction of the desired physical measurements with the desired accuracy, and in the shortest amount of time possible. Following that, the design has to be rationalised to meet the constraints of the system, ensuring the correct commanding is used, and that the state of the instrument is appropriate. Next, the investigation must be simulated on the mission simulator, first in isolation to verify the commanding, and then, ideally, as part of a larger time-line, to help evaluate any constraints that may be present due to previous investigations, or that may be imposed on following investigations.

Residual Force Noise Measurements	
1	Noise performance baseline and tracking
2	Noise correlations with other measured quantities
3	Assessment of low-frequency (sub-milliHertz) performance via extended (multi-day) noise measurements
4	Coupling of beam power fluctuations to test-mass motion via radiation pressure.
5	Contribution of test-mass charging and stray potentials to the overall force noise budget.
System Identification	
5	The estimation of the primary parameter which affect the two x-axis DFACS loops. These have most impact on our ability to estimate the residual differential test-mass acceleration and include such quantities as the thruster and capacitive actuation gains, the "stiffness" couplings and the system delays.
6	Assessment of the main cross-talk effects which couple motion along/around other degrees-of-freedom into the primary science measurement. Investigations will be carried out to quantify the tilt-to-length coupling of the optical metrology system, sensing cross-talk of the GRS, actuation cross-talk in both the thrusters and the capacitive actuators, among others.
Optimisation of the Optical Metrology System	
7	Optimisation of the test-mass position along x to reduce the coupling of various noise terms into the main x-axis measurements.
8	Optimisation of the test-mass position along y and z to minimise the coupling of angular test-mass jitter into the x-axis measurements.
9	Optimisation of the various OMS-related control loops to minimise the noise contributions to the primary measurement.
Optimisation of the Gravitational Reference System	
10	Determine the best 3D operating point for each test-mass to minimise sensing noise.
11	Measurement and compensation of actuation cross-talk.
Characterisation of the thermal & magnetic environment	
12	Activation of heaters on electrode housings to estimate force effects excited due to thermal gradients.
13	Activation of heaters on the struts which support the optical bench, allowing characterisation of the effect of thermally driven distortions on the measured phase in the interferometers.
14	Probe phases distortions of the optical beams arising from thermal fluctuations of the optical windows which isolate the test mass enclosure from the optical bench environment by activating the heaters mounted on the optical windows.
15	Assess the contribution of fluctuating magnetic fields to the overall force noise budget by generating oscillating magnetic fields with the coils mounted outside the test-mass housings.

Table 1. Examples of some of the investigations that will be carried out on LPF during the science operations phase.

4.2. The Operational Environment

A tight, real-time interaction of the scientists and operations engineers is required if we are to extract the maximum science from the available operational days. This in itself places many requirements on the operational environment, including the development of dedicated planning tools which seamlessly allow a design conceived at the science planning stage to flow all the way through to a valid commanding file which can be passed on to the Mission Operations Centre for uploading to the satellite. A series of software tools have been developed to facilitate this process, and the entire end-to-end chain is currently being tested in a series of simulation runs.

Following the upload and execution of an investigation, the relevant data needs to be retrieved and delivered to the science teams for prompt analysis. This allows a relatively quick re-planning

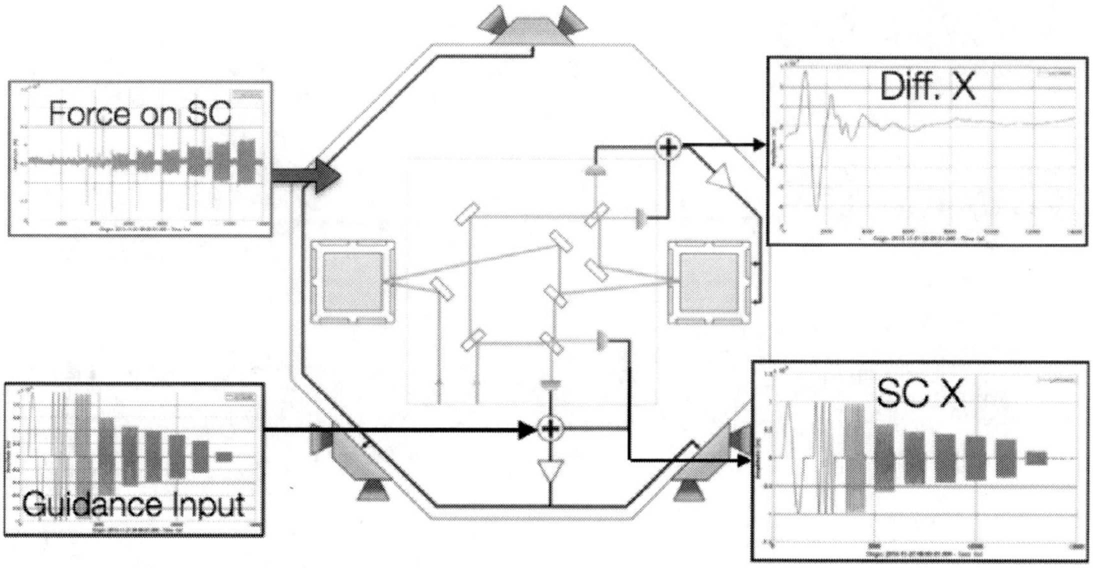

Figure 10. Schematic of LISA Pathfinder showing where, and what, signals are injected into the drag-free loop, and the expected output of the longitudinal interferometers.

cycle, giving the possibility of adapting the future mission time-line based on the results obtained to date.

The data analysis is carried out using a dedicated toolbox which is built on top of MATLAB. The LTPDA Toolbox provides an object-oriented data analysis environment with full history tracking of all processing steps which are performed on the data. The resulting analysis objects can be stored and retrieved from specially designed data repositories, giving the possibility for multiple analysis teams to access and share data concurrently. Figure 11 shows the flow of data through the various elements involved in analysing the data from LISA Pathfinder. In addition to the front-line analysis carried out at ESA's Science and Technology Operations Centre, off-line, in-depth analysis will be carried out at dedicated complementary data centres around Europe.

5. Conclusions

Throughout the history of the LISA/eLISA mission, the science return has never been in doubt - eLISA will observe the Universe in a way which has never been possible before. This has captured the imagination of the science community, but at the same time, mission studies have highlighted the challenging experimental metrology needed to realise such a revolutionary observatory. Together, these facts prompted the European Space Agency to adopt the LISA Pathfinder mission - the science return of a spaceborne gravitational wave detector easily justifies the technology development mission.

LISA Pathfinder is the first mission of its kind; the path from the original concept to the production of flight hardware has not been easy, however, it has been achieved. Production of the flight hardware is complete, with all units demonstrating that we can not only meet, but also exceed the performance requirements levied on the mission. The system is now entering the final tests, in preparation for the launch campaign.

LISA Pathfinder is scheduled to launch in mid-2015, with first performance results available

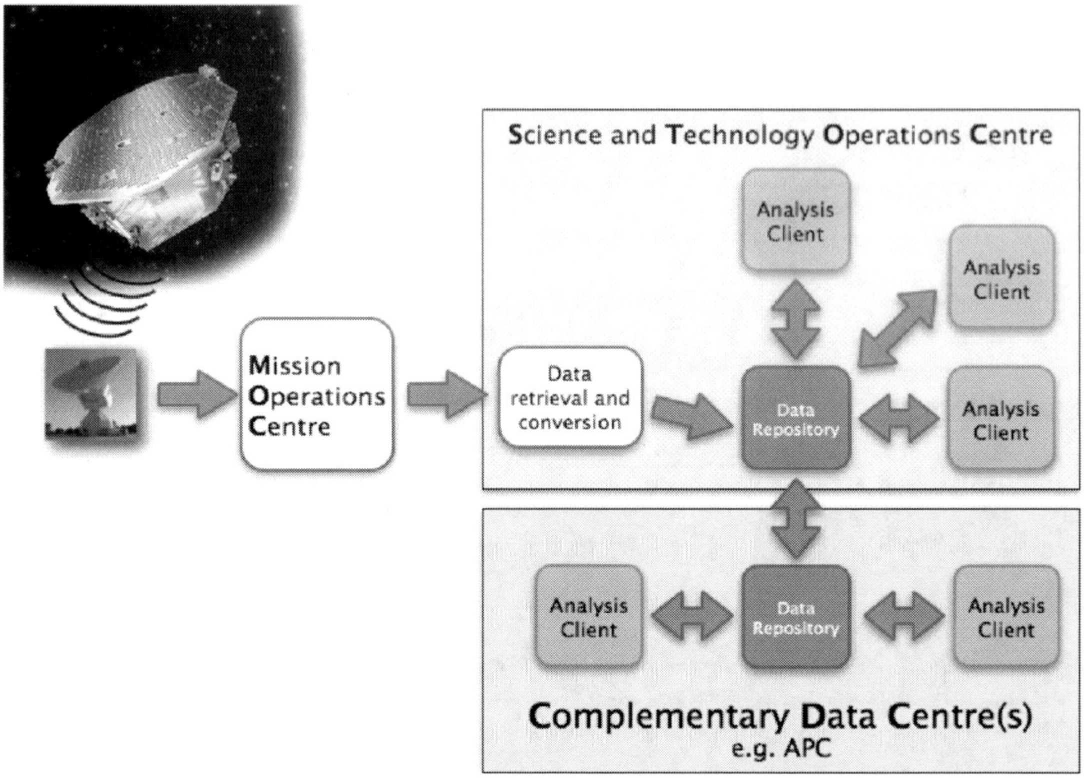

Figure 11. A schematic overview of the data downlink, showing the main elements of the science operations infrastructure and the route the telemetry takes through to the analysis by the on-duty scientists.

approximately three months thereafter.

Acknowledgments

The authors would like to thank the entire LISA Pathfinder team for their continued work in making the mission possible. The LISA Pathfinder mission is funded through the European Space Agency, while the LTP is funded by ESA, France, Germany, Italy, Netherlands, Spain, Switzerland, and the United Kingdom.

References

[1] Jennrich, O., *et al* , (2011) LISA - Unveiling a hidden Universe, *ESA*/SRE(2011)3 unpublished
[2] Jennrich, O., *et al* (2011) NGO - Revealing a hidden Universe: opening a new chapter of discovery, *ESA*/SRE(2011)19 unpublished
[3] Danzmann, K.D., *et al* (2013) The Gravitational Universe Whitepaper, available from *http*://elisascience.org/whitepaper
[4] Committee for a Decadal Survey of Astronomy and Astrophysics (2010) New Worlds, New Horizons in Astronomy and Astrophysics, **ISBN: 0-309-15800-1**
[5] Danzmann, K.D., *et al* (1998) European LISA Technology: Demonstration Satellite for the LISA Mission in ESA's Space Science Programme, unpublished proposal
[6] Congedo G, *et al* 2013 *Phys. Rev. D* **88** 082003
[7] Vitale S 2014 *Gen. Relativ. Gravit.* **46** 1730
[8] Rummel R, Yi W, and Stummer C 2011 *J. Geodesy* **85** 777

[9] Bortoluzzi D, *et al* 2004 *Class. Quantum Grav.* **21** S573
[10] Antonucci F, *et al* 2011 *Class. Quantum Grav.* **28** 094002
[11] Grynagier A, Fichter W, and Vitale S 2009 *Class. Quantum Grav.* **26** 094007
[12] Canizares P, *et al* 2011 *Class. Quantum Grav.* **28** 094004
[13] Dolesi R *et al* 2003 *Class. Quantum Grav.* **20** 99
[14] Weber W J, *et al* , 2002 in Cruise, M., Saulson, P. (Eds.), *Proceedings of the SPIE Astronomical Telescopes and Instrumentation Conference* vol. 4856, SPIE Optical Engineering Press, Bellingham, WA, pp. 31?42
[15] Robertson D I, Fitzimmons E D, Killow C J, Perreur-Lloyd M, Bogenstahl J, Ward H 2007 Optical Bench Interferometer Detailed Design Document *Internal project document* **S2-UGL-DDD-3003)**
[16] Robertson D I, Fitzimmons E D, Killow C J, Perreur-Lloyd M, Bryant J, Cruise A M, Dixon G, Hoyland D, Smith D and Bogenstahl J, 2013 Construction and testing of the optical bench for LISA pathfinder *Classical and Quantum Gravity* **30(8)**
[17] Guzman Cervantes F, *et al.*, 2012 LISA Technology Package Flight Hardware Test Campaign *APS Conference Series* **467**
[18] Cruise A M, Hoyland D, and Aston S M 2005 Implementation of the phasemeter for LISA LTP *Classical and Quantum Gravity* **22** S165 - S169
[19] Heine, F, et al., Space qualified laser sources, Optical Sensing II - Proceedings of SPIE, 6189, 61892I (2006)

Free-flight experiments in LISA Pathfinder

M Armano[a], H Audley[b], G Auger[c], J Baird[n], P Binetruy[c], M Born[b], D Bortoluzzi[d], N Brandt[e], A Bursi[t], M Caleno[f], A Cavalleri[g], A Cesarini[g], M Cruise[h], C Cutler[u], K Danzmann[b], I Diepholz[b], R Dolesi[g], N Dunbar[i], L Ferraioli[j], V Ferroni[g], E Fitzsimons[e], M Freschi[a], J Gallegos[a], C García Marirrodriga[f], R Gerndt[e], LI Gesa[k], F Gibert[k], D Giardini[j], R Giusteri[g], C Grimani[l], I Harrison[m], G Heinzel[b], M Hewitson[b], D Hollington[n], M Hueller[g], J Huesler[f], H Inchauspé[c], O Jennrich[f], P Jetzer[o], B Johlander[f], N Karnesis[k], B Kaune[b], N Korsakova[b], C Killow[p], I Lloro[k], R Maarschalkerweerd[m], S Madden[f], P Maghami[s], D Mance[j], V Martín[k], F Martin-Porqueras[a], I Mateos[k], P McNamara[f], J Mendes[m], L Mendes[a], A Moroni[t], M Nofrarias[k], S Paczkowski[b], M Perreur-Lloyd[p], A Petiteau[c], P Pivato[g], E Plagnol[c], P Prat[c], U Ragnit[f], J Ramos-Castro[qr], J Reiche[b], J A Romera Perez[f], D Robertson[p], H Rozemeijer[f], G Russano[g], P Sarra[t], A Schleicher[e], J Slutsky[s], C F Sopuerta[k], T Sumner[n], D Texier[a], J Thorpe[s], C Trenkel[i], H B Tu[g], D Vetrugno[g], S Vitale[g], G Wanner[b], H Ward[p], S Waschke[n], P Wass[n], D Wealthy[i], S Wen[g], W Weber[g], A Wittchen[b], C Zanoni[d], T Ziegler[e], P Zweifel[j]

[a] European Space Astronomy Centre, European Space Agency, Villanueva de la Cañada, 28692 Madrid, Spain
[b] Albert-Einstein-Institut, Max-Planck-Institut für Gravitationsphysik und Universität Hannover, 30167 Hannover, Germany
[c] APC UMR7164, Université Paris Diderot, Paris, France
[d] Department of Industrial Engineering, University of Trento, via Sommarive 9, 38123 Trento, and Trento Institute for Fundamental Physics and Application / INFN
[e] Airbus Defence and Space, Claude-Dornier-Strasse, 88090 Immenstaad, Germany
[f] European Space Technology Centre, European Space Agency, Keplerlaan 1, 2200 AG Noordwijk, The Netherlands
[g] Dipartimento di Fisica, Università di Trento and Trento Institute for Fundamental Physics and Application / INFN, 38123 Povo, Trento, Italy
[h] Department of Physics and Astronomy, University of Birmingham, Birmingham, UK
[i] Airbus Defence and Space, Gunnels Wood Road, Stevenage, Hertfordshire, SG1 2AS, UK
[j] Institut für Geophysik, ETH Zürich, Sonneggstrasse 5, CH-8092, Zürich, Switzerland
[k] Institut de Ciències de l'Espai (CSIC-IEEC), Campus UAB, Facultat de Ciències, 08193 Bellaterra, Spain
[l] Istituto di Fisica, Università degli Studi di Urbino/ INFN Urbino (PU), Italy
[m] European Space Operations Centre, European Space Agency, 64293 Darmstadt, Germany
[n] The Blackett Laboratory, Imperial College London, UK
[o] Physik Institut, Universität Zürich, Winterthurerstrasse 190, CH-8057 Zürich, Switzerland
[p] SUPA, Institute for Gravitational Research, School of Physics and Astronomy, University of Glasgow, Glasgow, G12 8QQ, UK
[q] Department d'Enginyeria Electrònica, Universitat Politècnica de Catalunya, 08034 Barcelona, Spain

[r] Institut d'Estudis Espacials de Catalunya (IEEC), C/ Gran Capità 2-4, 08034 Barcelona, Spain
[s] NASA Goddard Space Flight Center, 8800 Greenbelt Road, Greenbelt, MD 20771, USA
[t] CGS S.p.A, Compagnia Generale per lo Spazio, Via Gallarate, 150 - 20151 Milano, Italy
[t] NASA Jet Propulsion Laboratory, 4800 Oak Grove Drive, Pasadena, CA 91109

E-mail: james.i.thorpe@nasa.gov

Abstract. The LISA Pathfinder mission will demonstrate the technology of drag-free test masses for use as inertial references in future space-based gravitational wave detectors. To accomplish this, the Pathfinder spacecraft will perform drag-free flight about a test mass while measuring the acceleration of this primary test mass relative to a second reference test mass. Because the reference test mass is contained within the same spacecraft, it is necessary to apply forces on it to maintain its position and attitude relative to the spacecraft. These forces are a potential source of acceleration noise in the LISA Pathfinder system that are not present in the full LISA configuration. While LISA Pathfinder has been designed to meet it's primary mission requirements in the presence of this noise, recent estimates suggest that the on-orbit performance may be limited by this 'suspension noise'. The drift-mode or free-flight experiments provide an opportunity to mitigate this noise source and further characterize the underlying disturbances that are of interest to the designers of LISA-like instruments. This article provides a high-level overview of these experiments and the methods under development to analyze the resulting data.

1. Introduction

The basic operating principle of an interferometric gravitational-wave detector is the measurement of fluctuations in space-time curvature via the exchange of photons between pairs of geodesic-tracking references separated by large baselines [1]. A key challenge for implementing such a detector is the development of an object whose worldline approximates a geodesic – an *inertial* particle. The LISA Pathfinder (LPF) mission [2] will validate the technology of drag-free test masses for use as inertial references in a future space-based gravitational wave detector such as the Laser Interferometer Space Antenna (LISA).

The technique of drag-free flight for disturbance reduction [3, 4] can be briefly summarized as follows. A reference or 'test' mass is placed inside a hollow housing within the host spacecraft (SC). On orbit, the test mass is allowed to float freely inside the housing while a sensor system monitors the position and attitude of the test mass relative to the SC. This information is used by a control system which commands the SC to follow the orbit of the test mass. In doing so, the SC isolates the test mass from external disturbances. By actuating the SC rather than the test mass (as is done in a traditional inertial guidance system) to maintain the relative position and attitude, the test mass is isolated from noise associated with the actuation itself. The remaining residual acceleration noise of the test mass results from forces local to the SC as well as external forces that are not absorbed by the SC (e.g. magnetic). In practice, the "accelerometer" (feedback to the test mass) and "drag-free" (feedback to the SC) techniques are combined into a hybrid system where the feedback can be routed differently depending on the kinematic degree of freedom (DoF) and frequency band. For the LISA application, it is sufficient to have drag-free flight only along the linear DoF and only in the frequency band of desired sensitivity. This is analogous to the pendulum suspensions used for ground-based gravitational wave detectors that provide approximate free-fall in one DoF for frequencies sufficiently far above the resonance frequency.

In the LPF implementation [5], two 46 mm cubic Au-Pt test masses are contained in a single SC. One of the test masses is designated as the reference test mass (RTM) and the SC performs drag free flight along one linear DoF while controlling the remaining five DoFs (three angular

Journal of Physics: Conference Series **610** (2015) 012006 doi:10.1088/1742-6596/610/1/012006

and two linear) with an electrostatic suspension system. The second or non-reference test mass (NTM) is located $\sim 38\,\text{cm}$ away from the primary test mass along the drag-free axis and is used as a witness to assess the residual acceleration of the RTM. An interferometric metrology system [6], monitors relative displacements between the RTM and NTM with a precision of $\sim 10\,\text{pm}/\sqrt{\text{Hz}}$ in the measurement band $0.1\,\text{mHz} < f < 100\,\text{mHz}$

A key difference between the LPF configuration and the LISA configuration is that the LISA acceleration measurement is made between test masses on *separate* spacecraft whereas the LPF acceleration measurement is made between two test masses on the *same* spacecraft. As a result, the LPF NTM must be electrostatically suspended along the sensitive DoF because the SC cannot simultaneously follow the trajectories of both the RTM and NTM along the same DoF. This suspension force has a noise component which represents a disturbance in the LPF measurement that is not present in LISA-like configurations. With all noise sources at the design requirement levels for LPF, this residual suspension noise contribution is not a significant contribution to the overall measured differential acceleration noise in the LISA measurement band (see Figure 1a). The dominating term over this band are various types of spurious forces on the test masses, precisely the phenomena of interest to the designers of future gravitational wave instruments.

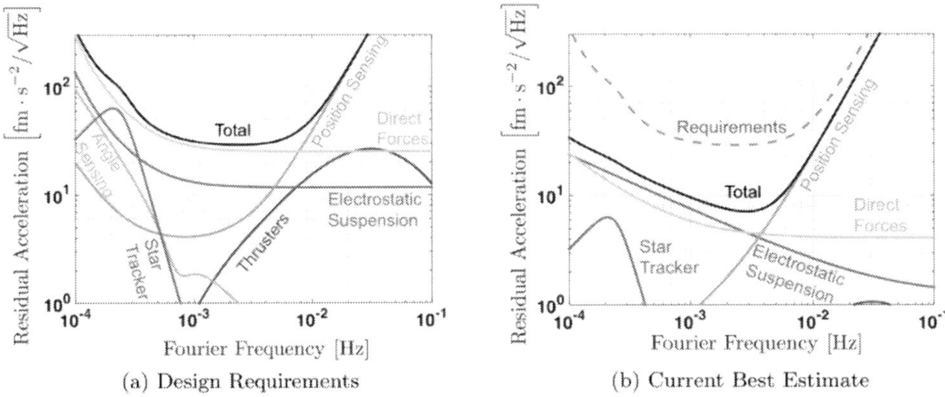

Figure 1: Breakdown of noise sources for measurement of acceleration of the reference test mass relative to the non-reference test mass along the sensitive DoF. The left panel shows the design requirement levels whereas the right panel shows the current best estimates based on ground test campaigns of flight hardware and system modeling. See [5] for more detail.

Figure 1b shows the current best estimate (CBE) for the differential acceleration noise in LPF. As would be expected from a conservative set of design requirements, the CBE levels for all terms are lower than the corresponding ones for the requirements. However, suspension noise is now expected to play a significant role in the measurement band. This expectation is supported by stringent limits placed on the magnitudes of unknown or unmodeled forces on the LPF inertial sensor using torsion pendulums [7, 8].

To be clear, LPF performance at either the requirement or CBE level would accomplish the goal of validating drag-free flight as a technique for realizing inertial reference sensors for LISA-like observatories. Nevertheless, a direct measurement of the forces acting on the test masses would provide additional valuable information to the designers of such observatories.

2. Free-Flight Experiments

The electrostatic suspension of the NTM in LPF is needed to counteract forces in the NTM-SC system that differ from those in the RTM-SC system, which are suppressed by the drag-free control loop. These include disturbance forces that are local to the test masses, such as residual gas disturbances, thermal disturbances, electrostatic forces, magnetic forces, etc. In addition, the static gravitational gradient differs at the RTM and NTM location, leading to a constant bias in the force that must be applied to the NTM along the sensitive axis. Although this bias will be minimized through the use of compensation masses designed based on pre-flight gravitational models, it is expected that residual gravitational accelerations along the sensitive DoF will be on the order of $10^{-10}\,\mathrm{m/s^2}$. This sets the amplitude, and consequently the noise, of the required suspension force.

In the *drift mode* or *free-flight* experiment, the compensation of the static field experienced by the NTM is performed with a series of discrete 'kicks' rather than with a continuously-applied force. Between the kicks, the electrostatic actuation of the NTM along the sensitive DoF is turned off, allowing the NTM to drift under the influence of the constant forces. In principle, this 'kick control' strategy could be employed on all DoFs of the NTM, which would suppress actuation noise from other DoFs from leaking into the sensitive DoF. In practice, this actuation cross-talk is expected to be sufficiently small that kick control is only required along the sensitive DoF.

2.1. The LTP Drift Mode Experiment

The drift mode controller designed for the LISA Test Package (LTP)[9] fixes both the length of the drift period ($\sim 350\,\mathrm{s}$) and the duration of the kicks ($\sim 1\,\mathrm{s}$). A Kalman-filter based observer tracks the motion of the NTM during the free flight and estimates the impulse required to maintain the NTM position. The amplitude of the subsequent kick is adjusted to deliver this impulse and the process is repeated. Figure 2a shows the displacement of the RTM relative to the NTM for a segment of simulated data from the LTP drift mode. It consists of a series of repeated quasi-parabolic flights (the trajectory of the NTM is not a true parabola due to the influence of other force terms such as a linear spring term that couples the NTM to the SC) with a duration of 350 s. Figure 2b shows the corresponding acceleration of the RTM relative to the NTM, which shows a series of discrete kicks separated by nearly constant acceleration during the free flight segments.

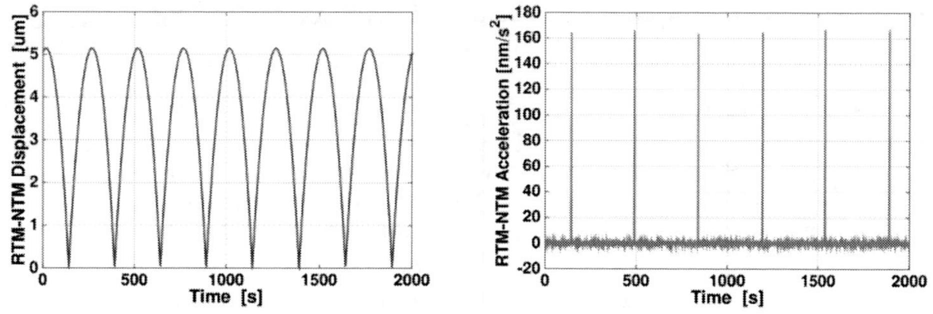

(a) RTM-NTM Relative Position (b) RTM-NTM Relative Acceleration

Figure 2: Example data for a LTP drift-mode simulation. The left panel shows the displacement of the RTM relative to the NTM whereas the right panel shows the acceleration.

2.2. A proposed ST7-DRS Free-Flight Experiment

An alternative drift-mode control design has recently been develop for possible use with the NASA-provided payload known as the Disturbance Reduction System (DRS)[1]. The ST7 design is based on a modified dead-band controller, where the dynamical state of the system determines the control mode. The two-dimensional phase space of NTM position and NTM velocity is used to describe the system state and a region centered around the nominal position and zero velocity is designated as the dead band. So long as the NTM position and velocity remain in this dead band region, no suspension control is applied along the sensitive DoF. When the NTM state drifts out of this deadband region, a suspension controller is engaged and remains on until the NTM state re-enters the dead-band region. The resulting trajectory of the NTM is more complex than that of the LTP drift mode experiment. In general, the duration of the free-flight segments vary from segment to segment. A 300 ks simulation yielded an approximately normal distribution of free-flights with a mean duration of $\sim 357\,\mathrm{s}$ and a standard deviation of $\sim 14\,\mathrm{s}$. While at first glance, this irregularity in the free flight durations might seem to be disadvantage, it has a potential advantage for data analysis because the few long segments can be used to estimate spectral information at lower frequencies. Additionally, unlike the LTP design, the DRS control design automatically adjusts to changes in the static gravity gradient and is robust to fairly significant changes. This means that the controller would not have to be re-optimized in-flight as the residual gravity gradient changes, for example due to fuel consumption.

2.3. Laboratory Experiments

Torsion pendulums with a LPF-like test mass suspended as a torsion member have been used to measure the small forces relevant to the free-fall purity in LISA and LPF. In this section, we describe an implementation of free-flight experiments in a torsion pendulum facility located at the University of Trento. The torsion pendulum consists of a hollow replica of the LISA Pathfinder test mass, suspended by a thin silica torsion fiber, and hangs inside a Gravitational Reference Sensor (GRS) prototype [8]. The sensitive degree of freedom is Φ, the rotation around the z-axis. A breadboard version of the front-end electronic chain provides torque authority of $\sim 200\,\mathrm{fN}\cdot\mathrm{m}$ by applying desired actuation voltages acrossa diagonal pair of electrodes. To simulate a large DC acceleration, the pendulum can be rotated by an angle $\Delta\Phi$ with respect to the inertial sensor, such that a DC torque $N_{DC} = -\,\Gamma\cdot\Delta\Phi$ is required to keep it centered. This is analogous to the bias on the NTM suspension in LPF. Unlike LPF, the bias for the torsion pendulum can be tuned by adjusting $\Delta\Phi$.

The equation of motion for the torsional degree of freedom for the test-mass is

$$I\ddot{\Phi} = -I\omega_0^2\Phi - \frac{I}{\tau}\dot{\Phi} + N(t) \tag{1}$$

where I is the moment of inertia, $\omega_0 = \sqrt{\frac{\Gamma}{I}} = \frac{2\pi}{T_0}$ is the pendulum resonance angular frequency (and T_0 the period), Γ is the pendulum rotational elastic constant and τ is the energy decay time. It is possible to soften electrostatically the pendulum by applying DC constant voltages to lengthen the pendulum period from roughly 465 s, without applied fields, to as much as $T_0 \approx 830\,\mathrm{s}$, to allow flight times comparable to those foreseen for LPF, $T_{fly} = 350\,\mathrm{s}$.

The pendulum torque sensitivity is around $1\,\mathrm{fN}\cdot\mathrm{m}/\sqrt{\mathrm{Hz}}$ at 1 mHz, corresponding to an equivalent acceleration of $50\,\mathrm{fm/s^2}\sqrt{\mathrm{Hz}}$, and thus near the LPF noise specification. The pendulum can thus allow a quantitatively significant test of the free-fall mode and our analysis algorithms, with real data exhibiting gaps as well as the large dynamic range needed in free-fall mode, and thus readout and dynamic system linearity challenges. The on-ground experiment

[1] the DRS is part of NASA's Space Technology 7 (ST7) mission

will also allow more flexibility to explore different control strategies, by varying flight and impulse time or control points, and different dynamic configurations made possible by having a variable stiffness.

Our free-fall test consists of three measurements. The first is the pendulum background torque noise level in absence of any applied force, which can be performed by rotating the pendulum such that the test mass is centered without applied torques. The measured angular displacement is then converted into torque

$$N_m = I\ddot{\Phi} + \Gamma\Phi + \frac{I}{\tau}\dot{\Phi} \tag{2}$$

In the second experiment, the pendulum is rotated by a large angle with respect to the GRS to simulate a large DC acceleration. In the current configuration, a rotation angle $\Phi_{EQ} \approx 2\,\text{mrad}$ requires a DC torque of roughly $13\,\text{pN}\cdot\text{m}$ to keep the test mass centered, a differential force of roughly $1.3\,\text{nN}$ (with electrostatic softening of the pendulum, this gives $\Phi_{EQ} \approx 5.3\,\text{mrad}$, see Figure 3b). The measured angular displacement is again converted into torque using (2) and the contribution from the noisy electrostatic actuation produces an excess in noise power relative to the first configuration.

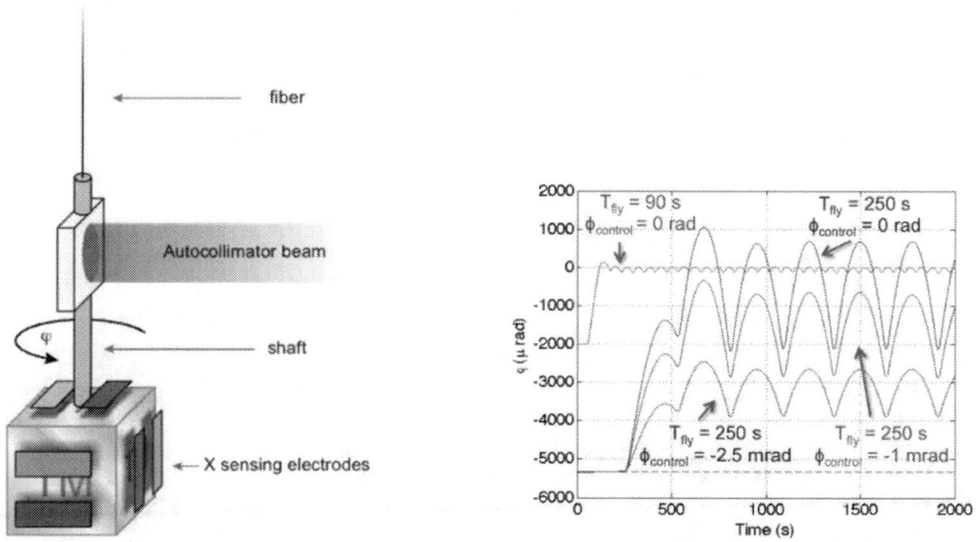

(a) Torsion-pendulum apparatus (b) Free-flight displacement data

Figure 3: Example data from a torsion pendulum facility used to simulate LPF free-flight experiments. The displacement timeseries is for an experiment performed with free-flight times $T_{fly} = 90\,\text{s}$ and $250\,\text{s}$ and angular set points $\Phi_{control} = 0,\ -1,\ -2.5\,\text{mrad}$.

In the final measurement, the pendulum rotation of the second experiment is maintained but a free-fall control scheme is employed to control the position of the TM. Torque impulses are applied periodically with a duty cycle χ, with average amplitude $N_{kick} \approx -\Gamma\Phi_{EQ}/\chi$. The free-fall torque noise can then be measured and compared with both the background and continuous-actuation cases.

The pendulum dynamics in between two impulses is a free oscillation around the equilibrium point. The torsional spring is small and positive, in contrast to the small negative spring expected in orbit. Because the stiffnesses are small and the flight time relatively short, 350 s flight compared to the 830 s free-oscillation period, the motion is similar to a parabolic flight in both cases.

The motion is periodically forced, by the impulses applied, to come back to a single initial position with a chosen velocity, to allow a periodic flight. To do that, a control scheme is implemented, where an observer estimates the pendulum position and velocity before each impulse with least squares fitting of the pendulum rotation data. Then a controller estimates the impulse intensity needed to reach the ideal initial point for the next cycle, using the pendulum dynamic constants and the flight and impulse times. We can also vary $\Phi_{control}$, the average position of the mass during the free flight, to change the actuation level.

Example preliminary data are shown in Figure 3b, with flight times of 90 and 250 s, using pendulum periods of 482 and 830 s, respectively, employing also different controller set points. The controller has been successfully employed with a variety of control configurations. The next step is that of debugging and understanding the torque noise in our on-ground free-fall model and beginning to test our flight data analysis algorithms on the experimental data.

3. Data Analysis Challenge

At their core, the goals of the LPF drift mode experiments are the same as those conducted in the standard science mode: measurement of the spectrum of the acceleration of the RTM relative to the NTM and estimation of system parameters such as the gravity gradients, stiffness terms, etc. The general analysis strategy for LPF experiments begins with models of the expected acceleration that are parametrized by relevant system parameters. Fitting these models to the data provides estimates of the system parameters and the fit residuals can be used to estimate the relative test mass acceleration.

The drift mode data poses two unique challenges to this approach. The first is the presence of the kicks, which represent a high-noise configuration of the NTM and can't be used for spectral estimation. Consequently, they must be excised in some way. The second is the size of the free-flight signals relative to the noise levels of interest. In displacement, the free flights have an amplitude of $\sim 10\,\mu$m, compared with a displacement sensitivity of $7\,\mathrm{pm}/\sqrt{\mathrm{Hz}}$. In acceleration, the force bias on the NTM is equivalent to an acceleration of $10^{-10}\,\mathrm{m/s^2}$, compared with expected noise levels of $\sim 10^{-14}\,\mathrm{m/s^2}/\sqrt{\mathrm{Hz}}$. Residuals in fitting the free-flight terms caused by small parameter or model errors can significantly impact the resulting estimate of the relative acceleration between the NTM and RTM. The effect of such residuals is exacerbated by the presence of the data gaps around the excised kicks, to which we now turn our attention.

After the 'deterministic' portion of the free-flight data has been removed, the next step is to estimate the spectrum of the residuals. The data from individual free-flights is well suited to this purpose and standard spectral estimation techniques can be applied. Unfortunately, the minimum Fourier frequency at which the spectrum can be estimated is determined by the length of the free flight segments. For the ~ 350 s flights for the LTP experiment, this limits the estimation to $f > 3$ mHz. To move to the lower end of the LTP band (1 mHz) and to the full LISA band (0.1 mHz), requires combining data from successive flights. Since the portions of data during the kicks cannot be used, this amounts to the not-uncommon problem of estimating the underlying spectrum in the presence of gaps. The most common technique for estimating spectra in unevenly-sampled timeseries is the Lomb-Scargle method [10, 11, 12], which is a mainstay in astronomical data analysis. However, Lomb Scargle is not ideally suited to the LPF free-flight problem because the data gaps are regular and periodic and the dynamic range of the spectra is larger than the typical simple power-law spectra encountered in astrophysics.

The general effect of data gaps is to introduce systematic biases into the estimated spectrum of the underlying continuous process. The precise nature of this bias depends both on the characteristics of the gaps (their duration, number, and grouping) as well as on the spectrum of the signal. The primary effect on LPF free-flight data is to 'fill in the bucket' in the LPF sensitivity. This is due to power from the upper end of the LPF band aliasing into lower frequencies and is exacerbated by the large range in noise power between the minimum noise of $\sim 3\times10^{-14}\,\mathrm{m/s^2/\sqrt{Hz}}$ at $f\sim 3\,\mathrm{mHz}$ and the noise of $\sim 5\times10^{-12}\,\mathrm{m/s^2/\sqrt{Hz}}$ at $f\sim 100\,\mathrm{mHz}$. To illustrate this effect, we generated a series of 'noise-only' mock LTP drift mode data using a linear state-space simulator of the LTP experiment. The simulator was configured to run in nominal science mode (no free-flights) with the capacitive actuation noise of the NTM along the sensitive axis artificially turned off to mimic the noise environment in the drift mode experiment. The red trace in Figure 4a shows the average power spectral density of the residual RTM-NTM acceleration noise for an ensemble of 100 runs of the simulator with different noise seeds, representing the expected level in a drift mode experiment. Gaps of 4 s duration separated by 350 s continuous segments were then artificially placed in the data to mimic the portion of the data that would be removed to avoid the influence of the kicks during drift mode. The increase in gap size over the 1 s kick duration allows the suppression of transients generated during downsampling of the data from 10 Hz to 1 Hz and an estimate of the acceleration from the measured displacement timeseries using finite differencing. As a zeroth-order estimate for the data in the gaps, a linear 'patch' is placed in each gap connecting the beginning point with the end point of the gap. The spectral estimate from this linearly patched data is the green trace in Figure 4a. It is clear that the linearly patched data gives a severely overestimated noise level over the entire LPF band. Figure 4b shows the fractional error of the spectral estimates over the lower portion of the LPF/LISA frequency band,

$$\mathrm{RE}(f) \equiv \frac{\left|\overline{S_{patched}(f) - S_{original}(f)}\right|}{\sigma_{original}(f)} \tag{3}$$

where $\overline{S_{original}(f) - S_{patched}(f)}$ represents the error in the power spectral density in each frequency bin for a given patching method, averaged over the ensemble of 100 simulator runs. $\sigma_{original}(f)$ is the standard deviation of the power spectral density in each frequency bin over the ensemble. $\mathrm{RE}(f)$ measures the error introduced by the patching method relative to the statistical uncertainty associated with a single simulator run. If $\mathrm{RE}(f) \lesssim 1$, then the error in the patching method is not significant for an individual free flight experiment. For the linear patching technique (green trace), $\mathrm{RE}(f) > 1$ for most of the LPF measurement band, with a maximum value of more than 10^2 at Fourier frequencies of a few mHz.

4. Approaches to Data Analysis

In this section, we briefly outline some of these techniques developed as part of this ongoing effort, with the details left to future publications.

4.1. *Windowing of data gaps*

One strategy for mitigating the adverse effects of data gaps on the estimation of spectra from timeseries is to employ windowing. When estimating spectra from timeseries without gaps, it is typical to first multiply the time series by a window function that smoothly tapers to zero at the beginning and end of the timeseries. This helps to suppress artifacts caused by mismatch between the beginning and the end of the timeseries. This same strategy can be applied to data with gaps by applying a window to each individual segment of continuous data. The cyan line in Figure 4a shows the result of the spectral estimate made using a sin^2 window in each free-flight segment. The estimate of the spectra made from this windowed data is a far better approximation of the original data than that made using the linear patches. However, the windowed estimate shows a large excess relative to the original spectrum near the inverse gap separation of $1/354\,\mathrm{s} \approx 2.8\,\mathrm{mHz}$[2]. With the exception of the inverse gap separation frequency,

[2] To clarify, 4 s gaps separated by 350 s free-flight segments gives a center-to-center gap spacing of 354 s.

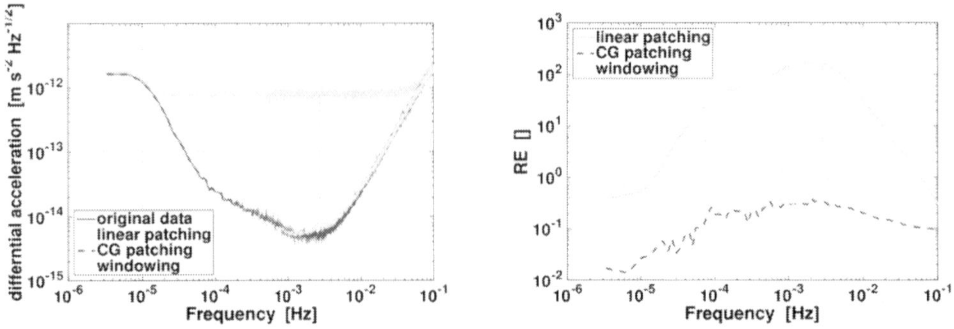

(a) Ensemble-averaged noise spectral densities

(b) Relative Error

Figure 4: Comparison of methods for spectral estimation of RTM acceleration relative to NTM with data gaps of 4 s duration separated by 350s free-flight segments. A linear state-space simulator of LTP was used to generate an ensemble of 100 noise realizations of LTP science mode data with no electrostatic suspension noise and data gaps were artificially inserted allowing for the comparison of the estimated power spectral densities with and without gaps. The left plot shows the power spectral densities of the original signal (red), the linearly-patched signal (green), the Constrained-Gaussian patched signal (dashed blue), and the power spectral density estimated with the window method (dashed cyan). The right plot shows the relative error as defined in (3) between the power spectral density estimated with gaps and that without gaps for the three approaches.

$\mathrm{RE}(f) < 1$ for most of the LPF measurement band.

To better compare the original acceleration noise power spectrum to the one recovered using the window method, fits were made to both versions of the power spectral density from each simulator run. The model was a four-component power law[3],

$$S_{mod}(f) = P_{-6} \cdot f^{-6} + P_{-2} \cdot f^{-2} + P_0 \cdot f^0 + P_4 \cdot f^{+4}, \tag{4}$$

where f is the Fourier frequency and $P_{-6,-2,0,+4}$ are the amplitudes of the four components. Figures 5a - 5d show histograms of the best-fit amplitudes for the spectral model over the ensemble of 100 simulator runs for both the original data (red) and the windowed method (cyan). Table 1 lists the mean and standard deviations for the ensemble of noise realizations for each of the four component amplitudes. As a rough measure of the statistical equivalence of the distributions of the best-fit coefficients to the original and recovered power spectral densities, we compute the difference between the means normalized by the quadrature sum of the deviations. The window method produces significant bias in the P_{-2} and P_0 coefficients (3.5 and 6 sigma, respectively) and slight bias (0.9 sigma, or a p-value of 36%) in the P_{+4} coefficient. The recovered P_{-6} coefficient is consistent but is not well-determined in the fits to either the original or recovered data.

4.2. Constrained-Gaussian Gap Patching

Another approach to dealing with data gaps and the difficulties they cause with estimating spectra is to fill the gaps with fabricated data. While the fabricated data in these patches will necessarily bias the spectral estimate, if the gaps are not too numerous and the patch data

[3] Note that the fit was performed on the *power* spectral density where as Figures 1a, 1b, and 4a plot the *amplitude* spectral density. As a result the spectral indices are a factor of two greater.

Table 1: Comparison of best-fit parameters for spectral model in (4) for original and recovered power spectral densities.

	original		window method			CG patching		
	μ	σ	μ	σ	$\frac{\mu-\mu_{orig}}{\sqrt{\sigma^2+\sigma_{orig}^2}}$	μ	σ	$\frac{\mu-\mu_{orig}}{\sqrt{\sigma^2+\sigma_{orig}^2}}$
$P_{-6}\times 10^{53}$	0.66	0.66	0.65	0.61	0.03	0.66	0.71	0.03
$P_{-2}\times 10^{35}$	1.14	0.18	0.44	0.09	-3.5	1.08	0.20	-0.2
$P_{0}\times 10^{29}$	2.25	0.16	4.29	0.30	6.1	2.31	0.19	0.3
$P_{+4}\times 10^{20}$	6.23	0.03	6.28	0.04	0.9	6.22	0.03	0.09

are chosen carefully, the bias can be far less than what arises from the gaps themselves. To minimize the bias resulting from the patches, the data in the patches should have the same spectral content as the missing data it is replacing. The first requirement is that the data points in the patches have proper correlations with *one another*, this is relatively straightforward problem of drawing random samples of (unevenly-sampled) data with a corresponding spectrum. The second requirement is that the data points in the patches have proper correlation with the *existing data*. In Appendix A, we demonstrate that enforcing the correlation with the existing data can be accomplished by biasing the mean values of the random data used to make the patches, a technique we call *Constrained-Gaussian Gap Patching* or CG patching. The values for these biases can be entirely computed from the existing data and a model of the underlying spectrum.

Once patched, standard approaches can be used to compute the spectrum of the entire data set as if it were continuous, without suffering from the bias introduced by the gaps. In principal this is a catch-22, one must know the spectrum in order to pick the patches that will allow the spectrum to be measured. In practice, an approximate *a priori* model of the spectrum will suffice. If necessary, an iterative procedure can be used in which an initial guess for S_y is used to generate the patches, a new value for S_y is estimated from the patched data series, and that spectrum is used to determine new patches.

The CG-patching approach was applied to the mock 'noise-only' drift-mode data in Figure 4a. The ensemble-average spectra is shown in the blue dashed trace in Figure 4a and matches the average spectra from the original data (red) quite closely for all frequencies. The value of RE(f) is less than unity across the entire measurement band and is significantly less than that of the windowed method for most frequencies. More significantly, when the spectra recovered with CG patching are fit to the power-law model in (4), the distribution in amplitudes is quite close to that of the original data, as shown in the blue dashed traces in the histograms in Figures 5a - 5d. Table 1 includes a statistical comparison between the fits to the power spectral densities of the original and CG-patched data. The normalized difference between the distributions in fit amplitudes is less than 0.3 sigma for all coefficients.

4.3. Other methods

While the window and CG-patching methods have received the most attention for potential application to data from free-flight experiments, other methods are also being studied. One approach is to recognize that if an approximate model for the underlying spectra is known (as is also required for the CG-patching method) and the location of the gaps are known, the spectral bias induced by the gaps can be analytically removed. This is a classic example of a deconvolution problem with the associated computational difficulties. However, if the data is aggressively low-pass filtered, the problem becomes tractable.

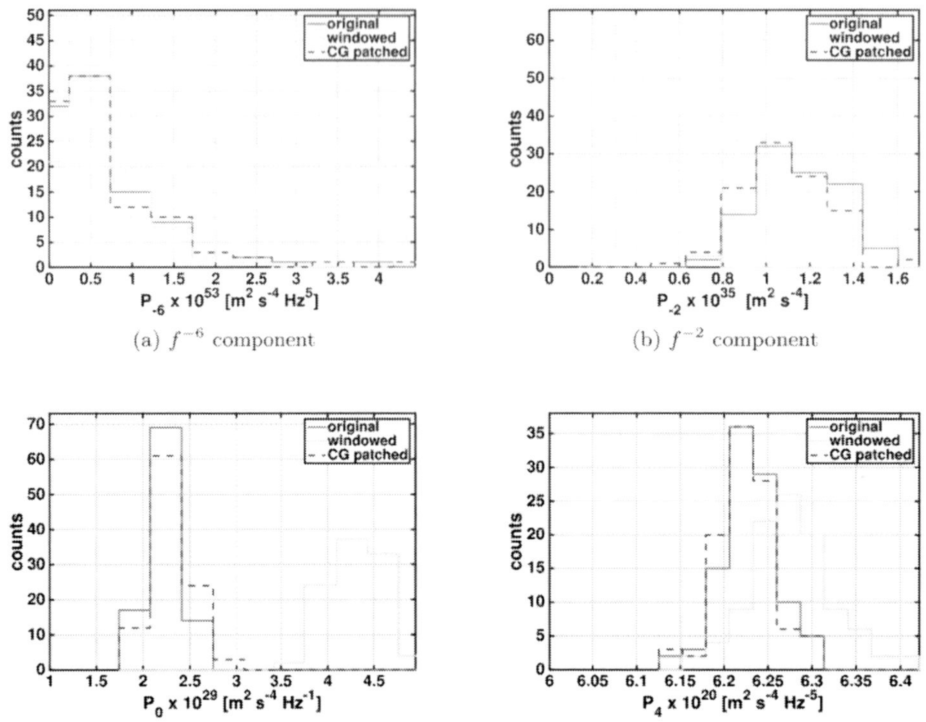

(a) f^{-6} component (b) f^{-2} component

(c) f^{0} component (d) f^{+4} component

Figure 5: Histograms of power-law amplitudes in a four-component fit to the spectrum of RTM acceleration relative to the NTM for the original data (red) and as recovered in the presence of gaps using the window method (cyan) and the Constrained Gaussian patching method (blue dashed). For each of the 100 simulator runs, a power spectrum was generated and fit using a least-squares algorithm to a four-component power law with indices -6,-2,0, and +4 as described in (4). The plots show the distribution of best-fit parameters over the ensemble of 100 simulator runs. For the case of the windowed method, the noise spike around 2.8 mHz was de-weighted to minimize it's bias on the fit.

An altogether unique approach is to consider only the data in the free-flights themselves and perform time-domain fits on the free-flight trajectories. Each of these fits will have parameters such as residual gravity gradient that are the same as those for the first stage, explained at the beginning of Section 3, of the methods outlined above. However, there will also be some variation in these parameters between successive flights that is caused by low-frequency (Fourier frequencies lower than the inverse free-flight time) acceleration noise in the system. By analyzing the variation in parameters recovered from fits to a series of free-flight trajectories, an estimate of acceleration noise at Fourier frequencies below the inverse free-flight frequency can be made without the need to deal with data gaps.

5. Conclusions

The free fight experiments planned for LPF will provide an opportunity to gather data that is even more representative of a full-scale LISA like instrument than the standard LPF science mode. This improvement will allow for better characterization of the small forces that ultimately limit the performance of LPF, LISA, and other future instruments requiring low-disturbance environments such as advanced geodesy or fundamental physics missions. The challenges in extracting this information from the free-flight data are significant, but progress in overcoming them is being made with a number of independent techniques. The challenge of estimating the relative RTM/NTM acceleration noise in the presence of data gaps has been successfully addressed by a number of techniques with the current limitation being the ability to combine this with accurate removal of the deterministic free-flight signal. It is fully expected that the remaining challenges will be resolved in time for launch and operations.

Appendix A: Generating Constrained-Gaussian Patches

In this appendix, we introduce the mathematical formalism of constrained-Gaussian gap patching. This method is similar to techniques used to patch gaps in 2D sky maps of cosmic microwave background data [13, 14]. In the time domain, spectral information is encoded in the two-point correlation function, C^{jk}, which measures the correlation between the samples y^j and y^k. For stationary processes, the two-point function depends only on the separation between the samples, $C^{jk} = \mathcal{C}(j-k)$, and is directly related to the spectral density,

$$C^{jk} = \int_{-\infty}^{+\infty} \frac{1}{2} S_y(f) e^{-2\pi f T(j-k)}. \tag{5}$$

Under the further assumption that y^j measures a Gaussian random process, the two-point function can be used to express the probability distribution function (PDF) for y^j,

$$P(\vec{y}) \propto \exp\left[-\frac{1}{2}\sum_{j,k} C_{jk} y^j y^k\right], \tag{6}$$

where C_{jk} is the inverse of C^{jk} such that $C_{jl}C^{lk} = \delta^j_k$.

For the case of data with gaps, it is useful to separate the data y_j into portions inside the gaps, $j \in \mathcal{G}$, and data between the gaps $j \in \bar{\mathcal{G}}$. The sum in (6) can then be divided into three parts,

$$\ln P(\vec{y}) = const - \frac{1}{2}\left[\sum^{\bar{\mathcal{G}},\bar{\mathcal{G}}'} C_{jk} y^j y^k + 2\sum^{\bar{\mathcal{G}},\mathcal{G}} C_{jk} y^j y^k + \sum^{\mathcal{G},\mathcal{G}'} C_{jk} y^j y^k\right]. \tag{7}$$

The first term in (7) is a fixed constant depending on measured data that can be absorbed into the overall normalization constant. The final term depends entirely on the data within the gaps, and just describes the PDF for (unevenly sampled) Gaussian noise with a particular spectrum. The middle term encodes the relationship between data inside the gaps and data outside the gaps. Faithfully reproducing that relationship is the key to generating good patches for filling gaps in measured data. Since we are interested in generating the data within the gaps, we can factor the summation of the middle term in (7) and define $\lambda_j \equiv \sum_j^{\bar{\mathcal{G}}} C_{jk} y^j$ such that that term becomes $2\sum^{\mathcal{G}} \lambda_j y^j$. The PDF can then be re-written as

$$\ln P(\vec{y}) = const - \frac{1}{2}\sum^{\mathcal{G},\mathcal{G}'} C_{jk}(y^j - \Delta^j)(y^k - \Delta^k), \tag{8}$$

where $\Delta^j \equiv -\sum_k^{\bar{\mathcal{G}}} C^{jk}\lambda_k$. Comparing (8) with the standard expression for multivariate Gaussian distributions, one finds that the data in the gap is described by a multivariate Gaussian with covariance C_{jk} and non-zero means Δ^k. Note that the values for Δ^k are entirely determined by the data outside the gap.

The prescription for *Constrained Gaussian Gap Filling* is thus as follows: make a guess for spectral density $S_y(f)$ describing the data; compute an estimated two-point correlation function, C_{jk}; compute the Δ^k for each point on the gap based on the data outside the gap and the estimated C_{jk}, draw data from a multivariate Gaussian distribution with covariance C_{jk} and mean Δ^k.

References

[1] Bondi H, Pirani F A E and Robinson I 1959 *Royal Society of London Proceedings Series A* **251** 519–533
[2] Armano M and et al 2009 *Class. Quant. Grav.* **26**
[3] Lange B 1964 *AIAA Journal* **2** 1590–1606 URL http://dx.doi.org/10.2514/3.55086
[4] deBra D B 1997 *Classical and Quantum Gravity* **14** 1549 URL http://stacks.iop.org/0264-9381/14/i=6/a=026
[5] Antonucci F, Armano M, Audley H, Auger G, Benedetti M, Binetruy P, Boatella C, Bogenstahl J, Bortoluzzi D, Bosetti P, Brandt N, Caleno M, Cavalleri A, Cesa M, Chmeissani M, Ciani G, Conchillo A, Congedo G, Cristofolini I, Cruise M, Danzmann K, Marchi F D, Diaz-Aguilo M, Diepholz I, Dixon G, Dolesi R, Dunbar N, Fauste J, Ferraioli L, Fertin D, Fichter W, Fitzsimons E, Freschi M, Marin A G, Marirrodriga C G, Gerndt R, Gesa L, Giardini D, Gibert F, Grimani C, Grynagier A, Guillaume B, Cervantes F G, Harrison I, Heinzel G, Hewitson M, Hollington D, Hough J, Hoyland D, Hueller M, Huesler J, Jeannin O, Jennrich O, Jetzer P, Johlander B, Killow C, Llamas X, Lloro I, Lobo A, Maarschalkerweerd R, Madden S, Mance D, Mateos I, McNamara P W, Mendes J, Mitchell E, Monsky A, Nicolini D, Nicolodi D, Nofrarias M, Pedersen F, Perreur-Lloyd M, Perreca A, Plagnol E, Prat P, Racca G D, Rais B, Ramos-Castro J, Reiche J, Perez J A R, Robertson D, Rozemeijer H, Sanjuan J, Schleicher A, Schulte M, Shaul D, Stagnaro L, Strandmoe S, Steier F, Sumner T J, Taylor A, Texier D, Trenkel C, Tombolato D, Vitale S, Wanner G, Ward H, Waschke S, Wass P, Weber W J and Zweifel P 2011 *Classical and Quantum Gravity* **28** 094002 URL http://stacks.iop.org/0264-9381/28/i=9/a=094002
[6] Audley H, Danzmann K, Marin A G, Heinzel G, Monsky A, Nofrarias M, Steier F, Gerardi D, Gerndt R, Hechenblaikner G, Johann U, Luetzow-Wentzky P, Wand V, Antonucci F, Armano M, Auger G, Benedetti M, Binetruy P, Boatella C, Bogenstahl J, Bortoluzzi D, Bosetti P, Caleno M, Cavalleri A, Cesa M, Chmeissani M, Ciani G, Conchillo A, Congedo G, Cristofolini I, Cruise M, Marchi F D, Diaz-Aguilo M, Diepholz I, Dixon G, Dolesi R, Fauste J, Ferraioli L, Fertin D, Fichter W, Fitzsimons E, Freschi M, Marirrodriga C G, Gesa L, Gibert F, Giardini D, Grimani C, Grynagier A, Guillaume B, Cervantes F G, Harrison I, Hewitson M, Hollington D, Hough J, Hoyland D, Hueller M, Huesler J, Jeannin O, Jennrich O, Jetzer P, Johlander B, Killow C, Llamas X, Lloro I, Lobo A, Maarschalkerweerd R, Madden S, Mance D, Mateos I, McNamara P W, Mendes J, Mitchell E, Nicolini D, Nicolodi D, Pedersen F, Perreur-Lloyd M, Perreca A, Plagnol E, Prat P, Racca G D, Rais B, Ramos-Castro J, Reiche J, Perez J A R, Robertson D, Rozemeijer H, Sanjuan J, Schulte M, Shaul D, Stagnaro L, Strandmoe S, Sumner T J, Taylor A, Texier D, Trenkel C, Tombolato D, Vitale S, Wanner G, Ward H, Waschke S, Wass P, Weber W J and Zweifel P 2011 *Classical and Quantum Gravity* **28** 094003 URL http://stacks.iop.org/0264-9381/28/i=9/a=094003
[7] Carbone L, Ciani G, Dolesi R, Hueller M, Tombolato D, Vitale S, Weber W J and Cavalleri A 2007 *Phys. Rev. D* **75**(4) 042001 URL http://link.aps.org/doi/10.1103/PhysRevD.75.042001
[8] Cavalleri A, Ciani G, Dolesi R, Hueller M, Nicolodi D, Tombolato D, Wass P J, Weber W J, Vitale S and Carbone L 2009 *Classical and Quantum Gravity* **26** 094012 URL http://stacks.iop.org/0264-9381/26/i=9/a=094012
[9] Grynagier A 2010 The drift mode for LISA Pathfinder, control description Tech. Rep. S2-iFR-TN-3005 Universität Stuttgart
[10] Lomb N R 1976 *Astrophysics and Space Science* **39** 447–462
[11] Scargle J D 1982 *The Astrophysical Journal* **263** 835–853
[12] Press W H and Rybicki G B 1989 *The Astrophysical Journal* **338** 277–280
[13] Hoffman Y and Ribak E 1991 *The Astrophysical Journal* **380** L5–L8
[14] Bucher M and Louis T 2012 *Monthly Notices of the Royal Astronomical Society* **424** 1694–1713

Journal of Physics: Conference Series **610** (2015) 012007 doi:10.1088/1742-6596/610/1/012007

Wide-Field InfraRed Survey Telescope (WFIRST) Mission and Synergies with LISA and LIGO-Virgo

N Gehrels[1], D Spergel[2], on behalf of the WFIRST SDT and Project
[1] NASA Goddard Space Flight Center, Greenbelt, Maryland 20771 USA
[2] Princeton University, Peyton Hall, Princeton, NJ, 08544 USA

[1] E-mail: neil.gehrels@nasa.gov

Abstract. The Wide-Field InfraRed Survey Telescope (WFIRST) is a NASA space mission in study for launch in 2024. It has a 2.4 m telescope, wide-field IR instrument operating in the 0.7 - 2.0 micron range and an exoplanet imaging coronagraph instrument operating in the 400 - 1000 nm range. The observatory will perform galaxy surveys over thousands of square degrees to J=27 AB for dark energy weak lensing and baryon acoustic oscillation measurements and will monitor a few square degrees for dark energy SN Ia studies. It will perform microlensing observations of the galactic bulge for an exoplanet census and direct imaging observations of nearby exoplanets with a pathfinder coronagraph. The mission will have a robust and well-funded guest observer program for 25% of the observing time. WFIRST will be a powerful tool for time domain astronomy and for coordinated observations with gravitational wave experiments. Gravitational wave events produced by mergers of nearby binary neutron stars (LIGO-Virgo) or extragalactic supermassive black hole binaries (LISA) will produce electromagnetic radiation that WFIRST can observe.

1. Introduction

WFIRST is the top-ranked large space mission of the 2010 NWNH Decadal Survey. The mission is designed to settle essential questions in dark energy, exoplanets, and infrared astrophysics. The mission will feature strategic key science programs plus a vigorous program of guest observations. The observatory includes a wide-field near infrared camera to perform surveys with HST-style imaging and sensitivity, but further in the IR and with hundreds of times the sky coverage. It also will have a pathfinder coronagraph instrument to perform sensitive direct imaging of exoplanets. A Science Definition Team and a Study Office at the Goddard Space Flight Center and Jet Propulsion Laboratory are studying the mission with reports [1, 2] in 2013 and 2014.

The Decadal Survey identified a tripod of focus science themes for WFIRST in the areas of cosmology, planetary systems and general astrophysics. The cosmology theme will highlight measurements of the expansion history of the universe and growth of structure to learn about dark energy. The observations will include near infrared galaxy imaging and slitless spectroscopic surveys over 2400 deg^2 in the baseline 6 year mission. The expansion history will also be studied by Type Ia supernova monitoring.

The planetary systems theme will highlight observations of exoplanets using the microlensing and coronagraphic techniques. Microlensing will be done by monitoring a 3 deg^2 field in the galactic bulge in the infrared to detect microlensing magnification events caused by the alignment of a background star and foreground star with a planetary system. The events will measure the masses and distances to the planets and their host star, and provide a census of the prevalence of different kinds of exoplanets. The direct imaging observations will be performed with a coronagraph, observing known nearby exoplanets and debris disks and searching for new ones. It will be sensitive to larger planets in the Neptune-Jupiter size class.

The general astrophysics theme will highlight survey and monitoring science of sources throughout the universe. The research will be done using archival data from the dark energy and exoplanet surveys, as well as new observations performed as part of a Guest Observer program. Science examples are solar system studies of asteroids, galactic studies of stellar populations and brown dwarfs, surveys to study galaxy evolution and star formation history of the universe, and cosmological studies of the earliest stars, black holes and galaxies.

The WFIRST observatory will have a target-of-opportunity capability that will make a valuable resource for follow-up of gravitational wave events detected by LIGO-Virgo and LISA. The combination of wide field of view, high sensitivity and precise imaging will allow deep searches and observations of electromagnetic counterparts.

The observatory is described in Section 2, observing program in Section 3, science in Section 4, gravitational wave interest in Section 5 and summary in Section 6.

2. WFIRST Observatory

WFIRST has recently been baselined with an existing 2.4 m telescope NASA acquired from the National Reconnaissance Office, compared with a 1.5 m telescope considered in NWNH. This configuration is referred to as WFIRST-AFTA. The 2.4 m obscured telescope feeds two different instrument volumes containing the wide-field instrument and the coronagraph. The wide-field instrument includes two channels, a wide-field channel and an integral field unit (IFU) spectrograph channel. The wide-field channel includes three mirrors (two folds and a tertiary) and a filter/grism wheel to provide an imaging mode covering 0.76 – 2.0 micron and a spectroscopy mode covering 1.35 – 1.95 micron. The wide-field focal plane uses 2.5 micron long-wavelength cutoff 4k x 4k HgCdTe detectors. The HgCdTe detectors are arranged in a 6x3 array, providing an active area of 0.281 deg^2. The pixel scale is 0.11 arcsec and filter bandpass in microns are z(0.76-0.98), Y(0.93-1.19), J(1.13-1.45), H(1.38-1.77), Ks(1.68-2.0) and Wide (0.93-2.00). The grism has a spectral resolving power of 461*lambda (1.35 microns < lambda < 1.95 microns). The IFU channel uses an image slicer and spectrograph to provide individual spectra of each 0.15 arcsecond wide slice covering the 0.6 – 2.0 micron spectral range over a 3.00 x 3.15 arcsec field. The spectral resolution is 75. The instrument provides a sharp PSF, precision photometry, and stable observations for implementing the WFIRST science.

The coronagraph instrument includes an imaging mode, an integral field spectrograph, and a low order wavefront sensor to perform exoplanet detection and characterization. The coronagraph covers a spectral range of 400 - 1000 nm, providing a contrast of 10^{-9} with an inner working angle of $3\lambda/D$ at 400 nm. The field of view is 2.5 arcsec and the pixel scale is 0.017 arcsec. The detector is a 1k x1k EMCCD. The coronagraph includes an Integral Field Spectrograph for low-background slit spectroscopy. It operates in the 600 - 1000 nm range and has a spectral resolution of 70.

3. WFIRST Observations

The WFIRST observing program will be determined in detail in the year before launch to give the best science return for the dark energy, exoplanets and highest-ranked GO programs. We present here an example of the observation strategy. The program has four main components, to be carried out over six years. The microlensing survey will monitor 2.81 deg^2 of the Galactic bulge, with 15-minute cadence, over six 72-day seasons, detecting thousands of exoplanets via the perturbations they produce on microlensing light curves. The high latitude survey (HLS) will carry out imaging and spectroscopy of a 2400 deg^2 area over 1.9 years of observing time, providing weak lensing shape measurements of 400 million galaxies and emission line redshifts of more than 20 million galaxies at redshifts $1 < z < 3$. The supernova survey will use 0.5 years of observing time over a 2-year period, with three tiers of imaging to discover supernovae and measure their light curve, IFU spectroscopy will be performed of more than 2700 Type Ia SNe at $0.2 < z < 1.7$ to measure redshifts, spectral diagnostics, and well calibrated fluxes in synthetic filter bands matched across redshift. The coronagraph direct imaging of exoplanet will take place over 1 year. The time will be divided between imaging and spectroscopy of known exoplanets from radial velocity and other programs and exoplanet searches of nearby stars. A Guest Observer program is allocated 1.5 years of observing time, spread across the 6-year mission.

Figure 1 shows the imaging depth achieved by the WFIRST HLS, in comparison to LSST (after 10 years of operation) and Euclid. In an AB-magnitude sense, WFIRST imaging is well matched to the i-band depth of LSST. The Euclid IR imaging is ~2.5 magnitudes shallower than WFIRST (and under sampled by its 0.3 arcsec near-IR pixels). The Euclid weak lensing survey uses a wide optical filter and reaches an AB-magnitude depth similar to what LSST achieves in each of its g, r, and i filters. The cumulative imaging depth in the WFIRST supernova survey fields, which will be observed many times over a 2-year interval, reaches 0.5 – 2.5 magnitudes fainter than the HLS.

Figure 2 shows the emission line sensitivity of the HLS spectroscopic survey. For moderately extended sources, the 7σ detection threshold is $\approx 1.0\text{–}1.5\times10^{-16}$ erg s^{-1} cm^{-2} depending on wavelength. The primary target is Hα at $1.05<z<2$, but it is also possible to extend to higher redshift, $1.7 < z < 2.9$, using galaxies with strong [OIII] emission. Forecasts based on recent estimates of the Hα and [OIII] galaxy luminosity functions predict ~ 20 million Hα galaxies, and ~ 2 million [OIII] galaxies at $z > 2$. At $z = 1.5$, the predicted comoving space density of the galaxy redshift survey is 1.5×10^{-3} Mpc^{-3}. This is 16 times higher than the predicted space density for Euclid (from the same luminosity function), providing better sampling of structure, over a smaller area of sky, for studies of baryon acoustic oscillations, redshift-space distortions, and higher order galaxy clustering. The "narrow/deep" strategy of WFIRST nicely complements the "shallow/wide" strategy of Euclid.

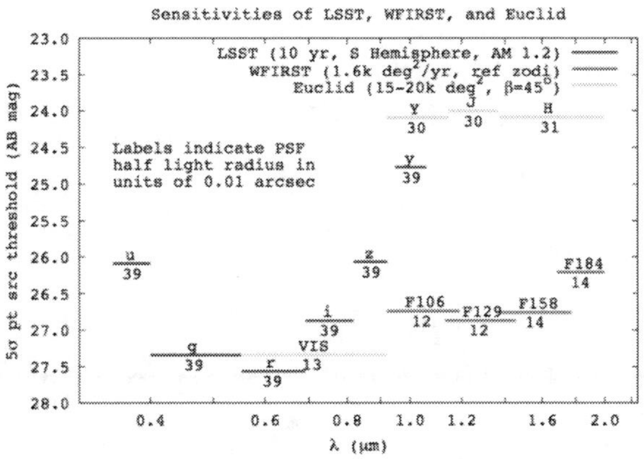

Figure 1. Imaging sensitivity of WFIRST compared to LSST and Euclid [1].

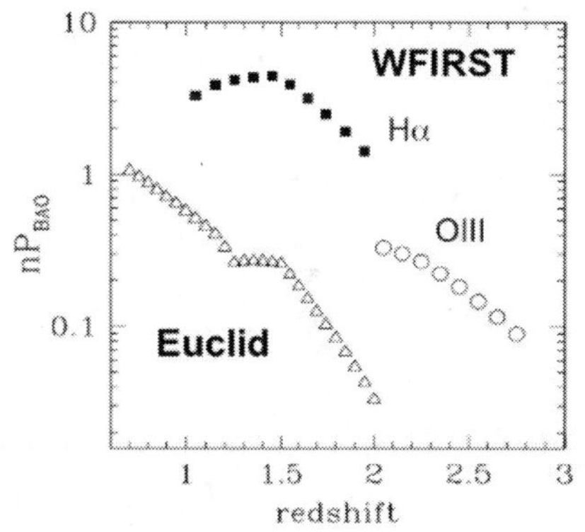

Figure 2. Spectroscopic sensitivity of WFIRST compared to Euclid [2].

4. WFIRST Science

4.1. Dark Energy

Figure 3 shows the various techniques that WFIRST will use to study dark energy. Observations of Type Ia SNe will determine "standard candle" distances out to $z = 1.7$, calibrated against a (ground-based) sample observed in the local Hubble flow. The aggregate precision of these measurements is 0.20% at $z < 1$ (error-weighted $\langle z\rangle = 0.50$) and 0.34% at $z > 1$ ($\langle z\rangle = 1.32$). The baryon acoustic oscillation (BAO) feature in galaxy clustering provides a "standard ruler" for distance measurement, calibrated in absolute units, independent of H_0. The galaxy redshift survey (GRS) enables measurements of the angular diameter distance $D_A(z)$ and the expansion rate $H(z)$ using $H\alpha$ emission line galaxies at z = 1 - 2 and [OIII] emission line galaxies at z = 2 – 3, with aggregate precision ranging from 0.40% to 1.8%.

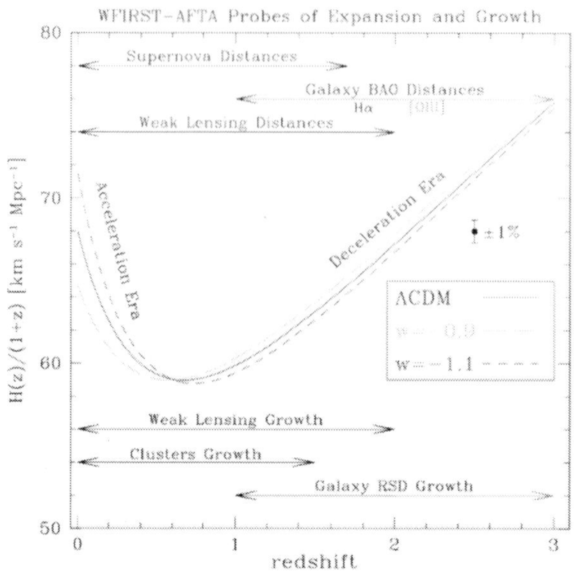

Figure 3. Dark energy techniques used by WFIRST [2].

The imaging survey will enable measurements of dark matter clustering via cosmic shear and via the abundance of galaxy clusters with mean mass profiles calibrated by weak lensing; we expect 40,000 $M \geq 10^{14} M_{Sun}$ clusters in the 2400 deg^2 area of the high- latitude survey. These data constrain the amplitude of matter fluctuations at $0 < z < 2$ and provide additional leverage on the redshift-distance relation. The expected aggregate precision on the fluctuation amplitude as an isolated parameter change is ≈ 0.15% at $z < 1$ and 0.3- 0.5% at $z > 1$. Redshift-space distortions in the GRS provide an entirely independent approach to measuring the growth of structure, with aggregate precision ≈ 1% at z = 1-2.

4.2. Microlensing

The WFIRST microlensing survey will continuously monitor a total of 2.81 deg^2 in the Galactic bulge for six 72-day campaigns, with a cadence of 15 minutes in a wide filter (0.927 – 2.00 μm) for planet discovery and 12 hours in z-band for characterization of source and lens stars. This survey will detect tens of thousands of stellar microlensing events and, in cases where the lensing star has an orbiting planet, can (when the alignment is favorable) produce a distinctive perturbation to the light curve. For a large fraction of these events, WFIRST will be able to completely characterize the lensing system, including the host star and planet mass, the planet orbital separation in physical units, and the distance to the system. It will do this through a combination of subtle deviations from the simplest light curve shapes, including finite-source effects, "parallax" caused by motion of the earth and the observatory's geosynchronous orbit, and source astrometric motion induced by the host lens, together with the isolation and direct measurement of the light from luminous host lenses afforded by the high angular resolution.

Figure 4 shows the parameter space for the WFIRST microlensing program compared to Kepler. Forecasts suggest that WFIRST will detect about 3000 planets overall, including 1000 "super-earths" (roughly 10 times the mass of earth), 300 earth-mass planets, and 40 Mars-mass planets. These detections would enable measurement of the mass function of cold exoplanets to better than ~10% per decade in mass for masses > 0.3 M_{Earth} and an estimate of the frequency of Mars-mass embryos accurate to ~15%. If there is one free-floating earth-mass planet per star in the Galaxy, then WFIRST will detect about 40 of them, and much larger numbers of more massive free-floating planets.

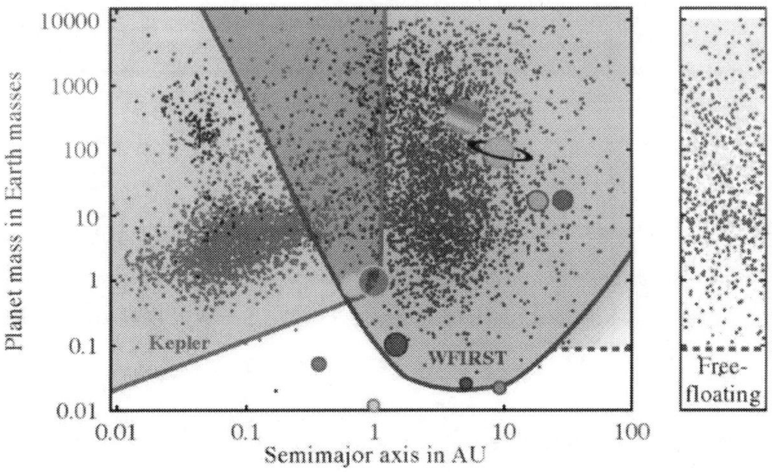

Figure 4. Parameter ranges for exoplanet detection for WFIRST and Kepler [2].

4.3. Exoplanet Direct Imaging

With a contrast level of 10^{-9} and an inner working angle of less than 0.2 arcsec, the WFIRST coronagraph will observe known nearby exoplanets and perform a survey of the nearest stars to search for exoplanets. These observations will directly image over a dozen known radial velocity planets and will discover an additional dozen previously unknown ice and gas giants as shown in Figure 5. The majority of these planets will also be characterized using R~70 spectra in the wavelength range 600-1000 nm via the IFS. These spectra will allow the detection of features expected due to methane, water, and alkali metals, reveal the signature of Rayleigh scattering, and easily distinguish between different classes of planets (i.e., Neptunes versus Jupiters), and planets with different metallicities. This survey will also be sensitive to debris disks with a few times the solar system's level of dust in the habitable zones and asteroid belts of nearby (~10 pc) Sun-like stars.

The high sensitivity and spatial resolution (0.05 arcsec is 0.5 AU at 10 pc) of WFIRST images will map the large-scale structure of these disks, revealing asymmetries, asteroid belts, and gaps due to unseen planets. WFIRST will make the most sensitive measurements yet of the amount of dust in or near the habitable zones of nearby stars. This is important for assessing the difficulty of imaging Earth-like planets with future missions as well as for understanding nearby planetary systems. Finally, spectrophotometry of these disks at the full range of available wavelengths from 400-1000 nm provides constraints on dust grain size.

4.4. Community Science

The combination of a 0.28 deg^2 field with the near-IR sensitivity and angular resolution afforded by a 2.4 m telescope gives WFIRST extraordinary discovery potential for general astrophysics. In the 6-year prime mission, 1.5 years would be allocated to GO programs. If WFIRST enters an extended mission phase, observing time would be competed, with the balance between large and small programs to be decided by the Time Allocation Committee. This will afford a large number of unique, high impact projects from characterizing nearby brown dwarfs to probing the epoch of reionization.

Journal of Physics: Conference Series **610** (2015) 012007 doi:10.1088/1742-6596/610/1/012007

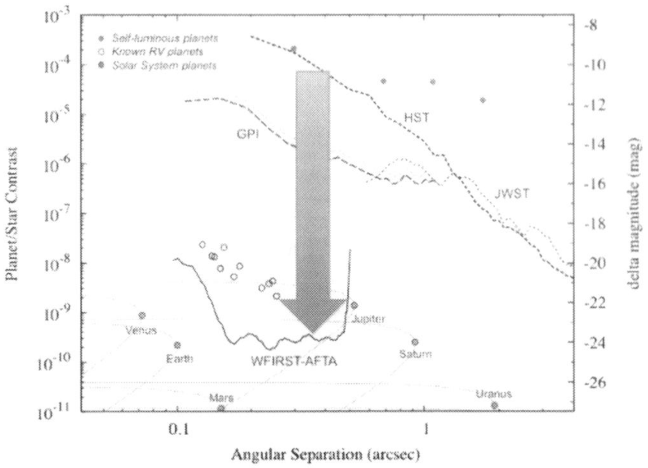

Figure 5. Exoplanet detection sensitivity of the WFIRST coronagraph compared with other instruments [2].

The WFIRST HLS will probe the low surface brightness structure around many thousands of nearby galaxies, and GO observations can extend these investigations to greater depth and to systematic coverage of different environments, from the centers of voids to the core of the Virgo cluster. For galaxies within a few Mpc, WFIRST will resolve individual red giant stars as well as imaging the low surface brightness background of main sequence stars. These investigations will show how recent accretion and minor mergers affect the growth of galaxies across a wide spectrum of mass, morphology, and environment. Regions of star formation, for example, the Taurus association, stretch across many degrees of the sky. The near-IR sensitivity of WFIRST will make possible a full census of stars in these systems to a depth far beyond what has been possible with ground-based and space-based telescopes. There are many other examples of community science, including white papers written by dozens of astronomers, in the 2013 SDT report [1].

5. WFIRST and Gravitational Wave Electromagnetic Counterparts

The next ten years will see major steps forward in the search and detection of gravitational waves. LIGO and Virgo will be in their advanced detector era with sensitivity by 2020 to detect NS binary mergers to 200 Mpc [3]. The detection rate will be 10's of events per year, with large uncertainties. In the next few years, the pulsar timing array [4] will be observing a sufficient number of millisecond pulsars with accurate enough timing observations to begin detecting binary supermassive black holes in very low frequency gravitational waves. The LISA Pathfinder mission [5] will launch in 2015 and demonstrate the drag-free control and laser interferometry technologies needed for LISA. For all of these current and future instruments, there is a great interest in detecting electromagnetic counterparts of their gravitational wave sources in order to determine distances and astrophysical properties. WFIRST can play a role in the counterpart observations.

For LIGO-Virgo, the most likely early gravitational wave detections will be NS-NS or NS-BH mergers within a range of 50 – 200 Mpc. They will be accompanied by bright gamma-ray bursts and afterglow if the jet axis is aimed at the observer. The chance of such alignment is about 1%. In the more common off-axis case, the event will be accompanied by a faint afterglow or possibly faint kilonova emission from radionuclides produced in the dense neutron-rich material surrounding the merger site. The kilonova emission is predicted to peak in the NIR, where WFIRST is the most sensitive. This is due to the opacity from the heavy elements formed in the ejecta which would be much more neutron rich than the ejecta in a type Ia supernova, where the opacity is dominated by Fe. [6]

The sky locations will not be accurate for early LIGO-Virgo events with error box sizes of 100's of deg^2 [3]. This will improve to a few deg^2 in the WFIRST time frame when LIGO-India is on-line. There will be several ways in which WFIRST can participate in follow-up observations as listed here:

1. Tile the error box with the WFI to search for a new source. With a 0.28 deg^2 field of view, 10's of pointing will be required. The afterglows could be in the one to two orders of magnitude brighter than the WFIRST J=27 AB sensitivity for days so detections are possible.
2. Perform pointed observations of galaxies within the error box. Since the mergers will occur in or near galaxies, WFIRST can point at the few brightest galaxies in the error box to search for emission [7]. This can will also be possible with the JWST.
3. Perform deep follow-up studies of candidate electromagnetic sources. WFIRST and JWST can observe candidate sources found by other telescopes and perform deep imaging and spectroscopic observation in the NIR.

For LISA, the most likely gravitational wave detections will be binary SMBH mergers at cosmic distances. Bare BH mergers have no electromagnetic radiation, but gas around the SMBHs will be stirred up and accrete forming quasar-like activity on years to decades time scales. Also, stars around the SMBHs will have their orbits perturbed and create tidal disruption events on years to decades time scales.

The sky location accuracy for LISA will like be a few deg^2 [8]. The primary way that WFIRST or a future similar telescope can make observations of LISA sources will be to perform monitoring observations of the fields to search for fading or tidal disruption flares. The error circle will either need to be tiled or candidate galaxies selected for pointed observations.

6. Conclusions

WFIRST will be a powerful mission for NIR surveys and exoplanets. It will perform Hubble quality and depth imaging and spectroscopy over 1000's of square degrees. It will advance our understanding of the expansion history of the universe and growth of structure driven by dark energy using several techniques. It will make a large step forward in exoplanet research by performing the first space microlensing census observing the dense star population in the galactic bulge and the first high-contrast coronagraphic direct-imaging observations of nearby exoplanets and debris disks. It will have a vigorous and well-fund guest observer program for community astrophysics. With its NIR coverage, depth and wide field of view, WFIRST will be a powerful tool for follow-up of LIGO-Virgo and LISA gravitational events.

7. References

[1] Spergel D, *et al.* 2013 *Wide-Field InfraRed Survey Telescope-Astrophysics Focused Telescope Assets (WFIRST-AFTA) Science Definition Team (SDT) Final Report* arXiv 1305.5422
[2] Spergel D, *et al.* 2014 *WFIRST-AFTA Science Definition Team Interim Report* (http://wfirst.gsfc.nasa.gov/)
[3] Aasi J, *et al.* 2014 Prospects for Localization of Gravitational Wave Transients by the Advanced LIGO and Advanced Virgo Observatories (preprint arXiv 1304.0670)
[4] Ravi V, Wyithe JSB, Shannon RM & Hobbs G 2014 Prospects for gravitational-wave detection and supermassive black hole astrophysics with pulsar timing arrays (preprint arXiv 1406.5297)
[5] Antonucci F *et al.*, 2012 Classical and Quantum Gravity 29, 124014.
[6] Kasen D, Badnell NR & Barnes J 2013, ApJ, 774, 25
[7] Kanner J, Camp J, Racusin J, Gehrels N, & White D 2012, ApJ, 759, 22
[8] Lang RN, Hughes SA, & Cornish NJ 2011, Phys Rev D, 84, 22002

Athena: the X-ray observatory to study the hot and energetic Universe

X Barcons[1], K Nandra[2], D Barret[3], J-W den Herder[4], A C Fabian[5], L Piro[6], M G Watson[7] and the Athena team[8]

[1] Instituto de Física de Cantabria (CSIC-UC), 39005 Santander, Spain
[2] Max-Planck Institut für Extraterrestriche Physik, 85741 Garching, Germany
[3] Institut de Recherche en Astrophysique et Planétologie, 31028 Toulouse, France
[4] SRON Netherlands Institute for Space Research, 3584 CA Utrecht, The Netherlands
[5] Institute of Astronomy, Cambridge CB3 0HA, United Kingdom
[6] INAF, Istituto Astrofisica e Planetologia Spaziali, 00133 Roma, Italy
[7] Dept of Physics & Astronomy, U of Leicester, Leicester LE1 7RH, United Kingdom
[8] www.the-athena-x-ray-observatory.eu

E-mail: barcons@ifca.unican.es

Abstract. Hot gas pervades the Universe: about half of the baryonic content in the Universe is expected to be at $T > 10^5$ K, and there are as many baryons at $T > 10^7$ trapped in galaxy clusters as there are locked into stars. There is an intimate relation between this hot gas, which delineates the large-scale structure of the Universe, and the most energetic phenomena occurring in the immediate vicinity of super-massive black holes, through a poorly known process called Cosmic Feedback. Studying the hot and energetic universe requires X-ray observatories in space, whose capabilities greatly exceed those of the current workhorse observatories: NASA's *Chandra* and ESA's *XMM-Newton*. *Athena* has been selected by ESA as the L2 mission (due for launch in 2028), to address the "Hot and Energetic Universe" science theme. It will be a large X-ray observatory capable of addressing the above topics, and many other fundamental questions in contemporary astrophysics. Here we present the *Athena* science objectives, the mission concept and its payload, including the X-ray telescope and its two baseline instruments: a Wide Field Imager (WFI) and an X-ray Integral Field Unit (X-IFU).

1. Introduction

ESA's Cosmic Vision 2015-2025 scientific programme will be implemented in terms of a number of missions of different sizes, with L-class missions being the flagships of the programme. Three L-class missions are contemplated. The first one was selected in 2012 and is JUICE, a mission to Jupiter's icy moons due for launch in 2022. For the selection of the L2 and L3 missions, ESA appointed a senior Science Survey Committee (SSC) in 2013 to recommend the scientific themes to be addressed by those two mission opportunities with nominal launch years of 2028 and 2034, respectively. The SSC reviewed over thirty community White Papers and recommended that L2 should address *The Hot and Energetic Universe* and L3 should address *The Gravitational Universe* [1]. ESA's governing bodies adopted that recommendation in November 2013.

The Hot and Energetic Universe [2] refers to the baryonic component of the Universe at such high temperatures which is largely undetectable by observational facilities at wavelengths longer than X-rays. There is a very substantial fraction of the baryonic Universe in this form: 50% of the baryons are thought to be at $T > 10^5$K and 30 to 40% at $T > 10^6$ K. The amount of hot baryons trapped in groups and clusters of galaxies at temperatures $T > 10^7$ K is of the same order than that locked into stars throughout the Universe. Hot gas is expected to trace the large-scale structures delineated by the dark matter potential wells, and it has an intimate relation with the very energetic phenomena arising in the immediate vicinity of black holes. Accretion onto Super-Massive Black Holes (SMBH) is accompanied by large releases of energy, both in the form of radiation and of mechanical energy. Massive winds and outflows transport this mechanical energy to galaxy scales, effectively terminating star formation and further SMBH growth, but also to larger (cluster) scales, excavating bubbles in the intra-cluster hot gas. This process is generically named Cosmic Feedback, it governs the life of the galaxies but its physical grounds are still poorly understood.

The intimate link between the hot and the energetic Universe calls for an X-ray observatory class mission to study these and other related topics. The *Athena* mission [3], selected by ESA in June 2014, will consist of a single large-aperture X-ray telescope equipped with two state-of-the-art focal plane instruments. *Athena*, being an observatory, will enable enormous science progress in many corners of astronomy, besides addressing the Hot and Energetic Universe theme.

This paper presents the *Athena* science goals in Section 2. Section 3 describes the science requirements and mission concept as submitted to ESA in response to the Call for Missions and selected in June 2014. In Section 4 we place *Athena* in a broad astrophysical context and note the opportunities for synergies with gravitational wave observatories, eLISA in particular.

2. The hot and energetic Universe

2.1. How do baryons assemble and form the large-scale structures in the Universe that we see today?

Hot gas pervades the Universe. On the largest scales, clusters of galaxies signal the locations where dark matter created the deepest potential wells. X-ray baryonic gas at temperatures above a million degrees is trapped in these potential wells and in an approximate local hydrostatic equilibrium. Determining how baryons accrete and dynamically evolve inside these potential wells which formed around z~2 is a question of high cosmological significance [4]. Quantifying the number of such first building blocks of the dark matter structure in the form of gas-filled groups of galaxies with mass > 5×10^{13} M_{Sun} will constrain large-scale structure formation, as their sky density is highly sensitive to the adopted model.

Measuring gas bulk velocities and turbulence can determine the physics on how baryons assemble and dynamically evolve in the cluster potential wells. How much gravitational energy is dissipated in to mechanical energy is not known, but models predict motions and turbulence at the level of 200 km/s. High-resolution X-ray spectroscopic imaging of cluster gas is essential to determine this [5].

It is also clear that astrophysical processes other than gravitational attraction drive the evolution of groups and clusters [4]. Supernovae and/or Active Galactic Nuclei (AGN) inject excess entropy during cluster history, which is reflected in deviations from a purely gravitational entropy profile. Determining when this excess energy was injected and what was the driving process is a must to understand large-scale structure and, indeed, assess the use of galaxy clusters as cosmological probes. Cluster entropy profiles need to be obtained out to z~1-2 to determine this.

Since clusters are the largest hot gas reservoirs in the Universe, they retain direct information about overall chemical evolution. Elements produced by stellar evolution and ejected by Supernovae and winds are trapped inside cluster potential wells. In nearby galaxy clusters elemental abundances and their spatial distribution across the cluster are needed to understand the relative contribution from type Ia and core-collapse Supernovae as well as from AGB stars to the chemical enrichment of the intra-cluster medium. In more distant clusters, determining abundances of trace elements like Fe, O and Si in cluster cores, spanning a broad range of masses and redshifts ($0<z<2$) will reveal the chemical evolution of the Universe [5].

AGN at cluster centres are responsible for preventing the gas from massively cooling to very low temperatures and result in massive amounts of star formation. Jets constitute the most powerful incarnations of AGN mechanical energy releases on cluster scales. Bubbles in the hot gas excavated by these jets measure the "instantaneous" mechanical power being released, while ripples in the distribution of hot gas across the whole cluster result from the integrated AGN action during its various evolution phases [6]. Detailed density and temperature diagnostics in cool cluster cores will reveal the delicate balance between heating and cooling.

Finally, the majority of hot baryons are not in clusters and their location and physical state is currently unknown [7]. More than 30% of the baryons in today's Universe are expected to reside in the Warm/Hot Intergalactic Medium (WHIM), following a filamentary structure tracing dark matter. Galaxy outflows and super-winds from starburst galaxies are further contributors to this tenuous intergalactic gas, whose mere detection is at the limit (if not beyond) the capabilities of current facilities. Determining the local baryon abundance requires detecting resonance absorption lines from various elements (O, Ne, C, etc) towards bright background X-ray sources. In addition, detecting this tenuous gas in emission will also reveal its ionization state and elemental abundances [7]. All this requires very sensitive weak line spectroscopy detection capability. As the number of bright X-ray sources in the sky is limited, Gamma-ray Burst (GRB) afterglows will also be used as X-ray lighthouses illuminating our line of sight towards the distant Universe.

2.2. How do black holes grow and shape the Universe?

SMBH growth and star formation are two fundamental astrophysical processes that drive galaxy evolution. SMBH grow by accretion, mergers, chaotic accretion (of smaller black holes) or a combination of all of them. All of these processes produce copious amounts of radiation and some of them are also expected to produce strong gravitational wave signals. X-rays are a particularly privileged part of the electromagnetic spectrum to use when looking at growing SMBH because they originate very close to the event horizon (closer than radiation at longer wavelengths), there is little contamination in X-rays from the host galaxy and finally X-rays can penetrate through significant amounts of obscuring gas.

Similar to star formation, accreting SMBH (AGN) are expected to be present from the earliest phases of galaxy evolution. During the re-ionisation (z~6-10) the number density of AGN depends strongly on the seeds of these SMBH, be those Population III star remnants (seed masses $\sim 10^3$ M_{Sun}) or the result of direct halo gravitational collapses (seed masses $\sim 10^5$-10^6 M_{Sun}). Galaxy formation models are not constrained enough by current observations as to being able to predict the SMBH distribution at that epoch. Optical surveys (e.g., with *Euclid*) will be able to find out the tip of the iceberg, i.e., the most massive and actively accreting SMBH at those epochs. But the average AGN population (X-ray luminosity $\sim 10^{43}$-10^{44} erg/s), which is needed to constrain SMBH seeds, will only be accessible through deep X-ray surveys [8]. Further information on the early universe can be obtained from X-ray observations of high-z GRB afterglows, using the fast response capability that will be implemented in *Athena*. Absorption features from heavy elements (or lack of them) will reveal whether the progenitors of these high-z GRBs are Population III stars or whether some chemical enrichment has already happened in the parent stellar populations.

SMBH accretion is known to happen very often in obscured environments, where there is very little (or no) radiation leakage. This has prevented so far to have a complete census of growing SMBH, including Compton-thick objects, where no direct X-rays escape from inside, but where a tiny fraction of reflected X-rays can be seen. *Athena* will perform such a complete census out to the epoch where most of the accretion and star formation in the Universe happened (z~2) [9]. In addition, the incidence of AGN winds and outflows including the most energetic ultra-fast outflows will be assessed and the mechanical energy released through these outflows will be measured in a number of case studies via high-resolution X-ray spectroscopy. This will deliver a complete picture of the SMBH growth at z~2, which together with the optical/IR galaxy surveys mapping star formation, will reveal the details on how galaxy evolution bursted at that particular epoch of cosmic history.

In nearby galaxies, the energy released by growing SMBH will be measured to exquisite detail. Significant amounts of mechanical energy are known to be associated to molecular outflows and to ionized gas. The current picture misses an essential ingredient, which is the massive outflows that are just beginning to be seen in X-rays. These could dominate in some cases the mechanical energy ejected by growing SMBH [10]. In addition, *Athena* will also measure the amounts of gas, metals and energy that AGN outflows and superwinds from starburst galaxies deposit in their surroundings.

Cosmic feedback is a truly remarkable phenomenon, where something happening at the scale of the event horizon of a SMBH impacts and largely drives the life of galaxies and clusters. Understanding the physics of matter under strong gravity conditions is essential to properly model this phenomenon. In the immediate vicinity of black holes, a number of relativistic effects take place and some of them have a direct observational impact [11]. *Athena* will measure, through a number of observational techniques, the spin of black holes in binaries. Specifically through X-ray spectroscopy, it will conduct a survey of AGN spins, whose distribution reflects the dominant SMBH growth mode (e.g., smooth accretion spins up SMBH, while chaotic accretion spins them down). Reverberation mapping, a technique at the edge of current observational capabilities, will in addition reveal the geometry of the material surrounding the SMBH.

2.3. *Additional science cases*

With the performance required to meet the above science goals, *Athena* will be able to tackle a number of other science goals, spanning virtually all corners of Astrophysical research.

Examples of such cases include the study of solar system body's atmospheres and exospheres as well as of the solar wind through charge exchange with these [12]. It also includes star formation, young stellar objects, ultra-cool dwarfs, outflows and magnetic fields in massive stars and planetary nebulae [13] as well as end points of stellar evolution such as neutron stars, black hole X-ray binaries, white dwarfs and ultra-luminous X-ray sources [14]. Furthermore, *Athena* allows to investigate the explosion mechanisms in Supernovae through their remnants [15] and study the cold and hot phases of the interstellar medium [15].

Last, but not least, the fast response capability of *Athena* will enable the study of all types of transient sources [16], including indeed GRBs. By 2028, astronomy might be driven –at least in part-by transient science, thanks in part to projects like LSST [17]. Athena should be able to observe transient sources a few hours after the trigger happens, and do that in a significant fraction of the transients.

3. The Athena mission

3.1. Athena Science Requirements

To address the Hot and Energetic Universe scientific goals, a large aperture X-ray telescope is needed together with two focal plane instruments: a Wide Field Imager (WFI) providing sensitive X-ray imaging over a large field of view [19] and an X-ray Integral Field Unit (X-IFU) delivering spatially resolved high-resolution X-ray spectroscopy [20]. These payload elements, and other mission systems, need to meet a number of rather stringent requirements to be able to address the above science goals. A description on the *Athena* science requirements is given in the mission proposal [3], and here only a qualitative description of these requirements on the most representative performance parameters is presented. Some key *Athena* performance parameters are illustrated in Figure 1.

The *Athena* X-ray telescope [18] will have an effective area of 2 m^2 at 1 keV and 0.25 m^2 at 6 keV. The former is driven by point source sensitivity (needed for surveys), and many spectroscopic measurements utilizing soft X-rays; the latter is driven by the need to use the Fe emission line to trace cluster gas properties as well as strong gravity effects around black holes. The angular resolution is specified to be 5 arcsec (half-energy width), largely driven by the needed point-source sensitivity and capability to match faint source positions to those at other wavelengths; 5 arcsec is also the scale of structures that develop in nearby cool cluster cores due to feedback from AGN.

The WFI [19] needs to have a large field of view (40' × 40') in order to provide sufficient survey grasp and efficiently find early galaxy groups, high-z AGN and significant numbers of obscured AGN at z~2. In addition, this enables observing most galaxy clusters in a single pointing. AGN reverberation mapping requires a moderate spectral resolution at 6 keV of 150 eV. A high count-rate capability for the WFI is also needed to study stellar-mass black holes.

The X-IFU [20] is required to have a spectral resolution of 2.5 eV at energies below 6 keV, along with an rms gain stability of a fraction of an eV. These values define respectively the ultimate weak line sensitivity needed to detect WHIM absorption lines, as well as the ultimate accuracy to which hot gas motions can be measured in clusters. At this resolution, OVII emission line triplets can be safely resolved and used to diagnose hot gas. The required field of view is 5' in diameter (the shape of the field of view is still under study), which ensures that distant clusters (z>0.2) can be observed in a single pointing. The X-IFU should be able to observe down to 0.2 keV in order to access important C lines from warm gas in local objects and at moderate redshifts.

The spacecraft itself should be rather agile, so it can react quickly to Target of Opportunity alerts, including GRBs. *Athena* will have a fast re-pointing capability (of the order of a few hours) to a large fraction of the sky, so enough fluence (flux integrated over time) can be obtained when observing GRBs. The absolute astrometric error needs to be below 3", which can be reduced in WFI observations to 1" using post-fact reconstruction.

3.2. The Athena mission concept

Athena is the result of maturing a concept that began with the X-ray Evolving Universe Spectroscopy (XEUS) mission [21] selected for study by ESA in 2007, then followed by the International X-ray Observatory (IXO) [22] and that later become the Athena-L1 mission [23]. The *Athena* mission will consist of a single spacecraft launched into an L2 halo orbit by an Ariane V – class launcher, containing as science payload a single aperture X-ray telescope with a focal length of 12 m, and two focal plane instruments whose position can be changed by a movable instrument platform. The WFI is

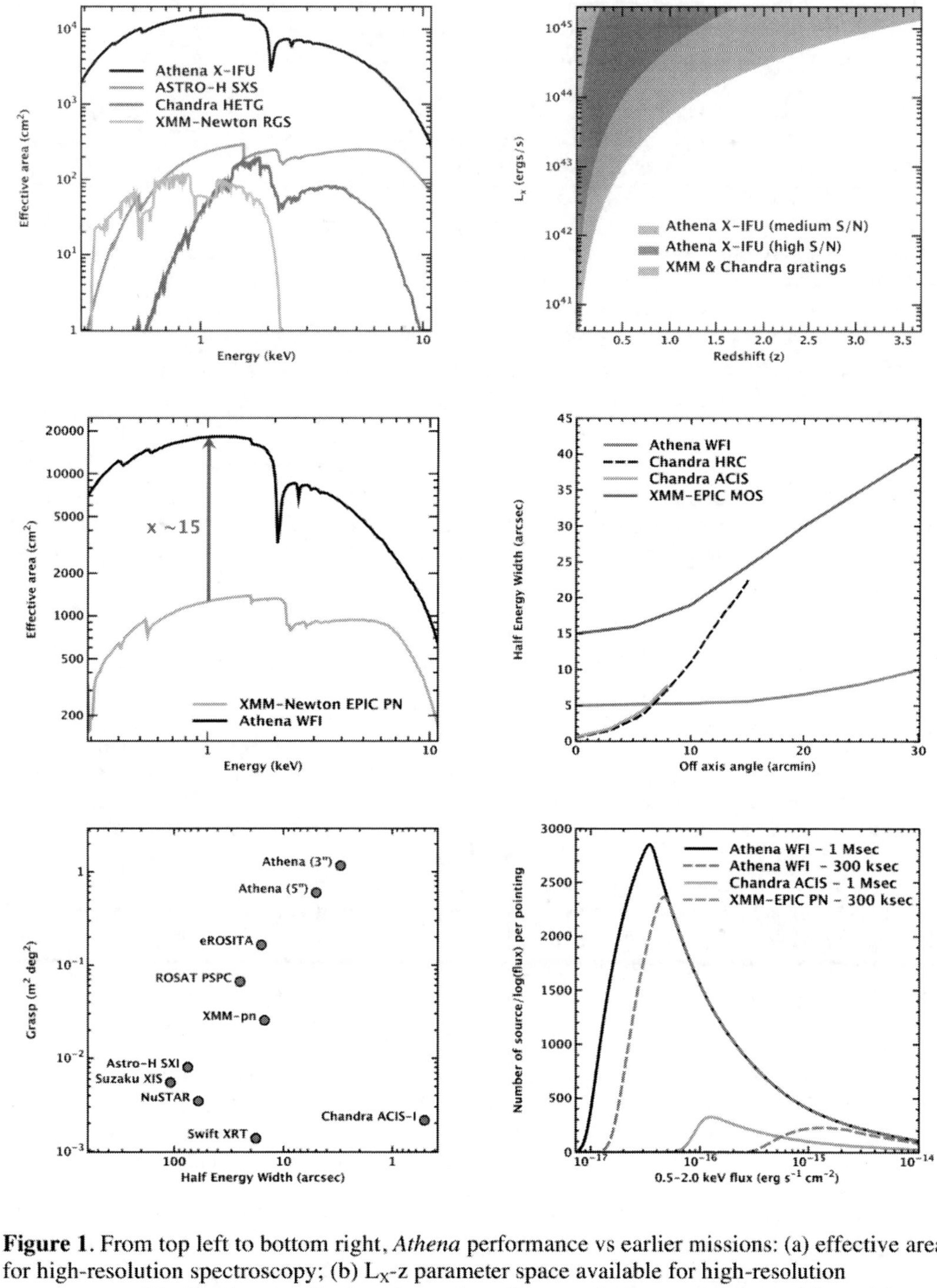

Figure 1. From top left to bottom right, *Athena* performance vs earlier missions: (a) effective area for high-resolution spectroscopy; (b) L_X-z parameter space available for high-resolution spectroscopy; (c) effective area for imaging and medium-resolution spectroscopy; (d) spatial resolution at various off-axis angles; (e) survey grasps and (f) survey speed in sources per pointing

a Si-based detector with DEPFET active pixel sensor readout [19], delivering sensitive imaging capabilities over a large field of view, modest energy resolution and high count-rate capability. The X-IFU is a calorimeter array, whose front-end sensor is made of absorbers thermally coupled to Transition Edge Sensors operating at 100 mK [20]. It will provide spatially resolved high-resolution X-ray spectroscopy for both point-like and extended sources.

The *Athena* mission proposed to ESA will last for 5 years, which will be sufficient to achieve the full set of Hot and Energetic Universe science goals, and accommodate a significant amount of additional science goals. It is expected, however, that the mission could be extended if there are strong scientific and programmatic reasons. The vast majority of the observations will be allocated following a traditional call for proposals and peer review process, although some time is expected to be guaranteed to consortia providing payload and other mission elements. The possibility of defining key programmes before launch, with the goal of securing some of the most demanding science goals, has been flagged in the mission proposal and will receive further consideration. It is envisaged that international partners, in particular NASA and JAXA, could participate in *Athena*, which will remain in any case a European flagship mission.

4. Athena in context

The Astrophysical landscape around 2028 will be significantly different to the one that we have today. A plethora of large astrophysical observatories will be available along the whole electromagnetic spectrum. Among those, ALMA, the ELTs (ESO's E-ELT, TMT and GMT) and JWST will be the dominant deep sky facilities at (sub)millimeter, optical and infrared wavelengths, in addition to LSST. The SKA could complement those facilities at cm-radio wavelengths and CTA will be the first true observatory at TeV energies.

Athena will be the X-ray element of the family. Its sensitivity has been designed to peer into the earliest growing SMBH at a comparable depth to which JWST and the ELTs will find the first star-forming galaxies and ALMA will see their cold Interstellar Medium. In the particular context of transient sources, as a number of gravitational wave sources are expected to be, the fast reaction capability of *Athena* might be instrumental in providing electromagnetic counterparts to the gravitational wave events. Cosmic X-rays arise either in extended gravitational potential wells (the Hot Universe) or in the deepest gravity potential wells (the Energetic Universe). In realistic conditions, it would be extremely surprising that a gravitational wave event, coming from a dramatic change of a strong gravitational field, would not come accompanied by copious X-ray emission. LIGO and VIRGO experiments first, and eLISA later will teach us how to identify gravitational wave events with electromagnetic signals. In the specific case of eLISA, where low-frequency gravitational waves from coalescing SMBH will be detectable, there will be a very important synergy with *Athena* in trying to discern what is the relative contribution from mergers and smooth accretion to SMBH growth. The hope is that both *Athena* and eLISA will be looking at the sky together for a number of years.

References

[1] Report of the Senior Survey Committee on the selection of the science themes for the L2 and L3 opportunities in the ESA Cosmic Vision programme, http://sci.esa.int/jump.cfm?oid=53260

[2] Nandra K, Barret D, Barcons X et al 2013 The Hot and Energetic Universe: A White Paper presenting the science theme motivating the Athena+ mission Athena Mission Proposal, arXiv:1306.2307

[3] Nandra K, Barcons X, Barret D, Fabian A C, den Herder J-W, Piro L and Watson M G 2014 Athena, the Advanced Telescope for High Energy Astrophysics. A mission addressing The Hot and Energetic Universe science theme, http://athena2.irap.omp.eu/spip.php?article17

[4] Poitecouteau E, Reiprich T et al 2013 The Hot and Energetic Universe: The evolution of galaxy groups and clusters arXiv:1306.2319

[5] Ettori S, Pratt G W et al 2013 The Hot and Energetic Universe: The astrophysics of galaxy

groups and clusters arXiv:1306.2322

[6] Croston J H, Sanders J S et al 2013 The Hot and Energetic Universe: AGN feedback in galaxy clusters and groups arXiv:1306.2323

[7] Kaastra J, Finoguenov A et al 2013 The Hot and Energetic Universe: The missing baryons and the warm and hot intergalactic medium arXiv:1306.2324

[8] Aird J, Comastri A et al 2013 The Hot and Energetic Universe: The formation and growth of the earliest super-massive black holes arXiv:1306.2325

[9] Georgakakis A, Carrera F J et al 2013 The Hot and Energetic Universe: Understanding the build-up of supermassive black holes and galaxies at the heyday of the Universe arXiv:1306.2328

[10] Cappi M, Done C et al 2013 The Hot and Energetic Universe: Astrophysics of feedback in local AGN arXiv:1306.2330

[11] Dovciak M, Matt G et al 2013 The Hot and Energetic Universe: The close environments of supermassive black holes arXiv:1306.2331

[12] Branduardi-Raymont G, Sciortino S et al 2013 The Hot and Energetic Universe: Solar system and exoplanets arXiv:1306.2332

[13] Sciortino S, Rauw G et al 2013 The Hot and Energetic Universe: Star formation and evolution arXiv:1306.2333

[14] Motch C, Wilms J et al 2013 The Hot and Energetic Universe: End points of stellar evolution arXiv:1306.2334

[15] Decourchelle A, Costantini E et al 2013 The Hot and Energetic Universe: The astrophysics of supernova remnants and the interstellar medium arXiv:1306.2335

[16] Jonker P, O'Brien P et al 2013 The Hot and Energetic Universe: Luminous extragalactic transients arXiv:1306.2336

[17] Kahn S 2015 These proceedings

[18] Willingale R, Pareschi G, Christensen F and den Herder J-W 2013 The Hot and Energetic Universe: The Optical Design of the Athena+ Mirror arXiv:1306.1709

[19] Rau A, Meidinger N et al 2013 The Hot and Energetic Universe: The Wide Field Imager (WFI) for Athena+ arXiv:1308.6785

[20] Barret D, den Herder J W, Piro L et al 2013 The Hot and Energetic Universe: The X-ray Integral Field Unit (X-IFU) for Athena+ arXiv:1308.6784

[21] Parmar A, Arnaud M, Barcons X et al 2006 XEUS: The x-ray evolving universe spectroscopy mission in *Space Telescopes and Instrumentation II: Ultraviolet to Gamma Ray*. Edited by Turner, Martin J. L.; Hasinger, Günther. Proceedings of the SPIE, Volume 6266, id. 62661R

[22] Barcons X, Barret D, Bautz M et al 2011 International X-ray Observatory (IXO) Assessment Study Report for the ESA Cosmic Vision 2015-2025 arXiv:1102.2845

[23] Barcons X, Barret D, Decourchelle A et al 2012 Athena (Advanced Telescope for High ENergy Astrophysics) Assessment Study Report for ESA Cosmic Vision 2015-2025 arXiv:1207.2745

Acknowledgments

We are grateful to the hundreds of scientists that helped preparing the successful *Hot and Energetic Universe* White Paper and *Athena* mission proposal. XB acknowledges partial financial support by the Spanish Ministry of Economy and Competitiveness through grants AYA2012-39767-C02-01 and AYA2012-31447.

Journal of Physics: Conference Series **610** (2015) 012009 doi:10.1088/1742-6596/610/1/012009

Status of MICROSCOPE, a mission to test the Equivalence Principle in space

Joel Bergé[1], Pierre Touboul[2], Manuel Rodrigues[1], for the MICROSCOPE team

[1] ONERA, the French Aerospace Lab, 29 avenue de la Division Leclerc, 92320 Châtillon, France

[2] ONERA, the French Aerospace Lab, chemin de la Hunière, 91123 Palaiseau, France

E-mail: joel.berge@onera.fr

Abstract. MICROSCOPE is a French Space Agency mission that aims to test the Weak Equivalence Principle in space down to an accuracy of 10^{-15}. This is two orders of magnitude better than the current constraints, which will allow us to test General Relativity as well as theories beyond General Relativity which predict a possible Weak Equivalence Principle violation below 10^{-13}. In this communication, we describe the MICROSCOPE mission, its measurement principle and instrument, and we give an update on its status. After a successful instrument's commissioning, MICROSCOPE is on track for on-schedule launch, expected in 2016.

1. Introduction

The 20th century gave birth to the Standard Model of Particle Physics (SM) and to General Relativity (GR). The former, based on quantum physics, describes the world at small scales and high energy, whereas the latter describes gravitation at large scales. Although superbly tested and confirmed in most of their respective regimes, SM and GR are still incompatible, and attempts to unify them have so far failed. Moreover, the discovery of the acceleration of the Universe's expansion put an extra pressure on current physics: whether it stems from dark energy or from a fifth force, or it is due to the failure of GR on cosmological scales, is still unknown. Thus, premiere tests of gravity are underway or planned, using such diverse probes as cosmological probes (e.g. weak lensing, baryon acoustic oscillations with surveys like Euclid[1], DES[2] or LSST[3]), gravitational waves detection and characterization (e.g. VIRGO[4], LIGO[5] and eLISA[6]), or search for Equivalence Principle violations. Looking for the latter is motivated by predictions from various theories aiming to unify GR and quantum physics, or to modify gravity in order to account for the accelerated expansion of the Universe. For instance, string-theory-inspired theories predict a violation of the Equivalence Principle at a level between 10^{-18} and

[1] http://sci.esa.int/euclid/
[2] http://www.darkenergysurvey.org/
[3] http://www.lsst.org
[4] http://www.ego-gw.it
[5] http://www.ligo.caltech.edu/
[6] https://www.elisascience.org/

10^{-13} [4, 9].

MICROSCOPE (Micro-Satellite à traînée Compensée pour l'Observation du Principe d'Equivalence [11, 12]) is a drag-free microsatellite which aims to test the WEP down to the 10^{-15} level, expected for launch in 2016. In this paper, we first motivate the science case for MICROSCOPE in Sect. 2; we then give a short overview of MICROSCOPE in Sect. 3 before stating on its current status in Sect. 4.

2. Weak Equivalence Principle

2.1. Weak Equivalence Principle and General Relativity

The Weak Equivalence Principle (WEP) states that two bodies in the same gravitational field experience the same acceleration, independently of their mass and composition. In other words, it states the universality of free fall. Together with the local position invariance (laws of physics do not depend on the position) and the local Lorentz invariance (laws of physics do not depend on the speed of the observer in an inertial frame), this principle constitutes the Einstein Equivalence Principle (EEP). The latter is at the basis of Einstein's General Relativity. Any violation of the WEP (and of the EEP thereof), would indicate that GR is not the ultimate theory of gravity.

2.2. Tests of the Weak Equivalence Principle

The WEP has been tested throughout the 20th century with an increasing precision. Deviations from WEP are usually described by the Eötvös parameter

$$\eta = 2\frac{(m_g/m_i)_A - (m_g/m_i)_B}{(m_g/m_i)_A + (m_g/m_i)_B} \quad (1)$$

where "A" and "B" are two bodies experiencing the same gravitational field, m_g is the gravitational mass and m_i the inertial mass. If the WEP holds, then $m_g = m_i$ for all bodies, and $\eta = 0$. Figure 1 shows how the upper limit on η has decreased during the 20th century, as more accurate experiments have been put forth. The first measurement was made by Eötvös with a torsion pendulum, and allowed him to constrain the WEP at a level of 10^{-8}. Most recently, the monitoring of the Earth and Moon system with the Lunar Laser Range (LLR – e.g. [8, 15]), and the measurement by the Eöt-Wash group with a torsion pendulum, reached the best limits $\eta \leqslant 10^{-13}$ [13]. On-ground measurement are however reaching their limits in terms of signal-to-noise ratio, making a better measurement more difficult. As a consequence, efforts are underway to perform WEP tests with atomic interferometers (e.g. [3]). Another solution to increase the precision on the WEP is to test it in space: this is the goal of the MICROSCOPE mission. Theories currently developed to explain the accelerated expansion of the Universe, or to unify GR and quantum physics, predict that WEP is violated at a level $10^{-18} \leqslant \eta \leqslant 10^{-13}$ [4, 9]. MICROSCOPE will be able to measure η at the level 10^{-15}: it will thus allow us to probe a significant part of the η-space under consideration by those new theories, and to start discriminating against theories.

3. MICROSCOPE

MICROSCOPE will test the WEP by comparing the acceleration experienced by two free-falling test masses in the Earth's gravity field. To this aim, it embarks two ultrasensitive electrostatic differential accelerometers. Each accelerometer consists of two coaxial cylindrical test masses whose motion is electrostatically constrained. In one (reference – 'REF') accelerometer, the test masses are made of the same material to demonstrate the experiment's accuracy; they are made of different materials in the second ('EP') accelerometer, which is used to test the WEP. The difference of electric potentials applied to keep the masses in equilibrium is a measure of the difference in the proof masses motion; hence, a non-zero difference of applied potentials is a measure of a WEP violation.

Journal of Physics: Conference Series **610** (2015) 012009 doi:10.1088/1742-6596/610/1/012009

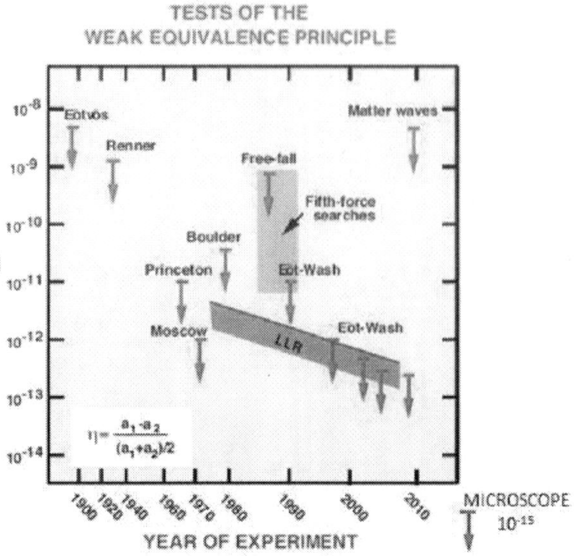

Figure 1. Tests of WEP throughout the 20th century. The arrow on the lower right corner shows the expectation for MICROSCOPE. Figure adapted from [14].

Figure 2 shows the measurement principle for the EP accelerometer: its two test masses, since they have the same center-of-mass, experience the same gravitational field (red arrows). If the WEP is violated and if, for example, the internal test mass falls faster than the external test mass, then the difference in accelerations along the EP test axis (horizontal black arrows, along which the test is performed) will be modulated by the instrument's motion around the Earth. We then expect to detect a sine wave corresponding to the modulation of the difference of the voltages applied in the two test-mass electrostatic configuration to keep them centered. Depending on the spacecraft's spin (either null for an inertial session as depicted by the figure, or non-null for a session where the satellite is forced to spin around the axis perpendicular to the orbital plane), the WEP violation signal will have a typical, expected frequency.

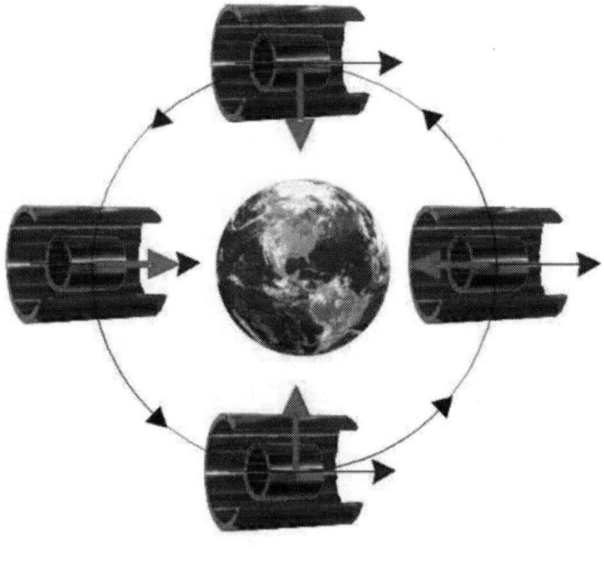

Figure 2. MICROSCOPE's measurement principle. A WEP violation is detected if the two cylindrical test-masses experience different accelerations (red arrows) as the satellite orbits the Earth; the difference in those accelerations is measured by the difference in the voltages applied to the test-masses to keep them in equilibrium. Black arrows show the sensitive axis along which a WEP violation is looked for.

The difference of acceleration deduced by the measured difference in applied voltages can be written as:

$$\overrightarrow{\Gamma}_{\mathrm{meas,d}} = [\mathcal{M}_c]\left(\eta\overrightarrow{g} + ([\mathcal{T}] - [\mathcal{I}n])\overrightarrow{\Delta} - 2[\Omega]\dot{\overrightarrow{\Delta}} - \ddot{\overrightarrow{\Delta}}\right) + \overrightarrow{K_{0,d}} + [\mathcal{M}_d]\overrightarrow{\Gamma}_{\mathrm{App,c}} + \overrightarrow{\Gamma}_{\mathrm{measquad,d}} + \overrightarrow{\Gamma}_{\mathrm{n,d}} + [\mathcal{C}_d]\dot{\overrightarrow{\Omega}} \quad (2)$$

where $\overrightarrow{g}$ is the Earth gravity field at the spacecraft's center-of-mass; $\overrightarrow{\Delta} = \overrightarrow{O_iO_j}$ is the distance between the center of the differential accelerometer's test masses; $\overrightarrow{K}_{0,d}$ is the accelerometer's bias; $[\mathcal{T}]$ is the Earth gravity gradient tensor; $[\mathcal{I}n]$ is the instrument's moment of inertia tensor, $[\Omega]\dot{\overrightarrow{\Delta}}$ is the Coriolis acceleration, with $[\Omega]$ the spacecraft's angular velocity matrix; $\overrightarrow{\Gamma}_{\mathrm{App,c}}$ is the common mode acceleration and includes non-gravitational external accelerations; $\overrightarrow{\Gamma}_{\mathrm{measquad,d}}$ quantifies the second-order term measured differential acceleration; $\overrightarrow{\Gamma}_{\mathrm{n,d}}$ is the instrument's noise; $[\mathcal{C}_d]$ is the differential angular to linear coupling matrix; $[\mathcal{M}_c]$ is the common mode sensitivity matrix and $[\mathcal{M}_d]$ is the differential mode sensitivity matrix. Those last two matrices contain combinations of the accelerometers' scale factor, coupling between axes, and axes alignments with respect to the satellite reference frame, for the common mode accelerations (the half-sum of the accelerations measured by the two accelerometers, which quantifies non-gravitational accelerations) and for the differential mode accelerations (the half-difference of the accelerations measured by the two accelerometers), respectively. They can be finely estimated by in-flight calibrations. All nuisance parameters are either corrected for through careful modeling (e.g. gravity gradient tensor) or calibrated in flight (e.g. bias [5, 6]), or minimized by design of the instrument and of the satellite (e.g. inertial tensor, instrument's noise).

MICROSCOPE's instrument (T-SAGE – Twin Space Accelerometer for Gravitation Experiment) and its performance have been described elsewhere (e.g. [11, 12]). As mentioned above, the instrument's mechanical core consists of two differential accelerometers (Sensor Units – SU), whose test masses are co-axial cylinders kept in equilibrium with electrostatic actuation. The test masses' materials were chosen carefully so as to maximize the scientific return of the experiment and to optimize their industrial machining: the EP test masses are made of alloys of Platinum-Rhodium (PtRh10 – 90% Pt, 10% Rh) and Titanium-Aluminium-Vanadium (TA6V – 90% Ti, 6% Al, 4% V), while the REF test masses are made of the same PtRh10 alloy. For each SU, the test masses are controlled electrostatically, through electrodes, without any mechanical contact; only a thin 7 μm-diameter gold wire, used to fix the masses' electrical potential to the electronics reference voltage, provides a mechanical contact (and associated, accounted for, damping noise) between the test masses and their cage. The test masses' control is performed by an electronic servo-loop. Two Front End Electronics Unit (FEEU) boxes (one per SU) include the capacitive sensing of masses, the reference voltage sources and the analog electronics to generate the electrical voltages applied to the electrodes; an Interface Control Unit (ICU) includes the digital electronics associated with the servo-loop digital control laws, as well as the interfaces to the satellite's data bus. Finally, the same electronics' output is used by the drag-free system of the satellite.

Performance analyses predict a noise Amplitude Spectral Density of 10^{-12} m/s^2/Hz$^{1/2}$ in the frequency band $10^{-3} - 0.03$ Hz, compatible with a test of the WEP at a 10^{-15} accuracy (see [11] for a detailed uncertainty analysis).

The spacecraft is derived from the CNES's Myriad series of microsatellites. With a mass of 325 kg and dimensions of 1380x1040x1580 mm^3, it has been designed to be as symmetric as possible, with the T-SAGE instrument sitting at its center-of-mass. No moving mechanical parts can contaminate the Equivalence Principle measurement. Its propulsion system, based on cold gas, is derived from the GAIA mission and supplied by ESA. It will be adapted and integrated by CNES on the satellite. The Equivalence Principle test will take place on a Sun-synchronous, very low eccentricity, 720 km orbit.

4. MICROSCOPE status

4.1. Instrument and satellite

As of August 2014, the T-SAGE flight model is assembled. Fig. 3 shows its Sensor Units, FEEU and ICU.

Figure 3. MICROSCOPE's Flight Model instrument. From left to right: Sensor Units, FEEU, ICU.

T-SAGE's Sensor Units have successfully undergone commissioning. In particular, we checked the metrology and verified that the test masses are well aligned. For the EP Sensor Unit, we measured an off-centering of the two test-masses' center of mass of 30.8 μm, under the required 34.6 μm. Although we found it slightly bigger than the specification for the REF Sensor Unit (35.7 μm vs 34.6 μm), we took it into account to update the error budget and found no appreciable impact on the WEP measurement. All environmental tests (thermal, vibrational) were successfully passed.

Extensive free-fall tests have been performed at the ZARM drop tower in Bremen (see [7] for their description). In the micro-gravity conditions afforded by those tests, we were able to check that the instrument performs as expected.

T-SAGE's electronics is currently undergoing commissioning. The instrument will be ready for delivery to CNES, where it will be integrated in the satellite, next fall.

4.2. Science Mission Center

MICROSCOPE's ground segment operations are centered around three entities: the Command Control Center (CCC) and the Center of Expertise for Compensation of Drag (CECT), hosted by CNES-Toulouse, are in charge of the satellite's operations and drag-free and attitude control respectively, and the Science Mission Center (CMSM), hosted by ONERA-Palaiseau, is in charge of the monitoring and operations of MICROSCOPE's instrument, and of the data processing and analysis. In particular, the CMSM is responsible for (1) ensuring all operational functions to maximize the instrument's operation, (2) day-to-day instrument management and monitoring, (3) weekly mission performance check, (4) proposing modifications of the mission scenario to the Science Working Group, and (5) releasing and archiving the data.

As of August 2014, we are developing the instrument's monitoring and data processing software. Compatibility tests with CNES (that provides preprocessed data) are planned for the fall and winter 2014.

4.3. Data analysis
As mentioned above (see also Fig. 2), a WEP violation signal will be detected as a sine wave corresponding to the modulation of the difference in accelerations experienced by the instrument's test-masses as the satellite orbits the Earth. The main task of the data analysis is therefore to extract this sine wave from the instrumental noise and to estimate its amplitude. Of foremost importance, we expect a small amount of data to be lost (e.g. because short mechanical crackles will saturate the instrument); as a consequence, the noise will leak from the high frequencies (where its Power Spectral Density increases) to the lower frequencies where the WEP violation signal is measured, therefore hampering the signal detection and estimation. Methods to deal with this problem are currently being implemented: Baghi *et al* [1] developed the KARMA (Kalman Auto-Regressive Model Analysis) technique, and Bergé *et al* [2] adapt an inpainting technique (see e.g. [10]) to the MICROSCOPE data analysis case.

Conclusion
MICROSCOPE will test the Weak Equivalence Principle in space down to an accuracy of 10^{-15}. This is two orders of magnitude better than the current constraints, and could allow us to rule out new theories that predict a WEP violation around 10^{-14}, or to complete General Relativity if a WEP violation is detected. Beside this science goal, MICROSCOPE will fulfill a technology objective by showing that the technology is ready for extremely fine satellite attitude control and precise drag-free system. Together with the results of LISA Pathfinder, this will be of interest for future ambitious missions like eLISA. As the expected launch date is coming, MICROSCOPE's instrument is fully integrated and is close to the end of its commissioning. It will be delivered in the fall 2014 to CNES for integration in the satellite, for a launch on time in 2016. The first scientific results should be available a few months after launch, while the mission is planned to last two years including partial eclipses when the scientific data are not available.

Acknowledgments
The authors are grateful to the MICROSCOPE mission staff of CNES in Toulouse, the Observatoire de la Côte d'Azur in Grasse and ONERA in Châtillon, for all exchanges about the mission, the satellite, the payload and the science mission center. This work was supported primarily by ONERA, where the T-SAGE instrument has been developped under a contract with CNES. MICROSCOPE is a CNES mission with ESA cooperation.

References
[1] Baghi, Q., Métris, G., Bergé, J., Christophe, B., Touboul, P., & Rodrigues, M., Phys. Rev. D. submitted
[2] Bergé, J. et al, in prep
[3] Bonnin, A., Zahzam, N., Bidel, Y., & Bresson, A. 2013, Phys. Rev. A, 88, 043615
[4] Damour, T. 2012, Classical and Quantum Gravity, 29, 184001
[5] Hardy, É., Levy, A., Métris, G., Rodrigues, M., & Touboul, P. 2013, Space Sci. Rev., 180, 177
[6] Hardy, É., Levy, A., Rodrigues, M., Touboul, P., & Métris, G. 2013, Advances in Space Research, 52, 1634
[7] Liorzou, F., Boulanger, D., Rodrigues, M., Touboul, P., & Selig, H. 2014, Advances in Space Research, 54, 1119
[8] Murphy, T. W., Jr., Adelberger, E. G., Battat, J. B. R., et al. 2012, Classical and Quantum Gravity, 29, 184005
[9] Overduin, J., Everitt, F., Worden, P., & Mester, J. 2012, Classical and Quantum Gravity, 29, 184012
[10] Pires, S., Starck, J.-L., Amara, A., et al. 2009, MNRAS, 395, 1265
[11] Touboul P. 2009, Space Sci. Rev., 148, 455
[12] Touboul P., Métris G., Lebat V., Robert A. 2012, Classical and Quantum Gravity, 29, 184010
[13] Wagner, T. A., Schlamminger, S., Gundlach, J. H., & Adelberger, E. G. 2012, Classical and Quantum Gravity, 29, 184002
[14] Will, C. M. 2006, Living Reviews in Relativity, 9, 3
[15] Williams, J. G., Turyshev, S. G., & Boggs, D. H. 2012, Classical and Quantum Gravity, 29, 184004

Journal of Physics: Conference Series **610** (2015) 012010 doi:10.1088/1742-6596/610/1/012010

GRACE-Follow On Laser Ranging Interferometer: German contribution

Klaus Abich[1], Christina Bogan[2], Claus Braxmaier[1,3], Karsten Danzmann[2], Marina Dehne[4], Martin Gohlke[1], Alexander Görth[2], Gerhard Heinzel[2], Mark Herding[4], Christoph Mahrdt[2], Vitali Müller[2], Kolja Nicklaus[4], Josep Sanjuan[1], Daniel Schütze[2], Benjamin Sheard[2], Gunnar Stede[2] and Kai Voss[4]

[1] Deutsches Zentrum für Luft- und Raumfahrt, Institute of Space Systems (DLR-RY), Robert-Hooke-Str. 7, 28359 Bremen, Germany

[2] Max-Planck-Institut für Gravitationsphysik (Albert-Einstein-Institut, AEI), Callinstrasse 38, 30419 Hannover, Germany

[3] Center of Applied Space technology and Microgravity (ZARM), University of Bremen, Am Fallturm, 28359 Bremen, Germany

[4] SpaceTech GmbH (STI), Seelbachstrasse 13, 88090 Immenstaad, Germany

E-mail: jose.sanjuan@dlr.de

Abstract. The Gravity Recovery and Climate Experiment (GRACE) is a joint US/German mission that has been mapping the Earth's gravity field since 2002 by measuring the distance variations between two spacecraft using a micro-wave link. GRACE is reaching the end of its lifetime. For this reason and in order to minimize data gaps, an almost identical mission will be launched in 2017. This mission is called GRACE-Follow On (GRACE-FO) and it will include an additional instrument as a technological demonstrator to monitor distance changes between the spacecraft. This instrument is the Laser Ranging Interferometer (LRI), which is based on heterodyne laser interferometry at 1064 nm and takes advantage of many technologies developed for LISA. In this paper a short overview of the current status of the German contribution is presented.

1. Introduction

The Gravity Recovery and Climate Experiment (GRACE) mission has been measuring changes in the gravity field of the Earth since 2002 and has allowed the quantification of mass variations, such as the mass loss of polar ice sheets, the contribution of mass influx to sea level rise, and changes in the hydrological cycle [1]. GRACE consists of two spacecraft (SC) in circular polar orbits at 400 km altitude. The distance between the two SC is kept between 170 km and 270 km and a two-way microwave link (K/Ka band ranging) is used to measure the distance changes between them at the micro-meter level [2]. This signal, together with the on-board accelerometer data [3] and GPS signals [4], is used to reconstruct the gravity field of the Earth. GRACE was designed to be a 5-year-lifetime mission but subsequently was extended until the end of its orbit (in 2015 or 2016). Continuous and long-term monitoring of the spatial and temporal variations of the gravity field of the Earth is of paramount importance for understanding the processes and physics behind the observed changes. Therefore, to minimize the data gap between GRACE and future gravity field missions an almost exact replicate called GRACE-Follow On (GRACE-FO) will be launched in 2017 by the joint collaboration between the US (NASA) and Germany (DLR and GFZ).

GRACE-FO offers a unique opportunity to accommodate a second ranging instrument (Laser Ranging Interferometer, LRI) based on laser interferometry as a technological demonstrator, which should surpass the microwave instrument performance due to its smaller wavelength; and be the first intersatellite interferometric mission in space —see figure 1 [5, 6]. The LRI measurement principle is based on the LISA interferometry concept: a master SC and a slave SC with its laser offset phase-locked to the beam coming from the master SC. The beam is sent back to the master SC where phase changes, which correspond to distance changes between the SC, are measured by interfering the local SC beam with the incoming beam from the slave SC. The LRI ranging noise goal is $\approx 80\,\mathrm{nm\,Hz^{-1/2}}$ for $f \geq 10\,\mathrm{mHz}$ and relaxing as f^{-2} towards lower frequencies.

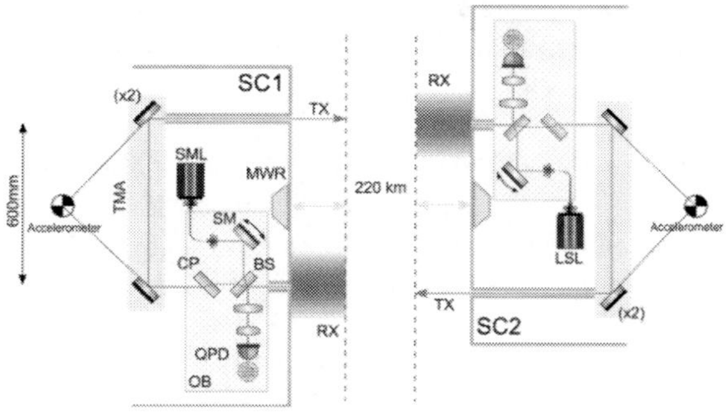

Figure 1. LRI concept. A detailed explanation can be found in [5]. SC1 sends light to SC2, which sends the light back to SC1 after being offset phase-locked to the incoming laser light from the distant SC. The change in the phase of the light is proportional to the SC distance change. The phase signal is measured by interfering the light from SC1 with the light coming from SC2. The steering mirrors correct the pointing of the beams using also the interferometric signal. SML: Stabilized Master Laser. LSL: Locked Slave Laser. SM: steering mirror. BS: Beam-splitter. QPD: Quadrant-photo Detector. CP: Compensation Plate. OB: Optical Bench. RX: Received Beam. TX: Transmitted Beam. MWR: Microwave Ranging. Cold gas tanks are not shown.

The German (AEI/GFZ/STI/DLR/Airbus) role in the LRI focuses on the optics, the triple mirror assembly, the beam steering mechanism and the photo-receivers. The US (JPL/Ball Aerospace/Tesat GmbH) side is responsible of the frequency stabilization system, phasemeter and payload processing electronics, and laser source [7]. Australia (ANU/EOS/CSIRO) contributes with triple mirror assembly prototyping and investigations on the link acquisition [8]. In this paper a short overview of the status of the German contribution is presented. In section 2 the Triple Mirror Assembly (TMA) is described. Section 3 focuses on the different subsystems within the so-called Optical Bench Assembly (OBA). Section 4 summarizes the German contribution status and future prospects.

2. Triple Mirror Assembly

The TMA is a large lateral offset (600 mm) retro-reflector composed of three orthogonal mirrors [9, 10] and is a key element for the LRI since: (i) it routes the beam around the micro-wave ranging instrument and the cold gas tanks; (ii) it keeps the incoming and outgoing beams anti-parallel and (iii) it allows to place its virtual vertex in the SC accelerometer reference point, and thus minimize SC jitter to path-length coupling, which is crucial due to the GRACE-FO low Earth orbit. The SC attitude jitter can be rather large: around $100\,\mu\text{rad}\,\text{Hz}^{-1/2}$ in pitch and yaw and about $1\,\text{mrad}\,\text{Hz}^{-1/2}$ in roll. The residual jitter couples due to misplacement of the TMA vertex on the accelerometer reference point (assuming this coincides with the SC center of mass) and non-idealities of the TMA itself, i.e., co-alignment[1] imperfections [11]. While the SC attitude stability is good enough for the microwave ranging, it is not sufficient for the laser interferometer due to its narrower receiving field of view and smaller beam divergence. For this reason, a steering mirror together with Differential Wavefront Sensing (DWS) [12] is used to steer the laser beam: the local SC measures the tilt between the local beam and the received beam from the distant SC and sends a signal to the steering mirror, which corrects the local beam pointing accordingly —see section 3. The local beam is then sent to the far SC after passing through the TMA. The TMA coalignment also plays an important role in the beam pointing because an error is added only in the transmit beam path and cannot be corrected by the beam steering mechanism. Consequently, poor TMA coalignment (above $\sim 100\,\mu\text{rad}$) can jeopardize the beam pointing control loop, due to the drop in power at the receiver due to the relative angle between the transmitted beam and the line-of-sight between the spacecraft. The received power for a relative angle of $100\,\mu\text{rad}$ is around 30% of the transmitted one. For $200\,\mu\text{rad}$ drops to $\sim 1\%$.

A Carbon Fibre Reinforced Polymer (CFRP) hollow tube (600 mm long and 39 mm in diameter) is used as the TMA mechanical frame due to its low coefficient of thermal expansion and its good strength/stiffness-mass ratio compared to other brittle materials such as Zerodur or Silicon Carbide. At the ends of the CFRP frame two Zerodur mirror assemblies (Zerodur pieces at both ends of the CFRP tube —see figure 2) were wrung and glued. The input Zerodur assembly holds one mirror while the output one holds the other two mirrors. The interface to the SC is a bracket made of CFRP. The TMA design (frame and mirrors), the materials processing, qualification and manufacturing of the CFRP frame has been carried out by STI, while the Zerodur mirror assemblies have been manufactured by Cassidian Optronics.

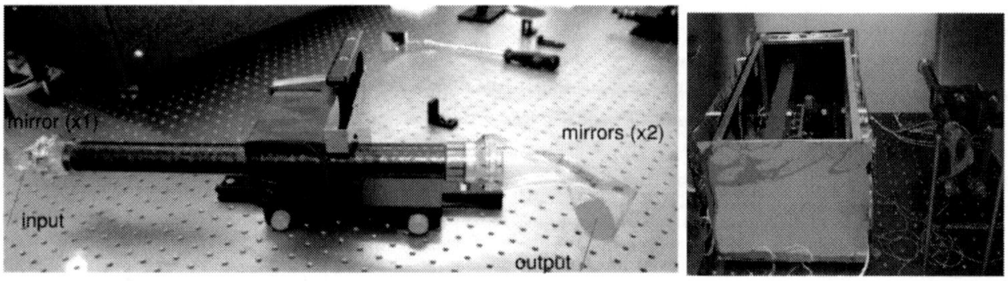

Figure 2. Left: QM TMA. Right: QM TMA co-alignment test campaign at STI facilities. The co-alignment is measured with the help of a Zerodur reference flat and an autocollimator (not shown).

The DM (Demonstrator Model) and QM (Qualification Model) have gone under performance (right panel on figure 2) and environmental tests in late 2013-early 2014. The TMA Critical Design Review (CDR) was passed successfully in March 2014. TMA Flight Models (FM) will go under performance and environmental tests in late 2014 and early 2015.

[1] Co-alignment is defined as the deviation from perfect anti-parallelism between the input and output beams of the TMA.

3. Optical Bench Assembly

The OBA includes all the necessary subsystems to

- launch the local beam onto the quadrant photo-receiver (QPR) and to the distant SC,
- provide the opto-electrical interfaces (photo-receivers and steering mirrors and their electronics) to the US-supplied Laser Ranging Processor (LRP), which ultimately
  - measures the phase difference change between the local beam and the incoming beam from the distant SC, which corresponds to the change in the relative distance between the SC,
  - measures the relative angle between the local beam and the incoming beam from the distant SC by means of DWS,
  - steers the local beam to align it to the beam coming from the far SC by zeroing the DWS signal since the SC pointing control is not sufficient to keep the heterodyne signal locked.

The OBA is shown in figures 1 and 3: the light from the laser is delivered to the OBA by means of a fibre injector that creates a collimated 5 mm beam in diameter. The output of the fibre is routed via the steering mirror, which is able to tip and tilt it. This is necessary to keep the beams from both SC aligned and maintain the heterodyne efficiency to operational levels. A tiny fraction of the light is transmitted through a beam splitter while most of it ($\approx$20 mW) is reflected and directed to the TMA and sent out to the far SC. Before the beam reaches the TMA it is guided through a compensation plate, which is used to reduce linear attitude jitter-pathlength coupling [5]. The transmitted part of the light ($\approx$2 mW) is directed to redundant QPRs. An imaging system is used to image both the local beam and the received beam, which is mostly reflected by the beam-splitter into the QPR aperture. The local beam is used as the local oscillator that generates the beat note when interfering with the beam coming from the far SC ($\approx$100 pW). The beat note heterodyne frequency is in the 4-16 MHz range[2] and is detected by a LISA-like phasemeter in the LRP [13] that also measures the relative angle between them using the DWS technique. This information is used to drive the steering mirror and keep both beams co-aligned. The subsystems needed to perform all these tasks are:

- Fibre injector assembly (FIA): it is a fully monolithic design developed by STI, with the fibre directly spliced to the aspheric lens body, thus allowing an extremely high thermal stability (both for pointing and wavefront planarity) while minimizing ghost reflections. The wavefront planarity error is less than $\lambda/15$ over the full operation temperature range.
- Quadrant photoreceivers: they are 1 mm in diameter InGaAs quadrant photo-diodes with suitable electronics to detect beat notes in the frequency range from 4 MHz to 16 MHz. The photo-receiver equivalent input current noise is $5\,\mathrm{pA\,Hz^{-1/2}}$ and has a responsivity around $0.6\,\mathrm{A\,W^{-1}}$. Two QPRs are present in each OBA for redundancy.
- Fast steering mirror: it has a large optical angle dynamic range ($\pm$8.1 mrad) in conjunction with a large bandwidth: $\pm$3 mrad at 150 Hz (equivalent to a speed of about $2\,\mathrm{rad\,s^{-1}}$). Both are especially needed for the initial acquisition procedure [14] where an uncertainty space of $\pm$3 mrad has to be quickly scanned. Such requirements are met by using voice coil actuators together with eddy current sensors (Kaman) in a closed-loop configuration.
- Imaging optics, beam-splitter and compensation plate.

The OBA CDR was successfully passed in May 2014. In early 2015 the FM OBA will be delivered and tested.

[2] Doppler shifts due to relative velocity of the SC are about $\pm$3 MHz.

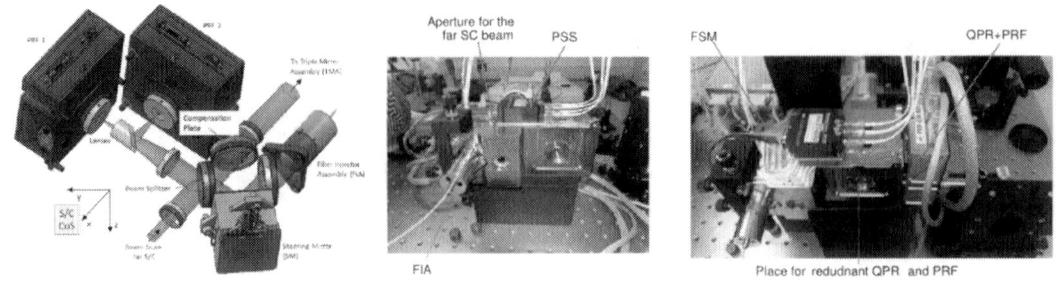

Figure 3. Left: OBA drawing. Middle and right: QM OBA during the last test campaign in 2014 at STI facilities. FSM: Fast Steering Mirror. PSS: Position Sensing System (eddy current sensors). QPR: Quadrant-photo-receiver. PRF: Photo-receiver Front-End.

4. Conclusion

The status of the German contribution for the LRI has been presented. All the systems have passed CDR in 2014 and the flight models will be delivered by the end of 2014 and beginning of 2015. The LRI performance will be about one order of magnitude better than the current GRACE microwave ranging and it will also provide SC attitude information via the DWS and the fast steering mirror Kaman sensors. The LRI is intended to pave the road for the next generation of gravity field missions based on intersatellite ranging. Moreover, it will also provide very valuable information for LISA since they have many elements in common, i.e., the frequency band (milli-Hertz), the received power by the far SC (pico-Watt) and the Doppler shifts (mega-Hertz).

References

[1] Tapley B D, Bettadpur S, Thompson P F, and Watkins M M 2004 *Sci.* **305** 503-505
[2] Davis E S, Dunn C E, Stanton R H and Thomas J B 1999 *50th Intern. Astronautical Congress*, The Netherlands
[3] Rodrigues M, Foulon B, Liorzou F and Touboul P 2003 *Class. Quantum Grav.* **20** S291-S300
[4] Dunn C, Bertiger W, Franklin G, Harris I, Kruizinga G, Meehan T, Nandi S, Nguyen D, Rogstad T, Thomas J B and Tien J 2002 *Proceedings of the 15th International Technical Meeting of the Satellite Division of The Institute of Navigation* (ION GPS), Portland, Oregon, USA
[5] Sheard B, Heinzel G, Danzmann K, Shaddock D A, Klipstein W and Folkner W M 2012 *J. Geod.* **86**, 1083
[6] Heinzel G, Sheard B, Brause N, Danzmann K, Dehne M, Gerberding O, Mahrdt C, Müller M, Schütze D, Stede G, Klipstein W, Folkner W, Spero R, Nicklaus K, Gath P and Shaddock D 2012. *International Conference on Space Optics* (ICSO), Corse, France
[7] Thompson R, Folkner W M, de Vine G, Klipstein W M, McKenzie K, Spero R, Yu N, Stephens M, Leitch, J, Pierce R, Lam T T-Y, Shaddock D A 2011, *Freq. Control and the European Freq. and Time Forum (FCS)*
[8] Wuchenich D M R. Inter-satellite laser interferometry. Ph.D dissertation. Australian National University, 2014
[9] Ward R L, Fleddermann R, Francis S, Mow-Lowry C, Wuchenich D, Elliot M, Gilles F, Herding M, Nicklaus K, Brown J, Burke J, Dligatch S, Farrant D, Green K, Seckold J, Blundell M, Brister R, Smith C, Danzmann K, Heinzel G, Schütze D, Sheard B S, Klipstein W, McClelland D E, and Shaddock D A, 2014 *Class. Quantum Grav.* **31**, 095015.
[10] Schütze D, Farrant D, Shaddock D A, Sheard B S, Heinzel G and Danzmann K 2014, *Rev. Sci. Instrum.* **85**
[11] Müller V, Simulations for LISA and GRACE-Follow-On. Master of Science Degree Thesis, Max-Planck-Institute for Gravitational Physics (Albert Einstein Institute, AEI) and Gottfried Wilhelm Leibniz Universtität Hannover, Germany (2013)
[12] Heinzel G, Wand V, García A, Jennrich O, Braxmaier C, Robertson D, Middleton K, Hoyland D, Rüdiger A, Schilling R, Johann U and Danzmann K 2004 *Class. Quantum Grav.* **21** S581
[13] Shaddock D, Ware B, Halverson P G, Spero R E, Klipstein W 2006 *AIP Conf.Proc.* **873** 654-660
[14] Mahrdt C, Laser link acquisition for the GRACE Follow-on laser ranging interferometer, Ph.D. dissertation. Hannover, 2013.

10th International LISA Symposium (LISAX) IOP Publishing
Journal of Physics: Conference Series **610** (2015) 012011 doi:10.1088/1742-6596/610/1/012011

Descope of the ALIA mission

Xuefei Gong[1], Yun-Kau Lau[1,2,3,4], Shengnian Xu[1], Pau Amaro-Seoane[5], Shan Bai[6], Xing Bian[1,7], Zhoujian Cao[1,3,4], Gerui Chen[8], Xian Chen[5], Yanwei Ding[9], Peng Dong[1], Wei Gao[1,7], Gerhard Heinzel[10], Ming Li[9], Shuo Li[11], Fukun Liu[12,13], Ziren Luo[14], Mingxue Shao[2], Rainer Spurzem[11,13,15], Baosan Sun[16], Wenlin Tang[1,7], Yan Wang[10], Peng Xu[2,17], Pin Yu[18], Yefei Yuan[19], Xiaomin Zhang[8] and Zebing Zhou[16]

[1] Institute of Applied Mathematics, Academy of Mathematics and Systems Science, Chinese Academy of Sciences, No.55 Zhongguancun Donglu, Beijing 100190, China
[2] Morningside Center of Mathematics, Chinese Academy of Sciences, No.55 Zhongguancun Donglu, Beijing 100190, China
[3] State Key Laboratory of Scientific and Engineering Computing, Academy of Mathematics and Systems Science, Chinese Academy of Sciences, No.55 Zhongguancun Donglu, Beijing 100190, China
[4] State Key Laboratory of Theoretical PhysicsInstitute of Theoretical Physics, Chinese Academy of Sciences, No.55 Zhongguancun Donglu, Beijing 100190, China
[5] Max-Planck-Institut für Gravitationsphysik (Albert Einstein Institut), Am Mfühlenberg 1, D-14476 Potsdam, Germany
[6] Theoretisch-Physikalisches Institut, Friedrich-Schiller-Universität Jena, Fröbelstieg 1, D-07743 Jena, Germany
[7] University of Chinese Academy of Sciences, No.19A Yuquan Road, Beijing 100049, China
[8] College of Applied Sciences, Beijing University of Technology, 100 Pingleyuan, Beijing 100124, China
[9] Dongfanghong Small Satellite Company, F12 Shenzhou Building, No.31 Zhongguancun Nandajie, Beijing 100081, China
[10] Max-Planck-Institut für Gravitationsphysik (Albert Einstein Institut), Callinstraße 38, D-30167 Hannover, Germany
[11] National Astronomical Observatories, Chinese Academy of Sciences, 20A Datun Road, Beijing 100012, China
[12] Department of Astronomy, School of Physics, Peking University, No.5 Yiheyuan Road, Beijing 100871, China
[13] The Kavli Institute for Astronomy and Astrophysics, Peking University, 5 Yiheyuan Road, Beijing 100871, China
[14] Institute of Mechanics, Chinese Academy of Sciences, 15 Beisihuanxi Road, Beijing 100190, China
[15] Astronomisches Rechen-Institut, Zentrum für Astronomie, Universität Heidelberg, Mönchhofstrasse 12-14, 69120 Heidelberg, Germany
[16] School of Physics, Huazhong University of Science and Technology, 1037 Luoyu Road, Wuhan 430074, China; Center for Gravitational Experiments, Huazhong University of Science and Technology, 1037 Luoyu Road, Wuhan 430074, China
[17] Institute of Theoretical Physics, Chinese Academy of Sciences, No.55 Zhongguancun Donglu, Beijing 100190, China
[18] Mathematical Science Center, Tsinghua University, Beijing 100084, China
[19] Center for Astrophysics, Department of Astronomy, University of Science and Technology of China, 96 Jinzhai Road, Hefei 230026, China

E-mail: lau@amss.ac.cn

Abstract. The present work reports on a feasibility study commissioned by the Chinese Academy of Sciences of China to explore various possible mission options to detect gravitational waves in space alternative to that of the eLISA/LISA mission concept. Based on the relative merits assigned to science and technological viability, a few representative mission options descoped from the ALIA mission are considered. A semi-analytic Monte Carlo simulation is carried out to understand the cosmic black hole merger histories and the possible scientific merits of the mission options in probing the light seed black holes and their coevolution with galaxies in early Universe. The study indicates that, by choosing the armlength of the interferometer to be three million kilometers and shifting the sensitivity floor to around one-hundredth Hz, together with a very moderate improvement on the position noise budget, there are certain mission options capable of exploring light seed, intermediate mass black hole binaries at high redshift that are not readily accessible to eLISA/LISA, and yet the technological requirements seem to within reach in the next few decades for China.

1. Introduction

The present work originated from a recently completed second phase of the feasibility study commissioned by National Space Science Center, Chinese Academy of Sciences of China to explore various possible mission options to detect gravitational waves in space alternative to that of (e)LISA mission concept. At the beginning of the study dating back to more than three years ago, the ALIA (Advanced Laser Interferometer Antenna) mission concept first proposed by Peter Bender [1, 2] was chosen as the starting point. It is conceivably in many ways the simplest adaptation of the LISA mission concept to a measurement band centered around a few hundredth Hz. A more detailed study [3] of the possible sciences of the mission further indicates that, apart from the known LISA sources, the mission also holds the promise in mapping out the mass and spin distribution of intermediate mass black holes (IMBHs) possibly present in dense star clusters at low redshift as well as in shedding important light on the structure formation in the early Universe. However, when the key technologies of the mission is further looked at, the sub-picometer interferometry requirement in the laser metrology part poses a major obstacle on the technological side of the mission. With a view that China will have a reasonable chance to realise the mission in the next few decades and to minimise possible risks in future R&D of the key technologies, the task of mission descope is inevitably forced upon us.

Upon further evaluation of the relative merits between science and technological viability of a few representative descope options descended from the ALIA mission, it emerges that there exists certain class of mission design that is technologically viable within the next few decades for China, and yet it contains significant science that goes beyond eLISA/LISA. The principal aim of this report is to outline this set of mission design parameters and briefly sketch the scientific case study.

The outline of the present work may be described as follows. In Section 2, we will state the mission design parameters, display the corresponding sensitivity curves and the detection ranges with respect to black hole binaries with various mass ratios. In Section 3, a semi-analytic Monte-Carlo simulation is carried out to understand the scientific potential of the prospective missions in probing the structure formation in early Universe.

2. Mission descope

After some careful considerations, the following set of baseline design parameters will be chosen for future study and development.

Table 1.

Armlength (m)	Telescope diameter (m)	Laser power(W)	1-way position noise ($\frac{\text{pm}}{\sqrt{\text{Hz}}}$)	Acceleration ($\frac{\text{m s}^{-2}}{\sqrt{\text{Hz}}}$)
3×10^9	0.45-0.6	2	5-8	3×10^{-15}(> 0.1mHz)
5×10^8 (ALIA)	1.0	30	0.1	3×10^{-16}(> 1mHz)
5×10^9 (LISA)	0.4	2	18	3×10^{-15}(> 0.1mHz)
1×10^9 (eLISA)	0.2	2	11	3×10^{-15}(> 0.1mHz)

For reference purpose, the baseline design parameters of ALIA, LISA/eLISA are also given in Table 1. The relevant sensitivity curves are displayed in Figure 1. Apart from the instrumental noises, confusion noise generated by both galactic and extra-galactic compact binaries are also taken into consideration. Relevant confusion levels are converted from estimations by Hils and Bender [4] and Farmer and Phinney [5].

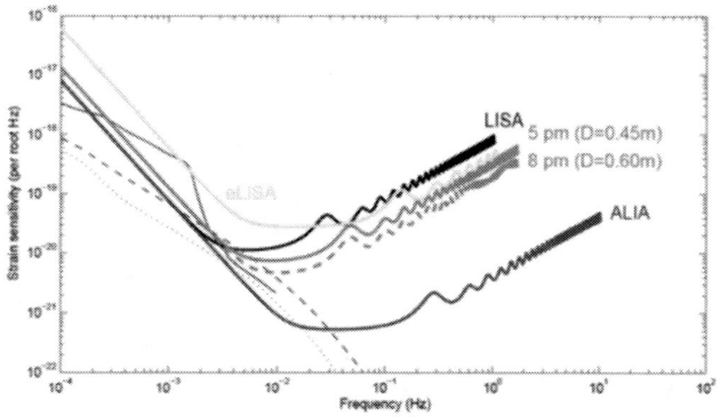

Figure 1.

For black hole binaries with mass ratio 1 : 4, typical of what one would expect from hierarchical black hole growth at high redshift, the all angle averaged detection range are plotted in Figure 2. Apart from galactic confusion noise [4], upper level (dashed curve) and lower level (dotted dashed curve) of confusion noise generated by extragalactic compact binaries as those estimated by [5] are also taken into account.

In calculating the averaged SNR, we have used hybrid waveforms in the frequency domain with black hole spin not taken into account [6, 7]. For one year of observation before merger, the contributions in SNR due to large spin is indeed negligible according to our calculations. Spin is relevant only in the parameter estimation stage, which will not be discussed in the present work. As may be seen from Figure 2, for a given redshift, the proposed mission concept is capable of detecting lighter black hole binaries in comparsion with eLISA/LISA and thereby provides better understanding of the hierarchical assembling process in early Universe.

Apart from IMBH binaries at high redshift, the designed sensitivity at around 0.01Hz measurement band means that the instrument is also capable of detecting IMRIs (Intermediate Mass Ratio Inspirals) harboured at globular clusters or dense young star clusters at low redshift

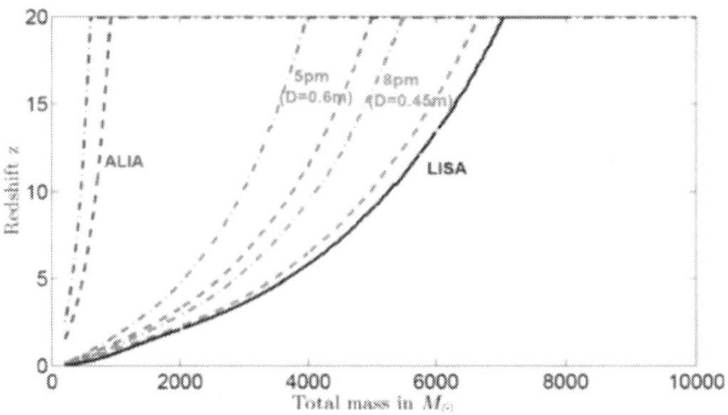

Figure 2. All-angle averaged detection range under a single Michelson threshold SNR of 7 for 1:4 mass ratio IMBH-IMBH binaries, one year observation prior to merger. For each mission option, both upper and lower confusion noise levels (represented by the dashed curve and dotted dashed curve respectively) due to extragalactic compact binaries are considered.

($z < 0.6$). See [8] for a further discussion of the capture dynamics of an IMRI in dense star clusters. Displayed in Fig 3 are the detection ranges of IMRIs with different mass ratios one year prior to merger. The stellar black hole inspiral into an IMBH is fixed to be $10M_\odot$, while the mass of the IMBH is subject to variation to generate different mass ratios in the figure.

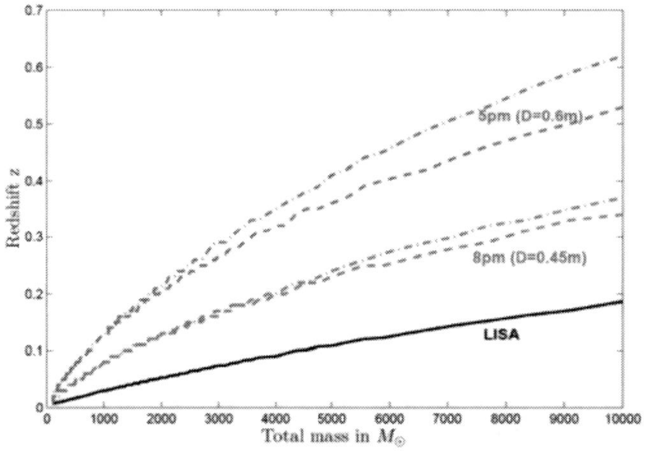

Figure 3. All-angle averaged detection range under a single Michelson threshold SNR of 7 for a stellar mass black hole spiralling into IMBHs with reduced masses of $10M_\odot$, one year observation prior to merger. For each mission option, both upper and lower confusion noise levels (represented by the dashed curve and dotted dashed curve respectively) due to extragalactic compact binaries are considered.

10th International LISA Symposium (LISAX) IOP Publishing
Journal of Physics: Conference Series **610** (2015) 012011 doi:10.1088/1742-6596/610/1/012011

3. Scientific case study

The primary science driver of the ALIA mission is to make direct detection of IMBH binaries at high redshift descended from the heavy Pop III stars [3, 9, 10] and thereby provide insight into the physics of the structural formation processes at early Universe. It would be of interest to find out to what extent this science objective is compromised in the descoped mission. For this purpose, we carry out a simple Monte Carlo simulation of black hole merger histories based on the realizations of EPS formalism and semi-analytical dynamics, in accordance with the prescription given in [11, 12, 13].

Pop III remnant black holes of $150M_\odot$ are placed in 3.5σ biased halos at z=20 with initial spins of the seeds generated randomly. By prescribing VHM-type dynamics [11, 12], we trace downwards the black hole merging history. The halo mass ratio criteria for major merger is set to be > 0.1. Both the prolonged accretion and the chaotic accretion scenario are considered. Black hole spins coherently evolve through both mergers and accretions processes and their magnitudes influence strongly the mass-to-energy conversion efficiency. We assume efficient gaseous alignment of the black holes so that the hardening time is short and only moderate gravitational radiation recoils take place. Numerical simulations [14] suggest that the hardening and merging times scales remain short even in gas free environment. In calculations relevant to GW observations, we assume a threshold SNR of 7 for detection in the sense of single Michelson interferometer and one year observation prior to merger. The results are schematically given in Figure 4 and Figure 5.

We assess our simulations by fitting the black hole mass functions and luminosity functions at six almost equally divided successive redshift intervals ranging from z=0.4 to 2.1. In the prolonged accretion scenario, the results deviate from the observational constrains given by Soltan type argument when going up to redshift $z > 1.5$. It may therefore underestimate the black holes growth rate and perhaps also the coalescence rate. Observationally the existence of very high redshift ($z > 6$) AGNs implies that feed back mechanisms may be very different at early epoch so that fast growth of the seed black hole could be possible.

In terms of coalescence rate, our result displayed in Figure 4 is in overall agreement with the results given by Sesana et al [13, 15] and Arun et al [16], though the coalescence counts given by their simulations are about two or three times higher. It is likely due to various numerical discrepancies in the simulations. Overall, our black hole mass growth is slower, particularly in the prolonged accretion scenario. At $z = 15$, the total mass of the black hole binaries typically are still less than $600M_\odot$ in the prolonged model and this may lead to a smaller counts in detectable sources. We expect our results yields a very conservative (pessimistic) estimate of black hole binaries merger event rate.

The astrophysics encapsulated in our simulation represents the state of art understanding of structural formation after the dark age. Due to our poor understanding of the evolution of the Universe at this epoch, it is likely that the simulation overlooks many details of the physical processes involved. The event rate count should be looked upon in a cautious way. Instead of reading into the precise numbers, it serves as an indication what spaceborne gravitational wave detector is capable of and in our case, the advantage of setting the most sensitive regime of the measurement band from a few mHz to 0.01Hz.

Globular cluster harbored IMRIs

Before we conclude, let us also briefly estimate the detection capability for IMRIs in dense star clusters. The calculations and underlying hypotheses are identical to that in [3](see also [17, 18]). The results are given in Table 2.

The above event rate estimate is subject to many uncertainties and perhaps we should not attach too much importance to the precise numbers. Instead, the calculations serves as an indication of the detection potential of the mission concept as far as IMRIs at low redshift are

*** Formatting Issue - Best Available Paper/Graphic ***

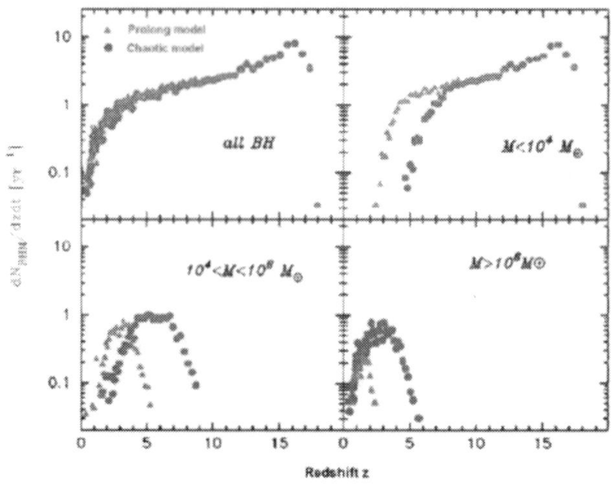

Figure 4. Coalescence rate predicted by the realized simulations.

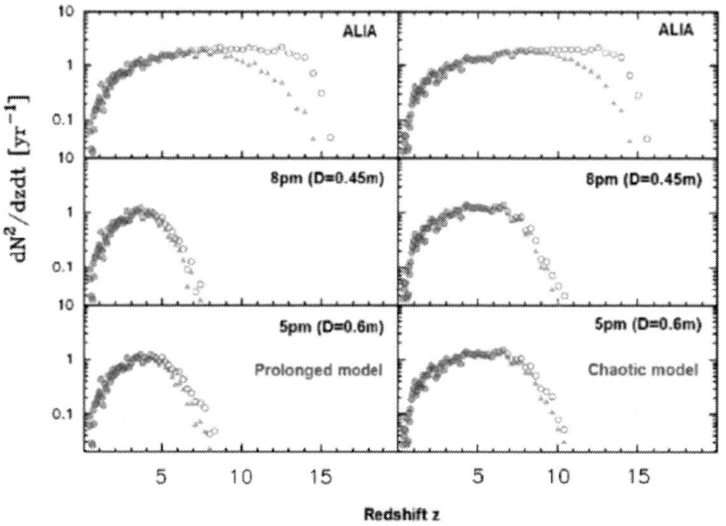

Figure 5. Event rates for descoped options.

concerned. Further, as event rate goes up as the cubic of the improvement in sensitivity, it also brings out the advantage of shifting slightly the most sensitive region of the measurement band to a few hundredth Hz, as far as detection of IMRIs is concerned. It should also be remarked that collision of dense star clusters [19] constitutes a possible IMBH gravitational wave sources, while the inspiral of massive black holes ($\sim 10^3 M_\odot$ to $\sim 10^4 M_\odot$) into the supermassive black hole at the center of a galaxy is also a promising IMRI source [20, 21]. However, the corresponding event rates would be difficult to estimate.

*** Formatting Issue - Best Available Paper/Graphic ***

Table 2. Prospective detection rate for IMRIs in globular clusters.

Mission option	Upper level of confusion	Lower level of confusion
ALIA	$z_c = 5$: ~ 8000 $z_c = 3$: ~ 6000	~ 12000 ~ 7000
5pm (D=0.6m)	~ 90	~ 130
8pm (D=0.45m)	~ 26	~ 32
LISA	~ 3	

4. Concluding remarks

With the second phase of the feasibility study of gravitational wave detection in space coming to a close, a preliminary mission design deemed suitable as blue print for future development of the project in the Chinese Academy of Sciences is put forward. Together with the roadmap to advance the project step by step, the mission design will serve as a guide for future developments on both the theoretical as well as technological fronts. We hope to report upon further progress of the project on various areas soon.

Acknowledgments

The present work is part of the report submitted to the National Space Science Center, Chinese Academy of Sciences for the project entitled "Feasibility study of gravitational wave detection in space" (project number XDA04070400). Prof. Wenrui Hu's relentless effort to promote gravitational wave detection in space in China motivated the feasibility study in the first place. We are also grateful to Prof. Shuangnan Zhang for his helpful suggestions and advices on many occasions throughout the course of our work. As always, Prof. Peter Bender generously shares with us his expertise and insights in this subject. Professors Shing-Tung Yau and Lo Yang have been very supportive and the Morningside Center of Mathematics provides a very conductive research environment to carry out the study. Partial support from the NSFC (contract numbers 11305255 and 11171329) and the Marine Public Welfare Project of China (contract number 201105032) are acknowledged. RS is supported by Chinese Academy of Sciences Grant Number $2009S1-5$ (RS), and through the "Thousand Talents" (Qianren) program of China.

References

[1] Bender P L 2004 *Class. Quantum Grav.* **21** S1203
[2] Bender P L and Begelman M C "Trends in Space Science and Cosmic Vision 2020" 2005 ESA SP-588 (Noordwijk:ESA Publications Division) pp 33-8
[3] Gong X *et al* 2011 *Class. Quantum Grav.* **27** 084010
[4] Bender P L and Hils D 1997 *Class. Quantum Grav.* **14** 1439
[5] Farmer A J and Phinney E S 2003 *Mon. Not. R. Astron. Soc.* **346** 1197
[6] Ajith P *et al* 2008 *Phys. Rev.* D **77** 104017
[7] Ajith P *et al* 2011 *Phys. Rev. Lett.* **106** 241101
[8] Konstantinidis S, Amaro-Seoane P and Kokkotas K D 2013 *Astron. and Astrophys.* **557** A135
[9] Madau P and Rees M J 2001 *Astrophys. J.* **551** L27
[10] Heger A and Woosley S E 2002 *Astrophys. J.* **567** 532
[11] Volonteri M, Haardt F and Madau P 2003 *Astrophys. J.* **582** 559
[12] Volonteri M, Haardt F and Madau P 2003 *Astrophys. J.* **593** 661
[13] Sesana A, Volonteri M and Haardt F 2007 *Mon. Not. R. Astro. Soc.* **377** 1711
[14] Khan F M, Holley-Bockelmann K, Berczik P and Just A 2013 *Astrophys. J.* **773** 100
[15] Sesana A, Gair J, Berti E and Volonteri M 2011 *Phys. Rev.* D **83** 044036

[16] Arun K G *et al* 2009 *Class. Quantum Grav.* **26** 094027
[17] Will C M 2004 *Astrophys. J.* **611** 1080
[18] Amaro-Seoane P and Santamaría L 2010 *Astrophys. J.* **722** 1197
[19] Amaro-Seoane P and Freitag M 2006 *Astrophys. J. Lett.* **653** L53
[20] Miller M C 2005 *Astrophys. J.* **618** 426
[21] Liu F K, Li S and Komossa S 2014 *Astrophys. J.* **786** 103
[22] Bender P L, Begelman M C and Gair J R 2013 *Class. Quantum Grav.* **30** 165017

Journal of Physics: Conference Series **610** (2015) 012012 doi:10.1088/1742-6596/610/1/012012

Status of advanced ground-based laser interferometers for gravitational-wave detection

K L Dooley[1] [1], T Akutsu[2], S Dwyer[3], P Puppo[4]

[1] Max-Planck-Institut für Gravitationsphysik (Albert-Einstein-Institut) und Leibniz Universität Hannover, Callinstr. 38, D-30167 Hannover, Germany.
[2] National Astronomical Observatory of Japan, 2-21-1 Osawa, Mitaka, Tokyo 181-8588
[3] LIGO Hanford Observatory, PO Box 159, Richland, WA 99352, USA
[4] INFN, Sezione di Roma, I-00185 Roma, Italy

E-mail: `kate.dooley@aei.mpg.de`

Abstract. Ground-based laser interferometers for gravitational-wave (GW) detection were first constructed starting 20 years ago and as of 2010 collection of several years' worth of science data at initial design sensitivities was completed. Upgrades to the initial detectors together with construction of brand new detectors are ongoing and feature advanced technologies to improve the sensitivity to GWs. This conference proceeding provides an overview of the common design features of ground-based laser interferometric GW detectors and establishes the context for the status updates of each of the four gravitational-wave detectors around the world: Advanced LIGO, Advanced Virgo, GEO 600 and KAGRA.

1. Introduction

This is an exciting time for the maturing field of gravitational wave (GW) physics. The network of ground-based laser interferometer GW detectors (GWD) is making rapid progress towards its goal of producing advanced instruments sensitive enough to make monthly GW detections in the 10 Hz to 10 kHz frequency range by the end of the decade [1]. Simultaneously, there has recently been a claim by the BICEP2 experiment that the measurement of the B-modes in the cosmic microwave background polarization can be the signature of the primordial GWs produced by inflation [2]. If confirmed, this would add to the original evidence for GWs from 1975 by Hulse and Taylor who observed that the rate of change of the orbital period of a binary star system precisely agrees with the predictions of GR [3, 4]. Moreover, the eLISA space-based gravitational wave detector [5] together with the promising pulsar timing technique [6] will be very important in the coming decades for the investigation of very massive objects and other GW sources in the milli- to micro-Hz frequency ranges, respectively.

GWs are dynamic strains in space-time that travel at the speed of light and are generated by non-axisymmetric acceleration of mass. They perturb the flat Minkowski metric describing space-time. The effect is the production of a dimensionless strain between two inertial masses located at a proper distance L from one another so that their distance changes as:

$$\Delta L(t) = \frac{1}{2}\, L\, h(t) \tag{1}$$

[1] Current address: The University of Mississippi, University, MS 38677, USA

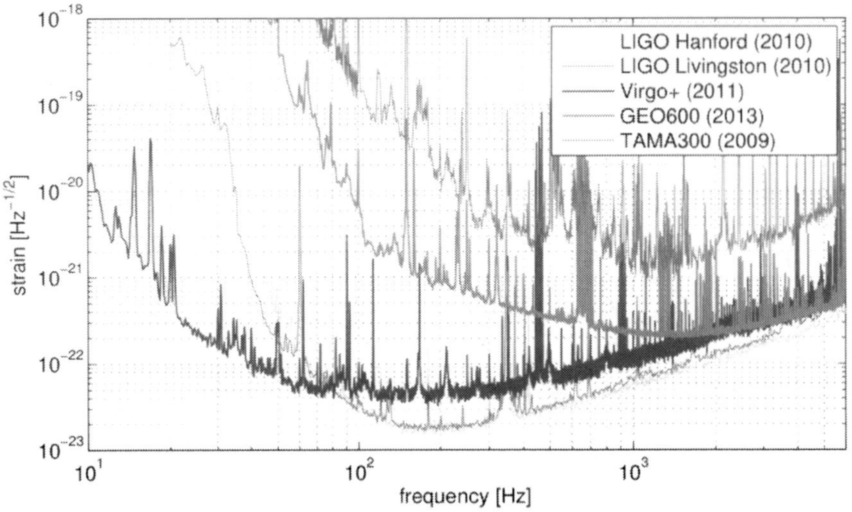

Figure 1. Strain sensitivities achieved by the first generation detectors.

In the late 20th century following the era of the operation of bar detectors for GW detection [7], large laser interferometers were identified as the promising route forward because of the very high strain sensitivities that could be achieved over a wide frequency band. R. Weiss produced the first detailed design study in 1972 of a large scale laser interferometer for GW detection, complete with calculations of fundamental noise sources [8]. Then, following the work of Forward [9] who built the first laser interferometric GW detector prototype, many groups around the world proceeded to study the benefits of laser interferometry, build new prototypes, perfect the design, and push technology development [10, 11, 12, 13].

The LIGO detectors in the U.S. [14], the Virgo detector in Italy [15], the GEO 600 detector in Germany [16] and the TAMA 300 [17] detector in Japan formed the first generation laser-interferometric GWD network. Construction of these projects began in the mid-1990s and then progressed in sequential commissioning and data-taking phases in the 2000s. These first-generation detectors achieved their original instrument sensitivity goals and are now undergoing major hardware upgrades to expand thousand-fold the observable volume of the cosmos. The strain sensitivities achieved by these initial detectors is shown in Figure 1. This paper describes the common design basics of the detectors and provides an overall summary of the current status of the worldwide network of GWDs, including a timeline for operations. The next four proceedings focus on the particular features and individual status of each of the advanced versions of these detectors [18, 19, 20, 21].

2. Background

A direct detection of gravitational waves themselves has not yet been made, but this is also not a surprise. The rate estimates for coalescing binary neutron stars, for instance, predict a detection probability of one event in 100 years for the initial detector sensitivities [1]. About two years' worth of double-coincidence data were collected. Nonetheless, the LIGO and Virgo collaborations have already produced astrophysical results from the data collected thus far. They have placed an upper limit that beats previous best estimates of the fraction of spin-down power emitted in GWs from the Crab Pulsar [22]. They have also placed an upper limit at 100 Hz on the normalized energy density of the stochastic GW background of cosmological and

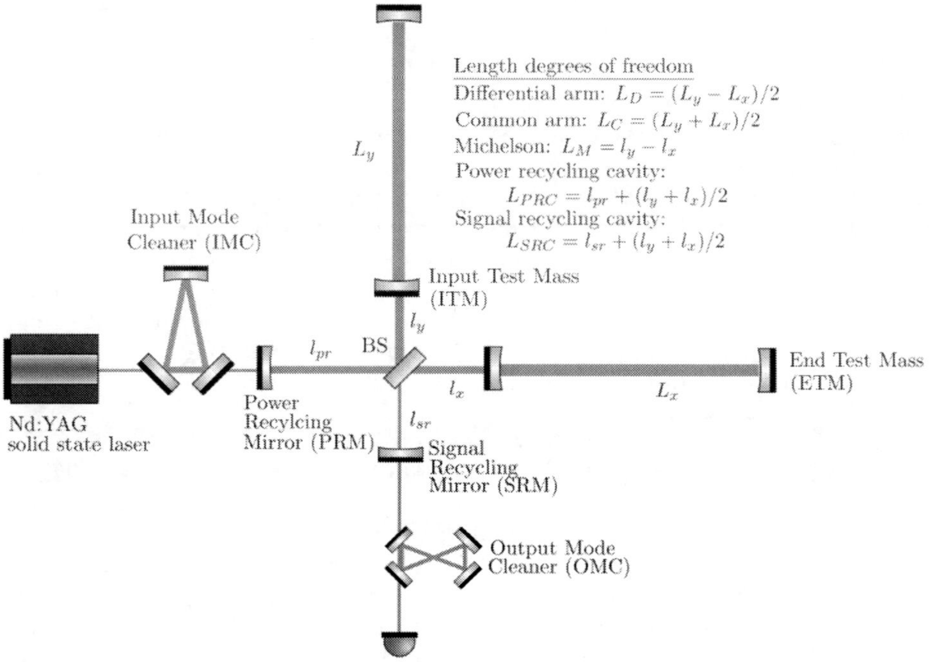

Figure 2. Layout of advanced ground-based interferometric gravitational wave detectors (not to scale). To keep the interferometer in its linear operating range, the length sensing and control system must control five interferometer length degrees of freedom and keep both mode cleaner cavities on resonance. The differential arm degree of freedom is sensitive to gravitational waves, and is sensed by the photo-detector in transmission of the output mode cleaner. GEO 600 is the exception in that it has no arm cavities.

astrophysical origin, a result otherwise inaccessible to standard observational techniques [23].

The last decade brought great advances in demonstrating the experimental feasibility of achieving the strain sensitivities required to witness astrophysical events and informed the design of today's generation of GWDs. Both LIGO and Virgo are re-using the infrastructure from the initial generation detectors and are replacing the hardware within the vacuum system. The upgraded detectors are called Advanced LIGO [24] and Advanced Virgo [25], respectively. TAMA 300 as well as CLIO, a prototype cryogenic laser interferometer [26], informed the design of a brand new detector called KAGRA, the first underground, cryogenic laser interferometer [27, 28]. GEO 600 is keeping its infrastructure and most of its initial generation hardware, but is carrying out upgrades to demonstrate advanced techniques in a program called GEO-HF [29]. In addition, a proposal to expand LIGO's baseline by building an interferometer in India is moving forward.

3. Detector design overview

The typical advanced detector configuration is that of a dual-recycled Fabry-Perot Michelson (DRFPM) laser interferometer as depicted in Figure 2. A power amplified, and intensity and frequency stabilized Nd:YAG solid state laser system injects linear-polarized 1064 nm light into a triangular input mode-cleaner (IMC) cavity. The IMC suppresses laser frequency noise and provides spatial filtering of the laser beam to reduce beam jitter that could otherwise couple to the GW readout. A beam splitter (BS) sends the beam to the two Fabry-Perot arms, which are made of an input test mass mirror (ITM) and an end test mass mirror (ETM). Both arms are of km-scale lengths and are set to maintain nearly perfect destructive interference of the recombined light at the anti-symmetric (AS) port which carries the GW information. Here, the beam is directed to an output mode-cleaning (OMC) cavity and then onto a photo-detector. The OMC transmits only the signal-carrying light to improve the signal-to-noise ratio.

A power recycling mirror at the symmetric port directs the constructively-interfered light back into the interferometer. The transmissivity of the power recycling mirror is set to match the losses of the main optics to create a nearly critically coupled cavity. A signal recycling mirror at the anti-symmetric port creates an additional cavity which can be used to adjust the storage time of the gravitational wave signal in the interferometer and thus the frequency response of the detector. The signal recycling mirror has a transmissivity selected to compromise between high and low frequency sensitivity based on thermal noise and laser power. Until a few years ago all detectors used heterodyne readout. Since approximately 2008 as part of intermediary upgrades to the initial detectors [30], homodyne (DC) readout was implemented together with the addition of an OMC [31, 32].

Nearly the entire interferometer is enclosed in an ultra high vacuum (UHV) system to render phase noise of residual gas unimportant and to keep the optics free of dust. The primary interferometer optics are suspended as pendulums to decouple them from ground motion so that they act like free masses in the horizontal plane at the frequencies in the GW detection band. To minimize the impact of thermal noise, the mirror suspensions are designed to minimize dissipation and, in the case of KAGRA, operated at cryogenic temperatures. Each mirror is equipped with actuators for coarse and fine control of the mirror position and orientation.

A feedback control system is implemented to hold the system sufficiently near the intended operating point so that the response to residual deviations remains linear. Calibration of the detector must take into account the action of the control system and the frequency response of the detector [33]. The various length (illustrated in Figure 2) and angular degrees of freedom are sensed through the use of radio-frequency sidebands on the carrier light that are created through phase modulation by electro-optic modulators. The differential arm length signal is sensitive to gravitational waves, and is sensed using homodyne readout in transmission of the OMC, where the GW signal is encoded as power variations of the light. The standard technique for locking optical cavities is the Pound–Drever–Hall method of laser frequency stabilization [34]. Although the interferometer is an analog instrument, it is interfaced through a digital control system which allows complex filters to be implemented and tuned from the control room.

The sources of noise that contaminate the detector's output can be grouped into two categories: displacement noise and sensing noise. Displacement noises are those that create real motion of the mirrors, while sensing noises are those that arise in the process of measuring the mirrors' position. The primary displacement noise that plagues terrestrial laser interferometers at very low frequencies is motion of the ground, i.e. seismic noise. Thermal motion of the mirrors, their dielectric coatings and suspensions as well as quantum radiation pressure noise are the other two types of displacement noise which dominate in the low- to mid-frequency range [35]. The primary sensing noise is shot noise that arises from the Poisson statistics of photon arrival at the photodetector.

Figure 3 shows the spectra of these ultimate limits to the performance of each of the GW detectors in the advanced detector network. Each curve reflects the incoherent sum of fundamental noise sources, which gives a likely best limit to performance. The actual sensitivity will depend also on technical noise sources. The narrow lines in each of the noise curves represent the thermally-excited violin modes of the test mass suspension fibers. Quantum noise and thermal noise play the dominant roles in limiting the sensitivity of each of the detectors. The frequency at which there is a dip in the noise together with the shapes of the noise curves are largely affected by the signal recycling parameters, which may be adjusted during the course of

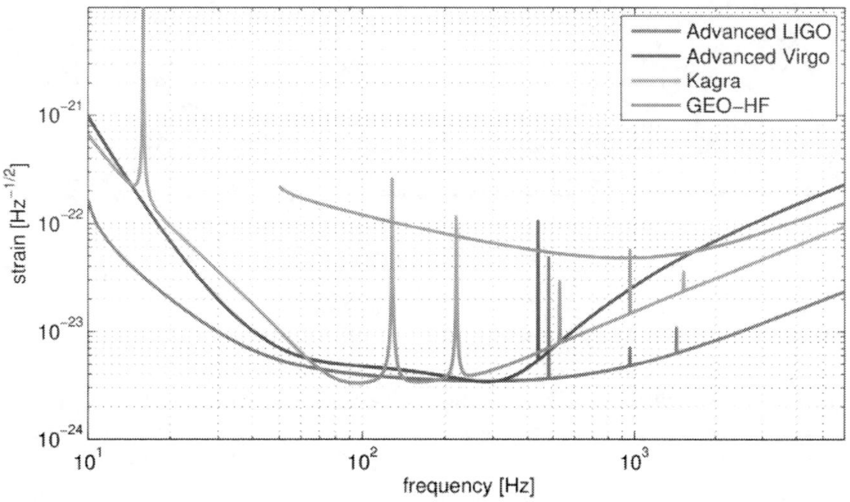

Figure 3. Prediction of the incoherent sum of fundamental noise sources which give a likely best limit to the performance of each of the instruments in the advanced detector network. Noise curves for both Advanced Virgo and KAGRA show a de-tuned signal recycling configuration optimized for sensitivity to neutron star binaries [25, 27, 36], whereas the Advanced LIGO and GEO-HF noise curves are shown with zero de-tuning [37, 38].

Table 1. Some design properties of each of the four GWDs. DRFPMI stands for dual-recycled Fabry-Perot Michelson.

	Advanced LIGO	Advanced Virgo	GEO-HF	KAGRA
arm length	4 km	3 km	2×600 m	3 km
power recycling gain	44	39	900	11
arm power	800 kW	700 kW	20 kW	400 kW
# of pendulum stages	4	8	3	6
mirror mass	40 kg	42 kg	6 kg	23 kg
mirror material	fused silica	fused silica	fused silica	sapphire
temperature	room	room	room	cryogenic
topology	DRFPMI	DRFPMI	DRMI	DRFPMI

operation of the advanced detectors. We postpone the detailed description of each noise curve to the proceedings dedicated to each detector. Table 1 provides an overview comparing some of the major properties of each of the detector designs.

4. Outlook

The advanced detectors form a four-site network which is crucial for GW signal characterization. The sky coverage depends on the detector locations and orientations. The two LIGO detectors are nearly aligned for maximum correlation, but they are relatively close together which results in signals that have largely redundant information about the source direction and character. For this reason, only the four-site network can provide full sky coverage [39].

Another important feature of a detector network is sky localization for electro-magetic follow ups and multi-messenger investigations. In the case of signals from coalescing binary neutron

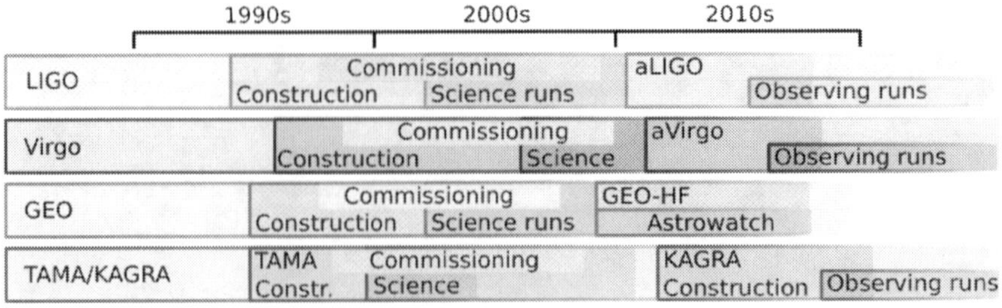

Figure 4. Timeline showing the approximate dates for construction, commissioning and science runs of each of the ground based GWDs. The first joint science runs in the advanced detector era are expected to begin taking place by late 2015.

star systems, the two Advanced LIGO detectors, Advanced Virgo, and KAGRA would provide a localization accuracy of 10 deg^2 for 50 % of the sources, while a five site network, including LIGO India, would improve the accuracy to 5 deg^2 for 30 % of the sources [40, 41]. This network of GW detectors together with the joint detection of the electro-magnetic counterparts is poised to open a promising window to gravitational wave astronomy.

Figure 4 shows a timeline of the construction, commissioning, and science run stages of each of the GWDs. The down times for each of the detectors is a bit staggered, with GEO 600 serving the role of being the sole detector online for the entirety of the period when the other detectors are offline. Commissioning and science run stages of the advanced detectors will be interspersed to allow for possible early results once astrophysically interesting sensitivities are reached, but before design sensitivity is reached [42]. The early science runs are likely to start in late 2015 with the Advanced LIGO detectors, and Advanced Virgo and KAGRA will join in due time.

Together with the promise of continuing to add to the collection of upper limits on GW emission in the era of advanced detectors, the first direct detection of GWs is expected to be made in only a couple years' time from now. A first detection is expected to witness an event such as a binary neutron star coalescence [1].

Acknowledgements

The authors gratefully acknowledge the support of the United States National Science Foundation for the construction and operation of the LIGO Laboratory; the Science and Technology Facilities Council of the United Kingdom, the Max-Planck-Society, and the State of Niedersachsen/Germany for support of the construction and operation of the GEO600 detector; the Italian Istituto Nazionale di Fisica Nucleare and the French Centre National de la Recherche Scientifique for the construction and operation of the Virgo detector; and the Japan Society for the Promotion of Science (JSPS) Core-to-Core Program A, Advanced Research Networks, and Grant-in-Aid for Specially Promoted Research of the KAGRA project. This document has been assigned LIGO document number LIGO-P1400153.

References

[1] Abadie J *et al.* 2010 *Classical and Quantum Gravity* **27** 173001+
[2] Ade P *et al.* (BICEP2 Collaboration) 2014 *Phys. Rev. Lett.* **112**(24) 241101
[3] Hulse R A and Taylor J H 1975 *Astrophys. J. Lett.* **195** L51–L53
[4] Weisberg J M and Taylor J H 2005 *Binary Radio Pulsars* (*Astronomical Society of the Pacific Conference Series* vol 328)
[5] Danzmann K and the LISA study team 1996 *Classical and Quantum Gravity* **13** A247
[6] Hobbs G *et al.* 2010 *Classical and Quantum Gravity* **27** 084013+

[7] Weber J 1960 *Physical Review Online Archive (Prola)* **117** 306–313
[8] Weiss R 1972 Electromagnetically Coupled Broadband Gravitational Antenna Tech. Rep. MIT
[9] Forward R L 1978 *Physical Review D* **17** 379–390
[10] Shoemaker D, Schilling R, Schnupp L, Winkler W, Maischberger K and Rüdiger A 1988 *Phys. Rev. D* **38**(2) 423–432
[11] Robertson D I, Morrison E, Hough J, Killbourn S, Meers B J, Newton G P, Robertson N A, Strain K A and Ward H 1995 *Review of Scientific Instruments* **66** 4447–4452
[12] Livas J, Benford R, Dewey D, Jeffries A, Linsay P, Saulson P, Shoemaker D and Weiss R 1985 (*Proc. of the Fourth Marcel Grossmann Meeting on General relativity*) 581+
[13] Abramovici A *et al.* 1996 *Physics Letters A* **218** 157–163
[14] Abbott B P *et al.* 2009 *Reports on Progress in Physics* **72** 076901+
[15] Acernese F *et al.* 2008 *Journal of Optics A: Pure and Applied Optics* **10** 064009+
[16] Grote H and the LIGO Scientific Collaboration 2010 *Classical and Quantum Gravity* **27** 084003+
[17] Takahashi R and the TAMA Collaboration 2004 *Classical and Quantum Gravity* **21** S403
[18] Dwyer S and the LIGO Scientific Collaboration *Journal of Physics: Conference Series* to be published in this issue.
[19] Acernese F *et al.* (the Virgo Collaboration) *Journal of Physics: Conference Series* to be published in this issue.
[20] Dooley K L and the LIGO Scientific Collaboration *Journal of Physics: Conference Series* to be published in this issue.
[21] Akutsu T and the KAGRA Collaboration *Journal of Physics: Conference Series* to be published in this issue.
[22] Abbott B *et al.* 2008 *The Astrophysical Journal Letters* L45+
[23] The LIGO Scientific Collaboration and the Virgo Collaboration 2009 *Nature* **460** 990–994
[24] Harry G M and the LIGO Scientific Collaboration 2010 *Classical and Quantum Gravity* **27** 084006+
[25] The Virgo Collaboration 2012 Advanced Virgo Technical Design Report VIR-0128A-12
[26] Uchiyama T *et al.* 2012 *Phys. Rev. Lett.* **108**(14) 141101
[27] Somiya K 2012 *Classical and Quantum Gravity* **29** 124007+
[28] Arai K *et al.* 2009 *Classical and Quantum Gravity* **26** 204020+
[29] Lück H *et al.* 2010 *Journal of Physics: Conference Series* **228** 012012+
[30] Adhikari R, Fritschel P and Waldman S 2006 Enhanced LIGO Tech. Rep. T060156 LIGO Laboratory
[31] Fricke T *et al.* 2012 *Classical and Quantum Gravity* **29** 065005+
[32] Hild S *et al.* 2009 *Classical and Quantum Gravity* **26** 055012+
[33] Hewitson M, Heinzel G, Smith J R, Strain K A and Ward H 2004 *Review of Scientific Instruments* **75** 4702–4709
[34] Drever R, Hall J, Kowalski F, Hough J, Ford G, Munley A and Ward H 1983 *Applied Physics B* **31** 97–105
[35] Saulson P R 1990 *Physical Review D* **42** 2437–2445
[36] URL `http://gwcenter.icrr.u-tokyo.ac.jp/researcher/parameters`
[37] URL `https://dcc.ligo.org/LIGO-T0900288/public`
[38] GEO-HF logbook, p.60 URL `https://intranet.aei.uni-hannover.de/geo600/geohflogbook.nsf/f5b2cbf2a827c0198525624b00057d30/4837a612ac990060c12575ce004e70fd`
[39] Schutz B F 2011 *Classical and Quantum Gravity* **28** 125023
[40] Fairhurst S 2011 *Classical and Quantum Gravity* **28** 105021
[41] Fairhurst S 2009 *New Journal of Physics* **11** 123006
[42] The LIGO Scientific Collaboration and the Virgo Collaboration 2013 Prospects for Localization of Gravitational Wave Transients by the Advanced LIGO and Advanced Virgo Observatories URL `http://arxiv.org/abs/1304.0670`

Journal of Physics: Conference Series **610** (2015) 012013 doi:10.1088/1742-6596/610/1/012013

Advanced LIGO status

S Dwyer for the LIGO Scientific Collaboration

Abstract. Advanced LIGO is currently in the final stages of installation and early commissioning. In the design of Advanced LIGO a key goal was the ability to detect gravitational waves from compact object binary inspirals, as these are thought to be the most likely candidates for early detections with ground based interferometers. Special emphasis has been placed on improving the low frequency sensitivity relative to the first generations of LIGO, in addition to improving the high frequency sensitivity by increasing the laser power. The interferometer in Livingston Louisiana has been locked (continuously held within the linear operating range) and noise investigations have begun, and the major installation activities for the interferometer at Hanford, Washington are completed.

1. Features of Advanced LIGO

During the successful science observation runs of Initial and Enhanced LIGO, several changes to the LIGO instruments were identified which were needed to build instruments with a high probability of detecting gravitational waves. A new interferometer, called Advanced LIGO, has been designed and built in the LIGO infrastructure based on these lessons learned during Initial LIGO. A major improvement in Advanced LIGO is in the reduction of seismic noise. The design calls for 10 orders of magnitude attenuation of ground motion at 10 Hz. Advanced LIGO relies on seven stages of isolation to achieve this: an external stage which relies on inertial sensors and hydraulic actuators for active control of large low frequency drifts, two internal stages which employ a combination of passive isolation and active isolation [2, 3, 4], and four pendulum stages used to suspend the test masses which provide passive isolation above the pendulum resonances from 0.5-2 Hz [5, 6]. These seven stages together are designed to improve the sensitivity at 30 Hz by 4 orders of magnitude compared to initial LIGO and to extend the low frequency limit of LIGO's observation band down to 10 Hz, as shown in the noise budget in Figure 1. This low frequency sensitivity will be critical for the detection of gravitational waves from compact binary inspirals.

In the intermediate frequency range, Brownian motion of the suspensions and the optical coatings are the dominant noise sources. To reduce mechanical dissipation in the suspensions, and therefore thermal motion of the test mass, the final suspension stage is one continuous assembly of fused silica elements. This creates a resonator with low mechanical losses and a quality factor estimated to be around 1 billion for the violin modes of the suspension fibers [7]. The beam size on the highly reflective end test masses has also been increased in Advanced LIGO compared to initial LIGO to reduce the impact of Brownian noise from optical coatings by averaging over a larger area of the coating surface.

At high frequencies, shot noise will be the dominant noise source. Advanced LIGO aims to improve the shot noise limited sensitivity by using up to 125 W of input power [8, 9], along with power and signal recycling [10]. With 125 W of input power, the arm cavity circulating power

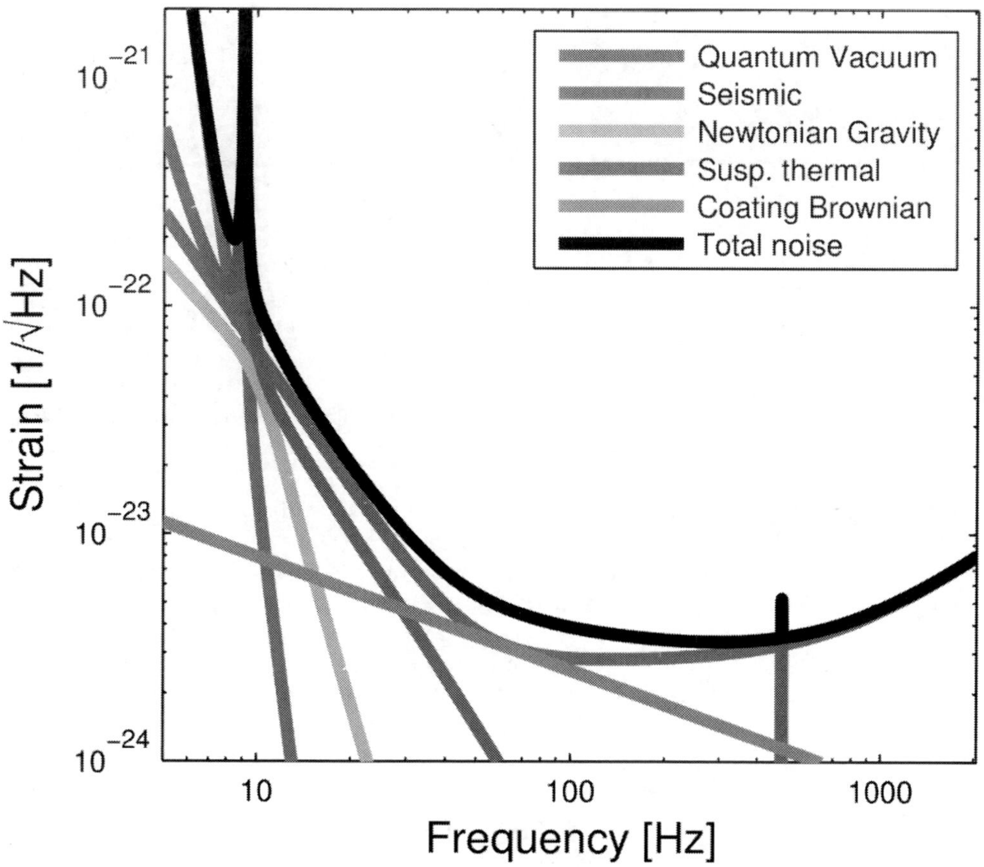

Figure 1. Noise sources for Advanced LIGO design sensitivity. Additional technical noise sources will make contributions to the noise performance, by design at the level of approximately 10 % [1]

is expected to be 800 kW, increased from 40 kW in Enhanced LIGO [11]. This high circulating power will pose several challenges, one of which is thermal lensing of optics due to absorption. A thermal compensation system has been designed to cancel the thermal lensing using a CO_2 laser projected onto a compensation plate and a heating element on the outer radius of the test masses. Hartmann sensors have been installed to sense the thermal lensing of each optic individually[12], which allows the possibility of locally controlling the curvature of each optic independently. A second challenge posed by high power operation is the increased radiation pressure force on the optics. The mass of the test masses in Advanced LIGO has been increased to 40 kg to reduce the displacement caused by this radiation pressure force.

Advanced LIGO will use a signal recycling mirror for broadband resonant sideband extraction. In this scheme the signal sidebands created by a gravitational wave in the arm cavities are extracted towards the interferometer dark port, so that the number of round trips the signal sidebands make in the arms is reduced [13]. For the next several years this will be used with no detuning, allowing for a broadband sensitivity and simplified operation. The reduction of shot noise with squeezed light injection has been demonstrated in Enhanced LIGO and at GEO [14, 15], and if implemented as anticipated in Advanced LIGO will further improve the shot noise limited sensitivity.

2. Lock Acquisition

The sensors used to measure the length degrees of freedom in ground based interferometers have very narrow linear operating ranges. This means that lock acquisition can be a difficult challenge for ground based interferometers, where five length degrees of freedom and many angular degrees of freedom must all be within a narrow range before the signals used to control them become available [17]. Each of these degrees of freedom are also coupled, so that vertex degrees of freedom disturb the sensors used to control the arm degrees of freedom and vice versa. Advanced LIGO has employed a new technique using a second frequency of laser light (doubled Yag, 532 nm green light) to stabilize the arm degrees of freedom and continue to stabilize them while holding them off resonance from the main laser [18, 19]. The vertex degrees of freedom are then stabilized using sensors that are relatively insensitive to the arm degrees of freedom. Once all degrees of freedom are controlled by these decoupled sensors, which have relatively poor noise performance, the arm cavities are brought on resonance and controls switched in a controlled way to the highly coupled, low noise sensors used for gravitational wave detection. This system was designed to speed up the lock acquisition process and improve the duty cycle by making the locking sequence more robust. The performance of this lock acquisition sequence has now been demonstrated in both Advanced LIGO interferometers [20].

The Advanced LIGO detector in Livingston, LA was fully locked for the first time in June 2014, a significant milestone for advanced gravitational wave detectors. These first locks have been followed by rapid progress in noise reduction. The last major installation activities for the interferometer at Hanford, WA are being completed in July 2014 with the first attempts at locking the full interferometer expected to take place starting in September.

3. Outlook

Now that Advanced LIGO has achieved the first locks of one full interferometer, the focus will be on improving noise performance in preparation for the first observing runs. In the coming years the schedule will be designed to increase the likelyhood of making an early detection of gravitational waves. While the chances of detection increase with the cube of improved detector performance, time spent commissioning the instruments must be balanced against time spent collecting data, especially in light of uncertainty in the estimates of event rates [21]. An early short observing run is planned for late 2015, when the range for binary neutron star inspirals is significantly better than Enhanced LIGO's range, between 40-80 Mpc [22]. This run will be followed by a series of longer observing runs at improved sensitivities with the goal of making a first direct detection of gravitational waves.

4. Acknowledgments

The authors gratefully acknowledge the support of the United States National Science Foundation for the construction and operation of the LIGO Laboratory and the Science and Technology Facilities Council of the United Kingdom, the Max-Planck-Society, and the State of Niedersachsen/Germany for support of the construction and operation of the GEO600 detector. The authors also gratefully acknowledge the support of the research by these agencies and by the Australian Research Council, the Council of Scientific and Industrial Research of India, the Istituto Nazionale di Fisica Nucleare of Italy, the Spanish Ministerio de Educacion y Ciencia, the Conselleria dEconomia, Hisenda i Innovacio of the Govern de les Illes Balears, the Royal Society, the Scottish Funding Council, the Scottish Universities Physics Alliance, The National

Aeronautics and Space Administration, the Carnegie Trust, the Leverhulme Trust, the David and Lucile Packard Foundation, the Research Corporation, and the Alfred P. Sloan Foundation.

References

[1] GWINC, Gravitational Wave Interferometer Noise Calculator, version 3 URL https://awiki.ligo-wa.caltech.edu/aLIGO/GWINC
[2] F Matichard et al 2014 URL arXiv:1407.6377
[3] F Matichard et al 2014 URL arXiv:1407.6324
[4] F Matichard et al 2013 *Proceedings of ASPE conference on Precision Control for Advanced Manufacturing Systems*
[5] Aston S M, Barton M A, Bell A S, Beveridge N, Bland B, Brummitt A J, Cagnoli G, Cantley C A, Carbone L, Cumming A V, Cunningham L, Cutler R M, Greenhalgh R J S, Hammond G D, Haughian K, Hayler T M, Heptonstall A, Heefner J, Hoyland D, Hough J, Jones R, Kissel J S, Kumar R, Lockerbie N A, Lodhia D, Martin I W, Murray P G, ODell J, Plissi M V, Reid S, Romie J, Robertson N A, Rowan S, Shapiro B, Speake C C, Strain K A, Tokmakov K V, Torrie C, van Veggel A A, Vecchio A and Wilmut I 2012 *Classical and Quantum Gravity* **29** 235004 URL http://stacks.iop.org/0264-9381/29/i=23/a=235004
[6] Carbone L, Aston S M, Cutler R M, Freise A, Greenhalgh J, Heefner J, Hoyland D, Lockerbie N A, Lodhia D, Robertson N A, Speake C C, Strain K A and Vecchio A 2012 *Classical and Quantum Gravity* **29** 115005 URL http://stacks.iop.org/0264-9381/29/i=11/a=115005
[7] Cumming A V, Bell A S, Barsotti L, Barton M A, Cagnoli G, Cook D, Cunningham L, Evans M, Hammond G D, Harry G M, Heptonstall A, Hough J, Jones R, Kumar R, Mittleman R, Robertson N A, Rowan S, Shapiro B, Strain K A, Tokmakov K, Torrie C and van Veggel A A 2012 *Classical and Quantum Gravity* **29** 035003 URL http://stacks.iop.org/0264-9381/29/i=3/a=035003
[8] Kwee P, Bogan C, Danzmann K, Frede M, Kim H, King P, Pöld J, Puncken O, Savage R L, Seifert F, Wessels P, Winkelmann L and Willke B 2012 *Opt. Express* **20** 10617–10634 URL http://www.opticsexpress.org/abstract.cfm?URI=oe-20-10-10617
[9] Winkelmann L, Puncken O, Kluzik R, Veltkamp C, Kwee P, Poeld J, Bogan C, Willke B, Frede M, Neumann J, Wessels P and Kracht D 2011 *Applied Physics B* **102** 529–538 ISSN 0946-2171 URL http://dx.doi.org/10.1007/s00340-011-4411-9
[10] Arain M A and Mueller G 2008 *Opt. Express* **16** 10018–10032 URL http://www.opticsexpress.org/abstract.cfm?URI=oe-16-14-10018
[11] Dooley K L, Barsotti L, Adhikari R X, Evans M, Fricke T T, Fritschel P, Frolov V, Kawabe K and Smith-Lefebvre N 2013 *J. Opt. Soc. Am. A* **30** 2618–2626 URL http://josaa.osa.org/abstract.cfm?URI=josaa-30-12-2618
[12] Brooks A F, Hosken D, Munch J, Veitch P J, Yan Z, Zhao C, Fan Y, Ju L, Blair D, Willems P, Slagmolen B and Degallaix J 2009 *Appl. Opt.* **48** 355–364 URL http://ao.osa.org/abstract.cfm?URI=ao-48-2-355
[13] Mizuno J, Strain K, Nelson P, Chen J, Schilling R, Rdiger A, Winkler W and Danzmann K 1993 *Physics Letters A* **175** 273 – 276 ISSN 0375-9601 URL http://www.sciencedirect.com/science/article/pii/037596019390620F
[14] LIGO Scientific Collaboration 2013 *N Photon* **7** 613–619
[15] Grote H, Danzmann K, Dooley K L, Schnabel R, Slutsky J and Vahlbruch H 2013 *Phys. Rev. Lett.* **110**(18) 181101 URL http://link.aps.org/doi/10.1103/PhysRevLett.110.181101
[16] Evans M, Barsotti L, Kwee P, Harms J and Miao H 2013 *Phys. Rev. D* **88**(2) 022002 URL http://link.aps.org/doi/10.1103/PhysRevD.88.022002
[17] Izumi K, Sigg D and Barsotti L 2014 *Opt. Lett.* **39** 5285–5288 URL http://ol.osa.org/abstract.cfm?URI=ol-39-18-5285
[18] Mullavey A J, Slagmolen B J J, Miller J, Evans M, Fritschel P, Sigg D, Waldman S J, Shaddock D A and McClelland D E 2012 *Opt. Express* **20** 81–89 URL http://www.opticsexpress.org/abstract.cfm?URI=oe-20-1-81
[19] Izumi K, Arai K, Barr B, Betzwieser J, Brooks A, Dahl K, Doravari S, Driggers J C, Korth W Z, Miao H, Rollins J, Vass S, Yeaton-Massey D and Adhikari R X 2012 *J. Opt. Soc. Am. A* **29** 2092–2103 URL http://josaa.osa.org/abstract.cfm?URI=josaa-29-10-2092
[20] A Staley et al *Under Review*
[21] Abadie J *et al.* 2010 *Classical and Quantum Gravity* **27** 173001+ ISSN 0264-9381 URL http://dx.doi.org/10.1088/0264-9381/27/17/173001
[22] LVC 2013 Prospects for localization of gravitational wave transients by the Advanced LIGO and Advanced VIRGO observatories URL arXiv:1304.0670v1

The Advanced Virgo detector

F. Acernese[35,11], T. Adams[19], M. Agathos[23],
K. Agatsuma[23], A. Allocca[36,14], P. Astone[15], G. Ballardin[5],
F. Barone[35,11], M. Barsuglia[1], A. Basti[31,14], Th. S. Bauer[23],
V. Bavigadda[5], M. Bejger[4], C. Belczynski[3],
D. Bersanetti[28,9], A. Bertolini[23], M. Bitossi[5,14],
M. A. Bizouard[18], S. Bloemen[23,24], M. Boer[2], G. Bogaert[2],
F. Bondu[32], L. Bonelli[31,14], R. Bonnand[19], V. Boschi[14],
L. Bosi[13], C. Bradaschia[14], M. Branchesi[38,8], T. Briant[20],
A. Brillet[2], V. Brisson[18], T. Bulik[3], H. J. Bulten[26,23],
D. Buskulic[19], C. Buy[1], G. Cagnoli[21], E. Calloni[29,11],
F. Carbognani[5], F. Cavalier[18], R. Cavalieri[5], G. Cella[14],
E. Cesarini[16], E. Chassande-Mottin[1], A. Chincarini[9],
A. Chiummo[5], S. Chua[20], F. Cleva[2], E. Coccia[34,10],
P.-F. Cohadon[20], A. Colla[33,15], M. Colombini[13],
A. Conte[33,15], J.-P. Coulon[2], E. Cuoco[5], S. D'Antonio[16],
V. Dattilo[5], M. Davier[18], R. Day[5], G. Debreczeni[25],
J. Degallaix[21], M. De Laurentis[29,11], S. Deléglise[20],
W. Del Pozzo[23], H. Dereli[2], R. De Rosa[29,11], L. Di Fiore[11],
A. Di Lieto[31,14], A. Di Virgilio[14], V. Dolique[21],
M. Drago[37,17], M. Ducrot[19], G. Endrőczi[25], V. Fafone[34,16],
S. Farinon[9], I. Ferrante[31,14], F. Ferrini[5], F. Fidecaro[31,14],
I. Fiori[5], R. Flaminio[21], J.-D. Fournier[2], S. Franco[18],
S. Frasca[33,15], F. Frasconi[14], L. Gammaitoni[30,13],
F. Garufi[29,11], A. Gatto[1], G. Gemme[9], B. Gendre[2],
E. Genin[5], A. Gennai[14], S. Ghosh[23,24], A. Giazotto[14],
R. Gouaty[19], M. Granata[21], G. Greco[8,38], P. Groot[24],
G. M. Guidi[38,8], J. Harms[8], A. Heidmann[20], H. Heitmann[2],
P. Hello[18], G. Hemming[5], D. Hofman[21], R.J.G. Jonker[23],
M. Kasprzack[18,5], F. Kéfélian[2], A. Królak[22,7], A. Kutynia[22],
C. Lazzaro[12], E. Lebigot[1], M. Leonardi[37,17], N. Leroy[18],
N. Letendre[19], M. Lorenzini[34,16], V. Loriette[6], G. Losurdo[8],
E. Majorana[15], I. Maksimovic[6], V. Malvezzi[34,16], N. Man[2],
V. Mangano[33,15], M. Mantovani[5,14], F. Marchesoni[27,13],
F. Marion[19], J. Marque[5], F. Martelli[38,8], L. Martinelli[2],
A. Masserot[19], D. Meacher[2], J. Meidam[23], F. Mezzani[15,33],
C. Michel[21], L. Milano[29,11], Y. Minenkov[16], A. Moggi[14],
M. Mohan[5], B. Mours[19], M. F. Nagy[25], I. Nardecchia[34,16],
L. Naticchioni[33,15], G. Nelemans[23,24], I. Neri[30,13],
M. Neri[28,9], F. Nocera[5], C. Palomba[15], F. Paoletti[5,14],
A. Pasqualetti[5], R. Passaquieti[31,14], D. Passuello[14],
M. Pichot[2], F. Piergiovanni[38,8], G. Pillant[5], L. Pinard[21],
R. Poggiani[31,14], M. Prijatelj[5], G. A. Prodi[37,17],

M. Punturo[13], P. Puppo[15], D. S. Rabeling[26,23], I. Rácz[25], P. Rapagnani[33,15], M. Razzano[31,14], V. Re[34,16], T. Regimbau[2], F. Ricci[33,15], F. Robinet[18], A. Rocchi[16], L. Rolland[19], R. Romano[35,11], P. Ruggi[5], B. Sassolas[21], D. Sentenac[5], V. Sequino[34,16], S. Shah[23,24], K. Siellez[2], N. Straniero[21], B. Swinkels[5], M. Tacca[1], M. Tonelli[31,14], F. Travasso[30,13], G. Vajente[31,14], N. van Bakel[23], M. van Beuzekom[23], J. F. J. van den Brand[26,23], C. Van Den Broeck[23], M. V. van der Sluys[23,24], J. van Heijningen[23], M. Vasúth[25], G. Vedovato[12], J. Veitch[23], D. Verkindt[19], F. Vetrano[38,8], A. Viceré[38,8], J.-Y. Vinet[2], H. Vocca[30,13], L.-W. Wei[2], M. Yvert[19], A. Zadrożny[22], J.-P. Zendri[12]

[1]APC, AstroParticule et Cosmologie, Université Paris Diderot, CNRS/IN2P3, CEA/Irfu, Observatoire de Paris, Sorbonne Paris Cité, 10, rue Alice Domon et Léonie Duquet, F-75205 Paris Cedex 13, France
[2]ARTEMIS, Université Nice-Sophia-Antipolis, CNRS and Observatoire de la Côte d'Azur, F-06304 Nice, France
[3]Astronomical Observatory Warsaw University, 00-478 Warsaw, Poland
[4]CAMK-PAN, 00-716 Warsaw, Poland
[5]European Gravitational Observatory (EGO), I-56021 Cascina, Pisa, Italy
[6]ESPCI, CNRS, F-75005 Paris, France
[7]IM-PAN, 00-956 Warsaw, Poland
[8]INFN, Sezione di Firenze, I-50019 Sesto Fiorentino, Firenze, Italy
[9]INFN, Sezione di Genova, I-16146 Genova, Italy
[10]INFN, Gran Sasso Science Institute, I-67100 L'Aquila, Italy
[11]INFN, Sezione di Napoli, Complesso Universitario di Monte S.Angelo, I-80126 Napoli, Italy
[12]INFN, Sezione di Padova, I-35131 Padova, Italy
[13]INFN, Sezione di Perugia, I-06123 Perugia, Italy
[14]INFN, Sezione di Pisa, I-56127 Pisa, Italy
[15]INFN, Sezione di Roma, I-00185 Roma, Italy
[16]INFN, Sezione di Roma Tor Vergata, I-00133 Roma, Italy
[17]INFN, Trento Institute for Fundamental Physics and Applications, I-38123 Povo, Trento, Italy
[18]LAL, Université Paris-Sud, IN2P3/CNRS, F-91898 Orsay, France
[19]Laboratoire d'Annecy-le-Vieux de Physique des Particules (LAPP), Université de Savoie, CNRS/IN2P3, F-74941 Annecy-le-Vieux, France
[20]Laboratoire Kastler Brossel, ENS, CNRS, UPMC, Université Pierre et Marie Curie, F-75005 Paris, France
[21]Laboratoire des Matériaux Avancés (LMA), IN2P3/CNRS, Université de Lyon, F-69622 Villeurbanne, Lyon, France
[22]NCBJ, 05-400 Świerk-Otwock, Poland
[23]Nikhef, Science Park, 1098 XG Amsterdam, The Netherlands
[24]Department of Astrophysics/IMAPP, Radboud University Nijmegen, P.O. Box 9010, 6500 GL Nijmegen, The Netherlands
[25]Wigner RCP, RMKI, H-1121 Budapest, Konkoly Thege Miklós út 29-33, Hungary
[26]VU University Amsterdam, 1081 HV Amsterdam, The Netherlands
[27]Università di Camerino, Dipartimento di Fisica, I-62032 Camerino, Italy
[28]Università degli Studi di Genova, I-16146 Genova, Italy
[29]Università di Napoli 'Federico II', Complesso Universitario di Monte S.Angelo, I-80126 Napoli, Italy
[30]Università di Perugia, I-06123 Perugia, Italy
[31]Università di Pisa, I-56127 Pisa, Italy
[32]Institut de Physique de Rennes, CNRS, Université de Rennes 1, F-35042 Rennes, France
[33]Università di Roma 'La Sapienza', I-00185 Roma, Italy

[34]Università di Roma Tor Vergata, I-00133 Roma, Italy
[35]Università di Salerno, Fisciano, I-84084 Salerno, Italy
[36]Università di Siena, I-53100 Siena, Italy
[37]Università di Trento, I-38123 Povo, Trento, Italy
[38]Università degli Studi di Urbino 'Carlo Bo', I-61029 Urbino, Italy

E-mail: paola.puppo@roma1.infn.it

Abstract. The Advanced Virgo interferometer is the upgraded version of the Virgo detector having the goal to extend by a factor 10 the observation horizon in the universe and consequently increase the detection rate by three orders of magnitude. Its installation is in progress and is expected to be completed in late 2015. In this proceeding we will present the scheme and the main challenging technical features of the detector and we will give an outline of the installation status and the foreseen time schedule which will bring Advanced Virgo to its full operation.

PACS numbers: 04.80.Nn, 95.55.Ym, 95.75.Kk

1. Introduction

Advanced Virgo (AdV) is the largest European gravitational-wave detector located at the European Gravitational Observatory (EGO) site in the Pisa countryside (Italy). In the last 10 years, the Virgo collaboration has been extended from Italy and France to three other european countries: Holland, Poland and Hungary, and it now includes up to 19 laboratories in Europe.

In March 2014 a Memorandum of Understanding for full data exchange, joint data analysis and publication policy was signed with the LIGO Scientific Collaboration, thus continuing a world wide network of second generation detectors (including Advanced Virgo, the two Advanced LIGO [1], GEO HF [2]).

2. Advanced Virgo main upgrades

Advanced Virgo (AdV) is the project to upgrade the Virgo detector [3] to a second generation instrument. It is designed with the aim to increase the sensitivity of about one order of magnitude and consequently extend the amount of probed Universe of a factor thousand. The Virgo detector upgrade passed through an intermediate step, named as Virgo+, in which some technical improvements were applied. They include the increase of the laser power up to 25 Watts and the introduction of the thermal compensation system to correct the thermal lensing effects. A second action was the use of silica fibres to suspend the mirrors for the reduction of the suspensions thermal noise. Finally, increasing the Finesse of the cavities from 50 to 150 completed this step. The Virgo+ detector has been run from 2010 up to the end of 2012 in coincidence with the Enhanced LIGO detectors[1].

The upgrade to AdV has been funded and started in 2009, and it is planned to finish in fall 2015 when the commissioning of the interferometer will start. This planning should allow a first science run data taking to be started in mid-2016.

2.1. Interferometer optical configuration and features

Advanced Virgo is a dual recycled interferometer as shown in fig. 2 of the introductory overview of this issue [6]. The detector will be illuminated by a 200 W laser beam delivering 125 W at the entrance of the interferometer. The optical power inside the power recycling cavity, at the beamsplitter will be 5 kW and up to 700 kW of light will be stored in the arms. Compared to Virgo, the light circulating in the arm cavities will be increased by a factor 350.

The choice of the arm cavities features deals with the aimed reduction of the mirror coating thermal noise which dominates in the mid-frequency range, of the thermal gradients on the mirror surface and of the alignment instabilities induced by the radiation pressure effect. To reduce the coating thermal noise the beam waist has been increased with respect to Virgo from 2 cm to 5 cm. With the 350 mm mirror diameters, such a value ensures negligible beam clipping losses (below 1 ppm). Moreover the beam waist will be placed in the center of the Fabry-Perot cavities nearer the input masses resulting in a larger beam size on the End Mirror (5.8 cm) than on the Input Mirror (4.87 cm) allowing the coating thermal noise of all the test masses to give an optimized contribution to the sensitivity curve.

Advanced Virgo will use marginally stable recycling cavities. The choice is driven by the construction schedule, the budget and the increased suspension complexity that would have to be required by a non degenerate-cavity solution like that one adopted in Advanced LIGO [1]. This configuration combined with the enlarged beam size is more sensitive to thermal effects on the mirror surfaces and to the aberrations due to their fabrication defects. However, a new mirror technology has been developed so that high performance optics are employed and a very sophisticated correction system of the thermal aberrations was set up to on purpose.

A new kind of substrate (Suprasil 3001/3002) is employed to make the cavity mirrors, so that the optical absorption is 3 times lower than in Virgo. Thanks to a new coating system, both the ITM and ETM mirrors are produced in pairs, as a consequence the asymmetry between the interferometer cavities is made nearly negligible. Moreover the use of Ti doped Ta_2O_5 as higher optical index material and the optimized layers thicknesses for the coating gives good results for the reduction of the mechanical and optical loss factors [4, 5].

With the signal recycling technique the detector response depends mainly on the properties of the signal recycling cavity and on the circulating power. The arm cavity finesse of about 450, the power recycling and signal recycling mirror transmissions respectively of 5% and 20% are chosen as a trade off between maximizing the circulating power and reducing the aberration effects, thus allowing the sensitivity to be optimized in a chosen sensitivity bandwidth.

To make a initial locking of the full interferometer an auxiliary green laser will be used, which will see lower cavity finesses easing the locking procedure. The mirror coatings are designed to have the optical properties suitable for this purpose.

2.2. Dealing with the high power stored in the cavities

Although the quantity of power absorbed by the highly improved quality mirror substrates and coatings is very small (0.3-0.4 ppm), the big amount of power in the cavity is sufficient to produce a non negligible thermal lensing effect which can spoil the laser beam matching in the cavities and degrade the interferometer performance

[7]. This effect is mainly due to the heat absorption in the mirror bulk and coatings and to reduce it a thermal compensation system (TCS), successfully used in Virgo+, has been designed and improved to deal with the new higher power [8].

In Advanced Virgo the TCS will need to correct the optical aberrations in the power recycling cavity and to correct the radius of curvature of all the test masses [9, 10]. Indeed, the inhomogeneous heating of the mirror surfaces will produce a change of their curvature radius and produce a distortion in the light beam wavefront via the thermo-elastic effect on the mirror coatings. A useful way to picture the optical distortion effect, is to use the fractional power scattered out from the TEM_{00} mode [11, 12], termed coupling losses, and the Gaussian-weighted RMS of the optical path length increase. To allow a correct operation of the detector, the coupling losses due to the thermal lensing must be reduced at least of a factor 10^3, this value should result in a reduction of the RMS fluctuations of the optical paths of 2 nm . The adjustment of the mirrors radius of curvature is provided with the use of heating rings surrounding the test masses on their back side. Moreover an appropriate heating pattern generated by a CO_2 laser which is completely absorbed by the fused silica substrates will be used. This last solution was tested during the Virgo+ operation by projecting a CO_2 laser beam in annulus shape on the mirror. However, in AdV the use of a higher power correcting CO_2 beam on the mirrors could introduce a displacement noise due to its intensity fluctuation, for this reason an additional transmissive optic (named Compensation Plates) is employed and placed on the mirrors back side, in the recycling cavities, avoiding any direct coupling effect of its motion to the dark fringe. The studies of the corrective patterns have been carried out with Finite Element Models so that it was possible to define an optimal corrective procedure that would allow the optical path RMS to be reduced by a factor 20 down to 0.35 nm, well within the requirements. The sensing system used in the TCS have been set up employing Hartmann wavefront sensors and phase cameras to detect the production of spurious light modes [**?**, 14].

2.3. Dealing with the stray light

A small amount of the power circulating in the interferometer is backscattered by all the illuminated surfaces. This effect becomes very important in the second generation interferometer with a bigger circulating power [15]. The stray light can increase the phase noise and spoil the sensitivity. To deal with such a problem diaphragm baffles are mounted around the suspended mirrors and placed at the vacuum tubes entrances and inside the links where the stray beams can be more noisy. They are designed to intercept and absorb as much as possible the stray light. For this reason they must have low scattering and low reflectivity features, and where required they must be suspended so that the residual injected strain noise is kept negligible. In AdV the baffles are stainless steel plates coated with Diamond-like Carbon, ensuring an acceptable damage threshold of 0.5 kW/cm^2, the choice was driven by a compromise between the requirements and the costs.

The scattered light that could arise from all the photodiodes used for the interferometer diagnostics and locking and for the signal detection and injection during the scientific runs will be reduced by suspending all the optical benches in vacuum. The five optical benches placed at the two cavity ends, at the power recycling pick off port and at the injection and detection systems are hung from five multistage vibration isolators (minitowers). This solution makes the interferometer less vulnerable to the

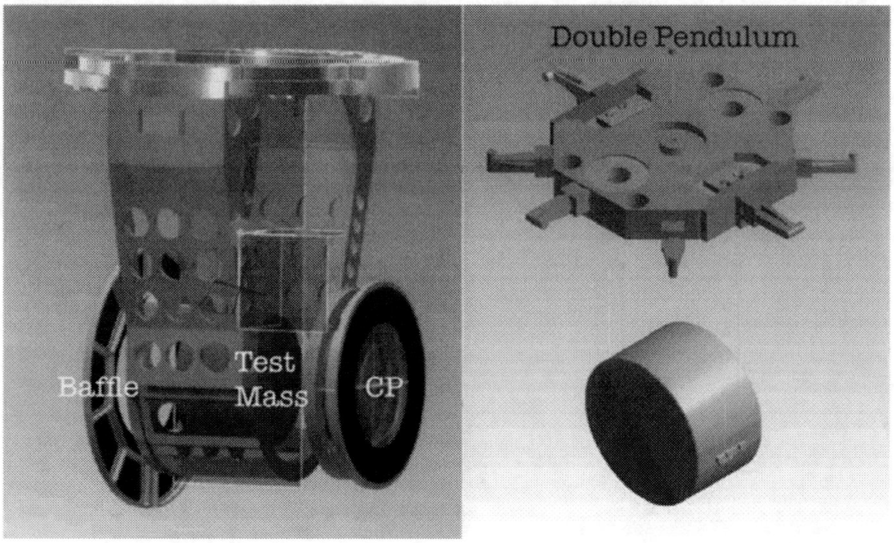

Figure 1. The new payloads in AdV must be compliant with the presence of the baffles and the TCS elements and with the double stage mirror monolithic suspension.

effect of the upconversion of the scattered light due to the seismic motion.

2.4. The mirror suspension system

In AdV heavier test masses of (42 kg) with the same diameter (350 mm) as in Virgo but a doubled thickness (20 cm) will be adopted. This solution is crucial to decrease the radiation pressure noise effect due to the high power circulating into the interferometer cavities. Moreover, the enlarged beam will require a larger beam splitter having a diameter of 550 mm with a total weigth of 34 kg.

All the test masses are suspended from a double pendulum stage (right picture in the figure 1) through the low loss silica fibers having a diameter of 400 μm whose profile has also been optimized to reduce the thermoelastic losses and consequently decrease the suspension thermal noise. Profiting from the experience gained in Virgo+[18, 19], the design of the fiber clamping systems both on the mirrors and on the intermediate suspension stages were improved to deal with the silica-steel interfaces, which can be a thermal noise source[16, 17]. All these clamping systems rely on the well known and tested silicate bonding technique which can reproduce the connection between materials at the molecular level and for this reason we talk of monolithic suspension system [20].

The mirror last stage suspension (payload), used to control the test masses and equipped with the sensors and actuators for the local and global control purposes, must be compliant with the monolithic suspension system, with the presence of the baffles and of the TCS elements. For this reason this part has been completely redesigned (left picture in the figure 1), and a big effort to improve the mechanical features has been made with help of the FEM. The study was crucial for defining the CP supporting

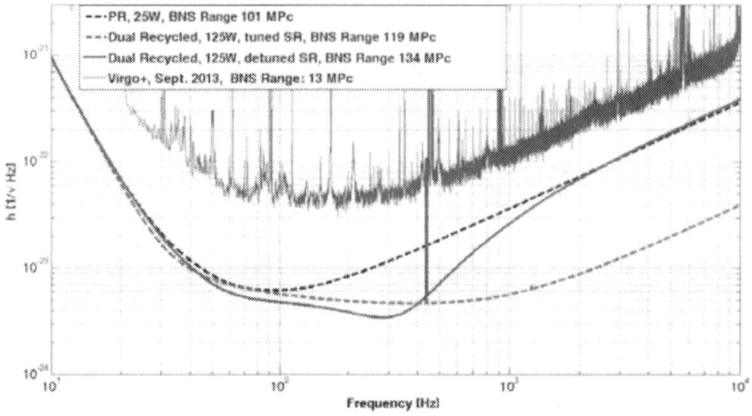

Figure 2. AdV sensitivity for the three configuration foreseen for the advanced detector step by step commissiong procedure in the Advanced Virgo technical design report (TDR) [23]. The early operation (black dashed line) with 25 W input power and without signal recycling, the mid-term operation with 125 W laser and tuned SRM (red dashed line), the early operation with 125 W laser and detuned SRM (solid blue line). For comparison the best noise sensitivity obtained in Virgo+ is shown.

system without adding extra displacement noise to the ITF output signal [21].

Moreover the superattenuators providing the seismic isolation whose performances have been widely demonstrated in Virgo [22], have been improved to be coupled with the new payloads. In particular new steering filters have been produced and the possibility to actively control the tilts induced by the seismic motion is being implemented.

2.5. Important infrastructure upgrades

The AdV infrastructure is essentially the same as for the Virgo detector. However some important improvements have been done in the central hall to host the minitowers, and to turn the laser and detection lab into acoustically isolated clean rooms. Large cryotraps are being installed at the end of the vacuum links to improve the residual pressure by a factor 100. Finally, several upgrades on the data acquisition and electronic systems will be done to handle the increased number of channels due to the new dual recycling control scheme.

3. The sensitivity of AdV

The peak sensitivity of Advanced Virgo is expected to be 3.4×10^{-24} $1/\sqrt{\text{Hz}}$ around 300 Hz. This value refers to the target sensitivity presented in the Advanced Virgo Technical Design Report (TDR) [23] approved by the funding agencies in 2012 and shown in figure 2 (solid blue line) and compared also with the other second generation detector sensitivities in the figure 3 of the introductory overview in this issue. The expected inspiral range which can be derived from this sensitivity, is 140 Mpc for

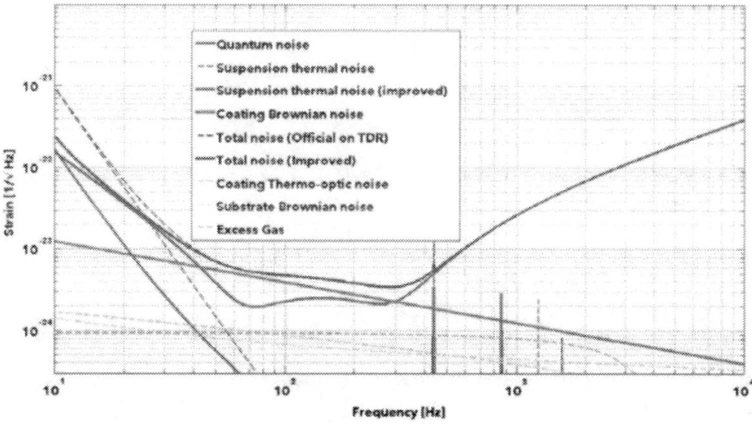

Figure 3. The official AdV sensitivity (blue dotted line) compared with the noise budget projection (blue solid line) based on the new improvements introduced in the detector design and modeling. With the new suspension design the thermal noise is improved by an order of magnitude in the low frequency range (green dotted line compared with the solid green line). In the mid-frequency range the noise is dominated by the coating thermal noise. Some other technical noise sources are sketched

binary neutron stars (BNS) of 1.5 $M_\odot$ and 1 Gpc for coalescing binaries black holes (BBH) of 30 $M_\odot$. In the figure 2 are also shown the two other sensitivity stages that Advanced Virgo pass through as mid-steps to reach its final operation. Indeed, AdV will not be operated in the final configuration at the beginning, and to face the problems arising from the new features, a step by step approach will be carried on. The planning is to have an early configuration with the power recycled detector using the initial Virgo+ 25 W laser, and no SR mirror (black dashed line) very similar to the former Virgo detector and easier to deal with. Then, after the installation of the new powerful laser, a mid-term dual-recycled configuration with a high power laser of 125 W and a tuned SR (red dashed line) will be followed by the configuration with a high power laser of 125 W and a detuned SR allowing the targeted sensitivity to be reached (solid blue line). The commissioning periods will be alternated with periods of data taking in coordination with Advanced LIGO in order to maximize the network capabilities.

3.1. Some improvement on the noise budget.

The main contributors to the sensitivity curve are the suspension thermal noise, the coating thermal noise up to 300 Hz, and then the quantum noise [23]. Since the TDR release some progress was made both in the design of the detector and in the modeling of the noise. The first tests of the new BS payloads, similar to the test mass payloads, showed a big improvement in the dissipation of the second stage suspension. Thanks to this result we are able to make a more optimistic projection of the AdV noise budget in this zone as it is sketched in the sensitivity curve shown in figure 3.

4. Status and Conclusions

At the time of writing, during the summer 2014, the installation of the interferometer optics has started. The injection bench and the mode cleaner far mirror were installed and their commissioning is started. Moreover the Beam Splitter payload assembly is succesfully finished and it is ready to be integrated in the tower by the end of September. The important infrastructure work was completed in October 2013 and the equipment installation is in progress.

The main mirrors of Advanced Virgo are being coated and characterized and the two input masses will be ready for their installation on site at the end of 2014. The commissioning of the new interferometer will start in mid-2015 and will last around 6 months with first a configuration similar to Virgo+ (25 W of input power and no signal recycling effect). The goal is to have an interferometer ready to participate in the first science run in coincidence with the Advanced LIGO interferometers in 2016. As a last step the input power will be raised gradually to 200 W and the signal recycling will be installed to allow the detector to reach its maximal sensitivity.

References

[1] S Dwyer and the LIGO Scientific Collaboration 2014 *Journal of Physics: Conference Series to be published in this issue.*
[2] K L Dooley and the LIGO Scientific Collaboration 2014 *Journal of Physics: Conference Series to be published in this issue.*
[3] The Virgo collaboration 2012 *JINST* **7** 3012
[4] C.Comtet et al., Proceedings of the 42th Rencontres de Moriond, 2007.
[5] Cimma, D. Forest, P. Ganau, B. Lagrange, J.M. Mackowski, C. Michel, J.L. Montorio, N. Morgado, R. Pignard, L. Pinard, A. Remillieux, Applied Optics, vol. 45 7 (2006) 1436-1439.
[6] K L Dooley, T Akutsu, S Dwyer, P Puppo 2014 *Journal of Physics: Conference Series to be published in this issue.*
[7] Hello P and Vinet J 1993 *Phys. Lett. A* **178** 351
[8] L. Pinard, R. Flaminio, D. Forest, *Bulk absorption measurement at 1064 nm of the new Heraeus ultra pure fused silica used for the Advanced Virgo Test Mass: effect of the annealing*, Virgo note VIR-0091A-10 (URL `https://tds.ego-gw.it/ql/?c=7215`) (2010).
[9] F. Acernese et al., J. Phys.: Conf. Ser. 120, 032007 (2008).
[10] T. Accadia et al., Proceedings of the 12th Marcell Grossmann Meeting ed T Damour, R T Jantzen and R Ruffini (Singapore: World Scientific) (2011).
[11] P. Hello, Eur. Phys. J. D 15, 373-383 (2001) .
[12] J.Y. Vinet, Living Rev. Relativity, 12, 5 (2009).
[13] A. Rocchi et al., J. P. Conf. Series 363, 012016 (2012).
[14] R. Day, Simulation of use of phase camera as sensor for correcting common high order aberrations in MSRC, VIR-0389A-11 (2011).
[15] A.Chiummo, R.Day, J.Marque, AdV - Stray Light Control: Requirements for wide-angle scattering in the arm cavity, VIR-0055A-13 URL `https://tds.ego-gw.it/ql/?c=9398`
[16] G. Cagnoli G. et al., Physical Letters A, vol. 255, p. 230, 1999
[17] Y. L. Huang, P. R. Saulson P. R., Review Scientific Instruments, vol. 69, p. 544, 1997
[18] T. Accadia et al., Class. Quantum Grav. 28 (2011) (doi:10.1088/0264-9381/28/11/114002)
[19] P. Puppo, M. Colombini, I. Nardecchia, E. Majorana, P. Rapagnani, F. Ricci: Virgo+ Thermal Noise Study, VIR-0074B-12, June 13 (2012), URL `https://tds.ego-gw.it/ql/?c=8878`
[20] G. Cagnoli et al., Physical Review Letters, vol. 85, p. 2442, 2000
[21] Conte, P. Puppo, P. Rapagnani: Thermal Noise evaluation of the Compensation Plate, Virgo internal note: VIR-0469A-13, October 24 (2013), URL `https://tds.ego-gw.it/ql/?c=9851`
[22] Braccini S *et al.* 2005 *Astroparticle Physics* **23** 557
[23] Collaboration T V 2012 Advanced Virgo technical design report, VIR-0128A-12 URL `https://tds.ego-gw.it/ql/?c=8940`

Journal of Physics: Conference Series **610** (2015) 012015 doi:10.1088/1742-6596/610/1/012015

Status of GEO 600

K L Dooley [1] **for the LIGO Scientific Collaboration**

Max-Planck-Institut für Gravitationsphysik (Albert-Einstein-Institut) und Leibniz Universität Hannover, Callinstr. 38, D-30167 Hannover, Germany

E-mail: `kate.dooley@aei.mpg.de`

Abstract. The German-British laser-interferometric gravitational wave detector GEO 600 is in its 13th year of operation since its first lock in 2001. After participating in science runs with other first generation detectors, GEO 600 has continued collecting data as an astrowatch instrument with a duty cycle of 62% during the time when the other detectors have gone offline to undergo substantial upgrades. Less invasive upgrades to demonstrate advanced technologies and improve the GEO 600 sensitivity at high frequencies as part of the GEO-HF program have additionally been carried out in parallel to data taking. We report briefly on the status of GEO 600.

1. Background

GEO 600 [1] is the German-British laser-interferometric gravitational wave (GW) detector that was built in the late 1990s and started operation in the early 2000s [2, 3] in conjunction with other first generation GW detectors. GEO 600 uses a dual-recycled Michelson configuration with 1200 m long arms folded inside 600 m long beam tubes. From the start, the project focused on researching, incorporating, and demonstrating riskier technologies compared to the designs of its longer-arm counterparts, Virgo (3 km) [4] and LIGO (4 km) [5]. Early design features that were unique to GEO 600 included multi-chain suspensions with a monolithic final stage and a reaction mass chain, electro-static actuators, and signal recycling [6].

Following the era of first generation detector science runs when the other observatories began going offline for substantial upgrades [7, 8], GEO 600 opted to serve the role of being the sole detector continuing to collect data through a program called *Astrowatch*. Simultaneously, starting in 2009, GEO 600 began a program called *GEO-HF* [9] which has the aim of carrying out a set of upgrades to improve the detector's high frequency sensitivity above approximately 600 Hz where quantum shot noise is the limiting noise source. The primary techniques implemented include the injection of squeezed vacuum, a change in signal recycling bandwidth and operating point, DC readout with an output mode cleaner, and higher laser power combined with a thermal compensation system.

2. GEO-HF status

Figure 1 shows the results to date of how the GEO-HF upgrades have affected the GEO 600 strain sensitivity. The most notable effect is above 600 Hz where as much as a factor of 4 improvement in sensitivity is achieved. The dominant contributing factor to this improvement

[1] Current address: The University of Mississippi, University, MS 38677, USA

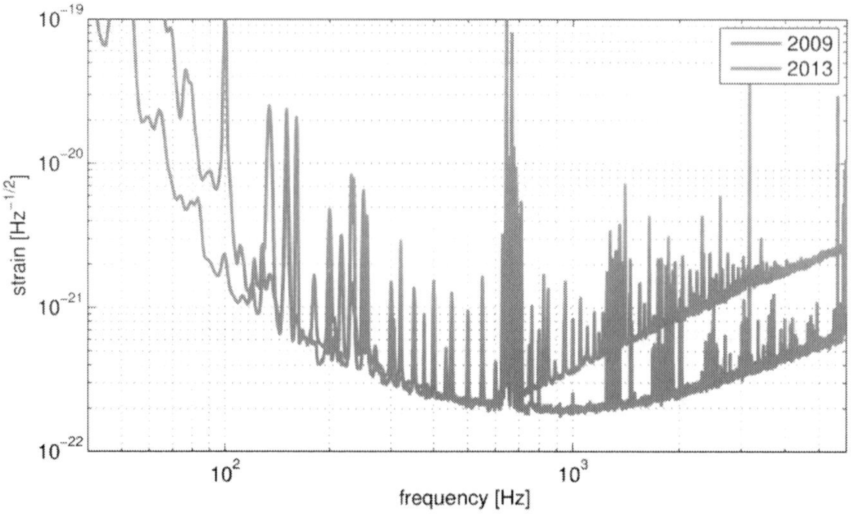

Figure 1. Strain sensitivity of GEO 600 today compared to the sensitivity at the start of the GEO-HF upgrade [3]. A new signal recycling mirror, squeezed vacuum injection, and DC readout contribute to the substantial improvement in high-frequency sensitivity. An effort to increase the laser power is ongoing. The large cluster of lines centered around 650 Hz are the violin modes of the suspension fibers.

is the decrease in the finesse of the signal recycling cavity (SRC). This was accomplished in 2010 by exchanging the signal recycling mirror from one with reflectivity of $R = 98\%$ to one with $R = 90\%$, thus reducing the storage time of the GW signal in the interferometer and therefore altering the detector's frequency response. The next most significant factor is squeezing, which creates (in its current state) a factor 1.5 improvement in sensitivity above approximately 1 kHz and was first demonstrated in 2010 [10]. Finally, a change in 2009 from heterodyne to homodyne (DC) readout of the GW signal [11] renders a fundamental factor $\sqrt{3/2}$ improvement in signal-to-noise ratio (SNR) [12]. An increase in laser power is not reflected in the 2013 sensitivity curve and is a topic of ongoing work.

The marked degradation in strain sensitivity below 100 Hz arises from technical side effects of both DC readout and the new signal recycling mirror. The limiting noise sources at these frequencies are alignment feedback noise and signal recycling length feedback noise. First, the dark-fringe offset required for DC readout creates a TEM_{00} carrier field on the Michelson alignment sensors. Jitter of this field caused by output optics creates an alignment sensing offset which is impressed on the mirrors. Second, the new signal recycling cavity's lower finesse decreased the SNR of the SRC length signal, creating higher feedback noise.

3. Ongoing research

Ongoing research at GEO 600 has several fronts. For one, there are still some technical challenges and open research questions related to realizing the full potential of DC readout and squeezing. Second, the GEO-HF goal of increasing the laser power by a factor of 4 – 6 must still be achieved and requires the design and commissioning of a new thermal compensation subsystem.

3.1. Squeezed vacuum integration
Injection of squeezed vacuum to the interferometer's output port is a novel technique employed to suppress quantum noise [13]. For nearly the last 3 years, GEO 600 has demonstrated the first long-term application of squeezed vacuum in a gravitational wave detector, achieving a 90% duty cycle during science times and improving the shot-noise-limited strain sensitivity up to a factor of 1.5 (3.7 dB) [10, 14]. In order to achieve even more stable and better squeezing levels, recent work at GEO 600 has focused on the development and commissioning of squeezer alignment and phase control systems [6]. Ongoing work includes the commissioning of a new wavefront sensing scheme in reflection of the OMC to ensure maximum overlap of the squeezed field with the interferometer output field to achieve higher squeezing levels. In addition, a new error signal for sensing the relative phase of the squeezed field with the interferometer output is being studied in order to eliminate lock point offsets and create more stable squeezing [15].

3.2. Higher laser power
One unique feature of GEO 600 is that it does not have Fabry-Perot arm cavities. Instead, a signal recycling mirror alone is used together with the power recycling mirror to amplify the GW signal. A side effect is that *all* of the stored power in the interferometer is transmitted through the substrate of the beam splitter (BS). The current operation of GEO 600 uses 2–3 kW at the BS, which leads to thermal lensing problems of the same magnitude that will be faced by advanced detectors when operated at full design power. GEO 600 currently does not have any thermal compensation at the BS and this limits the ability to operate with higher laser power. During commissioning periods, stable operation with as much as 7 kW can be achieved, but there is excess noise below 600 Hz in this high power state. Design of an array of heating elements to project optimized patterns of radiation onto the BS to compensate the lens represents ongoing work. Other challenges related to higher power which have already been addressed are the correction of astigmatism at one of the end mirrors using segmented thermal compensation [16] and the elimination of stray light coupling to shadow sensors on each of the mirrors through a modulation-demodulation technique [17].

3.3. Alternative OMC alignment techniques
The upgrade from RF to DC readout involved the installation of an output mode cleaner (OMC) in the interferometer's output port directly before the GW readout photodiode. Its purpose is to filter out all light that does not contribute to the GW signal. Higher order modes make it challenging to develop an alignment scheme with sufficient bandwidth that maximizes the SNR of the light transmitted through the OMC without introducing beam jitter coupling. Ongoing work at GEO 600 is the commissioning of a new alignment technique which uses wavefront sensors in reflection of the OMC. The wavefront sensors sense the relative alignment between fields representing the GW mode and the OMC eigenmode. The fields used are the radio frequency sidebands which are resonant in the interferometer and audio sidebands which are generated from a length dither of the OMC [6]. This scheme both eliminates low frequency dithering of the output steering optics and facilitates an increase in the alignment control bandwidth up to several Hz.

4. Astrowatch
Starting in 2007, GEO 600 has served as the community's Astrowatch detector. Astrowatch mode is a science run from the standpoint of detector operations such that GEO 600 is maintained at its highest possible sensitivity in a full interferometer lock and no experiments are allowed which could reduce its sensitivity. Furthermore, a realtime system with interferometer parameter tracking assures valid strain calibration [6]. A search in the Astrowatch data is triggered by an external event such as a gamma ray burst [18], neutrino detection, or an optical detection

of a supernova. Upon operating in science mode mostly on nights and weekends, GEO 600 has maintained an average science time duty cycle of 62% over the last several years. Active commissioning and upgrade activities are carried out during the remainder of the time.

5. Summary and outlook

GEO 600 continues to actively pursue the goals of the GEO-HF upgrade program and to dedicate significant time to collecting Astrowatch data during the time when the other GW detectors are being upgraded. The techniques of signal recycling, homodyne (DC) readout, and squeezed vacuum injection have been demonstrated to the benefit of the GEO 600 high frequency sensitivity. Research into better integration of squeezed vacuum and improved OMC alignment schemes are active areas of continued work on these topics. At the same time, a new thermal compensation system for the BS is being developed which will enable the use of higher laser powers. GEO 600 will continue to collect Astrowatch data and research advanced techniques in conjuction with furthering the improvement of its high-frequency sensitivity.

Acknowledgements

The authors are grateful for support from the Science and Technology Facilities Council (STFC), the University of Glasgow in the UK, the Bundesministerium für Bildung und Forschung (BMBF), and the state of Lower Saxony in Germany. This work was partly supported by DFG grant SFB/Transregio 7 Gravitational Wave Astronomy. This document has been assigned LIGO document number P1400172.

References

[1] Willke B *et al.* 2002 *Classical and Quantum Gravity* **19** 1377+
[2] Grote H 2008 *Classical and Quantum Gravity* **25** 114043+
[3] Grote H and the LIGO Scientific Collaboration 2010 *Classical and Quantum Gravity* **27** 084003+
[4] Accadia T *et al.* 2012 *Journal of Instrumentation* **7** P03012+
[5] Abbott B P *et al.* 2009 *Reports on Progress in Physics* **72** 076901+
[6] Affeldt C *et al.* 2014 *Classical and Quantum Gravity* **31** 224002+
[7] Harry G M and the LIGO Scientific Collaboration 2010 *Classical and Quantum Gravity* **27** 084006+
[8] The Virgo Collaboration 2009 Advanced Virgo Baseline Design Tech. Rep. VIR-027A-09
[9] Lück H *et al.* 2010 *Journal of Physics: Conference Series* **228** 012012+
[10] The LIGO Scientific Collaboration 2011 *Nat Phys* **7** 962–965
[11] Hild S *et al.* 2009 *Classical and Quantum Gravity* **26** 055012+
[12] Niebauer T M, Schilling R, Danzmann K, Rüdiger A and Winkler W 1991 *Physical Review A* **43** 5022–5029
[13] Caves C M 1980 *Physical Review Letters* **45** 75–79
[14] Grote H, Danzmann K, Dooley K L, Schnabel R, Slutsky J and Vahlbruch H 2013 *Physical Review Letters* **110** 181101+
[15] Dooley K L, Schreiber E, Vahlbruch H, Affeldt C, Leong J R, Wittel H, and Grote H 2014 *submitted to OSA* arXiv:1411.3454
[16] Wittel H *et al.* 2014 *Classical and Quantum Gravity* **31** 065008+
[17] Affeldt C 2014 *Laser Power Increase for GEO600* Ph.D. thesis Leibniz Universität Hannover
[18] The LIGO Scientific Collaboration and the Virgo Collaboration 2014 *Physical Review D* **89** 122004+

Journal of Physics: Conference Series **610** (2015) 012016 doi:10.1088/1742-6596/610/1/012016

Large-scale cryogenic gravitational-wave telescope in Japan: KAGRA

Tomotada Akutsu for the KAGRA collaboration

National Astronomical Observatory of Japan, 2-21-1 Osawa, Mitaka, Tokyo 181-8588, Japan

E-mail: tomo.akutsu@nao.ac.jp

Abstract. KAGRA, the large-scale cryogenic gravitational-wave telescope (formerly known as LCGT), is a laser interferometric detector under construction at the Kamioka mine in Japan. We report on the current status of the project as well as the overview of the main features and the schedule. The construction has been underway since 2012, and the tunnel excavation in the mine for two 3-km arms finished at the end of March 2014. As with the other advanced terrestrial interferometers, KAGRA will have two long Fabry-Perot cavities to sense gravitational waves, but they are to be installed underground for lower seismic noise. KAGRA will be the first large interferometer having cryogenic cavities, the mirrors for which are planned to be cooled down to around 20 K for reducing thermal noise.

1. Introduction

KAGRA is a Japanese large-scale cryogenic gravitational-wave telescope, formerly called LCGT, under construction at the Kamioka mine in Gifu prefecture [1]. Unlike other advanced ground-based interferometers [2, 3, 4], the KAGRA project is not a simple upgrade of existing facilities; it started from scratch for the construction of infrastructures underground such as a tunnel for the two 3-km length arms, drainage for ground water, and clean booths for installation works.

The KAGRA project is hosted by Institute for Cosmic Ray Resarch (ICRR) in the University of Tokyo together with two co-hosts, KEK (High Energy Accelerator Research Organization) and National Astronomical Observatory of Japan (NAOJ). The KAGRA collaboration has grown to include over 200 members from more than 60 universities and institutions in Japan and abroad.

Completing a long L-shaped tunnel in the Kamioka mine, Japan, is the first milestone of the KAGRA project, and finished on time at the end of March in 2014. The tunnel excavation started in 2012, which was delayed due to the impact of the Earthquake in 2011.

This article is part of a series of the conference proceedings looking over advanced ground-based gravitational-wave detectors, and here mainly covers KAGRA. If you wish to find common basic knowledges of ground-based laser interferometers, or an overall summary of the worldwide network of them, see the first one of the series. This short article briefly describes main features of KAGRA in section 2, the current status in section 3, and a summary in the last section.

2. Main features of KAGRA

KAGRA has two features that differ from other advanced ground-based interferometers [5]. One is that the whole interferometer will be built more than 200 m underground, below a mountain, for low seismic noise. The base rock of the mountain, in which the L-shaped tunnel

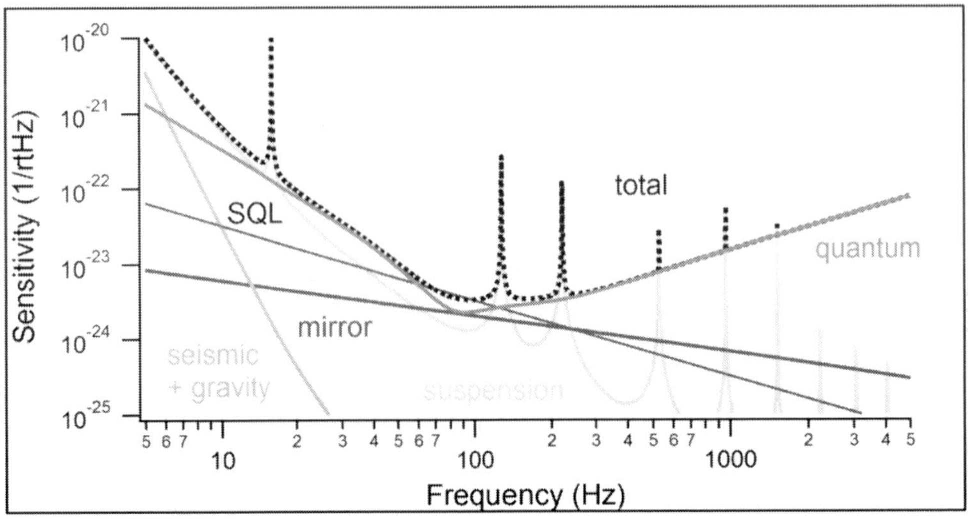

Figure 1. Estimated noise budget for bKAGRA [8]; see [5] for details.

for the interferometer is excavated, is expected to be quite stable with low seismic fluctuations. Seismic fluctuations in another tunnel but under the same mountain were measured to be about $10^{-9}/f^2$ m/Hz$^{1/2}$, where f is frequency, above 1 Hz frequency region; it was smaller by about 1/100 compared to TAMA300 in Tokyo in the same frequency region.

The other difference is that test masses for the 3-km Fabry-Perot cavities will be cooled down to 20 K for reducing thermal fluctuations, while a circulating power in each arm cavity will rise to the order of 100 kW. The reduction of mirror thermal noise with cryogenic mirrors has been already demonstrated in CLIO, a 100 m prototype interferometer in Kamioka [6]. The test masses for KAGRA (and also CLIO) are made of sapphire, which has better thermal conductivities than that of traditional silica mirrors at the low temperature. Thermal lensing effects in the test masses, which is due to the stored power in the arm cavities, can be reduced as well.

3. Overall plan and current states of KAGRA

3.1. Overall plan

Two major configuration phases are set in the overall schedule of KAGRA [5]. We are now in the first phase, called initial KAGRA (iKAGRA), and aim to develop basic technologies to operate a large interferometer as a gravitational-wave observatory in the underground; we will build a Michelson interferometer with two 3-km Fabry-Perot arm cavities, detector characterization system, data taking system, and data management system for an engineering run planned at the end of 2015. The test masses of the arm cavities are made of silica, suspended by rather simplified pendulums, and the whole interferometer is operated at room temperature.

The second phase, called baseline KAGRA (bKAGRA), will start from the beginning of 2016. In this phase, the iKAGRA interferometer will be upgraded, and it will reach the goal sensitivity (Figure 1). The silica test masses will be replaced by cryogenic ones made of sapphire, which will be cooled down to 20 K, and suspended by cryogenic suspensions. The interferometer will be modified with power recycling and signal recycling cavities, and operated with the resonant sideband extraction (RSE) technique to reach beyond the standard quantum limit [7].

Figure 2. Panorama photograph of the KAGRA tunnel in December 2012.

3.2. Construction status of iKAGRA

Tunnel excavation for KAGRA was completed at the end of March 2014. In the tunnel, two branches are separated by nearly 90 degrees with respect to one another and extend 3 km for the X and Y arm cavities (Figure 2). The branches are tilted by 1/300 for drainage of ground water; the highest altitude is at the X end, and the lowest at the Y end. The halfway point is at the corner of the branches, where a beam splitter will be installed. The corner station is 372 m above sea level. It is excavated by New Austrian Tunneling Method (NATM), and takes no more than two years. There are two teams for the tunneling; one starts from the corner of the branches toward both X and Y ends, and the other from the end of the Y arm branch. You can see a movie of the final blast to penetrate the Y arm at the halfway (1.5 km away from the corner) point here [9]. In bKAGRA phase, test masses will need long suspensions of 14 m height for their seismic attenuation system, so upper floors and pits to connect the upper and ground floors are also excavated above the chambers for the test mass.

Delivery of vacuum chambers and ducts for KAGRA is scheduled to start in August of 2014. In the iKAGRA phase, 23 chambers will be installed; 19 on the base floor and 4 on the upper. Each 3-km long arm duct is assembled from 12-m long ducts at the site. The inner surface of each vacuum duct is finished by electro-chemical buffing (ECB) method to reduce outgassing, so that the interferometer can be operated in vacuum of $\sim 10^{-7}$ Pa.

3.3. Research and development for bKAGRA

A cryogenic payload, the most challenging part in the KAGRA project, should support a sapphire test mass satisfying both low temperature and seismic noise. A conceptual design of this system is that the suspension wire from the upper floor at the room temperature comes downward via several pendulum stages to the cryogenic payload through an aperture on the top of the double inner shields, in which a sapphire test mass is suspended by sapphire wires or rods (Figure 3). Polishing of sapphire pathfinders (test pieces of substrates) produce satisfactory results so far, but the value of bulk absorption is a concern [10]. Mechanical loss of the coating on a sapphire is also investigated, and the result will be published later.

To conduct heat from a sapphire mirror, it is supported by sapphire wires from an upper stage (intermediate mass). Measurements show that heat conductivities of test pieces of sapphire fibers are more than 5000 W/m/K at 20 K, which is a promising result that they can be used as wires for the cryogenic payload [11].

The entire 3-km arm is not cooled, so a sapphire mirror will be exposed to the 300 K radiation from the room temperature parts. In order to reduce the entering radiation, two cryogenic duct shields of 5-m long each are installed on either sides of the sapphire mirror (Figure 3); they are also cooled to temperatures around 80 K and are designed to have an array of funnels inside to absorb or reflect this radiations [12].

Figure 3. Conceptual design of cryostat and cryogenic suspension for a sapphire test mass of KAGRA.

4. Summary

KAGRA is a Japanese large-scale cryogenic gravitational-wave telescope under construction at the Kamioka mine in Gifu prefecture, Japan. Unlike the other terrestrial advanced interferometers, the KAGRA interferometer is not a simple upgrade of previous one, and started with building a tunnel to make the underground site. The tunnel was finally completed at the end of March 2014. The first engineering run is planned at the end of 2015 with the initial configuration of the interferometer. The upgrade of the interferometer will start from the beginning of 2016 to reach the designed sensitivity of KAGRA.

Acknowledgments

The authors are grateful for support from the Japan Society for the Promotion of Science (JSPS) Core-to-Core Program A, Advanced Research Networks, and Grant-in-Aid for Specially Promoted Research of the KAGRA project.

References

[1] URL `http://gwcenter.icrr.u-tokyo.ac.jp`
[2] S Dwyer and the LIGO Scientific Collaboration 2015 Journal of Physics: Conference Series, to be published in this issue.
[3] F. Acernese et al (the Virgo Collaboration) 2015 Journal of Physics: Conference Series, to be published in this issue.
[4] K L Dooley and the LIGO Scientific Collaboration 2015 Journal of Physics: Conference Series, to be published in this issue.
[5] Somiya K 2012 *Classical and Quantum Gravity* **29** 124007
[6] Uchiyama T, Miyoki S, Telada S, Yamamoto K, Ohashi M, Agatsuma K, Arai K, Fujimoto M K, Haruyama T, Kawamura S, Miyakawa O, Ohishi N, Saito T, Shintomi T, Suzuki T, Takahashi R and Tatsumi D 2012 *Phys. Rev. Lett.* **108** 141101
[7] Aso Y, Michimura Y, Somiya K, Ando M, Miyakawa O, Sekiguchi T, Tatsumi D and Yamamoto H 2013 *Phys. Rev. D* **88** 043007
[8] URL `http://gwcenter.icrr.u-tokyo.ac.jp/researcher/parameters`
[9] URL `https://www.youtube.com/watch?v=tMjYdqI7G9o`
[10] Hirose E, Bajuk D, Billingsley G, Kajita T, Kestner B, Mio N, Ohashi M, Reichman B, Yamamoto H and Zhang L 2014 *Phys. Rev. D* **89** 062003

[11] Khalaidovski A, Hofmann G, Chen D, Komma J, Schwarz C, Tokoku C, Kimura N, Suzuki T, Scheie A O, Majorana E, Nawrodt R and Yamamoto K 2014 *Classical and Quantum Gravity* **31** 105004
[12] Sakakibara Y, Kimura N, Yamamoto K, Suzuki T, Tomaru T, Miyoki S, Uchiyama T and Kuroda K 2012 *Classical and Quantum Gravity* **29** 205019

Journal of Physics: Conference Series **610** (2015) 012017 doi:10.1088/1742-6596/610/1/012017

Pulsar Timing Arrays

Ryan S. Lynch
Department of Physics and Astronomy, West Virginia University, Morgantown, WV, 26506-6315, USA

Abstract.
Pulsars, and especially millisecond pulsars, can be used as precision clocks to explore a wide range of phenomena. An array of millisecond pulsars distributed across the sky can be used to detect low-frequency gravitational waves by searching for a correlated signature between pulsars, arising from the influence of gravitational waves at the Earth. These pulsar timing arrays complement other methods of gravitational wave detection in their frequency coverage and the physics they probe. I will introduce the basics of pulsar timing and discuss how it applies to the detection and study of the gravitational wave Universe. Pulsar timing array collaborations, current results, and future prospects will also be presented.

1. Pulsar Basics

A pulsar is a highly magnetized, rapidly rotating neutron star that emits beamed radio and high energy (X-ray and/or gamma-ray) radiation. A pulse of radiation is seen each time the beam crosses the line of sight of an observer, giving rise to a *pulse profile* that encodes information about the structure of the magnetosphere and the emission process (see Figure 1). Empirically, though there is substantial pulse-to-pulse variability, an average pulse profile emerges after $\sim$ 100–1000 rotations. This pulse profile, and the rotation of the pulsar itself, are observed to be very stable over long timescales (though see [1] for some caveats). These two factors allow pulsars to be used as precision astronomical clocks, suitable for studying a wide range of phenomena in exquisite detail.

Pulsars are born with periods $P \sim 10$s ms but quickly spin down due to the loss of rotational energy, to $P \sim 0.1$–1 s. Radio emission will continue for $\sim$ 10–100 Myr, after which the pulsar will become a quiet neutron star. However, if the pulsar exists in a binary system, it may find new life as a millisecond pulsar (MSP). In this scenario, mass and angular momentum are accreted from a companion star, spinning the pulsar back up to periods $P < 10$ ms while simultaneously decreasing the magnetic field to $B \sim 10^8$–10^9 G. Due to their lower magnetic fields, MSPs spin down much slower than non-recycled pulsars, having lifetimes greater than 10 billion years. To date, $\sim$ 260 MSPs have been discovered[1]. Figure 1 shows the known pulsars as a function of period and spin-down.

2. Pulsar Timing

The extremely high rotational stability of pulsars, and especially MSPs, make them precision clocks, suitable for measuring minute deviations in the expected *time of arrival* (TOA) of a pulse. This is accomplished through phase coherent *pulsar timing*, which unambiguously accounts for

[1] For an up-to-date list, see the ATNF pulsar catalog (`http://www.atnf.csiro.au/people/pulsar/psrcat/`)

Journal of Physics: Conference Series **610** (2015) 012017 doi:10.1088/1742-6596/610/1/012017

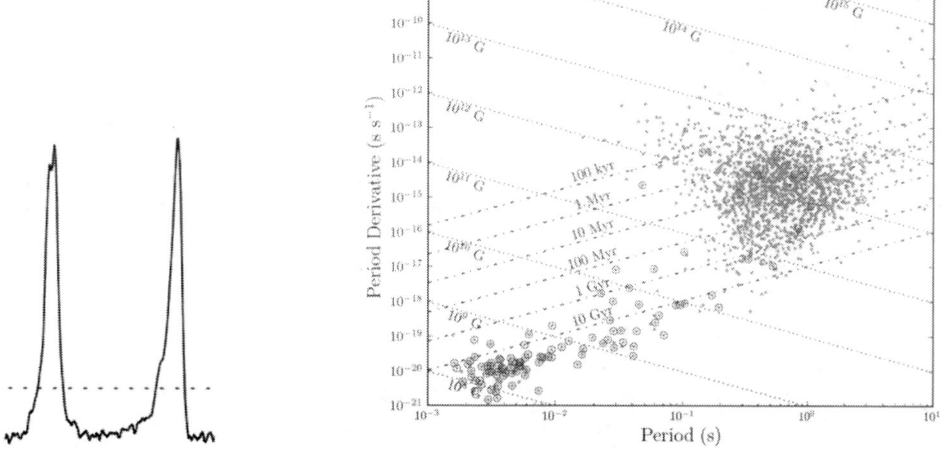

Figure 1. *Left:* A typical pulse profile, in this case of PSR J0741+6618, an MSP discovered in the Green Bank North Celestial Cap survey [2]. *Right:* The P-$\dot{P}$ diagram, showing pulsar spin period vs. spin-down. Circles indicate binary systems, and lines of constant magnetic field and age are indicated by dotted and dashed-dotted lines, respectively. Two distinct populations are clearly visible: the non-recycled pulsars, with $P \sim 0.1$–4 s and $\dot{P} \sim 10^{-17}$–10^{-13} s s^{-1}, and the fully recycled millisecond pulsars with $P \sim 3$ ms and $\dot{P} \sim 10^{-20}$ s s^{-1}, with partially recycled pulsars bridging the two populations. Data taken from the ATNF pulsar catalog.

every rotation of the pulsar over time-spans much greater than the star's rotational period. That is, the rotational phase of the pulsar at any time t (measured in an inertial reference frame with respect to the pulsar) is given by

$$\phi(t) = \phi(t_0) + f(t - t_0) + 1/2\dot{f}(t - t_0)^2 + \ldots \tag{1}$$

where t_0 is a reference epoch, $f = d\phi/dt$ is the rotational frequency, and $\dot{f}$ is the first time derivative of f. Higher order derivatives can sometimes be measured (particularly in young pulsars), but usually they are either too small (as is the case for most MSPs), or are contaminated by timing noise (see §3).

The first step in pulsar timing is to transform a TOA measured at the telescope to an approximately inertial reference frame with respect to the pulsar, typically the Solar system barycenter. Then,

$$t = t_{\rm topo} - t_{0,\rm topo} + \Delta_{\rm clock} - \Delta_{\rm DM} + \Delta_{\rm R\odot} + \Delta_{\rm E\odot} + \Delta_{\rm S\odot} + \Delta_{\rm R} + \Delta_{\rm E} + \Delta_{\rm S}. \tag{2}$$

Here, topo indicates time measured in the topocentric frame of the observatory, and the Δ's indicate various corrections and delays: $\Delta_{\rm clock}$ accounts for differences between the observatory clock and a terrestrial time standard, $\Delta_{\rm DM}$ corrects for a radio frequency-dependent dispersive delay as the signal travels through the ionized interstellar medium (ISM), $\Delta_{\rm R}$ is the classical Römer delay that accounts for light travel time across an orbit, $\Delta_{\rm E}$ is the Einstein delay, which takes into account time dilation due to orbital motion and the gravity of other bodies, and $\Delta_{\rm S}$ is the Shapiro delay. These orbital delays arise from the orbit of the Earth and the influence of the Sun and other bodies in the Solar system (indicated by $\odot$) and from the orbit of the pulsar and its companion when found in a binary system.

Two points are worth emphasizing. First, the various delays encode information that reveals a wealth of science, such as information about the pulsar orbit. Second, many of these corrections are time-dependent. The ISM is not static, so *dispersion measure* (DM, the column density of free electrons) must be measured at different epochs; clock corrections must be regularly updated; changes in the orbit of the pulsar owing to general relativistic effects or the presence of other bodies (planets or a third stellar companion) will manifest in the pulsar orbital terms.

The rotational, astrometric, and possible orbital parameters of a pulsar comprise a *timing model* that, along with the other corrections in Equation 2, predict the rotational phase of the pulsar at any given time as observed from the Solar system barycenter. The observed TOA of a pulse, or more precisely, a fiducial point on the pulse profile, can then be compared to this prediction. The difference between the observed and predicted arrival time is known as a *timing residual*. If we assume the fiducial point on the pulse profile always corresponds to the same rotational phase (i.e., that the profile does not drift or change), then an integer number of rotations must fit between any two TOAs. Model parameters are iteratively adjusted under this constraint to minimize the timing residuals. So long as the uncertainties in the model parameters do not propagate to a phase uncertainty $|\sigma_\phi| > 1$, phase coherence can be maintained, even though only a fraction of the total rotations of the pulsar are actually observed. This allows for extremely precise measurement of the model parameters, illustrating the power of pulsar timing.

3. The Limits and Challenges of Pulsar Timing

So far, we have not considered the influence of gravitational waves (GWs) on pulsar timing. Ultimately, we will be interested in measuring deviations in our timing residuals that are indicative of the influence of GWs on the Earth. Our sensitivity to GWs will depend in part on the uncertainty of our TOAs and our ability to model or otherwise account for so-called noise sources that will lead to non-zero residuals.

A TOA is calculated by cross-correlating a template and observed pulse profile. While this can be done in the time domain, far greater precision is achieved by working in the Fourier domain [3]. The TOA uncertainty is then $\sigma_{\mathrm{TOA}} \sim \mathrm{SNR}/\delta$, where SNR is the signal-to-noise of the observed profile and δ is the pulse width. It is common to obtain $\sigma_{\mathrm{TOA}} \sim 1$ milliperiod, which for MSPs corresponds to $\sim \mu$s. However, for the best MSPs $\sigma_{\mathrm{TOA}} \sim 10$ ns is possible.

Processes that lead to non-zero residuals can be broadly classified into two groups, depending on whether their power spectral density is flat or rises towards low frequency, i.e. *white* and *red noise* processes (see the contribution by Y. Wang in these proceedings). White noise processes include radiometer noise, pulse jitter, and interstellar scintillation, while red noise processes include timing noise, DM variations, and GWs. Each will be discussed in turn.

Radiometer noise arises from the observing system, atmosphere, cosmic microwave background, and Galactic synchrotron emission. Radiometer noise can be reduced by increasing observing time and radio bandwidth, observing with larger telescopes, and by using low-noise receivers and amplifiers. Interstellar scintillation (ISS) refers to scattering due to inhomogeneities in the ISM, analogous to the twinkling of stars. ISS can be mitigated by increasing observing bandwidths and integration times. Pulse jitter is a property of the pulsar itself [e.g. 4]. Although the average pulse profile is observed to be very stable over time, individual pulses vary widely in shape. Because the observed profile is comprised of a finite number of pulses, one never achieves a perfectly stable pulse profile. Pulse jitter can only be overcome by increasing observing time, i.e. summing more pulses. Pulse jitter is currently the limiting factor in timing stability for a few pulsars, and it will become more important as more sensitive observing systems reduce the impact of radiometer noise and ISS [5].

Timing noise is a well known process in non-recycled pulsars owing to torque variations on the neutron star. The effect is usually large enough that it contaminates the measurement of higher-order rotational frequency derivatives. It is observed to correlate with rotational period, and so is not prominent in most MSPs, though it may become so as timing precision improves and the length of data sets grow [6]. DM variations are more common (see the contribution from

Journal of Physics: Conference Series **610** (2015) 012017 doi:10.1088/1742-6596/610/1/012017

L. Levin in these proceedings). The ISM is turbulent, and both the pulsar and the Earth are in motion with respect to the ISM, so that over time the pulsar signal travels through a different column density of free electrons, i.e. different DMs. As such, Δ_{DM} in Equation 2 becomes $\Delta_{\mathrm{DM}}(t)$. High-precision measurements of DM must be made for each observing epoch.

4. Using Pulsars to Detect Gravitational Waves

Assuming that we have accurately calculated all timing parameters and effectively mitigated all noise sources, we are left with timing residuals that may contain a measurable effect of GWs [7, 8]. The nature of the signature depends on the GW source. Continuous GWs emitted by a single supermassive binary black hole (SMBBH) will lead to sinusoidal deviation in timing residuals, while the superposition of all GW-emitting SMBBH in the Universe will have a stochastic signature (see the contribution by C. Moore in these proceedings). The goal is to detect and characterize this signature with a high degree of confidence. To do this, we should search for characteristics that are uniquely attributable to GWs, as opposed to other processes, namely the quadripolar signature of GWs.

GWs will influence the pulsars themselves, as well as the telescopes we use to observe them. The so-called *pulsar term* will not be correlated between different pulsars, and although it is recoverable in principle if the distance to the pulsar is accurately known, in practice it is not likely to be accessible using current analysis techniques. The *Earth-term*, however, will be correlated between different pulsars, and with a unique dependence on the angular separation between pulsar pairs owing to the quadripolar nature of GWs. This forms the basis for pulsar timing arrays (PTAs). Correlated signatures in timing residuals can arise from other processes, but not with this quadripolar nature. For example, an error in Δ_{clock} will have a monopolar signature, while an error in $\Delta_{\mathrm{S}\odot}$ (due to an incorrect planetary mass) will have a dipolar signature.

The precise form of the correlation between pulsars is the famous Hellings-Downs curve, or overlap reduction function [9] (see Figure 2). Conceptually, it can be formed by simply cross-correlating timing residuals from pulsars distributed across the sky. The frequency of GWs that can be detected using this technique is bounded by the cadence of pulsar observations and the time-span of the timing data-set. This is currently ~ 1/years–1/weeks, or $\sim 10^{-9}$ Hz–10^{-7} Hz. At these frequencies, the most likely sources are believed to be SMBBH in the early stages of inspiral. Cosmic strings, if they exist, are also expected to emit at these frequencies. Inflationary GWs may also exist in this frequency range, but are likely to be well below PTA sensitivity. As such, PTAs are an excellent complement to CMB polarization experiments and space or ground based laser interferometers (see Figure 2).

In practice, the small strain of GWs is expected to manifest in timing residuals at the level of ~ 10 ns, and it is technically challenging to obtain the necessary timing precision and theoretically challenging to model the relevant noise sources, while timing noise could pose a floor to the achievable residuals. Computationally, timing model parameters and noise processes are modeled simultaneously in a GW detection pipeline, which requires the inversion of large matrices and advanced analysis techniques. Nonetheless, several teams around the world are making excellent progress addressing all of these challenges.

5. PTA Collaborations

There are currently three major PTA collaborations. The North American Nanohertz Observatory for Gravitational Waves (NANOGrav) is made up of researchers at institutions in the United States and Canada [10]. NANOGrav primarily uses the 100-m Green Bank Telescope (GBT) and the 305-m Arecibo Observatory to time 42 MSPs with approximately monthly cadence. Arecibo is the most sensitive radio telescope in the world but has a limited field-of-view and can only see a subset of the pulsars. The GBT, which is fully steerable and can observe to a declination of $\delta \approx -46°$, observes the MSPs that Arecibo cannot.

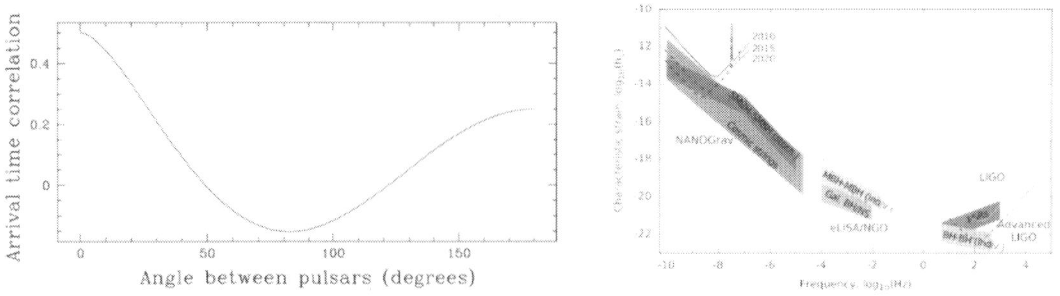

Figure 2. *Left:* The Hellings-Downs curve showing the expected angular correlation between pulsar timing residuals as a function of angular separation. The maximum correlation factor is 0.5, because only the Earth-term is correlated, while the pulsar-term is not. *Right:* A schematic representation of expected sources in the frequency range of PTAs and space and ground based interferometers. Approximate sensitivity curves are shown for various experiments. Figures courtesy of the NANOGrav collaboration.

The European Pulsar Timing Array (EPTA) [11] uses five telescopes: The 100-m Effelsberg Radio Telescope in Germany, the 76-m Lovell Telescope in England, the 64-m Sardinia Radio Telescope in Italy, the Westerbork Synthesis Radio Telescope (made up of 14, 25-m dishes) in the Netherlands, and the Nançay Radio Telescope in France, a Kraus-type design consisting of a flat primary and cylindrical secondary surface. The EPTA is able to use these telescopes to observe with a higher cadence than other PTAs. The LEAP (Large European Array for Pulsars) project is working to use all these telescopes as a phased array. When completed, the array will have a collecting area equivalent to a 194-m telescope.

The Parkes Pulsar Timing Array (PPTA) uses the 64-m Parkes Observatory in Australia [12]. As the southernmost telescope currently used for PTA observing, Parkes is able to see many southern hemisphere MSPs that are not visible to NANOGrav and EPTA telescopes, including PSR J0437−4715, the closest, brightest, and best timed MSP. PPTA members are primarily found in Australia and China.

All three major PTAs collaborate under the umbrella of the International Pulsar Timing Array (IPTA) [13]. The IPTA is a "collaboration of collaborations" with the goal of facilitating GW science through data and code sharing and greater cooperation. An IPTA data-set consisting of timing data from all three PTAs is currently being assembled. When complete, it will provide the greatest sensitivity to nanohertz-frequency GWs. But activities in all three PTAs and the IPTA are not limited to timing observations and data analysis. Surveys for new MSPs, instrumentation development, theoretical modeling of the nanohertz GW Universe, multi-messenger astronomy, and education and public outreach are all major projects in each PTA and the IPTA as a whole. Furthermore, PTA data sets have utility far beyond detecting GWs. Individual pulsars themselves are often excellent astrophysical laboratories, and there are currently IPTA-wide projects to develop a pulsar-based time standard (independent of terrestrial time standards) and to refine Solar system ephemerides.

6. Current Results and Future Prospects

The best limit on the amplitude of the stochastic GW background currently comes from the PPTA [14] (see Figure 3). These results are putting significant constraints on certain SMBBH merger models. The current best limit on the amplitude of single sources come from NANOGrav [15], and constrains the chirp mass of any SMBBHs in the Virgo cluster.

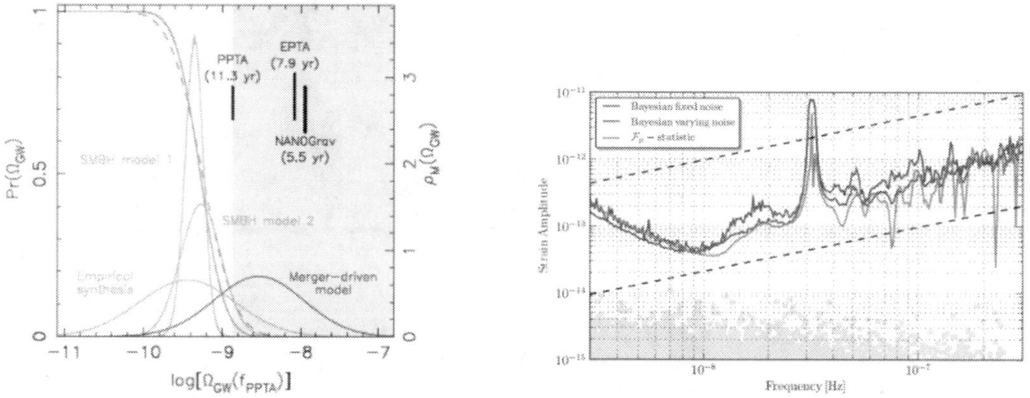

Figure 3. The current best results from PTAs. *Left:* The PPTA 11.3-year 95% confidence upper limit on the density of GWs, $\Omega_{\rm GW}$ due to a stochastic background, from [14]. Probability density functions for the predictions of various models for SMBBH inspirals are also shown, with the PPTA results excluding the shaded region. Earlier NANOGrav and EPTA results are also shown. *Right:* NANOGrav 95% confidence upper limits on strain amplitude due to individual, continuous GW sources from [15]. Both a Bayesian (blue and purple lines) and frequentist (red line) analysis are shown, along with the amplitude of the strongest sources in 1000 Monte Carlo realizations based on [16]. Dashed lines indicate the expected strain amplitude for a source in Virgo with chirp mass of 10^9 $M_\odot$ and 10^{10} $M_\odot$.

These limits are expected to improve substantially over the next 5–10 years as the length of the data sets increase and new pulsars are added to the arrays. The upcoming 9-year data release from NANOGrav and the combined IPTA data release will improve upon the PPTA result. Based on current estimates, a GW detection is likely in as little as three years, and almost certainly by 2022. The biggest variable is the unknown amplitude of the stochastic GW background, which will likely be the first detection, though the red noise properties of the data set (mostly arising from possible timing noise) are also important.

Detection is merely the first step in studying the low-frequency GW Universe. The frequency dependence of the stochastic background will shed light on the merger history of galaxies. PTAs will also contribute to multi-messenger astronomy: they are sensitive to SMBBHs with separations $\sim 10^{-1}$–10^{-4} pc, and electromagnetic counterparts at these stages of merger include circumbinary disk emission, dual nuclei, and periodic changes in radio jets, among others. However, alternative explanations for some of these phenomena are available, so a PTA single-source detection can provide supporting evidence for a SMBBH interpretation.

New telescopes will significantly contribute to PTA science. The Five Hundred Meter Aperture Spherical Telescope being built in China will provide Arecibo-like sensitivity over a wider range of declinations, significantly increasing the timing precision that can be achieved for many pulsars. The increased sensitivity will also allow pulsars that are currently too weak to achieve the requisite timing precision to be added to the arrays. No current or planned telescopes will be able to match the sensitivity expected from the Square Kilometer Array, which will time hundreds of pulsars with sub-microsecond precision. This will greatly increasing PTA sensitivity, especially to single sources as this is currently dominated by the two or three most precise MSPs. In the meantime, new receivers and back-ends will improve upon the telescopes that are already in use.

Over the next several years and decades, PTAs will provide insight into a new and unique regime of the GW Universe, nicely complementing ground and space based interferometers.

7. Additional Resources

The Handbook of Pulsar Astronomy [17] provides an excellent introduction to pulsar timing and many other areas of pulsar astronomy. More information about all areas of PTA science can be found in the focus issue of *Classical and Quantum Gravity* (2013, Volume 30, Issue 22).

References

[1] Lyne A, Hobbs G, Kramer M, Stairs I and Stappers B 2010 *Science* **329** 408–412 (*Preprint* 1006.5184)

[2] Stovall K, Lynch R S, Ransom S M, Archibald A M, Banaszak S, Biwer C M, Boyles J, Dartez L P, Day D, Ford A J, Flanigan J, Garcia A, Hessels J W T, Hinojosa J, Jenet F A, Kaplan D L, Karako-Argaman C, Kaspi V M, Kondratiev V I, Leake S, Lorimer D R, Lunsford G, Martinez J G, Mata A, McLaughlin M A, Roberts M S E, Rohr M D, Siemens X, Stairs I H, van Leeuwen J, Walker A N and Wells B L 2014 *ApJ* **791** 67 (*Preprint* 1406.5214)

[3] Taylor J H 1992 *Royal Society of London Philosophical Transactions Series* A **341** 117–134

[4] Liu K, Keane E F, Lee K J, Kramer M, Cordes J M and Purver M B 2012 *MNRAS* **420** 361–368 (*Preprint* 1110.4759)

[5] Shannon R M, Osłowski S, Dai S, Bailes M, Hobbs G, Manchester R N, van Straten W, Raithel C A, Ravi V, Toomey L, Bhat N D R, Burke-Spolaor S, Coles W A, Keith M J, Kerr M, Levin Y, Sarkissian J M, Wang J B, Wen L and Zhu X J 2014 *MNRAS* **443** 1463–1481 (*Preprint* 1406.4716)

[6] Shannon R M and Cordes J M 2010 *ApJ* **725** 1607–1619 (*Preprint* 1010.4794)

[7] Foster R S and Backer D C 1990 *ApJ* **361** 300–308

[8] Jenet F A, Hobbs G B, Lee K J and Manchester R N 2005 *ApJL* **625** L123–L126 (*Preprint* astro-ph/0504458)

[9] Hellings R W and Downs G S 1983 *ApJL* **265** L39–L42

[10] McLaughlin M A 2013 *Classical and Quantum Gravity* **30** 224008 (*Preprint* 1310.0758)

[11] Kramer M and Champion D J 2013 *Classical and Quantum Gravity* **30** 224009

[12] Hobbs G 2013 *Classical and Quantum Gravity* **30** 224007 (*Preprint* 1307.2629)

[13] Manchester R N and IPTA 2013 *Classical and Quantum Gravity* **30** 224010 (*Preprint* 1309.7392)

[14] Shannon R M, Ravi V, Coles W A, Hobbs G, Keith M J, Manchester R N, Wyithe J S B, Bailes M, Bhat N D R, Burke-Spolaor S, Khoo J, Levin Y, Oslowski S, Sarkissian J M, van Straten W, Verbiest J P W and Wang J B 2013 *Science* **342** 334–337 (*Preprint* 1310.4569)

[15] Arzoumanian Z, Brazier A, Burke-Spolaor S, Chamberlin S J, Chatterjee S, Cordes J M, Demorest P B, Deng X, Dolch T, Ellis J A, Ferdman R D, Garver-Daniels N, Jenet F, Jones G, Kaspi V M, Koop M, Lam M, Lazio T J W, Lommen A N, Lorimer D R, Luo J, Lynch R S, Madison D R, McLaughlin M, McWilliams S T, Nice D J, Palliyaguru N, Pennucci T T, Ransom S M, Sesana A, Siemens X, Stairs I H, Stinebring D R, Stovall K, Swiggum J, Vallisneri M, van Haasteren R, Wang Y and Zhu W W 2014 *ArXiv e-prints* (*Preprint* 1404.1267)

[16] Sesana A 2013 *MNRAS* **433** L1–L5 (*Preprint* 1211.5375)

[17] Lorimer D R and Kramer M 2012 *Handbook of Pulsar Astronomy*

Journal of Physics: Conference Series **610** (2015) 012018 doi:10.1088/1742-6596/610/1/012018

The sensitivity of pulsar timing arrays

C J Moore

Institute of Astronomy, Madingley Road, Cambridge, CB3 0HA, United Kingdom

E-mail: cjm96@ast.cam.ac.uk

Abstract. Simple analytic expressions are derived for the sensitivity curve of a pulsar timing array (PTA) to both a monochromatic source of gravitational waves and an isotropic stochastic background of gravitational waves. These derivations are performed in both frequentist and Bayesian frameworks and the results verified with numerical injections and recovery in mock PTA datasets. The consistency of the frequentist and Bayesian approaches is demonstrated and the results are used to emphasise the fact that the sensitivity curve of a PTA depends not only on the properties of the PTA itself, but also on the properties of the gravitational wave source being observed.

1. Introduction

Pulsar timing arrays (PTAs) are ongoing attempts to detect low frequency gravitational waves (GWs) by using the timing of a network of galactic millisecond pulsars. The chosen millisecond pulsars are extremely stable rotators which allow a timing model of sufficient accuracy to be constructed to account for every single rotation of the pulsar over the many years of observation. After subtracting the timing model from the pulse arrival times the remaining timing residuals include any unmodeled effects; e.g. timing noise and (possibly) GWs. All of the pulsars reside in the same GW field and hence the induced timing residuals in different pulsars are correlated by an amount that depends on their angular separation. It is this correlation which allows a network of pulsars spread across the sky to be exploited as a GW detector.

PTAs are sensitive to GWs in the range $10^{-9}\,\mathrm{Hz} \lesssim f \lesssim 10^{-6}\,\mathrm{Hz}$ and are typically most sensitive to the lowest frequencies in this range. Potential sources of GWs in this frequency band are individual supermassive black hole binaries with chirp masses around $\mathcal{M} \approx 10^9 M_\odot$ out to redshifts of $z \approx 2$ ((1; 2)). If there are many sources per unit frequency then they cease to be individually resolvable and instead overlap to form a stochastic background of GWs ((3; 4; 5)). It is currently unclear whether the first detection of GWs by PTAs will be an individual binary or a stochastic background. Stochastic backgrounds may also be caused by other mechanisms such as cosmic string networks (6), or primordial GWs ((7; 8)).

There are currently three PTAs operating: the European Pulsar Timing Array (EPTA[1], (9)), the Parkes Pulsar Timing Array (PPTA[2], (10)) and the North American Nanohertz Observatory for Gravitational Waves (NANOGrav[3], (11)). There are plans to merge the seperate PTAs to form the International Pulsar Timing Array (IPTA[4], (12)).

These proceedings are based on a talk given at the 10th International LISA Symposium, this talk in turn was based largely on the work done in (13). In Sec. 2 of this paper the properties of the PTA which will be assumed throughout the remainder of the paper are summarised, Sec.

[1] `http://www.epta.eu.org/`
[2] `http://www.atnf.csiro.au/research/pulsar/ppta/`
[3] `http://nanograv.org/`
[4] `http://www.ipta4gw.org/`

Journal of Physics: Conference Series **610** (2015) 012018 doi:10.1088/1742-6596/610/1/012018

3 describes the sensitivity of this PTA to an individual binary, Sec. 4 describes the sensitivity to a stochastic background and final concluding remarks are given in Sec. 5.

2. Our canonical PTA

In order to produce definite values for the sensitivity curves it is necessary to assume particular values for the PTA parameters. For all calculations in this paper the PTA was assumed to consist of 36 pulsars randomly placed on the sky all timed with a cadence of 1/(2 weeks) for a total of 5 years with a precision of 100 ns. The choice of the cadence of the measurements is important for the numerical calculations as it determines the maximum frequency that can be detected; however, it also enters in the results for the analytic sensitivity curves through the noise power spectral density defined in Eq. 2. These choices of PTA parameters are comparable to the OPEN1 dataset in the recent IPTA data challenge[5]. It should be noted that the performance of this canonical PTA is significantly better than any of the current PTAs. The IPTA will have a dataset of roughly this number of pulsars timed for this long in the next few years, however only a subset of these will be timed to an accuracy of 100ns.

3. Sensitivity to an individual binary

Binary black holes, in contrast to the stochastic backgrounds considered in Sec. 4, are a GW source which can be accurately modelled over the entire duration of observations; hence matched templates may be constructed for the purpose of detection and parameter estimation. PTAs are sensitive to individual binary black holes with very large chirp masses ($10^8 M_\odot \lesssim \mathcal{M} \lesssim 10^{10} M_\odot$) in the early, adiabatic stage of their inspiral. These systems are typically in circular orbits evolving on timescales longer than the total length of our observations and appear in the data as quasi-monochromatic signals. Therefore, for simplicity, in this section we consider the sensitivity of our PTA to a monochromatic source of GWs as a proxy for an individual binary.

3.1. Frequentist detection

The frequentist procedure for claiming a detection of a signal in the PTA dataset is to define a scalar detection statistic, $\mathcal{S}$, which is a function of the dataset. The signal to noise ratio (SNR) of this statistic, ϱ, is then defined as the expectation of this statistic in the presence of a signal divided by the square root of the variance of the statistic in the absence of a signal. A detection is claimed if the value of the detection statistic evaluated on the actual measured data exceeds some predetermined threshold, ϱ_{th} (the value of this threshold may be tuned to fix the false alarm rate).

Here the detection statistic was chosen to be the cross correlation of two data streams with the optimal (or "matched") filter function, $\tilde{\mathbf{K}}(f)$; see Eq. 1. This statistic has the advantage that it gives an SNR which is quartic in the signal amplitude (see Eq. 1, we tentatively venture to call this the "power SNR" in contrast to the usual "amplitude" SNR) and hence may be directly compared with the stochastic background SNR (see Sec. 4). It should be noted that in the case of a single binary it is also possible to use the usual matched filter statistic and associated amplitude SNR for the purposes of detection, but in the case of the stochastic background one is forced to use the cross correlation statistic because the signal in each pulsar is swamped by the intrinsic pulsar red-noise. The sensitivity of cross correlation statistic is suboptimal when compared with the matched filter, but it becomes equal in the limit of large signal amplitude. As we here impose a detection threshold equivalent to a Bayes factor of 1000 this approximation seems reasonable. Further justification for the accuracy of this approach come from comparison with the numerical results in Fig. 2 which were derived from an optimal (Bayesian) approach.

$$\mathcal{S} = \int \mathrm{d}t \int \mathrm{d}t' \, \mathbf{s}(t)^T \mathbf{K}^\dagger(t-t') \mathbf{s}(t') = \int \mathrm{d}f \, \tilde{\mathbf{s}}(f)^T \tilde{\mathbf{K}}^\dagger(f) \tilde{\mathbf{s}}(f) \tag{1}$$

$$\tilde{\mathbf{K}}(f) = \left. \frac{\tilde{\mathbf{h}}(f)\tilde{\mathbf{h}}(f)^T}{S_n^2} \right|_{\text{diag}\to 0} \Rightarrow \varrho^2 = \sum_y \sum_{x>y} \frac{8}{T} \int \mathrm{d}f \, \frac{|\tilde{h}_x(f)|^2 |\tilde{h}_y(f)|^2}{S_n^2} .$$

[5] `http://www.ipta4gw.org/?page_id=89`

In evaluating the SNR in Eq. 1 the noise terms in the timing residuals from pulsars x and y have been assumed to be white, Gaussian and uncorrelated between pulsars. If pulsar x is timed with precision σ_x and with a cadence $1/\delta t_x$ then this leads to the familiar expression

$$\langle \tilde{n}_x^*(f)\tilde{n}_y(f')\rangle = \frac{1}{2}S_n(f)\delta_{xy}\delta(f-f') \quad \text{where} \quad S_n = 2\sigma_x^2\delta t_x. \tag{2}$$

The red curve in Fig. 1 shows the sensitivity curve obtained by setting $\varrho = \varrho_{\rm th}$ in Eq. 1 and rearranging to find the amplitude as a function of frequency. This sensitivity curve has the obvious problem that it tends to a constant at low frequencies. This is because the need to use the measured data to fit for the free parameters in the pulsar timing model hasn't been accounted for. Heuristically the effect of this fitting may be understood by approximating the timing model as a quadratic in time. To motivate the choice of a quadratic notice that a constant offset in the timing residuals is degenerate with mis-estimating the distance to the pulsar, a linear drift in the timing residuals is degenerate with the pulsar spin and a quadratic trend in the timing residuals is degenerate with the pulsar spin-down rate. Expanding the sinusoidal signal at low frequencies ($ft \ll 1$) as a Taylor series in powers of ft allows us to include the effect of fitting for the quadratic timing model (see (13)) by subtracting off the first three terms in this series (up to and including $\mathcal{O}(ft)^2$). The resulting sensitivity curve is shown in Fig. 1. This procedure is rather *ad-hoc*, a much more satisfactory way of accounting for the timing model is provided by the Bayesian approach, which we now discuss.

3.2. Bayesian detection

The Bayesian procedure for claiming a detection of a signal in the PTA dataset is to consider two competing hypotheses which attempt to describe the measured data, $\mathbf{s}$: the noise hypothesis, $\{\mathcal{H}_n : \mathbf{s} = \mathbf{n} + \mathbf{m}\}$; and the signal hypothesis, $\{\mathcal{H}_h : \mathbf{s} = \mathbf{n} + \mathbf{m} + \mathbf{h}\}$. Here $\mathbf{n}$ denotes the random noise, $\mathbf{m}$ the timing model and $\mathbf{h}$ the signal; any of $\mathbf{n}$, $\mathbf{m}$ or $\mathbf{h}$ may depend on a number of free parameters, $\vec{\theta}$. For each hypothesis $i \in \{n, h\}$ the likelihood, $\mathcal{L}_i$, may be evaluated for a given choice of the parameters. The evidence for each hypothesis, $\mathcal{O}_i$, may be calculated by integrating, or *marginalising*, the likelihood over all of the free parameters in that hypothesis (weighted by any prior knowledge of each parameter). The odds ratio (or Bayes factor) is then defined as the ratio of the evidences $\mathcal{B} = \mathcal{O}_h/\mathcal{O}_n$ and a detection is claimed if the Bayes factor exceeds some predetermined threshold, $\mathcal{B}_{\rm th}$. The threshold Bayes factor is the Bayesian equivalent of the frequentist threshold SNR, and $\mathcal{B}_{\rm th}$ may be chosen to fix a desired false alarm rate.

Here the timing model was taken to be a quadratic in each pulsar, $m_x(t) = \alpha_x + \beta_x t + \gamma_x t^2$, where α_x, β_x and γ_x are free parameters with flat priors. These parameters are then marginalised over when evaluating the evidence for each hypothesis, this provides a rigorous method for accounting for the loss of sensitivity at low frequencies caused by fitting for the timing model. In reality the timing model is significantly more complex than a simple quadratic; it must account for effects as diverse as the Roemer delay, dispersion of the pulse by the interstellar medium and the orbital motion of the pulsar (if it happens to be in a binary) ((14; 15)). However the simple quadratic timing model considered here captures well the qualitative effect of fitting for the pulsar parameters and has the attractive feature that the free parameters (α_x, β_x and γ_x) may be marginalised over analytically.

The likelihood functions for both hypotheses may be written in terms of the noise covariance matrix, $\mathbf{\Sigma}_n = \sigma\mathbb{1}$, where it has been assumed for simplicity that $\sigma_x = \sigma$, $\forall x$,

$$\log\mathcal{L}_n(\vec{\Theta}) = \log A - \tfrac{1}{2}\left(\mathbf{s} - \mathbf{m}(\vec{\Theta})\right)^{\rm T}\mathbf{\Sigma}_n^{-1}\left(\mathbf{s} - \mathbf{m}(\vec{\Theta})\right)\,, \tag{3}$$

$$\log\mathcal{L}_h(\vec{\Theta},\vec{\Psi}) = \log A - \tfrac{1}{2}\left(\mathbf{s} - \mathbf{m}(\vec{\Theta}) - \mathbf{h}(\vec{\Psi})\right)^{\rm T}\mathbf{\Sigma}_n^{-1}\left(\mathbf{s} - \mathbf{m}(\vec{\Theta}) - \mathbf{h}(\vec{\Psi})\right)\,.$$

The evidence for each hypothesis may be evaluated by marginalising over the free parameters; for the pulsar parameters, $\vec{\Theta}$, flat priors were used, for the source parameters, $\vec{\Psi}$, delta function priors at the true values were used. The choice of delta function priors may seem unreasonable at first, but in reality it is simply a convenient way of imposing a large signal approximation. A detection of GWs is only claimed if the Bayes factor exceeds some threshold, this threshold must be chosen to be large if we are to have high confidence in the detection. In the limit of a large Bayes factor the prior becomes uninformative compared to the data and the value of the Bayes factor obtained in the experiment is independent of the choice of prior. Therefore a delta-function prior is chosen to ease the analytic calculations. For a further discussion of this point see (13). The evidence for hypothesis $\mathcal{H}_i$ is given by

$$\mathcal{O}_i(\mathbf{s}) = \int \mathrm{d}\vec{\lambda}_i \; \mathcal{L}_i(\mathbf{s}, \vec{\lambda}_i) P_i(\vec{\lambda}_i) \,, \quad \text{where } \lambda_n = \vec{\theta} \text{ and } \lambda_h = \left\{ \vec{\theta}, \vec{\psi} \right\} \,. \tag{4}$$

The Bayes factor, $\mathcal{B}$, may now be evaluated and the expectation over noise realisations calculated,

$$\overline{\mathcal{B}} = \int \mathrm{d}\mathbf{n} \; P(\mathbf{n})\mathcal{B} \quad \text{where } P(\mathbf{n}) = \frac{\exp\left(-\frac{1}{2}\mathbf{n}^{\mathrm{T}}\mathbf{\Sigma}_n^{-1}\mathbf{n}\right)}{\sqrt{(2\pi)^{N_p T/\delta t}\det\left(\mathbf{\Sigma}_n\right)}} \,. \tag{5}$$

The integral in Eq. 5 can be performed analytically and inverted to give an analytic expression for the minimum detectable amplitude as a function of frequency. This expression is too lengthy to reproduce here (see (13)) but is plotted in Fig. 1. Both of the analytic curves shown in Fig. 1 should be compared with the results of the numerical injections and recovery shown in Fig. 2.

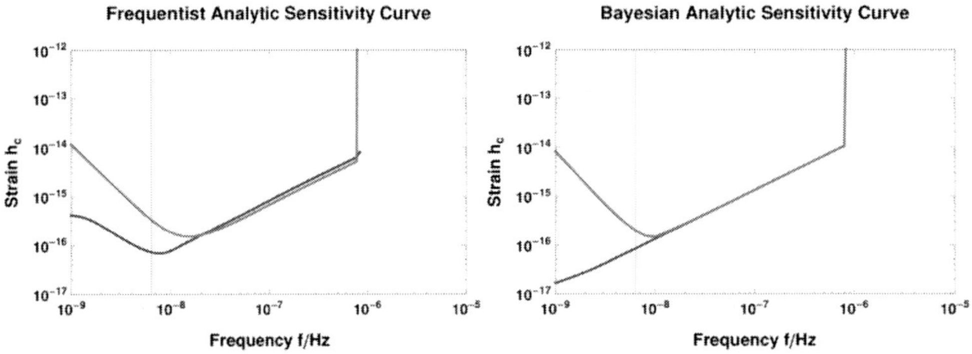

Figure 1: The sensitivity of our canonical PTA to a monochromatic source. The left-hand panel shows the prediction of the frequentist analytic formula in Sec. 3.1 with (red) and without (blue) the effect of fitting for the timing model; a threshold SNR of $\varrho = 3$ was used. The right-hand panel shows the prediction of the Bayesian analytic formula in Sec. 3.2 with (red) and without (blue) the effect of marginalising over the timing model parameters. The vertical black line indciates the frequency of 1/year. In both panels it can be seen that the effect of including the timing model is a reduction in sensitivity at low frequencies. (Figure reproduced with modifications from (13).)

4. Sensitivity to a stochastic background of GWs

A stochastic background of gravitational waves is completely specified by the power spectral density, $S_h(f)$, which is usually taken to be a power law in frequency,

$$\left\langle \tilde{h}_x(f)\tilde{h}_y^*(f') \right\rangle = \frac{1}{2}\delta(f-f')\Gamma_{xy}S_h(f) \,, \quad \text{where } S_h(f) = \frac{A^2}{12\pi^2 f_0^3}\left(\frac{f}{f_0}\right)^{-\gamma} \,. \tag{6}$$

The constant Γ_{xy} is called the overlap reduction function and, for an isotropic background, depends only on the relative sky positions of pulsars x and y. As can be seen from Eq. 6 the signal

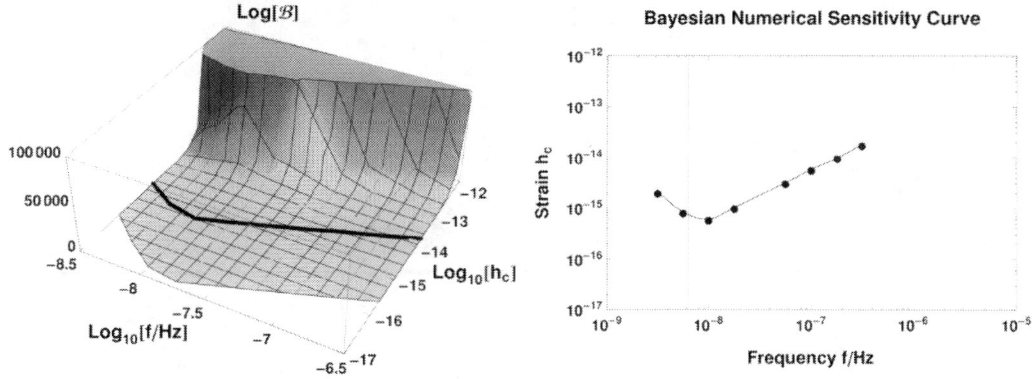

Figure 2: The results of the recovery of numerical injections of a monochromatic signal into the mock dataset. The left-hand panel shows a surface plot of the Bayes factor against the amplitude and frequency of a monochromatic source with the contour $\mathcal{B} = \mathcal{B}_{\text{th}}$ indicated by a solid black line. The right-hand panel shows the contour $\mathcal{B} = \mathcal{B}_{\text{th}}$ in a form more comparable to Fig. 1. For more detail on the numerical calculations used to produce these plots see (13). (Figure reproduced from (13).)

has contributions from all frequencies, in contrast to the monochromatic signal in Sec. 3. When calculating the sensitivity curve it no longer makes sense to calculate the minimum detectable amplitude as a function of frequency, $h_c(f)$. Instead the minimum detectable amplitude of the background in terms of the power-law slope may be determined, $A(\gamma)$. Unfortunately the interpretation of this quantity is somewhat opaque, so an approach suggested by (16) can be used. The parameter γ is gridded with n values selected between some large-negative and large-positive limits. For each value of γ the value $A(\gamma)$ may be calculated, giving the set of ordered pairs (γ_i, A_i) for $i \in \{1, 2, \ldots n\}$. For each pair the power-law curve for characteristic strain, $h_c = A(\gamma)\,(f/f_0)^{(\gamma-3)/2}$, may be drawn. This gives a set of power-law curves (straight lines on a log-log plot) for characteristic strain, the envelope of this set of curves is then interpreted as the *power-law-integrated* sensitivity curve (16), See Figs. 3 and 4.

4.1. Frequentist detection

Similar calculations to those performed in Sec. 3.1 may now be performed for the stochastic background. The principal difference is that in the expression for the filter function $\tilde{\mathbf{K}}(f)$ (Eq. 1) the product $\tilde{h}_x(f)\tilde{h}_y(f)$ is replaced by the expectation value $\langle \tilde{h}_x(f)\tilde{h}_y(f) \rangle$. The resulting expression for the SNR is (see (13))

$$\rho^2 = \sum_{x>y} \sum_{y} 8T \int \mathrm{d}f \, \frac{\Gamma_{xy}^2 S_h^2(f)}{S_n^2(f)} \,. \tag{7}$$

Setting $\rho = \rho_{\text{th}}$ in Eq. 7 and inverting gives the expression $A(\gamma)$ (13) and the power-law-integrated sensitivity curve may be drawn following the procedure outlined above, see Fig. 3.

4.2. Bayesian detection

Similar calculations to those performed in Sec. 3.2 may now be performed for the stochastic background. This yields an analytic expression for the expectation value of the Bayes factor, $\overline{\mathcal{B}}$. Setting $\overline{\mathcal{B}} = \mathcal{B}_{\text{th}}$ and inverting gives an analytic expression for $A(\gamma)$. The analytic expressions are too lengthy to reproduce here, but can be found in (13). Using the expression $A(\gamma)$ the power-law-integrated sensitivity curve may be drawn following the procedure outlined above, see Fig. 3.

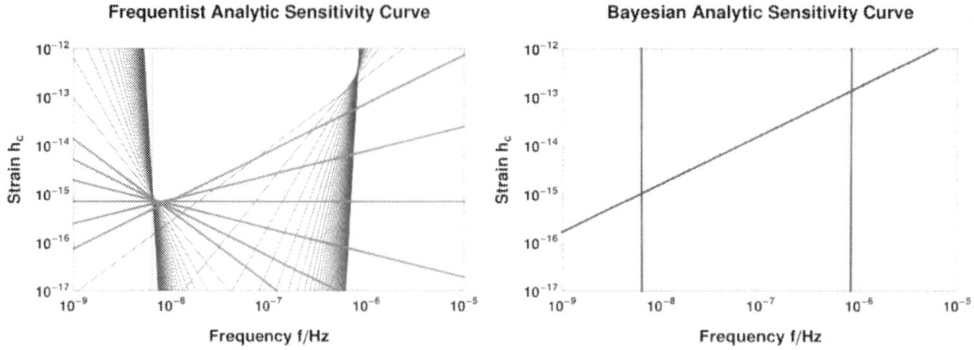

Figure 3: The sensitivity of our canonical PTA to a stochastic background. The left-hand panel shows the prediction of the frequentist analytic formula in Sec. 3.1 (with $\rho_{\rm th} = 3$), the curves drawn in red are for comparison with Fig. 4. The right-hand panel shows the prediction of the Bayesian analytic formula in Sec. 3.2. (Figure reproduced from (13).)

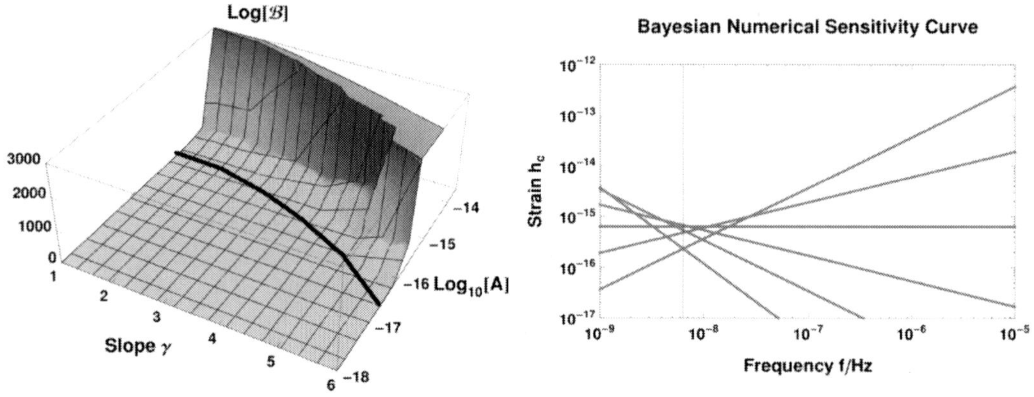

Figure 4: The results of the recovery of numerical injections of a stochastic background signal into the mock dataset. The left-hand panel shows a surface plot of the Bayes factor against the amplitude and power-law index of a stochastic background with the contour $\mathcal{B} = \mathcal{B}_{\rm th}$ indicated by a solid black line. The right-hand panel shows a power-law-integrated sensitivity curve formed by using points off the contour $\mathcal{B} = \mathcal{B}_{\rm th}$. The range of values of the power-law index was limited for reasons of numerical stability however this plot may be compared with the curves drawn in red in Fig. 3. For more detail on the numerical calculations used to produce these plots see (13). (Figure reproduced from (13).)

5. Conclusions

The sensitivity of a PTA depends not only on the properties of the PTA itself, but also on the properties of the source being observed. This statement is also true for more traditional interferometric detectors such as aLIGO and eLISA ((17; 18)) however the effect is particularly important for PTAs because their sensitivity is determined by the sampling properties of the dataset rather than properties of the noise. For example, in the case of LIGO the lowest frequency which may be detected is determined to a large extent by the properties of the seismic noise, whereas in the case of PTAs it is determined by the total length of the observations.

The sensitivity curve of a canonical PTA was evaluated for both a monochromatic source and a stochastic background, and this was done in both the frequentist and Bayesian frameworks, with consistent results found. The analytic calculations were also verified by numerical injections into a mock dataset. In the simplified case of regularly sampled data with white, Gaussian, uncorrelated noise, simple analytic expressions for the sensitivity of a PTA have been derived. It is hoped that these expressions will be useful going forward in order to assess the prospects for detection and astronomical inference using PTA observations.

Acknowledgments
The author is supported by the STFC. This talk was a partial summary of work done in collaboration with S. R. Taylor and J. R. Gair. The author would like to thank the organisers of the 10th international LISA symposium for the opportunity to give this talk.

References

[1] Jaffe A H and Backer D C 2003 *ApJ* **583** 616–631 (*Preprint* arXiv:astro-ph/0210148)
[2] Wyithe J S B and Loeb A 2003 *ApJ* **590** 691–706 (*Preprint* arXiv:astro-ph/0211556)
[3] Sesana A, Vecchio A and Colacino C N 2008 *MNRAS* **390** 192–209 (*Preprint* 0804.4476)
[4] Sesana A, Vecchio A and Volonteri M 2009 *MNRAS* **394** 2255–2265 (*Preprint* 0809.3412)
[5] Sesana A and Vecchio A 2010 *Classical and Quantum Gravity* **27** 084016 (*Preprint* 1001.3161)
[6] Ölmez S, Mandic V and Siemens X 2010 *Phys. Rev. D* **81** 104028 (*Preprint* 1004.0890)
[7] Grishchuk L P 1976 *Pis ma Zhurnal Eksperimental noi i Teoreticheskoi Fiziki* **23** 326–330
[8] Grishchuk L P 2005 *Physics Uspekhi* **48** 1235–1247 (*Preprint* arXiv:gr-qc/0504018)
[9] Kramer M and Champion D J 2013 *Classical and Quantum Gravity* **30** 224009
[10] Hobbs G 2013 *Classical and Quantum Gravity* **30** 224007 (*Preprint* 1307.2629)
[11] McLaughlin M A 2013 *Classical and Quantum Gravity* **30** 224008 (*Preprint* 1310.0758)
[12] Manchester R N and IPTA 2013 *Classical and Quantum Gravity* **30** 224010
[13] Moore C J, Taylor S R and Gair J R 2014 *ArXiv e-prints* (*Preprint* 1406.5199)
[14] Hobbs G B, Edwards R T and Manchester R N 2006 *MNRAS* **369** 655–672 (*Preprint* arXiv:astro-ph/0603381)
[15] Edwards R T, Hobbs G B and Manchester R N 2006 *MNRAS* **372** 1549–1574 (*Preprint* arXiv:astro-ph/0607664)
[16] Thrane E and Romano J D 2013 *Phys. Rev. D* **88** 124032 (*Preprint* 1310.5300)
[17] Harry G M and LIGO Scientific Collaboration 2010 *Classical and Quantum Gravity* **27** 084006
[18] Amaro-Seoane P, Aoudia S, Babak S, Binétruy P, Berti E, Bohé A, Caprini C, Colpi M, Cornish N J, Danzmann K, Dufaux J F, Gair J, Jennrich O, Jetzer P, Klein A, Lang R N, Lobo A, Littenberg T, McWilliams S T, Nelemans G, Petiteau A, Porter E K, Schutz B F, Sesana A, Stebbins R, Sumner T, Vallisneri M, Vitale S, Volonteri M and Ward H 2012 *cqg* **29** 124016 (*Preprint* 1202.0839)

Noise in pulsar timing arrays

Yan Wang

School of Physics, Huazhong University of Science and Technology, 1037 Luoyu Road, Wuhan, Hubei Province 430074, China

Center for Advanced Radio Astronomy, University of Texas at Brownsville, 1 West University Boulevard, Brownsville, Texas 78520, USA

E-mail: ywang12@hust.edu.cn

Abstract. To successfully detect gravitational waves with pulsar timing arrays, we need to have a comprehensive understanding of the physical origins and statistical characteristics of the noise in pulse arrival times and identify mitigation methods to reduce the noise. In this paper we will review radiometer noise, phase jitter noise and timing noise in the noise budget of pulsar timing and show various efforts used to reduce them. We will briefly discuss the results of an overall assessment of the components and physical causes of the timing residuals for millisecond pulsars in the North American Nanohertz Observatory for Gravitational Waves (NANOGrav).

1. Introduction

Pulsar timing arrays (PTAs) are striving to detect very low frequency ($10^{-9} - 10^{-6}$ Hz) gravitational waves (GWs) by observing a set of extremely stable millisecond pulsars (MSPs). Detection can be achieved by observing 20–40 pulsars over 5–10 years, assuming monthly observation cadence and 100 ns for root mean square (RMS) of timing residuals which is presumably dictated by white Gaussian noise [1]. However, detection can be delayed by up to about 10 years depending on the level of red noise and the possible errors (e.g., ephemeris, polarization calibration, time transfer) in the highest timing precision [2]. So far, upper limits have been reported for the amplitude of the stochastic GW background [3–5] and for continuous GW sources [6; 7] by three major PTAs (NANOGrav [4], PPTA [8] and EPTA [9]). Results from the International Pulsar Timing Array (IPTA) [10–12] which combine the data sets from all PTAs will also become available in the near future.

As in any gravitational wave detection experiment (e.g., LIGO [13], Virgo [14], eLISA [15]), noise characterization and mitigation are the central issues that need to be addressed in order to confidently detect and characterize the relatively weak GW signals, and carry out detailed astrophysical interpretations of them. For PTAs, a comprehensive understanding of the noise error budget of pulse arrival time is crucial to guide us to improve instruments, design experiments, and develop algorithms.

To achieve this goal, we need to build a complete measurement model that accounts for the end to end errors in the highest timing precision. Here, one end is a rotating pulsar, which produces regular radio pulses. The spacially coherent radiation carrying these pulses will be distorted by turbulent plasma in interstellar medium (ISM), causing dispersion, scattering, refraction, diffraction, etc. of the radio waves [16], along the line of propagation to the other end, a telescope. The faint EM signal is subsequently collected and focused by the large reflector, received by the radiometer and recorded by the backend system at the observatory. From emission to reception, there are various errors that can be introduced into the final estimation of pulse time of arrivals (TOAs).

Here, we shall not discuss the propagation effects pertaining to the interstellar medium [17; 18], since it is covered in the account by L. Levin in this volume. Instead, we focus on the other errors such as radiometer noise, phase jitter noise and timing noise, which are currently under intensive scrutiny of the pulsar timing community [19–21]. The rest of this paper is organized as follows. In Sec. 2 we provide an overview of these noise sources and methods of mitigation, and in Sec. 3 we discuss briefly the efforts and results in the NANOGrav collaboration on noise analysis. The paper is concluded in Sec. 4.

2. Noise

In timing analysis, it is a common practice to transform TOAs measured in the topocentric reference frame centered at a telescope to the inertial reference frame centered at the Solar System Barycenter [16; 22]. This transformation includes terms representing deterministic effects such as clock correction, time delay due to interstellar medium dispersion, geometric time delay (Römer delay), relativistic time delay (Shapiro delay, Einstein delay), etc. In addition, various error terms should be also added into the time transformation. A comprehensive list of timing errors can be found in Table 1 of [23].

In general, we can classify the errors into the ones pertinent to the time tagging of pulses (i.e. TOA measurement by template fitting) and the ones pertinent to the physical properties of pulsar or ISM. An example of the latter is the timing of a pulse with infinite SNR and known profile which can be measured to an arbitrary precision (no time tagging error), but the irregularity of pulsar rotation or the stochastic fluctuation of dispersion measure may introduce additional random components. The error can be achromatic which means its influence is independent of the observation frequency (timing noise), weakly chromatic (radiometer noise, jitter noise), or strongly chromatic (effects rooted from interstellar medium).

The power spectrum of the noise can be white or red. The TOA fluctuations caused by the stochastic GW background have red spectra for individual pulsars, which are angularly correlated between pairs of pulsars [24]. Thus, for detecting this background, it is imperative to assess and reduce, if possible, the confusing red noise components rooted from other sources, for example, pulsar and ISM.

2.1. Radiometer noise

Radiometer noise is instrumental in origin (thermal electron fluctuation) combining the contribution from the sky background (dominated by the synchrotron-radiating electrons in the plane of the Galaxy). Radiometer noise with a Gaussian probability density function is additive to the pulse profile, of any of the Stokes parameters. Usually, the Stokes I (intensity) is used to measure the TOAs. Radiometer noise is weakly chromatic if the declination of pulse flux density with increasing of observation frequency cancels out much of the variation of the sky background temperature.[1] The resulting TOA is estimated from template fitting of integrated pulse profile with theoretical or integrated template. The minimum RMS error for TOA estimation due to finite SNR and sampling rate is [23]:

$$\sigma_{\rm SNR} = 1\,\mu s \left(\frac{W}{1\,\rm ms}\right)\left(\frac{N}{10^6}\right)^{-1/2}\left(\frac{1}{\rm SNR_1}\right)\left(\frac{\Delta}{W}\right)^{1/2}, \tag{1}$$

$$= 0.71\mu s \left(\frac{W}{1\,\rm ms}\right)^{3/2}\left(\frac{P}{1\,\rm ms}\right)^{-1}\left(\frac{f}{1.4\,\rm GHz}\right)^{-\alpha}\left(\frac{\Delta f}{1\,\rm GHz}\right)^{-1/2}\left(\frac{N}{10^6}\right)^{-1/2}\left(\frac{S_{\rm sys}}{S_{1400}}\right). \tag{2}$$

[1] It could be strongly chromatic if the pulsar spectrum is flat in the observation band.

Here W is the effective pulse width which equals to $W_{\rm FWHM}/\sqrt{2\pi \ln 2}$ for a Gaussian pulse profile, $W_{\rm FWHM}$ is the full width at half maximum of the pulse, N is the number of pulse averaged synchronously to yield an integrated profile, $\rm SNR_1$ is the single pulse SNR, Δ is the sampling interval, and P is the spin period. Radio frequency f and bandwidth Δf are in GHz, $S_{\rm sys}$ and S_{1400} are the system equivalent flux density in Jy and mean flux density of the pulsar at 1.4 GHz in mJy. Note that optimal result can be archived by increasing the bandwidth and integration time, and choosing bright pulsars with short duty cycle. The second equality is obtained by using the radiometer equation [25]:

$$\Delta T_{\rm sys} = \frac{T_{\rm sys}}{\sqrt{n_{\rm p} t \Delta f}}, \tag{3}$$

and assuming that the flux density of pulsar follows a power law with spectral index α. $T_{\rm sys}$ and $\Delta T_{\rm sys}$ is the system temperature of the receiver and its RMS fluctuation, $n_{\rm p}$ is the number of polarization, and t is the integration time.

Eq. 1-2 are obtained under the assumption that there is no variance for the integrated pulse template over the frequency band, i.e. the profile has the same shape at high frequency as at low frequency. However, the profile evolution can arise from phenomena intrinsic to pulsar emission beam or ISM (DM change, scattering and scintillation). Ignoring it will degrade the timing precision promised by the modern broad band receivers and backends (see Table 1). This so called "the large-bandwidth problem" [26] is demonstrated in a recent 24-hour global observation for the millisecond PSR 1713+0747 (c.f. Fig. 4 in [20]). Current treatment of this problem is to generate templates for each frequency channel, add additional fitting parameters (JUMPs) to handle the possible offsets between them, and allow the standard TOA analysis packages such as `tempo` and `tempo2` to find out the best fit value for them [4]. A more consistent and efficient method may be to create a two-dimensional pulse portrait (rather than one-dimensional profiles at different frequencies) which takes into account of differential profile evolution and time offset as a smooth function of frequency. This strategy has been explored for broad band data in [27; 28].

Table 1. Wide band receivers and backends currently used or planned for pulsar observation.

Telescope	D(m)	Receiver	f(GHz)	$T_{\rm sys}$	Backend	Δf(MHz)	Ref.
GBT	100	L-band	1.15-1.73	20	GUPPI	800	[29]
Arecibo	305	L-band	1.15-1.73	25	PUPPI	800	[30; 31]
Parkes	64	Multibeam	1.23-1.53	28	APSR	1000	[8]
Effelsberg	100	UBB	0.6-3	24	ASTERIX	512	[32; 33]
FAST	500	L-band	1.15-1.72	25	–	–	[34]

2.2. Pulse phase jitter and amplitude modulation noise

Increasing the SNR of the integrated profile will reduce the radiometer noise, however it is not the only justification for using integrated pulse profile in TOA measurement. Individual pulse usually jitters in phase at the level of single pulse width, and its amplitude can change more than 100% from pulse to pulse (c.f. Fig. 4 of [35] for PSR J1740+1000). Thus measuring individual pulses even with large $\rm SNR_1$ will result in a TOA uncertainty in the order of the pulse width. Phase jitter and amplitude modulation are related to the stability of the integrated pulse shape which is determined by the shape of individual pulse and the probability density function of the phase jitter. A stable pulse profile can be obtained by summing over at least several hundreds of individual pulses, and the associated TOA uncertainty due to jitter roughly scales inversely as square root of the number of the individual pulses [36].

Pulse phase jitter and amplitude modulation appear in all well studied pulsars. It is weakly chromatic. Jitter noise is not additive in nature as the radiometer noise. In fact, it changes the integrated pulse profile in a statistical manner [21]. For a simple case in which we assume the pulse profile is Gaussian shape and the phase jitter follows a Gaussian distribution, the RMS error caused by jitter noise can be written as [23]

$$\sigma_{\rm J} = 0.28\,\mu s \left(\frac{W_i}{1\,{\rm ms}}\right)\left(\frac{N}{10^6}\right)^{-1/2}\left(\frac{f_{\rm J}}{1/3}\right)\left(\frac{1+m_{\rm I}^2}{2}\right)^{1/2}. \quad (4)$$

Here, $m_{\rm I}$ is the amplitude modulation index defined as the ratio of the standard deviation of the amplitude to the mean at different pulse phases and $m_{\rm I} \approx 1$. $f_{\rm J}$ is the dimensionless jitter parameter defined as the ratio of the standard deviation of the phase of single pulse to the intrinsic pulse width W_i of the template and $f_{\rm J} \approx 1/3$. The RMS of total error $\sigma_{\rm t}$ for estimated TOA is the quadratic summation of radiometer noise and jitter noise, i.e. $\sigma_{\rm t}^2 = \sigma_{\rm SNR}^2 + \sigma_{\rm J}^2$.

By equating Eq. 1 with Eq. 4, we find that jitter noise will become more important than radiometer noise when $\rm SNR_1$ exceeds only a few tenths. This sensitivity is accessible by the future radio telescopes such as FAST [37] and SKA [38] that have larger collecting areas and lower system temperatures. In this scenario, the noise will not be reduced by increasing the observation bandwidth. Increasing the observation time will become an inevitable choice. As a result, PTAs will need to request more observation time of radio telescopes.

2.3. *Timing noise*

Timing noise, also known as spin noise, appears as the structures with temporal correlation in timing residuals that depart greatly from the measurement error alone. It may be caused by the irregularity of rotational spin rate which could root from the changes in internal structure and/or magnetosphere of neutron star. It has been found in a number of canonical pulsars and a few MSPs (e.g., B1937+21, B1821-24) [39]. The RMS error of timing noise can be characterized by a scaling law [19]:

$$\sigma_{\rm TN} = C\nu^{\alpha}|\dot{\nu}|^{\beta}T^{\gamma}, \quad (5)$$

where C, α, β and γ are the fitting parameters determined from the whole populations of pulsars with measurable timing noise or upper limits, ν and $\dot{\nu}$ is the pulsar spin frequency and frequency derivative, T is the total span of observations. The best-fit values and $\pm 2\sigma$ confidence limits calculated from canonical pulsar and MSP population are $\ln C = 1.6 \pm 0.4$, $\alpha = -1.4 \pm 0.1$, $\beta = 1.1 \pm 0.1$, and $\gamma = 2.0 \pm 0.2$ [19]. This suggests that MSPs with high spin frequency and low spin frequency derivative would have a lower level of timing noise. Eq. 5 can serve as a criterion to select excellent timers that potentially have less timing noise from newly discovered MSPs for the PTAs. Despite the timing noise is not yet detected in most MSPs, it can be a latent phenomenon that may emerge from future data with longer observation span and higher timing precision.

Timing noise is the red noise intrinsic to pulsar. For $\gamma \approx 2$, it has a power spectral index between -4 and -6 based on the random walk model of pulsar rotation. As a comparison, the characteristic strain h_c of the stochastic GW background scales as $h_c = A(f/{\rm yr}^{-1})^{\alpha}$, where $\alpha = -2/3$ for a background generated by the incoherent superposition of supermassive black hole binaries. The RMS residuals induced by this background grows as $T^{5/3}$ associated with a spectral index of $2\alpha - 3 = -13/3$ [40]. One of the consequences of the red noise is that the time of GW detection predicted by white noise dominated model [1] may be delayed up to 10 years depending on the level of red noise [2]. Besides, more MSPs will be required than previously expected in order to distinguish between the red noise from the GW stochastic background and the other red noise sources by cross-correlating the data from different pulsars, and the number of pulsars becomes a more important factor than the observation cadence and timing quality of individual pulsar when the PTA enters a regime where the lowest frequencies of the timing residuals are dominated by GWs, a likely case for the current PTAs [2]. Currently, the NANOGrav are adding 3-4 new MSPs discovered from the ongoing major pulsar surveys at Arecibo Observatory and Green Bank Telescope (e.g., PALFA and GBNCC) in the observation campaign each year.

3. Noise assessment

As discussed above, the characterization of noise in PTA plays a central role in hunting for GWs. Due to its importance, the NANOGrav has formed a noise budget working group to use complementary methods to assess the constituents of timing residuals and their physical causes. The white noise and Gaussian statistics are two essential aspects of the assessments, especially the latter is a common assumption in forming the GW detection strategies [41–44]. Blindly applying these strategies without checking the presumption may lead to unreliable results.

Methods used include autocorrelation analyses, Bayesian inference, zero-crossing tests, Gaussianity tests, etc. A memorandum to consolidate the results from the overall assessment of the NANOGrav 5-yrs ASP/GASP data set for 17 MSPs [4] is in preparation. Further studies extended to the NANOGrav 9-yrs (including 4-yrs PUPPI/GUPPI) data set for more than 30 MSPs will be also carried out once the data are available.

Initial results show that for the 5-yrs data most of the pulsars are consistent with the white noise assumption [45; 46], although it is possible that the red timing noise can appear in the 9-yrs observations which have longer span and higher precision. Different levels of departure from Gaussian statistics are shown in most of the pulsars [46], it is suspected that the diffractive interstellar scintillation is the root cause. This suggests that the robust signal detection and characterization methods that are not sensitive to the non-Gaussianity should be implemented in GW data analysis.

4. Conclusions

Noise characterization and mitigation are the central issues in detecting GWs by pulsar timing arrays. In this article, we provide an overview of the features of radiometer noise, phase jitter noise and timing noise as well as the efforts to reduce them.

The radiometer noise is dominant in the current timing precision, it will be continuously mitigated with the developments of instrument and algorithm to a level smaller than the jitter noise. The jitter noise is intrinsic to a pulsar, it can only be mitigated by extending integration time, therefore affects the strategies on telescope time application and allocation.

Timing noise is latent for most MSPs, it can potentially postpone the detection of GWs by PTAs. Finding more MSPs with excellent timing performance in the ongoing and future surveys is imperative in the competition between the red noise from GWs and pulsars themselves.

Acknowledgments

We wish to acknowledge the invitation from Prof. Andrea Lommen for the 10th LISA symposium, and the assistance from the local organizer at University of Florida, especially Prof. Guido Mueller for his help in conference registration. We would like to thank the NANOGrav members for helpful discussion and the Center for Gravitational Wave Astronomy at UTB for partial support under NASA grant NNX09AV06A. The NANOGrav project is supported by the National Science Foundation under PIRE award number 0968296.

References

[1] Jenet F A, Hobbs G B, Lee K J and Manchester R N 2005 *Astrophysical Journal Letter* **625** L123–L126 (*Preprint* arXiv:astro-ph/0504458)

[2] Siemens X, Ellis J, Jenet F and Romano J D 2013 *Classical and Quantum Gravity* **30** 224015 (*Preprint* 1305.3196)

[3] van Haasteren R, Levin Y, Janssen G H, Lazaridis K, Kramer M, Stappers B W, Desvignes G, Purver M B, Lyne A G, Ferdman R D, Jessner A, Cognard I, Theureau G, D'Amico N, Possenti A, Burgay M, Corongiu A, Hessels J W T, Smits R and Verbiest J P W 2011 *Monthly Notices of the Royal Astronomical Society* **414** 3117–3128 (*Preprint* 1103.0576)

[4] Demorest P B, Ferdman R D, Gonzalez M E, Nice D, Ransom S, Stairs I H, Arzoumanian Z, Brazier A, Burke-Spolaor S, Chamberlin S J, Cordes J M, Ellis J, Finn L S, Freire P, Giampanis S, Jenet F, Kaspi V M, Lazio J, Lommen A N, McLaughlin M, Palliyaguru N, Perrodin D, Shannon R M, Siemens X, Stinebring D, Swiggum J and Zhu W W 2013 *Astrophysical Journal* **762** 94 (*Preprint* 1201.6641)

[5] Shannon R M, Ravi V, Coles W A, Hobbs G, Keith M J, Manchester R N, Wyithe J S B, Bailes M, Bhat N D R, Burke-Spolaor S, Khoo J, Levin Y, Oslowski S, Sarkissian J M, van Straten W, Verbiest J P W and Want J B 2013 *Science* **342** 334–337 (*Preprint* 1310.4569)

[6] Arzoumanian Z, Brazier A, Burke-Spolaor S, Chamberlin S J, Chatterjee S, Cordes J M, Demorest P B, Deng X, Dolch T, Ellis J A, Ferdman R D, Garver-Daniels N, Jenet F, Jones G, Kaspi V M, Koop M, Lam M T, Lazio T J W, Lommen A N, Lorimer D R, Luo J, Lynch R S, Madison D R, McLaughlin M A, McWilliams S T, Nice D J, Palliyaguru N, Pennucci T T, Ransom S M, Sesana A, Siemens X, Stairs I H, Stinebring D R, Stovall K, Swiggum J, Vallisneri M, van Haasteren R, Wang Y, Zhu W W and NANOGrav Collaboration 2014 *Astrophysical Journal* **794** 141 (*Preprint* 1404.1267)

[7] Zhu X J, Hobbs G, Wen L, Coles W A, Wang J B, Shannon R M, Manchester R N, Bailes M, Bhat N D R, Burke-Spolaor S, Dai S, Keith M J, Kerr M, Levin Y, Madison D R, Osłowski S, Ravi V, Toomey L and van Straten W 2014 *Monthly Notices of the Royal Astronomical Society* **444** 3709–3720 (*Preprint* 1408.5129)

[8] Manchester R N, Hobbs G, Bailes M, Coles W A, van Straten W, Keith M J, Shannon R M, Bhat N D R, Brown A, Burke-Spolaor S G, Champion D J, Chaudhary A, Edwards R T, Hampson G, Hotan A W, Jameson A, Jenet F A, Kesteven M J, Khoo J, Kocz J, Maciesiak K, Oslowski S, Ravi V, Reynolds J R, Sarkissian J M, Verbiest J P W, Wen Z L, Wilson W E, Yardley D, Yan W M and You X P 2013 *Publications of the Astronomical Society of Australia* **30** e017 (*Preprint* 1210.6130)

[9] Ferdman R D, van Haasteren R, Bassa C G, Burgay M, Cognard I, Corongiu A, D'Amico N, Desvignes G, Hessels J W T, Janssen G H, Jessner A, Jordan C, Karuppusamy R, Keane E F, Kramer M, Lazaridis K, Levin Y, Lyne A G, Pilia M, Possenti A, Purver M, Stappers B, Sanidas S, Smits R and Theureau G 2010 *Classical and Quantum Gravity* **27** 084014 (*Preprint* 1003.3405)

[10] Hobbs G, Archibald A, Arzoumanian Z, Backer D, Bailes M, Bhat N D R, Burgay M, Burke-Spolaor S, Champion D, Cognard I, Coles W, Cordes J, Demorest P, Desvignes G, Ferdman R D, Finn L, Freire P, Gonzalez M, Hessels J, Hotan A, Janssen G, Jenet F, Jessner A, Jordan C, Kaspi V, Kramer M, Kondratiev V, Lazio J, Lazaridis K, Lee K J, Levin Y, Lommen A, Lorimer D, Lynch R, Lyne A, Manchester R, McLaughlin M, Nice D, Oslowski S, Pilia M, Possenti A, Purver M, Ransom S, Reynolds J, Sanidas S, Sarkissian J, Sesana A, Shannon R, Siemens X, Stairs I, Stappers B, Stinebring D, Theureau G, van Haasteren R, van Straten W, Verbiest J P W, Yardley D R B and You X P 2010 *Classical and Quantum Gravity* **27** 084013 (*Preprint* 0911.5206)

[11] Manchester R N and IPTA 2013 *Classical and Quantum Gravity* **30** 224010 (*Preprint* 1309.7392)

[12] McLaughlin M A 2014 *ArXiv e-prints* (*Preprint* 1409.4579)

[13] Abbott B P, Abbott R, Adhikari R, Ajith P, Allen B, Allen G, Amin R S, Anderson S B, Anderson W G, Arain M A and et al 2009 *Reports on Progress in Physics* **72** 076901 (*Preprint* 0711.3041)

[14] Accadia T, Acernese F, Antonucci F and et al 2011 *Classical and Quantum Gravity* **28** 114002

[15] Seoane P A, Aoudia S, Audley H, Auger G and et al 2013 *ArXiv e-prints* (*Preprint* 1305.5720)

[16] Lorimer D R and Kramer M 2004 *Handbook of Pulsar Astronomy*

[17] Rickett B J 1977 *Annual review of astronomy and astrophysics* **15** 479–504

[18] Rickett B J 1990 *Annual review of astronomy and astrophysics* **28** 561–605

[19] Shannon R M and Cordes J M 2010 *Astrophysical Journal* **725** 1607–1619 (*Preprint* 1010.4794)

[20] Dolch T, Lam M T, Cordes J, Chatterjee S, Bassa C, Bhattacharyya B, Champion D J, Cognard I, Crowter K, Demorest P B, Hessels J W T, Janssen G, Jenet F A, Jones G, Jordan C, Karuppusamy R, Keith M, Kondratiev V, Kramer M, Lazarus P, Lazio T J W, Lee K J, McLaughlin M A, Roy J, Shannon R M, Stairs I, Stovall K, Verbiest J P W, Madison D R, Palliyaguru N, Perrodin D, Ransom S, Stappers B, Zhu W W, Dai S, Desvignes G, Guillemot L, Liu K, Lyne A, Perera B B P, Petroff E, Rankin J M and Smits R 2014 *Astrophysical Journal* **794** 21 (*Preprint* 1408.1694)

[21] Shannon R M, Osłowski S, Dai S, Bailes M, Hobbs G, Manchester R N, van Straten W, Raithel C A, Ravi V, Toomey L, Bhat N D R, Burke-Spolaor S, Coles W A, Keith M J, Kerr M, Levin Y, Sarkissian J M, Wang J B, Wen L and Zhu X J 2014 *Monthly Notices of the Royal Astronomical Society* **443** 1463–1481 (*Preprint* 1406.4716)

[22] Edwards R T, Hobbs G B and Manchester R N 2006 *Monthly Notices of the Royal Astronomical Society* **372** 1549–1574 (*Preprint* astro-ph/0607664)

[23] Cordes J M and Shannon R M 2010 *ArXiv e-prints* (*Preprint* 1010.3785)

[24] Hellings R W and Downs G S 1983 *Astrophysical Journal Letter* **265** L39–L42

[25] Wilson T L, Rohlfs K and Hüttemeister S 2009 *Tools of Radio Astronomy* (Springer-Verlag)

[26] Lommen A N and Demorest P 2013 *Classical and Quantum Gravity* **30** 224001 (*Preprint* 1309.1767)

[27] Pennucci T T, Demorest P B and Ransom S M 2014 *Astrophysical Journal* **790** 93 (*Preprint* 1402.1672)

[28] Liu K, Desvignes G, Cognard I, Stappers B W, Verbiest J P W, Lee K J, Champion D J, Kramer M, Freire P C C and Karuppusamy R 2014 *Monthly Notices of the Royal Astronomical Society* **443** 3752–3760 (*Preprint* 1407.3827)

[29] The Proposer's Guide for the Green Bank Telescope 2014

[30] http://www.naic.edu/~astro/RXstatus/Lwide/Lwide.shtml

[31] http://www.naic.edu/puppi-observing/

[32] http://www.mpifr-bonn.mpg.de/research/fundamental/ubb

[33] http://www.mpifr-bonn.mpg.de/research/fundamental/asterix

[34] Nan R, Li D, Jin C, Wang Q, Zhu L, Zhu W, Zhang H, Yue Y and Qian L 2011 *International Journal of Modern Physics D* **20** 989–1024 (*Preprint* 1105.3794)

[35] McLaughlin M A, Arzoumanian Z, Cordes J M, Backer D C, Lommen A N, Lorimer D R and Zepka A F 2002 *Astrophysical Journal* **564** 333–342 (*Preprint* astro-ph/0106371)

[36] Rathnasree N and Rankin J M 1995 *Astrophysical Journal* **452** 814

[37] Hobbs G, Dai S, Manchester R N, Shannon R M, Kerr M, Lee K J and Xu R 2014 *ArXiv e-prints* (*Preprint* 1407.0435)

[38] Smits R, Kramer M, Stappers B, Lorimer D R, Cordes J and Faulkner A 2009 *Astronomy and Astrophysics* **493** 1161–1170 (*Preprint* 0811.0211)

[39] Hobbs G, Lyne A G and Kramer M 2010 *Monthly Notices of the Royal Astronomical Society* **402** 1027–1048 (*Preprint* 0912.4537)

[40] Jenet F A, Hobbs G B, van Straten W, Manchester R N, Bailes M, Verbiest J P W, Edwards R T, Hotan A W, Sarkissian J M and Ord S M 2006 *Astrophysical Journal* **653** 1571–1576 (*Preprint* astro-ph/0609013)

[41] Babak S and Sesana A 2012 *Phys. Rev. D* **85** 044034 (*Preprint* 1112.1075)

[42] Ellis J A, Siemens X and Creighton J D E 2012 *Astrophysical Journal* **756** 175 (*Preprint* 1204.4218)

[43] Ellis J A 2013 *Classical and Quantum Gravity* **30** 224004 (*Preprint* 1305.0835)

[44] Wang Y, Mohanty S D and Jenet F A 2014 *Astrophysical Journal* **795** 96 (*Preprint* 1406.5496)

[45] Perrodin D, Jenet F, Lommen A, Finn L, Demorest P, Ferdman R, Gonzalez M, Nice D, Ransom S and Stairs I 2013 *ArXiv e-prints* (*Preprint* 1311.3693)

[46] Wang Y and et al 2014 *in preparation*

Journal of Physics: Conference Series **610** (2015) 012020 doi:10.1088/1742-6596/610/1/012020

Interstellar Medium Mitigation Techniques in Pulsar Timing Arrays

L Levin

Department of Physics & Astronomy, West Virginia University, Morgantown, WV 26506, USA

E-mail: Lina.Levin@mail.wvu.edu

Abstract. Pulsar Timing Arrays use a set of millisecond pulsars in an attempt to directly detect nanohertz gravitational waves. For this purpose, high precision timing of the pulsars is essential and ultimately a precision of the order of ~100 ns is required. Propagation effects in the interstellar medium cause the radio emission from a pulsar to be dispersed and scattered, introducing time variable delays of the pulses on their way to Earth. If these delays are not properly corrected for, they may cause significant errors in the timing analysis of a pulsar. These proceedings will review the effects of the interstellar medium on pulse arrival times and present some of the techniques used to mitigate the associated time delays from the pulsar signal. Correcting for these delays is essential to providing a higher timing precision and hence to increasing the array's sensitivity to gravitational waves.

1. Introduction

Pulsar Timing Arrays (PTAs) are observing a set of high-precision timing millisecond pulsars (MSPs) with the aim to detect nano-Hz gravitational waves (GWs). MSPs are extremely stably rotating neutron stars, with times of arrival (TOAs) of the pulses measurable to a few μs precision. In the timing procedure of a pulsar, residuals are created by making a timing model based on the parameters of the pulsar, and accounting for each rotation of the star. By observing MSPs for a long time span, the timing model is improved and the scatter of the residuals can go down to an RMS of ~100 ns. To search for GWs, PTAs look for correlations in the timing residuals of several MSPs. Since PTAs are sensitive to nanohertz frequency GWs, which make up a part of frequency space not covered by ground based and space based GW detectors, the efforts of various GW projects are complementary. The main GW sources expected in the PTA frequency band are super massive black hole binaries and cosmic strings, and searches are being carried out both for single sources and for a GW background. Currently, three collaborations are undertaking these efforts, namely the European Pulsar Timing Array[1] (EPTA; [1]), the Parkes Pulsar Timing Array[2] (PPTA; [2]), and the North American Nanohertz Observatory for Gravitational Waves[3] (NANOGrav; [3]). Together they make up the International Pulsar Timing Array[4] (IPTA; [4]).

[1] http://www.epta.eu.org
[2] http://www.atnf.csiro.au/research/pulsar/ppta
[3] http://nanograv.org
[4] http://www.ipta4gw.org

To succeed in detecting GWs, it is necessary to achieve as high timing precision as possible, over a long time span. As the signal from a pulsar travels through the ionized plasma of the interstellar medium (ISM), the pulses get scattered and the TOAs of the pulses are delayed. The two dominating ISM effects are dispersion and scattering, and both these effects introduce perturbations to the TOAs and will increase the timing error if not properly corrected for. Monitoring of ISM parameters and development of correction techniques is on-going within several pulsar groups [5–7], and ISM mitigation is increasingly recognized as one of the most important parts of the noise budget in PTAs [8–10].

This document is written as part of the 10th International LISA Symposium conference proceedings, and is based on a talk reviewing the challenges of ISM propagation effects on pulsar timing data as well as the different techniques used to mitigate these effects from the pulsar signals.

2. Dispersion

As the pulsar signal travels through the ISM, the overall column density of free electrons has dispersive effects on the broadband pulses. We can write the group velocity of the pulse train as

$$v_{\rm g} = c\sqrt{1-\frac{\nu_{\rm p}^2}{\nu^2}} \quad (1)$$

where c is the speed of light, ν is the observing frequency and $\nu_{\rm p}$ is the plasma frequency given by

$$v_{\rm p} = \sqrt{\frac{n_{\rm e}e^2}{\pi m_{\rm e}}} \approx 10^4{\rm Hz}\sqrt{\frac{n_{\rm e}}{1{\rm cm}^{-3}}}. \quad (2)$$

Here, e is the electron charge, $m_{\rm e}$ is the electron mass and $n_{\rm e}$ is the electron density in the relevant direction in the sky, all given in cgs units. If looking into the Galactic plane, the value of $n_{\rm e} \approx 0.03\,{\rm cm}^{-3}$, and it gets lower at higher Galactic latitudes. The definition of the Dispersion Measure (DM) as the integrated column density of free electrons along some line of sight can be written as

$${\rm DM} = \int_0^d n_{\rm e}\, dl, \quad (3)$$

where d is the distance to the pulsar. From this, we can derive the time delay of the radio wave due to the DM as

$$t_{\rm DM}(\nu) = \frac{e^2}{2\pi m_{\rm e} c\,\nu^2}{\rm DM} \approx 4.15\,{\rm ms}\,\frac{\rm DM}{\nu_{\rm GHz}^2}. \quad (4)$$

The observational consequence of the dispersive delay is that a pulsar observed with a broad frequency band will show pulses that arrive later at lower frequency than at higher frequency, following the ν^{-2} law. By observing pulsars at a wide enough frequency band, or at two or more separate frequency bands, the DM can be measured at each observing epoch. An example of an observation that has not been DM corrected can be found in Fig. 1.

One practical use of the DM is as a tool to infer the distance to the pulsar. This can be done with the use of models of the free electrons in the Galaxy. Since there are more free electrons in certain directions in the Galaxy, such as along the Galactic plane, a higher DM does not necessarily mean a larger distance, but also the sky position of a pulsar need to be considered. Currently, the NE2001 model [11] is the most comprehensive and widely used model.

Since the pulsar and the ISM have different relative velocities, the DM of a pulsar is not constant in time, and the variations of the DM need to be tracked and corrected for in the data. This can be achieved by observing pulsars at two or more separate frequencies at a maximum of a few days apart. Different methods on how to optimally correct for the DM variations are

Journal of Physics: Conference Series **610** (2015) 012020 doi:10.1088/1742-6596/610/1/012020

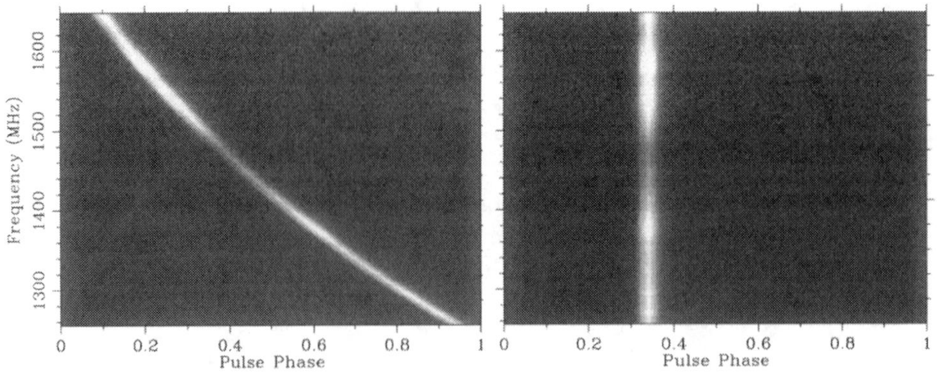

Figure 1. Wide-band observation of PSR J1744–1134, before (*left*) and after (*right*) DM correction. In this case, the dispersion delay over a 400 MHz band is almost equal to one full rotation of the pulsar. Due to scintillation, some intensity modulation over the band is also visible in this observation. These plots were created from raw data, contributing to the weak interference signals visible in the background.

being used by the different PTAs. Currently, the method used by NANOGrav involves fitting the multifrequency data for a DM within a ~10 day window, in a simultaneous fit with all pulsar parameters [12]. The PPTA technique is very similar to this, but removes the dispersive delay from the timing residuals after the timing solution has been subtracted. This method includes a frequency independent "common mode" delay [9], that is introduced to make sure that no GW signal is absorbed into the DM correction. Finally, the EPTA method uses the fact that the DM variations are correlated between adjacent observations, and applies a time-domain spectral analysis to measure the DM power-law spectrum and then extract the DM waveform by constructing a linear optimal filter [10].

3. Multi-path scattering

Spatial inhomogeneities in the ISM give rise to scattering of the radio waves from pulsars. Such multi-path scattering manifests itself in several ways, including refractive and diffractive intensity scintillations as well as pulse broadening.

Direct detection of pulse broadening is predominantly possible for pulsars with high DM values, i.e. for sources with more free electrons along their line of sight. Strong scattering cause large delays in the TOAs and hence high-DM pulsars are usually not included in PTAs. One exception is PSR J1903+0327, which because of its high flux density is useful for PTAs even though it has a large DM value. An observation of this pulsar can be seen in Fig. 2, showing a large scattering tail in its folded profile. For pulsars with more modest scattering, the pulse broadening can be measured indirectly through analyses of the diffractive scintillation pattern of the pulsar, which will be described later in this section.

The most commonly used model to describe the ISM consists of a thin disk of turbulent plasma midway between the pulsar and the observer [13]. When the pulses travel through the disk, inhomogeneities in the plasma introduce phase perturbations to the signal. In combination with the different relative velocities of the pulsar, the ISM, and the Earth, this contributes to a diffraction pattern being observed. The perturbations are correlated over a scintillation bandwidth ($\Delta\nu_{\rm d}$), which is inversely proportional to the scattering delay ($\tau_{\rm d}$), and over a scintillation timescale ($\Delta t_{\rm d}$). Such diffractive scintillation effects were first observed in pulsars

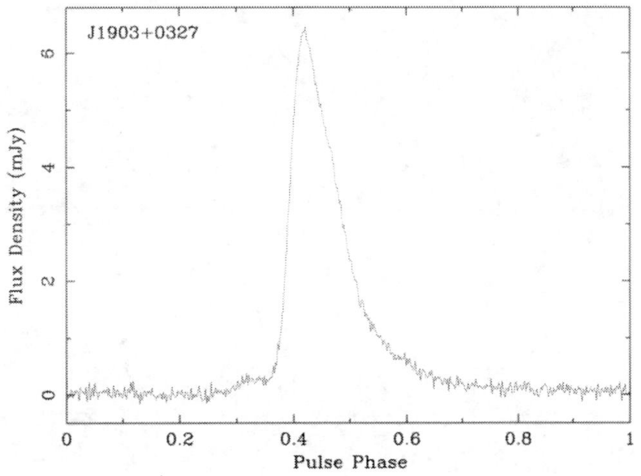

Figure 2. Pulse profile of PSR J1903+0327 at an observing frequency of 1.4 GHz, showing an exponential scattering tail. This pulsar has a large DM = 298 pc cm^{-3}, contributing to strong scattering of the profile even at this relatively high observing frequency.

by Lyne & Rickett [14], and originate in interstellar density structure of scale size $\sim 10^6 - 10^8$ cm. The ISM plasma is often considered to be composed as a Kolmogorov medium, which implies scaling of the scintillation parameters as $\Delta\nu_\mathrm{d} \propto \nu^{4.4}$, and $\Delta t_\mathrm{d} \propto \nu^{1.2}$ [15]. The diffractive scintillation pattern for a pulsar can change drastically on short time scales of weeks to months [16]. Similar to the case of DM variations, the relative velocities of the pulsar and the ISM give rise to the time variable scattering delays.

Larger structures in the ISM, with a size scale $\sim 10^{10} - 10^{12}$ cm, give rise to refractive scintillation effects. These effects arise through wavefront curvature, by causing the angle of arrival to diverge from the straight line of sight, and are observed through flux density variations in pulsars [17].

The multi-path broadening of pulses from diffraction give rise to the largest ISM delays, next to dispersive delays [5]. These can be measured indirectly through dynamical spectrum analyses of the pulsar data. A dynamic spectrum displays the intensity of the on-pulse part of the pulse profile over observing time and frequency. In the spectrum, scintillation maxima show up as bright islands on the map (so called "scintles", see Fig. 3). The scintles are limited in time since the interference pattern has a characteristic length scale, which becomes a characteristic time scale due to the relative velocity of the system, and they are limited in frequency because coherent radio beams have scattering delays of order τ_d.

To get enough signal to noise to measure the scintillation parameters from the scintles, a 2-dimensional autocorrelation function (ACF) is created from the dynamic spectrum. This 2D ACF is then summed in time and frequency separately, and the two resulting 1D ACFs are fitted with a Gaussian function, centered at zero lag. From the Gaussian fit, Δt_d is defined as the half-width at e^{-1} of the summed frequency lag and $\Delta\nu_\mathrm{d}$ is the half-width at half-maximum of the summed time lag [18]. The scattering delay, τ_d, can be calculated from $\Delta\nu_\mathrm{d}$ through

$$2\pi\Delta\nu_\mathrm{d}\tau_\mathrm{d} = C_1 \quad (5)$$

where C_1 is a constant that depends on the geometry and spectrum of the electron density, and with expected value ranging from 0.6–1.5 [19].

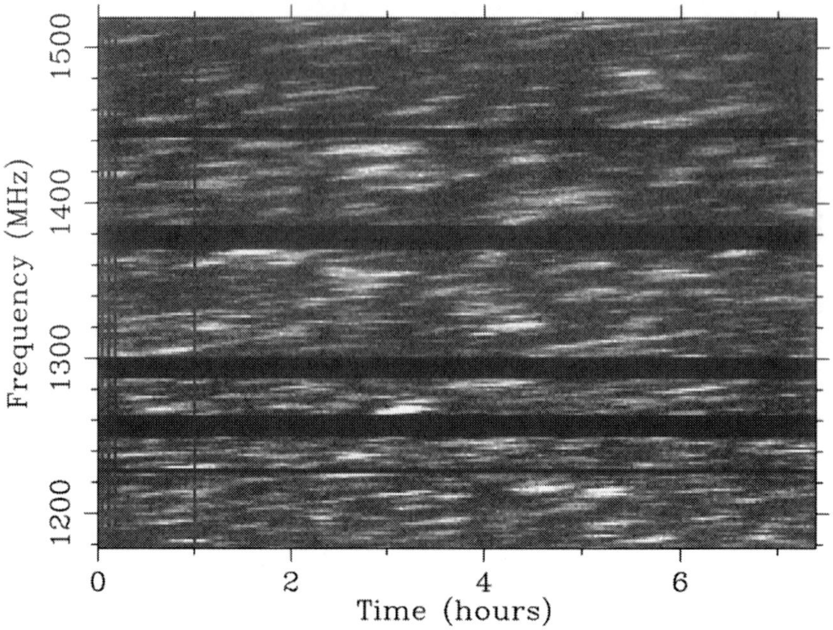

Figure 3. Dynamic spectrum of a ~7-hr long observation of PSR J1614–2230, after interference signals have been removed from the data (showing up as dark lines in the plot). Lighter color indicates higher intensity of the pulsar signal, with separate bright islands referred to as scintles. Analyses of the scintles in the dynamic spectrum can be used to estimate scintillation parameters for a pulsar.

By creating dynamic spectra of each observation for each of the PTA pulsars, the scintillation parameters can be monitored, and in particular the variation of the scattering delays can be analyzed. This is important, since a constant delay would not cause any additional errors in timing precision, no matter its magnitude, but a highly variable delay would contribute largely to the timing noise. Therefore, the maximum variation of the delays need to be monitored, and can be compared to the individual TOA error to see if the scattering is one of the dominating sources of noise in the data. At the commonly used observing frequency band centered at ~1.4 GHz, for most PTA pulsars, the scattering delay variations are in general much smaller than the current TOA uncertainty [7, 20], and hence are currently not limiting the timing precision. With longer data spans and new developments in instrumentation, the timing precision will improve, and hence a detailed scattering analysis may become increasingly important in the future.

The latest developments in ISM mitigation include a technique to estimate pulse broadening times and scattering delays using the signal-processing tool Cyclic Spectroscopy (CS; [21, 22]). This technique allows determination of the ISM impulse response function and extraction of the intrinsic pulse profile for highly scattered pulsars with a large flux density. CS backends have been developed at some of the radio telescopes [23], providing observations with much higher frequency resolution than possible with the filter banks used in conventional pulsar backends. Hence, very narrow scintillation bandwidths of high DM pulsars, that would not have been possible to measure with other pulsar backends, may be resolved in observations with a CS backend. This makes CS a very powerful tool for ISM mitigation and it will surely be used increasingly in the future.

4. Conclusions

PTAs strive to time MSPs with the highest possible precision, to allow for direct detection of nano-Hz GWs. Mitigating the delays that originate from radio wave propagation through the ISM is an essential part of the high-precision timing procedure.

The largest ISM delays are due to dispersion, and time-variable DM values are currently corrected for in the timing data. Optimal DM correction is a highly topical issue among the PTA members and development of new techniques in this area is on-going.

Further, interstellar density fluctuations also cause time-variable delays in the pulse time of arrival. For most PTA pulsars, scattering is currently not a limiting factor. However, with new wide-band receivers and higher timing accuracy, ability to correct for scattering delays may become an increasingly important part of high-precision timing. In addition, development of new techniques to properly mitigate scattering delays from the pulsar signal may allow for observations at lower frequencies as well as inclusion of more distant pulsars to the array in the future.

Acknowledgements

The NANOGrav project receives support from the National Science Foundation (NSF) PIRE program award number 0968296. The author would like to thank the NANOGrav collaboration for very useful advise and support, and for providing the pulsar data contributing to the figures in this document. She is also grateful to the organizers of the 10th International LISA Symposium for inviting her to present this talk.

References

[1] Kramer M and Champion D J 2013 *Classical and Quantum Gravity* **30** 224009
[2] Hobbs G 2013 *Classical and Quantum Gravity* **30** 224007 (*Preprint* 1307.2629)
[3] McLaughlin M A 2013 *Classical and Quantum Gravity* **30** 224008 (*Preprint* 1310.0758)
[4] Manchester R N and IPTA 2013 *Classical and Quantum Gravity* **30** 224010 (*Preprint* 1309.7392)
[5] Cordes J M and Shannon R M 2010 *ArXiv Astrophysics e-prints* (*Preprint* 1010.3785)
[6] Archibald A M, Kondratiev V I, Hessels J W T and Stinebring D R 2014 *ApJ* **790** L22 (*Preprint* 1407.0171)
[7] Levin L e a 2015 *ApJ, in prep*
[8] Stinebring D 2013 *Classical and Quantum Gravity* **30** 224006 (*Preprint* 1310.8316)
[9] Keith M J, Coles W, Shannon R M, Hobbs G B, Manchester R N, Bailes M, Bhat N D R, Burke-Spolaor S, Champion D J, Chaudhary A, Hotan A W, Khoo J, Kocz J, Osłowski S, Ravi V, Reynolds J E, Sarkissian J, van Straten W and Yardley D R B 2013 *MNRAS* **429** 2161–2174 (*Preprint* 1211.5887)
[10] Lee K J, Bassa C G, Janssen G H, Karuppusamy R, Kramer M, Liu K, Perrodin D, Smits R, Stappers B W, van Haasteren R and Lentati L 2014 *MNRAS* **441** 2831–2844
[11] Cordes J M and Lazio T J W 2002 *ArXiv Astrophysics e-prints* (*Preprint* 0207156)
[12] Demorest P B, Ferdman R D, Gonzalez M E, Nice D, Ransom S, Stairs I H, Arzoumanian Z, Brazier A, Burke-Spolaor S, Chamberlin S J, Cordes J M, Ellis J, Finn L S, Freire P, Giampanis S, Jenet F, Kaspi V M, Lazio J, Lommen A N, McLaughlin M, Palliyaguru N, Perrodin D, Shannon R M, Siemens X, Stinebring D, Swiggum J and Zhu W W 2013 *ApJ* **762** 94 (*Preprint* 1201.6641)
[13] Scheuer P A G 1968 *Nature* **218** 920–922
[14] Lyne A G and Rickett B J 1968 *Nature* **219** 1339–1342
[15] Cordes J M, Weisberg J M and Boriakoff V 1985 *ApJ* **288** 221–247
[16] Hemberger D A and Stinebring D R 2008 *ApJL* **674** L37–L40
[17] Rickett B J 1990 *Annual review of astronomy and astrophysics* **28** 561–605
[18] Cordes J M 1986 *ApJ* **311** 183–196
[19] Lambert H C and Rickett B J 1999 *ApJ* **517** 299–317
[20] Coles W A, Rickett B J, Gao J J, Hobbs G and Verbiest J P W 2010 *ApJ* **717** 1206–1221 (*Preprint* 1005.4914)
[21] Demorest P B 2011 *MNRAS* **416** 2821–2826 (*Preprint* 1106.3345)
[22] Walker M A, Demorest P B and van Straten W 2013 *ApJ* **779** 99 (*Preprint* 1310.3535)
[23] Jones G e a 2015 *ApJ, in prep*

The LIGO Open Science Center

Michele Vallisneri[1], Jonah Kanner[2], Roy Williams[2], Alan Weinstein[2] and Branson Stephens[3]

[1] Jet Propulsion Laboratory, California Institute of Technology, Pasadena CA 91109, USA
[2] LIGO – California Institute of Technology, Pasadena CA 91125, USA
[3] University of Wisconsin – Milwaukee, Milwaukee WI 53201, USA

E-mail: michele.vallisneri@jpl.nasa.gov

Abstract. The LIGO Open Science Center (LOSC) fulfills LIGO's commitment to release, archive, and serve LIGO data in a broadly accessible way to the scientific community and to the public, and to provide the information and tools necessary to understand and use the data. In August 2014, the LOSC published the full dataset from Initial LIGO's "S5" run at design sensitivity, the first such large-scale release and a valuable testbed to explore the use of LIGO data by non-LIGO researchers and by the public, and to help teach gravitational-wave data analysis to students across the world. In addition to serving the S5 data, the LOSC web portal (losc.ligo.org) now offers documentation, data-location and data-quality queries, tutorials and example code, and more. We review the mission and plans of the LOSC, focusing on the S5 data release.

1. Introduction

The emerging *Open Science* movement holds that the unprecedented ease of sharing information and collaborating remotely on the internet will foster a new era of greatly accelerated and amplified discovery, *if* data, publications, and results are made available openly and broadly [1]. The 2010 "Panton principles" open-data manifesto [2] characterizes science as based on "building on, reusing and openly criticising the published body of scientific knowledge," and argues that "for science to effectively function, and for society to reap the full benefits from scientific endeavours, it is crucial that science data be made open." The U.S. government accepted these arguments, and in 2013 it issued a general call to its agencies to make "information resources accessible, discoverable, and usable by the public" to "help fuel entrepreneurship, innovation, and scientific discovery" [3]. This memorandum characterizes *open data* as "fully discoverable and usable by end users," which it spells out further as public, accessible, described, reusable, complete, timely, and managed in post-release.

LIGO, the Laser Interferometer Gravitational-wave Observatory [4], is a facility dedicated to measuring cosmic gravitational waves (GWs); it is funded by the U.S. National Science Foundation (NSF) and operated by the California Institute of Technology and the Massachusetts Institute of Technology. LIGO has been collecting science data since 2002. Analyzing these data is one of the responsibilities of the LIGO Scientific Collaboration (LSC), which now comprises $\sim$ 90 institutions and 900 members worldwide. While the LSC is open to any contributor from the global community, any astrophysical publications based on LIGO data must currently be approved by the collaboration as a whole, and then published with a full list of authors. In 2008, the NSF asked LIGO and the LSC to devise a roadmap to open the LIGO data to professional researchers outside the LSC and to a broader public that includes amateur scientists, students, and more.

There are many good reasons why the LIGO data should be open. First, since LIGO is publicly funded, *the public owns the data.* Next and most important, opening the LIGO data will *maximize discovery.* It will benefit the LSC data-analysis efforts, enabling reproducibility, getting more eyes on LIGO data to recognize potential problems and solutions more rapidly, and improving the communication of LSC results to colleagues. It will also benefit collaborations with other GW and non-GW projects, facilitating electromagnetic followups and other multimessenger astronomy, and harnessing new perspectives and insights to advance LSC science. Last, open data access will serve to interest and engage a *much wider community,* bringing the excitement of GW science and multimessenger astronomy to students, amateur and professional astronomers and scientists in related fields.

LIGO's data-release roadmap [5], which later became part of LIGO's data management plan [6], identifies two phases to the release. During the early *discovery phase,* which will start as the upgraded Advanced LIGO detectors [7] begin to collect science data (as early as 2015), the understanding of the LIGO measurements is evolving as the instrument is tuned toward design sensitivity, and detections are rare. Use of the full data is restricted to the LSC, although early *transient alerts* of candidate GW events are broadcast to astronomers who have signed on to search for electromagnetic counterparts. The LSC releases to the public segments of GW strain data around validated discoveries, or astrophysically significant nondetections (such a release happened already for short-hard GRB 070201 after LIGO was able to exclude that the burst originated from a neutron-star binary coalescence in M31 [8]).

The *observational phase* begins after one of three conditions occurs: LIGO has obtained plentiful detections; it has probed a space-time volume of $3 \times 10^7\,\mathrm{Mpc}^3\,\mathrm{yr}$; or 3.5 years have elapsed since its formal acceptance. In this phase, the sensitivity of the observatory and the mature understanding of the data allow the exploration of the astrophysical content of the measured GWs. The entire dataset is released to the public in six-month segments with 24-month latency, as necessary to correct the data for instrumental and environmental issues [6]. In this phase also transient alerts are issued publicly.

2. The LIGO Open Science Center

Archiving and serving scientific data requires a dedicated service, characterized formally by the International Organization for Standardization (ISO) as an *Open Archival Information System* (OAIS, [9]). An OAIS includes elements that perform data ingestion, storage, annotation, preservation, operations, and access. Such a system exists already in the form of the *LIGO Data System,* which has archived one petabyte of LIGO data (1% of which in the GW channel, the rest in a myriad of environmental and instrument-monitoring channels) and has made it available within seconds of data taking to hundreds of LSC scientists.

However, the functions of an OAIS need to be fulfilled with respect to the needs and competences of one or more *designated communities* of data consumers. Consider for instance data preservation, which extends to the *bits* as well as the *meaning* of the data: achieving this goal for LSC members, who can rely the support of their peers and on the tacit knowledge obtained in their professional training, is clearly a very different task than doing so for non-LSC scientists and for the general public. To extend the LIGO Data System to these new designated communities, LIGO created the LIGO Open Science Center (LOSC), defined [6] as "a fabric to enable robust scientific discourse about the LIGO data for somebody who has never met any LSC member." Who may be the users of the LOSC data? Scientists with interests in GWs, relativity, astrophysics, and multimessenger astronomy; undergraduate and graduate students in the sciences; even K-12 STEM students and their teachers.

The LOSC team (comprising the named authors of this article) began work in 2012, focusing on an initial set of milestones for documenting, distributing, and curating the LIGO data:

- releasing the full GW strain data from the 2005–2007 "S5" science run, which achieved design sensitivity for Initial LIGO;
- validating the GW strain data and annotating it with easily interpreted data-quality information and with information about the simulated GW signals present in the data (a.k.a. *hardware injections*);
- setting up the hardware and software for a LOSC web portal, `losc.ligo.org`, offering access to open data as well as additional services;
- providing a database service for the available data and its quality;
- producing a suite of tutorials and example codes to acquire, load, manipulate, filter, and plot the data.

As we write this article, we have reached all these milestones. Most notably, in August 2014 the LOSC released an open dataset for LIGO's flagship fifth science run S5 (which took place between November 2005 and October 2007), the first in which LIGO performed at design sensitivity. The LSC, in partnership with the Virgo Collaboration [10], searched the data for a variety of sources, and published a number of papers limiting the rates of GW-emitting events in the local universe ([11, 12, 13] and several more, see [14] for a complete list of LSC publications). The S5 dataset includes data from the three LIGO instruments (the 4-km H1 and 2-km H2 in Hanford, WA, and the 4-km L1 in Livingston, LA), which collected 537, 546, and 457 days of science data, respectively. Although part of the S5 dataset was analyzed together with concurrent Virgo data, the purview of the LOSC does not include releasing Virgo datasets.

We intend for the S5 dataset to serve as a foundation for future data releases (including the "S6" data, which is currently under internal review); we anticipate that it will create a realistic testbed to engage the scientific community, the general public and the LSC itself in a discussion about the most useful treatments and representations of the LIGO data products; we trust that it will provide a valuable tool to teach GW data analysis to students at different levels. We plan to continue managing and supporting access to the S5 dataset as we prepare for further releases.

In the remainder of this proceeding, we describe the nature and technical details of the data archived by the LOSC, and the main services provided by the LOSC portal.

3. The LOSC data archive

The LIGO detectors generate thousands of channels of data ($\sim$ 6,600 for Initial LIGO [4]) corresponding to many different types of measurements and instrument monitors. The LOSC offers three simpler data products, which (we expect) will fulfill the scientific and pedagogical requirements of its designated communities. These products are the 4,096-Hz time series of the calibrated GW observable (known as *strain*, or $h(t)$); a 1-Hz time series that characterizes the *quality* of the strain data, by summarizing the data cuts and vetoes made by the LSC scientists who searched the S5 data for GWs; and another 1-Hz time series that flags the simulated GWs *injections* that were inserted in the data by actuating the LIGO test masses, to test search pipelines and characterize their performance.

The three LOSC time series are packaged together in files that each cover 4,096 contiguous seconds (an arbitrary but convenient choice). In designing and creating these files, we followed the principles discussed in the LIGO data-management plan [6], requiring LOSC data to be:

- *Interoperable*: The LOSC files are available in the broadly used and supported HDF5 format [15], and also in the *frame* format [16] used internally by the GW experimenters; free and open software tools to read both formats are available or linked on the LOSC portal.
- *Self-describing*: The LOSC files contain metadata describing their content, as well as "keys" to interpret the data-quality and injection time series.

- *Published online*: The LOSC files can be located and downloaded easily from the LOSC portal *by click* (i.e., through a web browser) and *by code* (i.e., by automated scripts that use a well-defined interface).
- *Contextualized*: The LOSC portal offers a "getting started" page about the files and their content, example code and tutorials, extensive documentation and references.
- *Supported*: The LOSC is organizing workshops and webinars about the data, will provide "help desk" functions as resources allow, and will continue to refine the datasets and correct problems if any are found.

This section continues with more details about the three time series contained in the LOSC data files.

3.1. GW strain

This time series represents the dimensionless GW strain $h(t)$, conceptually the fractional change induced by GWs in the LIGO arm length, as measured with laser interferometry. The $h(t)$ data were created by LSC analysts by *calibrating* [17] the LIGO interferometric measurements with a variety of techniques to infer actual arm-length changes from optical observables. These data record the GWs impinging on the detectors plus instrumental (strain-equivalent) noise; since no GW events were found in the S5 run, it is fair to say that the S5 $h(t)$ data are, to our knowledge, essentially all noise.

We downsample $h(t)$ to 4 kHz from the original 16 kHz to reduce the size of LOSC files while (arguably) losing no science content, since the higher frequencies are dominated by optical noise. Thus, a 4,096-second file contains 16,777,216 GW-strain samples from a single detector, represented as floating-point "doubles" (eight bytes each). If samples are missing because the detector was not taking science data at that time, they are represented as "NaN" (not a number) values. Each file has a starting GPS time[1] corresponding to an integer multiple of 4,096. Such a file takes up to 120 MB, allowing for compression; the whole S5 dataset (data from three detectors over two years, with roughly 65% duty cycle) has a size of $3 \times 2 \times 365 \times 24 \times 3600 \times 4096 \times 0.65 \times 8\,\text{bytes} =$ almost 4 TB.

3.2. Data quality

The LIGO instruments did not continuously collect data, since operation was frequently interrupted by commissioning work, seismic events, intermittent construction or forest logging nearby, and other causes. As a result, the S5 dataset contains many data gaps, $\sim 30\%$ of the total data-taking period. Most LSC searches would therefore begin from a *segment list* indicating intervals of existing science data, usually aligned to whole seconds. To capture this information, we label every second of usable science data with the marker `DATA` (and conversely, all the GW-strain samples in intervals not marked by `DATA` are NaNs in the LOSC files).

In addition to this minimal characterization, the GW searches performed by LSC scientists adopted further judgments about the quality of science data, ranking every second of observation in one of four *categories*. They performed the ranking by developing hundreds of *vetoes* that identify instrumental and environmental conditions (ranging from passing trains, saturated photodiodes, and powerline fluctuations to high winds, elevated seismic noise, and magnetometer spikes) found to be correlated to glitches and increased noise in the GW-strain channel. They then assigned each veto to a category, with slightly different assignments for different astrophysical searches (such as the search for GW signals from low-mass binary inspirals, known as `CBCLOW` [11], or the all-sky search for GW bursts, known as `BURST` [12]). Overall, the meaning of the four categories can be summarized as follows:

[1] Time as obtained from an accurate GPS receiver, given as the number of seconds since the January 6th, 1980 epoch, not including leap seconds.

- Data passing CAT1 vetoes is free of severe problems, so the detector is considered to be in observation mode.
- Data passing CAT2 (in addition to CAT1) vetoes is also free of problems resulting from well understood couplings between the GW-strain and instrumental channels. In LIGO searches, data passing CAT2 is typically used to identify GW event candidates.
- Data passing CAT3 (in addition to CAT1 and CAT2) vetoes is also free of problems resulting from less well established couplings between the GW and instrumental channels. In LIGO searches, data passing CAT3 is typically used to derive GW upper limits.
- Data passing CAT4 (in addition to CAT1–CAT3) vetoes is also free of problems resulting from poorly established couplings between the GW and instrumental channels.

The LOSC data-quality time series is given at 1 Hz cadence, and it consists of short *bitmasks* where setting an individual bit denotes that for that whole second (corresponding to 4,096 GW-strain samples) the LIGO data passed "CATn" vetoes according to the veto definitions of a certain search (for S5, one of CBCHIGH [18], CBCLOW [11], BURST [12], CONTINUOUS [13], and STOCHASTIC [19]). In addition, setting bit 0 corresponds to data possessing the DATA marker discussed above. The meaning of the data-quality bits is described in the metadata contained in each LOSC file.

Using this data-quality time series, users of the LOSC data may construct segment lists to analyze according to the most fitting veto definitions. For example, scientists mounting a search for short-duration GWs may use the segments of data for which the BURST_CAT2 bit is set, while those looking for longer transients may use CBCHIGH_CAT2, or perhaps CBCHIGH_CAT3 if setting upper limits. As we describe below, the LOSC website contains example code to form these segment lists.

3.3. Injections

Thousands of simulated GW signals, known as *hardware injections* (see, e.g., [20]), are routinely embedded in the LIGO strain measurements by actuating the LIGO test masses at scheduled (and carefully catalogued!) times. In S5, such injections were performed by the four LIGO search groups (those looking for compact binary coalescences, burst signals, continuous waves, and stochastic signals [11, 12, 13, 19]), to test the efficiency and characterize the statistics of their search pipelines, and to validate the calibration of the detector response to GWs. Search pipelines and detection protocols were further validated with a small number of *blind* injections that were not disclosed to data analysts until the searches had been completed (see, e.g., [21]). Binary and burst injections are short transients (several seconds to several minutes); continuous injections are longer-lived (potentially hours), but almost monochromatic, so they do not disturb the detection of transient signals occurring alongside them; by contrast, stochastic injections are both long-lived and broadband, so the loudest ones may invalidate, for the purpose of other searches, the stretches of data where they occur.

The LOSC data files include a 1-Hz bitmask timeline that identifies the seconds during which a hardware injection was active, as well as the type of the injection (when recorded in the LSC segment database). Thus, users of the LOSC data can build segment lists that exclude injection times, or focus on those very times to develop and test their search algorithms. In addition, separate listings of the times and parameters of all S5 hardware injections are available on the LSC portal.

4. LOSC data services

In addition to serving the HDF and frame files that contain the LOSC GW-strain, data-quality, and injection time series, the LOSC portal losc.ligo.org provides a variety of services and contextual information. We describe some highlights in this section.

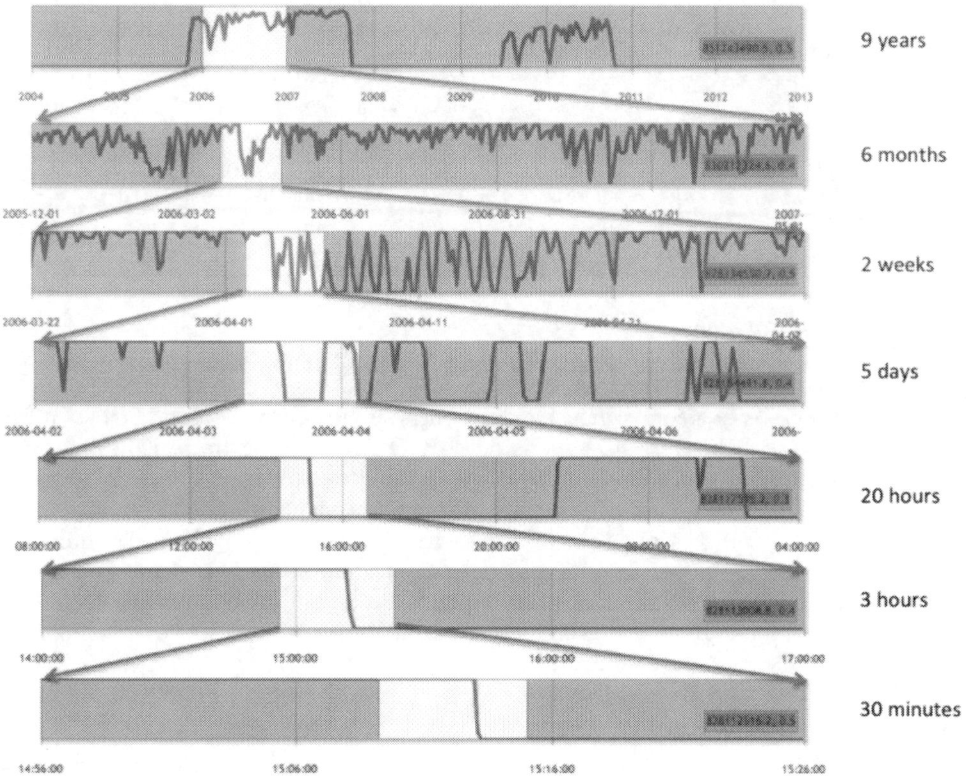

Figure 1. The LOSC Timeline tool offers instantaneous access to *duty-cycle* information for data-quality and injection bits over intervals of minutes to years.

4.1. Archive database

The LOSC portal offers access to a catalog of the LOSC files, which reproduces their basic metadata and a set of useful statistics: the minimum, maximum, standard deviation, and band-limited root-mean-square of the GW strain, as well as the number of seconds for which each data-quality and injection bit is set. The catalog can be queried to display the files that contain data between arbitrary GPS times, either as a table (color-coded by data quality and by the presence of injections), or in textual form, or in machine-readable JSON format [22].

4.2. Timeline

The LOSC Timeline tool presents a visual representation of the *duty cycle* of each data-quality (or injection) bit across a prescribed time interval: if the interval is short enough that each second can be shown onscreen, then the timeline for each bit is always either 0 or 1; for longer intervals, Timeline shows the average of the bit over nonoverlapping 2^n-second subintervals. The averages are cached at various levels to offer immediate access to data-quality or injection history over hours, weeks, years, and decades (see Fig. 1). Multiple bits can be explored together interactively, or the underlying information can be fetched in machine-readable JSON format or text-based segment lists.

In addition to all data-quality and injection bits for the S5 run, the LOSC Timeline tool currently visualizes science-mode duty cycles for LIGO's international partners: the German–British GEO600 [23] and the French–Italian Virgo [10]. Thus from a single web page we may explore the operation of five detectors, in three countries, over 12 years.

4.3. MySources

The LOSC MySources tool (`losc.ligo.org/mysources`) allows users to find out which GW detectors (among the three LIGO detectors, GEO, and Virgo) were taking data at the times of a user-provided collection of timed events, such as gamma-ray bursts, magnetar flares, or high-energy-particle measurements. The tool accepts the times (either UTC or GPS), event names, and (if desired) the sky positions of the events, and returns a listing of the observatories that have science data for those times, as well as the antenna patterns that quantify detector geometric response to a source at a given sky position.

4.4. Tutorials and examples

The LOSC portal also includes a set of tutorials and example codes (mostly in Python, but also in MATLAB and C) to learn how to download, parse, and manipulate the LOSC data files. The intention is to provide not a comprehensive data-analysis suite, which already exists in the public domain as the comprehensive (but challenging) LIGO Algorithm Library [24], but rather small and transparent samples that a beginner can use to build up to more sophisticated tools, or just to explore and learn. The tutorials cover software-environment prerequisites; searching for and downloading the files; inspecting them; working with the data-quality bitmasks to build segment lists; plotting time series, spectra, and spectrograms; identifying injections; and searching for (simple) chirp-like signals. A final tutorial explains how to use web services to automate the search of open data, how to navigate the metadata and catalogs, and how to download the entire S5 dataset file-by-file in a robust way.

5. Summary

The creation of the LOSC is an important step in aligning the GW community with the broad social movement towards open science, and in supporting the continuing trajectory of the international network of GW observatories toward a model of shared data analysis. The LOSC provides public access to LIGO data products and to value-added documentation, tutorials, and (free and open) software, with the aim of making the LIGO data truly comprehensible and usable by a broad audience. Significant resources have been committed to ensuring that the data are presented accurately, and that the contextual information is sufficient to allow reuse by scientists, students, and others inside and outside the LIGO Scientific Collaboration.

The LOSC has already released the first large dataset from Initial LIGO's flagship S5 run; more are planned to follow. The S5 release creates several concrete benefits. It will promote broad participation in the advancement of GW physics and astrophysics from professional and amateur scientists, graduate students, undergraduates, and secondary students; all are invited to help verify and improve the quality of LIGO's scientific results. The web services and tutorials created to enable GW research by scientists outside the LSC are already being used by researchers inside the collaboration, and are streamlining training and access to data for new LSC students. The public dataset will provide excellent opportunities for outreach and education at many levels, and for many different audiences. Finally, the effort to document and curate the S5 dataset has helped ensure that this archive will be a usable and valuable resource for many years to come.

LIGO will continue to release datasets through the LOSC, including data from the higher-sensitivity Advanced LIGO detectors that will begin operating in 2015 [6]. Any GW signals identified by the LSC will be included in these releases. When regular GW detections begin to occur, public participation in LIGO data analysis will add an exciting dimension to GW astronomy. Numerical relativists, relativity theorists, astrophysicists, and others will use LIGO data to understand the dynamics of strongly curved spacetime, as well as the origins and properties of GW sources.

Acknowledgements

We thank Stuart Anderson, Juan Barayoga, Patrick Brady, Duncan Brown, Laura Cadonati, Vuk Mandic, Szabi Marka, Greg Mendell, Keith Riles, Peter Shawhan, David Shoemaker, and Eric Thrane for helpful discussions and interactions. Beverly Berger, Marie-Anne Bizouard, Vicki Kalogera, Andrew Lundgren, Vivien Raymond, and John Whelan participated in the LSC review of the S5 dataset. We are grateful to the Virgo and GEO collaborations for allowing the LOSC Timeline tool to report observation-segment metadata for their detectors. Alexander Cole, Ashley Disbrow, Gary LaMotte, and Shannon Wang performed early tests of the data. LIGO was constructed by the California Institute of Technology and Massachusetts Institute of Technology with funding from the National Science Foundation, and operates under cooperative agreement NSF-0757058. The LIGO Open Science Center (LOSC) was partially funded by grant NSF-1210172. Part of this work was carried out at the Jet Propulsion Laboratory, California Institute of Technology, under contract to the National Aeronautics and Space Administration. This is LIGO document LIGO-P1400186. Copyright 2014.

References

[1] Nielsen M 2011 *Reinventing Discovery: The New Era of Networked Science* (Princeton University Press)
[2] Murray-Rust P, Neylon C, Pollock R and Wilbanks J 2010 Panton principles: Principles for open data in science retrieved 12 Sep 2014 from `http://pantonprinciples.org`
[3] Executive office of the US President, Office of Management and Budget 2013 Open data policy—managing information as an asset `http://www.whitehouse.gov/sites/default/files/omb/memoranda/2013/m-13-13.pdf`
[4] Abbott B P *et al.* 2009 *Reports on Progress in Physics* **72** 076901 (*Preprint* 0711.3041)
[5] Lazzarini A, Whitcomb S, Marx J, Weiss R, Reitze D, Barish B and P S 2009 A proposal for providing open access of LIGO data to the broader research community report LIGO-M080072-v2, `http://dcc.ligo.org/LIGO-M080072/public`
[6] Anderson S and Williams R 2014 Ligo data management plan, january 2014 report LIGO-M1000066-v19, `http://dcc.ligo.org/LIGO-M1000066/public`
[7] Harry G M and LIGO Scientific Collaboration 2010 *Classical and Quantum Gravity* **27** 084006
[8] Abbott B *et al.* 2008 *Astrophys. J.* **681** 1419–1430 (*Preprint* 0711.1163)
[9] Consultative Committee for Space Data Systems 2012 Space data and information transfer systems – Open archival information system (OAIS) – Reference model ISO 14721:2012
[10] Accadia T *et al.* 2012 *Journal of Instrumentation* **7** 3012
[11] Abbott B P *et al.* 2009 *Phys. Rev. D* **79** 122001 (*Preprint* 0901.0302)
[12] Abbott B P *et al.* 2009 *Phys. Rev. D* **80** 102001 (*Preprint* 0905.0020)
[13] Abbott B P *et al.* 2009 *Phys. Rev. Lett.* **102** 111102 (*Preprint* 0810.0283)
[14] LIGO Scientific Collaboration 2014 List of LSC Observational and Instrument Publications `https://www.lsc-group.phys.uwm.edu/ppcomm/Papers.html`
[15] The HDF Group 1997-2014 Hierarchical Data Format, version 5 http://www.hdfgroup.org/HDF5
[16] LIGO and Virgo collaborations 2014 Common Data Frame Format for Interferometric Gravitational Wave Detector http://lappweb.in2p3.fr/virgo/FrameL
[17] Abadie J *et al.* 2010 *Nuclear Instruments and Methods in Physics Research A* **624** 223–240 (*Preprint* 1007.3973)
[18] Abadie J *et al.* 2010 *Phys. Rev. D* **82** 102001 (*Preprint* 1005.4655)
[19] Abbott B P *et al.* 2009 *Nature* **460** 990–994 (*Preprint* 0910.5772)
[20] Brown D A and LIGO Scientific Collaboration 2004 *Classical and Quantum Gravity* **21** 797 (*Preprint* gr-qc/0312031)
[21] LIGO and Virgo collaborations 2011 GW100916 blind injection data release `http://www.ligo.org/scientists/GW100916`
[22] ECMA 2013 JSON (JavaScript Object Notation) `http://json.org`
[23] Grote H and LIGO Scientific Collaboration 2010 *Classical and Quantum Gravity* **27** 084003
[24] LIGO Scientific Collaboration 2014 LSC Algorithm Library Suite `https://www.lsc-group.phys.uwm.edu/daswg/projects/lalsuite.html`

Journal of Physics: Conference Series **610** (2015) 012022 doi:10.1088/1742-6596/610/1/012022

Summary of the results of the LISA-Pathfinder Test Mass release

C Zanoni[1,2,0], D Bortoluzzi[1,2], J W Conklin[3], I Köker[4], B Seutchat[1], S Vitale[5,2]

[1] Dept. of Industrial Engineering, University of Trento, Via Sommarive 9, I-38123 Trento, Italy
[2] INFN, Trento associated group, Via Sommarive 6, I-38123 Trento, Italy
[3] Dept. of Mechanical and Aerospace Engineering, University of Florida, FL 32611-6250, USA
[4] Airbus Defence and Space, Claude-Dornier-Strasse, 88090 Immenstaad, Germany
[5] Dept. of Physics, University of Trento, Via Sommarive 6, I-38123 Trento, Italy

E-mail: carlo.zanoni@unitn.it, daniele.bortoluzzi@unitn.it

Abstract. The challenging goal of LISA-Pathfinder in terms of maximum non-gravitational forces applied on the test mass poses tight constraints on the design of the Gravitational Reference Sensor. In particular, large gaps (3-4 mm) must exist between the test mass and its housing and any system there located must be either gold coated or made of a gold-based material. As a consequence, a significant adhesion may arise between the test mass and the mechanism designed to cage it during the spacecraft launch and to release it to free-fall. The criticality of the latter phase is enhanced by the control force authority exerted to the test mass by the surrounding electrodes. Such a force is limited by the large gaps (order of μN). Since the expected adhesion force between the test mass and its holding devices is much larger than the force authority, a dynamic release must be realized. However, following this procedure adhesion converts into test mass velocity, which can be controlled by the capacitive force only if it is smaller than 5 μm/s. At the University of Trento (Italy) the Transferred Momentum Measurement Facility has been designed and developed to measure the impulse produced by metallic adhesion upon quick rupture, in representative conditions of the LISA-Pathfinder test mass release to free-fall. Large sets of data have been collected and a mathematical model of the in-flight release dynamics has been developed, in order to estimate the test mass release velocity. A summary of the results is presented, together with an overview of the recent developments and a prediction of the in-flight performance.

1. Introduction

In the LISA-Pathfinder mission, the main sensor -the Gravitational Reference Sensor (GRS)- measures the position of a 1.96 kg Au/Pt cubic test mass (TM). This object is nominally shielded within the required limits [1] from all forces except gravity and spacecraft actuation. The motion ruled by gravity alone is called Drag-Free and follows a Geodesic trajectory.

In order to survive launch each TM must be carefully constrained during Launch and Early Orbit Phase. Such a constraint is then removed, allowing the beginning of the scientific phase. This release phase is a critical feature of LISA-Pathfinder. In fact, due to the limited available electrostatic actuation (0.5 μN), the initial state of each TM must be within the boundaries of Tab. 1.

The device in charge of holding the TM during launch is called CVM (Caging and Vent Mechanism) [2]. It acts along the z axis of the TM and applies a high load of about 300 N

[0] Current address: EN-MME, CERN, Route de Meyrin 385, 1217 Meyrin, Switzerland

Table 1. TM state requirement

State	Value
offset along x, y and z	$\pm$ 200 μm
linear velocity along x, y and z	$\pm$ 5 μm/s
angle around x, y and z	$\pm$ 2 mrad
angular rate around x, y and z	$\pm$ 100 μrad/s

on each TM corner. Once in space, the CVM hands the TM over to the GPRM (Grabbing, Positioning and Release Mechanism), [3].

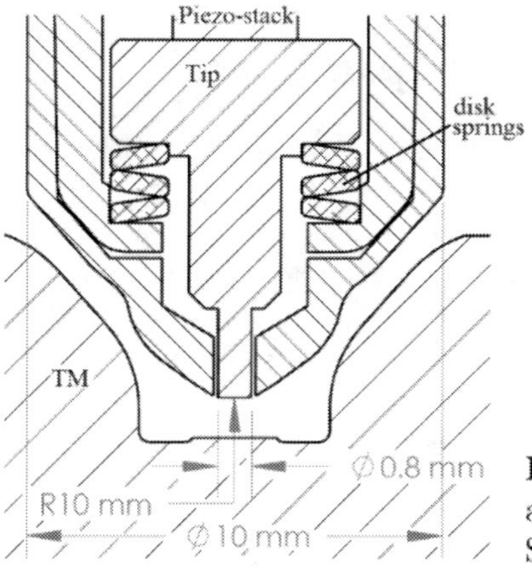

Figure 1. Cut view of the GPRM and TM recess (courtesy of RUAG Schweiz AG, RUAG Space)

The GPRM consists of two subsystems: the plunger (or grabbing finger) and the release tip. See Fig. 1. The first one is in charge of centering and aligning and the second is a small pin moved by a piezo-stack that removes the last mechanical action on the TM by means of a limited contact area and a low contact load (0.3 N). Both GPRM and CVM consist by two symmetrical parts on two TM faces.

The criticality of the release is the result of two effects. First of all, the mechanical action must be really symmetrical on both TM sides. Second, and more important, adhesion acts joining release device and TMand has a limited repeatability. Therefore, the removal of the constraint will transfer a momentum to the TM. More precisely, a pull on the TM is expected along + or - z. Adhesion is enhanced by the fact that both TM and GPRM are Au or Au coated for mitigation of stray forces due to non uniform surface work-functions.

An experimental-analytical on-ground qualification activity has been conceived for assessing the performance and the risk of the release. In the following, we will provide an overview of the main test results and an estimation of the TM release velocity.

2. Testing facility and experimental data

At the University of Trento (Italy), a facility has been designed for studying adhesion in dynamic conditions and in a space-like environment. The Transferred Momentum Measurement Facility (TMMF) is constituted by a 1.15 m pendulum. At its end is suspended a mock-up of the TM. Such a mock-up is a frame in which is included an Au/Pt insert, reproducing the flight TM contact surface. Similarly, a mock-up of the release tip is moved by an ultrasonic piezo stage. This last mock-up is loaded up to 300 mN against the pendulum, which is blocked by a set of 3 needles. The load is then relaxed until a nominal 0 mN value. Such a value is inferred from the commanded motion of the tip and the value of the stiffness of its blades, Fig. 2. At this point the tip is quickly retracted pulling the TM by means of adhesion. An interferometer measures the motion of the pendulum from which the applied force can be derived.

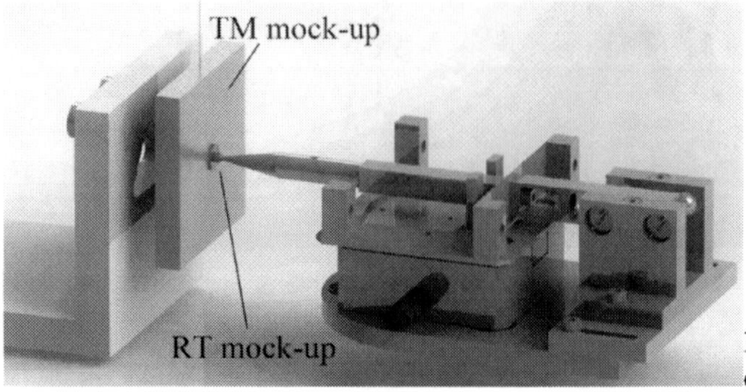

Figure 2. Rendered view of the core of the TMMF.

A more detailed description of the experiment can be found in [4, 5]. Indeed, a large set of actuators are also present for alignment and positioning of TM and release tip mock-ups. Also, the facility is supported by a damping table that mitigates ground vibrations effects.

Three main experimental campaigns have been performed, each one with a different mass attached to the pendulum: 0.0096 kg, 0.089 kg and 0.878 kg. The flight TM is equal to 1.96 kg, but adhesion arises only at the contact patch, whose representativeness is guaranteed on ground. At the same time, budget and organizational reasons make harder the use of a flight TM.. For each of these campaigns the peak force, the time-length of force action and the transferred momentum have been derived from the TM laws of motion. The histograms in Fig. 3 summarize the results in terms of peak force.

Unfortunately, adhesion is only one of the two forces that contribute to the measured motion of the TM. In order to avoid the rupture of the adhesive bond due to microseismic noise and for small positioning errors, a load of a few mN is always left between TM and tip. The effect of such a load is a net impulse on the TM, that can be estimated using:

$$I_{load} = F_{load}\sqrt{\frac{M_{TM}}{k_{nd}}} \tag{1}$$

where F_{load} is the residual load, M_{TM} the mock-up mass and k_{nd} the stiffness of the blocking system (the set of 3 needles).

3. Estimation of in-flight behavior

A first rough estimation of the release velocity is done by observing that the force applied on the TM in the TMMF lasts for about 2 ms. At the same time, the flight release mechanism moves 10 times quicker than the actuator used on-ground. The impulse is roughly equal to the area of the triangle $F_{peak}\delta t/2(=v_{TM}M_{TM})$, the maximum allowed peak force is 100 mN. We conservatively assume that the totality of the force is due to adhesion. The worst case measured peak adhesion value is 48 mN, which still leaves a margin factor of 2. However, this margin factor does not take into account the effect of a possible asymmetry in the mechanisms action.

A finer estimation requires a mathematical model for adhesion, such as [6]:

$$F_{adh} = A_1(\Delta l)e^{-\frac{\Delta l}{\lambda}} \tag{2}$$

where Δl is the elongation. This function, together with a linear model (i.e. spring) for the needles push, can be fitted in the TM measured acceleration in order to estimate adhesion, disentangling the needles effect. The result is a set of force-to-elongation functions.

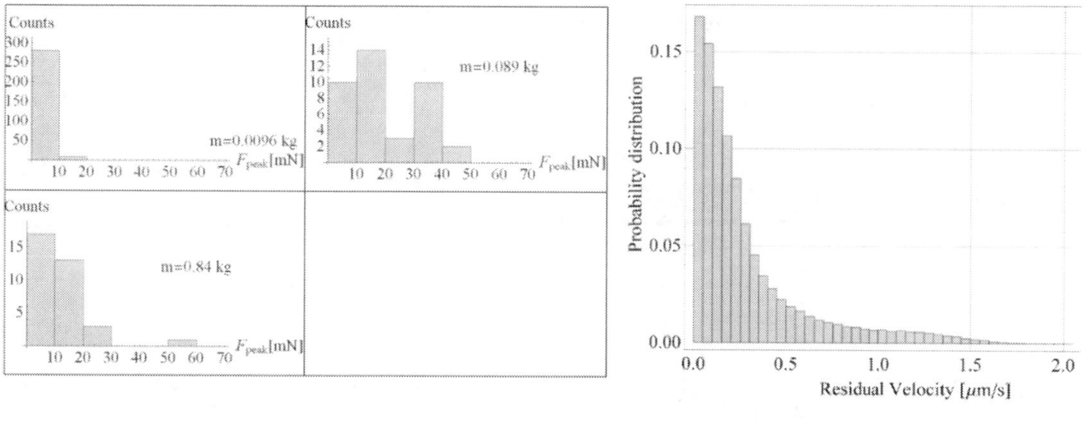

Figure 3. On the left: histograms of the experimental peak forces. On the right: release velocity probability distribution obtained from Montecarlo simulations

A model of the release tip has also been proposed [7] that has a voltage as input and the motion of the tip as output. A mathematical description of the release phase is possible by joining this last model with the adhesion functions and the TM equations of motion.

The output of the simulation is a random variable function of a large set of input random variables:

- Adhesion functions: derived from experimental data, as explained.
- Parameters of the release mechanism model: the model of the release tip is identified on the basis of measurements of its motion. The identification procedure provides both an optimal set of values and their covariance matrix.
- Contact geometry: the uncertainty on the contact radius of the tip is estimated on the basis of the design tolerance.
- Input voltage: the real input is not an ideal step, but has a small drop time whose statistical distribution is extracted from a set of measurements.
- Residual load before release: the nominal value is 300 mN with an uncertainty suggested by Airbus Defence and Space, responsible of all the Caging Assembly.

The release velocity distribution resulting from 100000 simulations is shown in Fig. 3. The 99.73 % upper value is 1.5 μm/sec.

This activity provides key information in the scope of future Drag-Free missions and eLISA especially. One parameter that could improve the performance is the tip's head curvature radius. As a first order approximation, adhesion is proportional to the contact area. In the Hertz model [8], the radius a of the contact area and the force F are:

$$a = \frac{\sqrt[3]{3}\sqrt[3]{\frac{FR}{E_m}}}{2^{2/3}} \qquad F = \frac{4}{3}d^{3/2}E_m\sqrt{R} \tag{3}$$

where R is the radius of the hemisphere in contact, d the indentation and E_m the equivalent elastic modulus. A larger R means a larger a and contact area. At the same time a small R increases the indentation d. The larger the contact area, the larger will be adhesion. Conversely, a larger indentation means a longer contact time and an enhancement of the effects of asymmetrical mechanisms actions. A trade-off value can therefore be found and it can be shown that a radius of about 1 mm would provide much better worst-case performance as shown in Fig. 4. Even in an estimation in which adhesion is low, the current value of 10 mm (Fig. 1) appears large.

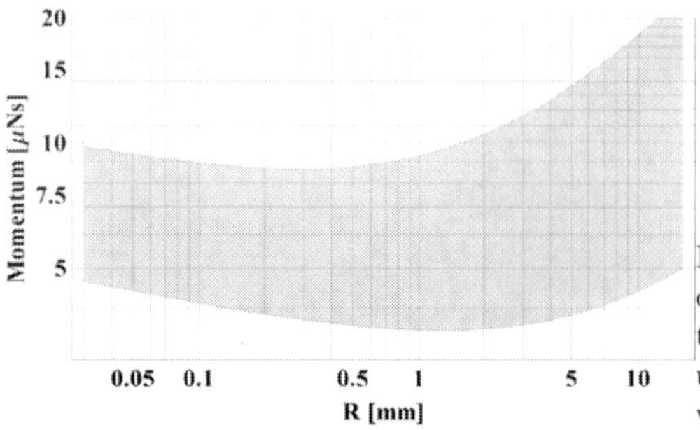

Figure 4. Estimation of the worst case velocity as a function of the tip radius. The two curves are an upper and bottom estimation of the worst-case.

4. Conclusion

The release of the test mass represents a critical phase for the LISA-Pathfinder space mission. In order to declare it successful, it must result in a test mass moving with a velocity below 5 μm/sec. An experimental-analytical procedure has been defined to evaluate the risk of this phase. Most of this risk is due to the presence of adhesion between the Au/Pt test mass and the release mechanism. An experimental facility was built to estimate adhesion effect in dynamic conditions in a space-like environment. The results of on-ground testing activity together with simulations suggest that the worst-case test mass velocity is compliant with the requirement with a margin factor larger than 2.

References

[1] Antonucci F, Armano M, Audley H, Auger G, Benedetti M, Binetruy P, Bogenstahl J, Bortoluzzi D, Bosetti P, Brandt N, Caleno M, Canizares P, Cavalleri A, Cesa M, Chmeissani M, Conchillo A, Congedo G, Cristofolini I, Cruise M, Danzmann K, Marchi F D, Diaz-Aguilo M, Diepholz I, Dixon G, Dolesi R, Dunbar N, Fauste J, Ferraioli L, Ferrone V, Fichter W, Fitzsimons E, Freschi M, Marin A G, Marirrodriga C G, Gerndt R, Gesa L, Gilbert F, Giardini D, Grimani C, Grynagier A, Guillaume B, Guzmn F, Harrison I, Heinzel G, Hernndez V, Hewitson M, Hollington D, Hough J, Hoyland D, Hueller M, Huesler J, Jennrich O, Jetzer P, Johlander B, Karnesis N, Killow C, others, Llamas X, Lloro I, Lobo A, Maarschalkerweerd R, Madden S, Mance D, Mateos I, McNamara P W, Mendes J, Mitchell E, Monsky A, Nicolini D, Nicolodi D, Nofrarias M, Pedersen F, Perreur-Lloyd M, Plagnol E, Prat P, Racca G D, Ramos-Castro J, Reiche J, Perez J A R, Robertson D, Rozemeijer H, Sanjuan J, Schleicher A, Schulte M, Shaul D, Stagnaro L, Strandmoe S, Steier F, Sumner T J, Taylor A, Texier D, Trenkel C, Tu H B, Vitale S, Wanner G, Ward H, Waschke S, Wass P, Weber W J, Ziegler T and Zweifel P 2012 *Classical and Quantum Gravity* **29** 124014

[2] Zahnd B, M Zimmermann M and Spörri R 2013 LISA-Pathfinder cage and vent mechanism - development and qualification *Proceedings 15th European Space Mechanism and Tribology Symposium (ESMATS)*

[3] Neukom A, Romano R and Nellen P M 2009 Testing and lesson learnt of LISA GPRM *Proceedings 13th European Space Mechanism and Tribology Symposium (ESMATS)*

[4] Bortoluzzi D, Benedetti M, Baglivo L, Cecco M D and Vitale S 2011 *Review of Scientific Instruments* **82** 125107 (pages 11)

[5] Bortoluzzi D, Benedetti M, Conklin J W, Baglivo L and Lio M D 2011 Testing the LISA-Pathfinder test mass injection into a geodesic *20th Annual Conference of the Italian Association of Theoretical and Applied Mechanics*

[6] Bortoluzzi D, Benedetti M and Conklin J W 2013 *Mechanical Systems and Signal Processing* **38** 384 – 398 ISSN 0888-3270

[7] Bortoluzzi D, Conklin J W and Zanoni C 2013 *Advances in Space Research* **51** 1145 – 1156 ISSN 0273-1177

[8] Maugis D D 2000 *Contact, adhesion, and rupture of elastic solids* (*Springer series in solid-state sciences* vol XIV) (Berlin, New York: Springer) ISBN 3-540-66113-1

Journal of Physics: Conference Series **610** (2015) 012023 doi:10.1088/1742-6596/610/1/012023

In-flight thermal experiments for LISA Pathfinder: Simulating temperature noise at the Inertial Sensors

F Gibert[a], M Nofrarias[a], M Armano[b], H Audley[c], G Auger[d], J Baird[n], P Binetruy[d], M Born[c], D Bortoluzzi[e], N Brandt[f], A Bursi[t], M Caleno[g], A Cavalleri[h], A Cesarini[h], M Cruise[i], K Danzmann[c], I Diepholz[c], R Dolesi[h], N Dunbar[j], L Ferraioli[k], V Ferroni[h], E Fitzsimons[f], M Freschi[b], J Gallegos[b], C García Marirrodriga[g], R Gerndt[f], Ll Gesa[a], D Giardini[k], R Giusteri[h], C Grimani[l], I Harrison[m], G Heinzel[c], M Hewitson[c], D Hollington[n], M Hueller[h], J Huesler[g], H Inchauspé[d], O Jennrich[g], P Jetzer[o], B Johlander[g], N Karnesis[a], B Kaune[c], N Korsakova[c], C Killow[p], I Lloro[a], R Maarschalkerweerd[m], S Madden[g], P Maghami[s], D Mance[k], V Martín[a], F Martin-Porqueras[b], I Mateos[a], P McNamara[g], J Mendes[m], L Mendes[b], A Moroni[t], S Paczkowski[c], M Perreur-Lloyd[p], A Petiteau[d], P Pivato[h], E Plagnol[d], P Prat[d], U Ragnit[g], J Ramos-Castro[q r], J Reiche[c], J A Romera Perez[g], D Robertson[p], H Rozemeijer[g], G Russano[h], P Sarra[t], A Schleicher[f], J Slutsky[s], C F Sopuerta[a], T Sumner[n], D Texier[b], J Thorpe[s], C Trenkel[j], H B Tu[h], D Vetrugno[h], S Vitale[h], G Wanner[c], H Ward[p], S Waschke[n], P Wass[n], D Wealthy[j], S Wen[h], W Weber[h], A Wittchen[c], C Zanoni[e], T Ziegler[f], P Zweifel[k]

[a] Institut de Ciències de l'Espai (CSIC-IEEC), Campus UAB, Facultat de Ciències, 08193 Bellaterra, Spain
[b] European Space Astronomy Centre, European Space Agency, Villanueva de la Cañada, 28692 Madrid, Spain
[c] Albert-Einstein-Institut, Max-Planck-Institut für Gravitationsphysik und Universität Hannover, 30167 Hannover, Germany
[d] APC UMR7164, Université Paris Diderot, Paris, France
[e] Department of Industrial Engineering, University of Trento, via Sommarive 9, 38123 Trento, and Trento Institute for Fundamental Physics and Application / INFN
[f] Airbus Defence and Space, Claude-Dornier-Strasse, 88090 Immenstaad, Germany
[g] European Space Technology Centre, European Space Agency, Keplerlaan 1, 2200 AG Noordwijk, The Netherlands
[h] Dipartimento di Fisica, Università di Trento and Trento Institute for Fundamental Physics and Application / INFN, 38123 Povo, Trento, Italy
[i] Department of Physics and Astronomy, University of Birmingham, Birmingham, UK
[j] Airbus Defence and Space, Gunnels Wood Road, Stevenage, Hertfordshire, SG1 2AS, UK
[k] Institut für Geophysik, ETH Zürich, Sonneggstrasse 5, CH-8092, Zürich, Switzerland
[l] Istituto di Fisica, Università degli Studi di Urbino/ INFN Urbino (PU), Italy
[m] European Space Operations Centre, European Space Agency, 64293 Darmstadt, Germany
[n] The Blackett Laboratory, Imperial College London, UK
[o] Physik Institut, Universität Zürich, Winterthurerstrasse 190, CH-8057 Zürich, Switzerland
[p] SUPA, Institute for Gravitational Research, School of Physics and Astronomy, University of Glasgow, Glasgow, G12 8QQ, UK

[q] Department d'Enginyeria Electrònica, Universitat Politècnica de Catalunya, 08034 Barcelona, Spain
[r] Institut d'Estudis Espacials de Catalunya (IEEC), C/ Gran Capità 2-4, 08034 Barcelona, Spain
[s] NASA Goddard Space Flight Center, 8800 Greenbelt Road, Greenbelt, MD 20771, USA
[t] CGS S.p.A, Compagnia Generale per lo Spazio, Via Gallarate, 150 - 20151 Milano, Italy

E-mail: gibert@ieec.cat

Abstract. Thermal Diagnostics experiments to be carried out on board LISA Pathfinder (LPF) will yield a detailed characterisation of how temperature fluctuations affect the LTP (LISA Technology Package) instrument performance, a crucial information for future space based gravitational wave detectors as the proposed eLISA. Amongst them, the study of temperature gradient fluctuations around the test masses of the Inertial Sensors will provide as well information regarding the contribution of the Brownian noise, which is expected to limit the LTP sensitivity at frequencies close to 1 mHz during some LTP experiments. In this paper we report on how these kind of Thermal Diagnostics experiments were simulated in the last LPF Simulation Campaign (November, 2013) involving all the LPF Data Analysis team and using an *end-to-end* simulator of the whole spacecraft. Such simulation campaign was conducted under the framework of the preparation for LPF operations.

1. Introduction

LISA Pathfinder (LPF) [1] is a shared ESA/NASA mission to demonstrate the technology for future space-basedspace-based gravitational wave observatories, to be launched along the second half of 2015. LPF aims to prove the feasibility of space-borne gravitational wave observatories, such as eLISA [2], by measuring a residual differential acceleration between free-falling test masses below $3 \times 10^{-14}\,\mathrm{m\,s^{-2}Hz^{-1/2}}$ at 1 mHz. The main instrument on board is the LISA Technology Package (LTP), which includes essentially the two Inertial Sensors with the test masses [4] and the Optical Measurement System [3] that provides the high precision relative distance measurements between the test masses.

The acceleration sensitivity of the LTP along its bandwidth of measurement ($1-30$ mHz) is limited at the low edge of the band by actuation noise while, at high frequencies the readout noise sets the limit. Other noise sources like magnetic field fluctuations, test mass random charging or temperature noise have a smaller impact to the sensitivity, but still the LTP needs to be equipped with a diagnostics subsystem to measure and characterise their contribution [5].

In order to characterise those thermal related effects that can have an impact in the instrument sensitivity, the LTP counts with a series of (a) thermistors to provide high precision measurements of temperature fluctuations and (b) heaters to apply controlled heat signals to the system. Specific temperature experiments will be carried out on board in order to characterise the temperature noise perturbations on the LTP performance.

Amongst the different kind of thermal perturbations [6], the LTP is specially sensitive to residual temperature gradient fluctuations on the Electrode Housings (EH) of the Inertial Sensors –the structures that host the test masses (TM) and support the layout of electrodes for electrostatic sensing and actuation. Asymmetric temperature distributions on these parts induce forces and torques on the test masses through three different thermal effects:

(i) The **radiometer effect**, significant in rarefied environments where the particles have a mean free path longer than the cavity dimensions.

(ii) The **radiation pressure**, as a consequence of the radiation emitted by a surface, following the Stefan-Boltzmann Law.

(iii) The temperature-dependent flow of particles emitted by a surface, inducing the asymmetric **outgassing** effect.

Their contribution to the LTP performance has been investigated by means of torsion pendulums [7] and will be measured in-flight through a dedicated thermal experiment. In order to assist the design of the experiment, a State-Space Model (SSM) of the thermal-induced forces and torques has been developed [8, 9] and, in a recent LPF simulation campaign we had the opportunity to test both the experiment simulator and the data analysis tools by integrating it to a global satellite simulator.

In this paper we report the results of such a simulation in the following way: the models for the thermal effects are presented in Section 2, the thermal experiment is described in Section 3, the simulation campaign in Section 4 and the analysis in presented in Section 5. The final discussion is found in Section 6.

2. Thermal effects model

The model implemented is based on the linearised expressions for the different thermal effects showed in Table 1. Each F_{ij} is the force on a surface j by the presence of another surface i at a different temperature. K^{ij} are the coefficients with information on the geometry, visibility conditions and constants. The ΔT_{ij} and T_0 terms represent the temperature difference between surfaces and the absolute temperature respectively, while p is the system pressure. Θ_{OG} is the activation temperature for the outgassing effect. More details of the model can be found in [8].

Table 1. Simplified expressions of the thermal effects [8].

Radiometer effect (RM)	$F^{ij}_{RM} = K^{RM}_{ij}\, p\, T_0^{-1}\, \Delta T_{ij}$
Radiation pressure (RP)	$F^{ij}_{RP} = K^{RP}_{ij}\, T_0^3\, \Delta T_{ij}$
Outgassing effect (OG)	$F^{ij}_{OG} = K^{OG}_{ij}\, T_0^{-2}\, e^{-\Theta_{OG} T_0^{-1}}\, \Delta T_{ij}$

The system surfaces are discretised in order to allow the calculation of net forces and torques on all the TM axis by computing the different node-to-node contributions. The different coefficients K^{ij}, Θ_{OG} and p are considered constant parameters while ΔT_{ij} and T_0 are simulated by means of a thermal model of the spacecraft.

3. The Electrode Housing thermal experiment

The purpose of the experiments is double: first, to determine the thermal coupling of the temperature gradient fluctuations with the interferometer readouts and, secondly, to disentangle the contribution of the different thermal effects. The relevance of the latter is due to the dependence of the radiometer effect on the pressure, that may allow to estimate the contribution of Brownian noise [10], a pressure-dependent effect expected to be the first noise contributor at the lower edge of the bandwidth –close to 1 mHz– after the actuation noise.

By applying alternate sequences of pulses to Heaters 1 and 2 –see Figure 1– a temperature gradient signal is induced on the x axis. Such a signal is measured by combining the readouts of the different thermistors on the EH walls, as:

$$\Delta T_x = \frac{T_3 + T_4 - T_1 - T_2}{2} \tag{1}$$

The thermal coupling coefficient on the X axis is then defined as

Journal of Physics: Conference Series **610** (2015) 012023 doi:10.1088/1742-6596/610/1/012023

$$\alpha_x = \frac{\mathrm{d}\Delta F_x}{\mathrm{d}\Delta T_x} \tag{2}$$

where the F_x is the x component of the force applied to the test mass. Since the different thermal effects have a different dependence on the absolute temperature, $\alpha_x(T)$ can be used as figure of merit to disentangle the different effects described above. This discrimination is achieved by repeating the same alternate heating sequence at different absolute temperatures.

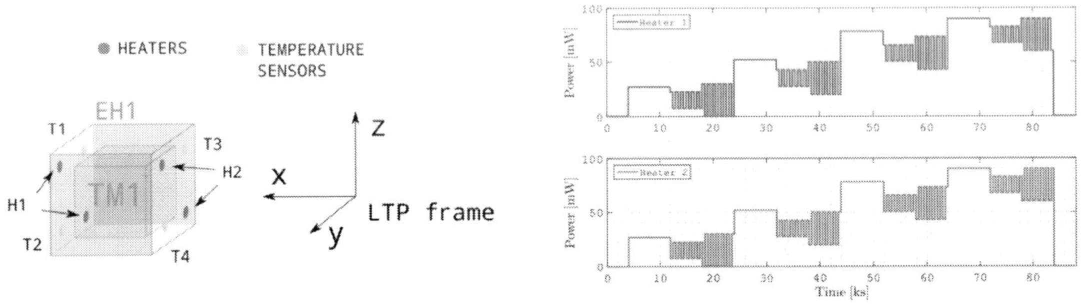

Figure 1. *Left*: Layout of temperature sensors (T1-T4) and heaters (H1, H2) on one of the EH. *Right:* Signals applied to H1 and H2 during the thermal experiment.

4. Simulation campaign

The 4th LPF Simulation Campaign took place in ESAC [1] on November 2013 and lasted for eight days, involving the whole LPF data analysis team. Every day, the telemetry corresponding to the experiments simulated the previous day was received and post-processed and a different group of data analysts was responsible of the analysis. The signal applied to the EH during one day of the campaign consists of the repetition of a same pattern at four different absolute temperature levels, quickly achieved by applying constant voltages to the heaters. At each temperature level, series of pulses of 500 s are applied alternatively to H1 and H2, resulting in a 1 mHz modulated temperature gradient signal along the x axis. For linearity checks, two different modulation amplitudes are applied at each temperature level –see Figure 1.

5. Analysis

The data analysis presented here has been carried out by means of the LTP Data Analysis toolbox, which is a dedicated MATLAB toolbox developed by the LTP scientific community to provide a common working framework during operations [11].

The relevant data for the experiment is presented in Figure 2. A final increment of temperature of 2.5 K is achieved, with the modulation patterns keeping close to stable temperatures. The interferometer response, as expected, is sensitive to the injection only on its differential channel x_{12}, which measures the relative distance between TMs. The control loop that regulates the spacecraft position with respect to the primary TM (x_1) has a high gain at 1 mHz, practically impeding the sensitivity of x_1 to the heat injections, while the control loop on x_{12} actuates on the x axis of the secondary TM with very low gain in the same band.

The temperature gradient signal shows an oscillating pattern with peak-to-peak amplitude levels of 100 and 200 mK as a consequence of the two different modulation amplitudes applied –see Figure 2. The thermal-induced force on the TM is recovered from the telemetry and, as

[1] ESAC: *European Space and Astronomy Centre*, Villanueva de la Cañada, Madrid, Spain

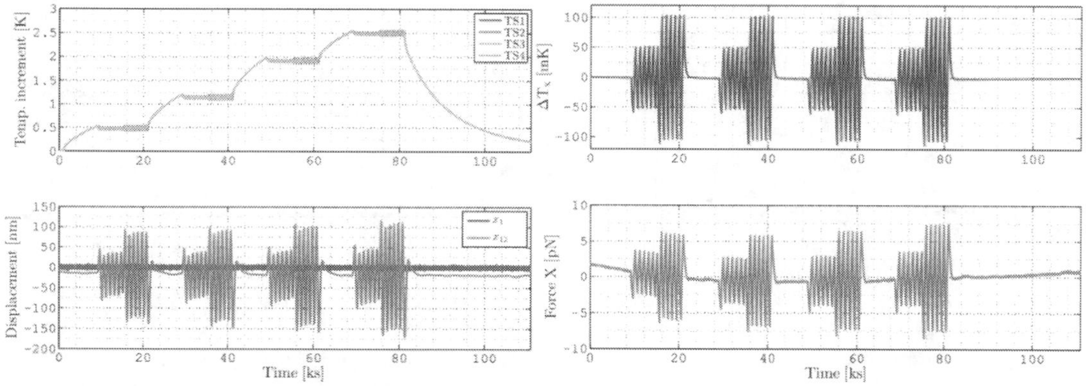

Figure 2. *Top left:* Temperature sensors readouts. *Lower-left* plot: IFO measurements x_1 and x_{12}. *Top right*: Temperature gradient on x axis. *Lower-right* plot: force induced to the TM.

expected, presents the same shape than the temperature gradient signal induced. A delay of about 52 s between the temperature gradient and the force is attributed to intrinsic delays in the thermal model, since the surface nodes used to model the forces are in average geometrically further from the heaters than the thermistors used to measure the temperature gradient.

5.1. Thermal coefficients

The thermal coefficients are obtained by demodulating and dividing the temperature gradient and the force signals. Table 2 show the individual values for each temperature modulation.

Table 2. Thermal coefficients estimated for the different temperature modulations. *A1* and *A2* correspond to the small and large modulation amplitudes respectively.

Temperature [K]	$\alpha_{x,A1}$ [pN/K]	Error [pN/K]	$\alpha_{x,A2}$ [pN/K]	Error [pN/K]
293.48	54.4	0.5	54.4	0.4
294.14	57.8	0.7	57.7	0.6
294.92	62.9	0.7	62.9	0.5
295.50	67.9	0.9	67.9	0.7

The final temperature coefficients are obtained by averaging the values at each temperature level –see Figure 3. The increment with the absolute temperature –much higher than $\propto T^3$– points to an outgassing-dominated scenario. Despite such variation, the reduced increment of absolute temperature available (around 2K between modulation points) limits the identification of the radiometer effect contribution, leaving only the option to set an upper limit of 7.4×10^{-15} m s^{-2} Hz$^{-1/2}$ to the Brownian effect, by considering the worse case of a totally radiometer-dominated environment at the lowest temperature.

5.2. Noise projection to interferometer

In addition to the previous analysis, transfer functions between the temperature gradient and the force can be calculated providing a tool to project the temperature noise to x_{12} equivalent acceleration noise. Figure 3 shows this projection during a specific noise measurement with no thermal injections, showing that the temperature noise to the acceleration sensitivity is kept about a factor 30 below the actual noise level.

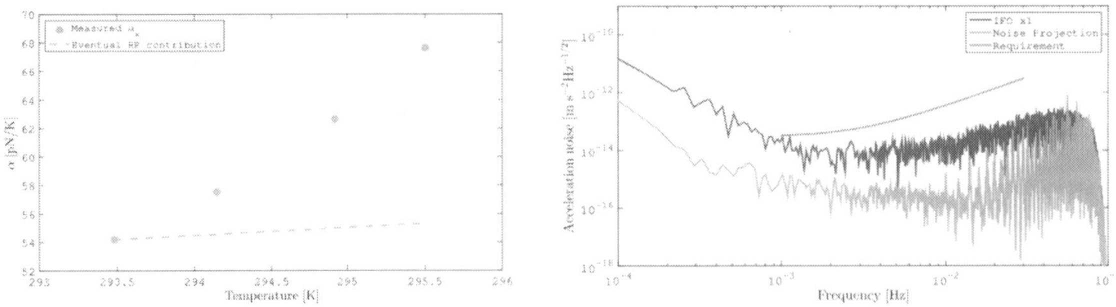

Figure 3. *Left*: Thermal coefficients (α_x) vs. absolute temperature. The comparison of the curve against the third power of the temperature –an eventual dominating radiation pressure contribution at the lowest temperature– suggests an outgassing-dominated scenario. *Right*: Temperature noise projection on an acceleration noise measurement, where the temperature contribution keeps about a factor 30 below the noise level. For $f > 0.02\,\mathrm{mHz}$ the projection is no longer significant since the readout noise of the thermistors is dominant in this band.

6. Discussion

The Electrode Housing thermal experiments to be carried out on board LISA Pathfinder have been simulated and analysed in a dedicated campaign, combining a global spacecraft simulator together with a SSM simulator of the thermal effects. Results show how the current procedure allows to estimate the thermal coefficients with enough precision in an outgassing-dominated environment, but the absolute temperature range is too small to identify the individual contribution of each thermal effect under these circumstances. Even though, an upper limit to the Brownian noise has been obtained from this experiment.

Acknowledgements

The author wants to thank the whole LTP Data Analysis team for the effort in preparing and carrying out the 4th LPF Simulation Campaign. In addition, the author acknowledges support from Project AYA2010-15709 of *Plan Nacional del Espacio* of the Spanish Ministry of Science and Innovation (MICINN).

References

[1] McNamara P *et al* 2013 The LISA Pathfinder Mission, *ASP Conference Series* **467**
[2] Amaro-Seoane P *et al* 2013 The Gravitational Universe, *Document submitted to the European Space Agency for the L2/L3 selection of ESA's Cosmic Vision program*
[3] Heinzel G *et al* 2004 The LTP interferometer and Phasemeter *Class. Quantum Grav.* **21** 581
[4] Dolesi R *et al* 2002 Gravitational sensor for LISA and its technology demonstration mission, *Class. Quantum Grav.* **20** S99
[5] Cañizares P *et al* 2008 The diagnostics subsystem on board LISA Pathfinder and LISA, *Class. Quantum Grav.* **26** 094005
[6] Nofrarias M *et al* 2008 Thermal diagnostic of the Optical Window on board LISA Pathfinder *Class. Quantum Grav.* **24** 5103
[7] Carbone L *et al* 2007, Thermal gradient-induced forces on geodetic reference masses for LISA, *Phys. Rev. D.* **76** 102003
[8] Gibert F *et al* 2012 State-space modelling for heater induced thermal effects on LISA Pathfinder's Test Masses, *J. Phys.: Conf. Ser.* **363** 012044
[9] Gibert F *et al* 2013 Closed Loop Simulations of the Thermal Experiments in LISA Pathfinder, *ASP Conference Series* **467**
[10] Cavalleri A *et al* 2009, Increased Brownian Force Noise from Molecular Impacts in a Constrained Volume, *Phys. Rev. L.* **103** 140601
[11] Hewitson M *et al* 2009 Data analysis for the LISA Technology Package, *Class. Quantum Grav.* **26** 094003

Journal of Physics: Conference Series **610** (2015) 012024 doi:10.1088/1742-6596/610/1/012024

Disentangling the magnetic force noise contribution in LISA Pathfinder

M Armano[a], H Audley[b], G Auger[c], J Baird[n], P Binetruy[c], M Born[b], D Bortoluzzi[d], N Brandt[e], A Bursi[t], M Caleno[f], A Cavalleri[g], A Cesarini[g], M Cruise[h], K Danzmann[b], I Diepholz[b], R Dolesi[g], N Dunbar[i], L Ferraioli[j], V Ferroni[g], E Fitzsimons[e], M Freschi[a], J Gallegos[a], C García Marirrodriga[f], R Gerndt[e], LI Gesa[k], F Gibert[k], D Giardini[j], R Giusteri[g], C Grimani[l], I Harrison[m], G Heinzel[b], M Hewitson[b], D Hollington[n], M Hueller[g], J Huesler[f], H Inchauspé[c], O Jennrich[f], P Jetzer[o], B Johlander[f], N Karnesis[k], B Kaune[b], N Korsakova[b], C Killow[p], I Lloro[k], R Maarschalkerweerd[m], S Madden[f], D Mance[j], V Martín[k], F Martin-Porqueras[a], I Mateos[k], P McNamara[f], J Mendes[m], L Mendes[a], A Moroni[t], M Nofrarias[k], S Paczkowski[b], M Perreur-Lloyd[p], A Petiteau[c], P Pivato[g], E Plagnol[c], P Prat[c], U Ragnit[f], J Ramos-Castro[qr], J Reiche[b], J A Romera Perez[f], D Robertson[p], H Rozemeijer[f], G Russano[g], P Sarra[t], A Schleicher[e], J Slutsky[s], C F Sopuerta[k], T Sumner[n], D Texier[a], J Thorpe[s], C Trenkel[i], H B Tu[g], S Vitale[g], G Wanner[b], H Ward[p], S Waschke[n], P Wass[n], D Wealthy[i], S Wen[g], W Weber[g], A Wittchen[b], C Zanoni[d], T Ziegler[e], P Zweifel[j]

[a] European Space Astronomy Centre, European Space Agency, Villanueva de la Cañada, 28692 Madrid, Spain
[b] Albert-Einstein-Institut, Max-Planck-Institut für Gravitationsphysik und Universität Hannover, 30167 Hannover, Germany
[c] APC UMR7164, Université Paris Diderot, Paris, France
[d] Department of Industrial Engineering, University of Trento, via Sommarive 9, 38123 Trento, and Trento Institute for Fundamental Physics and Application / INFN
[e] Airbus Defence and Space, Claude-Dornier-Strasse, 88090 Immenstaad, Germany
[f] European Space Technology Centre, European Space Agency, Keplerlaan 1, 2200 AG Noordwijk, The Netherlands
[g] Dipartimento di Fisica, Università di Trento and Trento Institute for Fundamental Physics and Application / INFN, 38123 Povo, Trento, Italy
[h] Department of Physics and Astronomy, University of Birmingham, Birmingham, UK
[i] Airbus Defence and Space, Gunnels Wood Road, Stevenage, Hertfordshire, SG1 2AS, UK
[j] Institut für Geophysik, ETH Zürich, Sonneggstrasse 5, CH-8092, Zürich, Switzerland
[k] Institut de Ciències de l'Espai (CSIC-IEEC), Campus UAB, Facultat de Ciències, 08193 Bellaterra, Spain
[l] Istituto di Fisica, Università degli Studi di Urbino/ INFN Urbino (PU), Italy
[m] European Space Operations Centre, European Space Agency, 64293 Darmstadt, Germany
[n] The Blackett Laboratory, Imperial College London, UK
[o] Physik Institut, Universität Zürich, Winterthurerstrasse 190, CH-8057 Zürich, Switzerland
[p] SUPA, Institute for Gravitational Research, School of Physics and Astronomy, University of Glasgow, Glasgow, G12 8QQ, UK

[q] Department d'Enginyeria Electrònica, Universitat Politècnica de Catalunya, 08034 Barcelona, Spain
[r] Institut d'Estudis Espacials de Catalunya (IEEC), C/ Gran Capità 2-4, 08034 Barcelona, Spain
[s] NASA Goddard Space Flight Center, 8800 Greenbelt Road, Greenbelt, MD 20771, USA
[t] CGS S.p.A, Compagnia Generale per lo Spazio, Via Gallarate, 150 - 20151 Milano, Italy

E-mail: nofrarias@ice.cat

Abstract. Magnetically-induced forces on the inertial masses on-board LISA Pathfinder are expected to be one of the dominant contributions to the mission noise budget, accounting for up to 40%. The origin of this disturbance is the coupling of the residual magnetization and susceptibility of the test masses with the environmental magnetic field. In order to fully understand this important part of the noise model, a set of coils and magnetometers are integrated as a part of the diagnostics subsystem. During operations a sequence of magnetic excitations will be applied to precisely determine the coupling of the magnetic environment to the test mass displacement using the on-board magnetometers. Since no direct measurement of the magnetic field in the test mass position will be available, an extrapolation of the magnetic measurements to the test mass position will be carried out as a part of the data analysis activities. In this paper we show the first results on the magnetic experiments during an end-to-end LISA Pathfinder simulation, and we describe the methods under development to map the magnetic field on-board.

1. Introduction

LISA Pathfinder (LPF) [1] will test key technologies required for gravitational wave detection in space. The main goal of the mission is to achieve a residual differential acceleration noise between two free-falling test masses of $3 \times 10^{-14}\,\mathrm{m/s^2/\sqrt{Hz}}$ in the $1\,\mathrm{mHz} \leq \mathrm{f} \leq 30\,\mathrm{mHz}$ frequency band, expressed in terms of noise spectral density. The satellite is equipped with the Gravitational Reference Sensor [2] hosting the test mass and controlling its position inside the spacecraft through the so called drag-free control loop, and the Optical Metrology Subsystem [3] achieving pico-meter sensitivities in the measurement of the relative displacement of the test masses.

An equally important goal of the LPF mission is to disentangle the different contributions in the measured noise. Indeed, future space gravitational wave observatories [4] will benefit from the detailed noise model that LPF will provide. With that aim, the Diagnostic Subsystem [5] contains sensors and actuators to perform experiments during flight operations.

The rationale behind these experiments is to induce a high signal-to-noise ratio calibration signal that allows a good estimate of the coupling between the system and the perturbation under study. Once the calibration signal is off, we will use the derived coupling to estimate the level of noise induced in the main measurement by the perturbation. For the magnetic case, this means that a set of coils will induce a controlled magnetic field in the test mass volume, resulting in a force that will enable determining the magnetic properties of the test mass. These will be used to compute the force noise contribution due to the background magnetic field.

The tri-axial fluxgate magnetometers used in LPF are located far from the test mass due to the electromagnetic fields they emit, hence a precise interpolation is required in order to know the background magnetic field affecting the test masses. Hence, there are two different problems that need to be addressed to characterise the magnetic contribution to the total force noise budget:

- To estimate the magnetic parameters of the test mass through the observed response to an injected magnetic signal.
- To interpolate the measured magnetic field at the magnetometers to the position of the test masses.

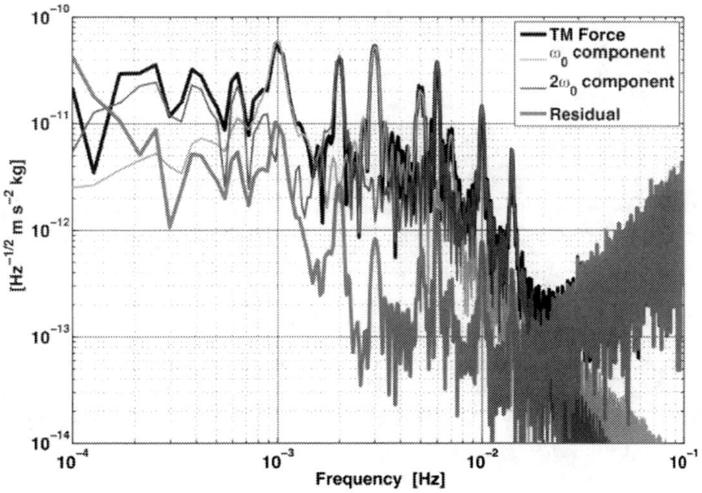

Figure 1. Reconstruction of the force in the x direction during the magnetic experiment in 4th operational exercise.

In the following we report on the latest developments for both

2. Estimation of test mass magnetic parameters

The inertial sensors displacement sensitivity at low frequencies is limited by actuation noise. After this, the main contribution limiting the sensor is due to the coupling of the test mass motion with background magnetic fields, which could represent up to 40% of the overall force noise budget. Although this coupling is greatly reduced because of the magnetic properties of the Au/Pt alloy from which the test mass is built, there is still a small remnant that needs to be taken into account.

If we assume an homogenous and isotropic test mass, the problem can be reduced to the estimation of the moment vector, $\vec{M} = (M_x, M_y, M_z)$, and susceptibility, $\chi = \chi_0 + i\,\chi_e$, in the following system of linear equations [6]

$$\begin{aligned} F_x &= \chi_0\, f_{DC} + M_x\, f_{x_{1\omega 0}} + \chi_0\, f_{x_{2\omega 0}} + \chi_e\, f''_{x_{2\omega 0}} \\ N_y &= M_z\, n_{y_{1\omega 0}} \\ N_z &= M_y\, n_{z_{1\omega 0}} \end{aligned} \quad (1)$$

where $f_{x_{1\omega 0}}$ is a force term at the injected frequency, ω_0; $f_{x_{2\omega 0}}$ is a force term at $2\,\omega_0$ and $f''_{x_{2\omega 0}}$ is a force term at $2\,\omega_0$ but out of phase when compared with the injected signal. All these terms depend on the magnetic field and the magnetic field gradient in the test mass volume and hence, are computed numerically and assumed known. The same applies to $n_{y_{1\omega 0}}$ and $n_{z_{1\omega 0}}$ which correspond to torques around the y and z axes due to the applied magnetic field.

This test mass characterisation scheme was put into test during the 4th operational exercise, which took place in November 2013. During this simulation, the science team was divided into small groups, each in charge of a different experiment. These simulations train the team for in-flight operations, so the data is received and processed in a mission-like scenario in terms of scheduling, satellite commanding and telemetry reception. In this occasion, the magnetic characterisation of the test took 19 h where the coil was commanded to apply a sinusoidal field

Table 1. Test mass magnetic parameters obtained during the 4th operational exercise.

Parameter	Value
χ_0	$(5.81 \pm 0.09) \times 10^{-5}$
M_x [Am^{-1}]	$(1.18 \pm 0.01) \times 10^{-4}$
M_y [Am^{-1}]	$(1.29 \pm 0.02) \times 10^{-4}$
M_z [Am^{-1}]	$(1.40 \pm 0.03) \times 10^{-4}$

injection at 1, 3, 5, and 7 mHz commanding in each case a 1 mA current into the coils. The set of frequencies is selected to maximize the SNR of the response signal [6].

The telemetry obtained during the simulation was processed using the LTPDA [7] magnetic data analysis pipeline. The main steps implemented in this procedure can be summarised as (1) pre-processing the data to detect possible irregularities as, for instance, missing data packets, (2) to translate the measured test mass displacement into measured force, based on the system dynamic parameters, (3) to compute the magnetic field at the test mass position based on the geometry of the problem and the applied current, (4) to compute the force contribution at the relevant frequencies in order to (5) solve the linear system of equations in Eq. (1).

The result of the experiment, shown in Fig. 1, is a clear signal when measuring the applied force on the test mass. The plot shows as well that the pipeline was able to retrieve the magnetic parameters characterising the test mass and reconstruct the applied force. Table 1 shows the values obtained by the pipeline. Notice that the imaginary part of the susceptibility was not estimated because this feature was not implemented in the simulator at the time of the simulation.

3. Magnetic field interpolation

The feed-back field from the tri-axial fluxgate magnetometers requires that they be located far from the desired measurement location. Although there exist proposals to alleviate the problem in future designs [8], the situation in LPF calls for a precise interpolation process. Several techniques have been proposed, among them the most promising is the usage of neural network algorithms [9] which learn from a set of given magnetic field configurations to interpolate the field in the real case.

Since the number of sources in the satellite greatly exceeds the number of available magnetometers, the interpolation becomes a degenerate problem. In all proposed methods the interpolation is usually strongly dependent on the *a priori* knowledge of the sources and therefore a clear detailed analysis of the distribution of the magnetic sources in the satellite is necessary for a successful analysis.

In Figure 2 we show the latest update on the distribution of magnetic sources in LISA Pathfinder. The figure graphically describes the complexity of the magnetic field due to the diversity of sources around the main experiment, located here in the position of the two test masses. It is important to notice that the recent replacement of the the field-emission electric propulsion (FEEP) thrusters by the cold gas ones has an important impact in the magnetic environment of the satellite. Indeed, the three main structures in Figure 2 are due to the cold gas latch valves (CGLV).

Again, this recent change stresses the importance of a good characterization on the ground. In Figure 3 we report a first study to evaluate the contributions of each identified magnetic source to the measured magnetic field for each magnetometer. These estimates show that the CGLV units are enough to explain nearly 60% of the contribution as measured for some magnetometers.

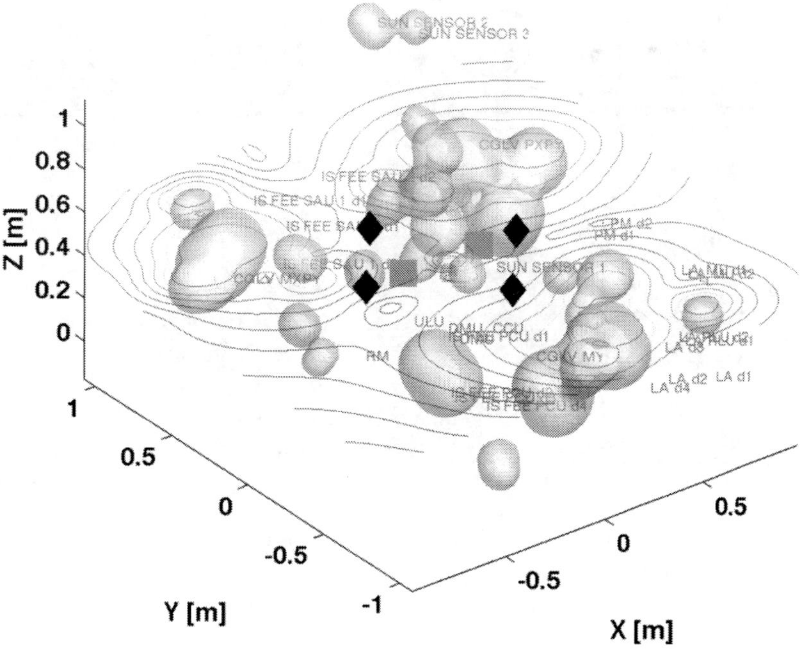

Figure 2. A magnetic map of LISA Pathfinder with some of the main magnetically active components. The shaded shapes show iso-surfaces of magnetic field while the contour lines display the magnetic field distribution in the XY plane defined by the test masses. The masses are shown as cubes in the center, the magnetometers appear as rhombus. Some of the acronyms in the plot are: cold gas latch valves (CGLV), phasemeter (PM), laser assembly (LA), radiation monitor (RM), ultraviolet lamp unit (ULU) and inertial sensor front-end electronics sensing and acutuation unit (IS FEE SAU)

Apart from these, each sensor is more sensitive to a given set of magnetic sources which explains the remaining contributions of the magnetic field.

As previously stated, *a priori* knowledge is crucial to solve the magnetic interpolation process. Hence, the ground characterisation of the units will be a valuable input for the in-flight characterisation since it will allow to exploit the particular spatial distribution of the magnetic sources in the LPF to help better characterise the magnetic field map on board.

4. Conclusions

Magnetically induced forces on the LISA Pathfinder test masses will be an important contribution to the final mission noise budget. Understanding the magnetic environment on board and characterising the magnetic properties of the test mass is crucial for the mission. During the 4th operational exercise, the experiments to induce a magnetic signal and thereby measure the magnetic moment and susceptibility of the test masses were simulated and the data analysis pipeline successfully tested, providing a successful characterisation.

The interpolation of the magnetic field from the magnetometers to the test mass position is

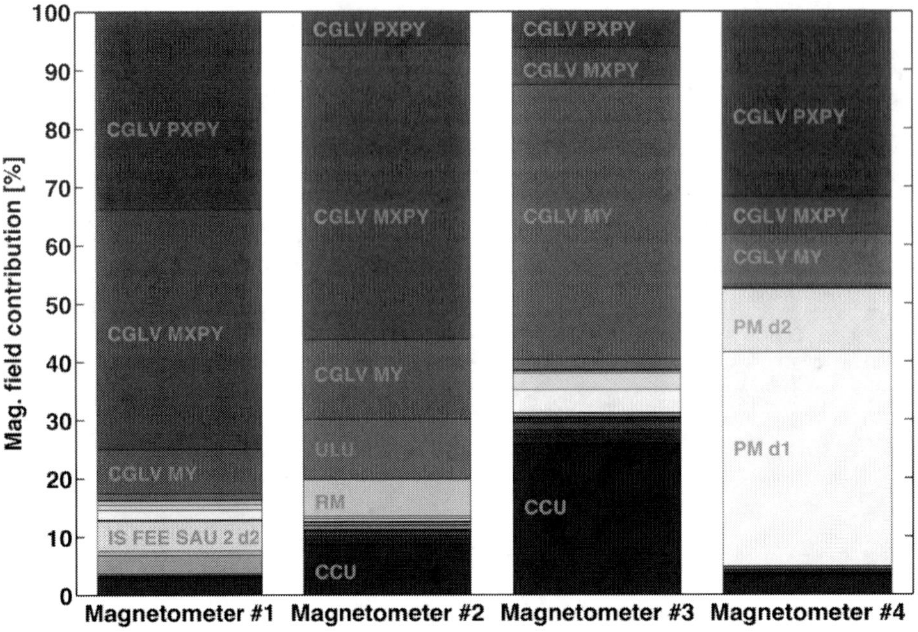

Figure 3. Main contributions to the total measured magnetic field in each magnetometer. The cold gas valve is dominating in each magnetometer.

a difficult task given the distribution of magnetic sources in the satellite surrounding the test masses. We have reported a first characterisation of the impact of individual magnetically active components to the total measured magnetic field. This information will be a valuable input for the interpolation methods being proposed, which relay on *a priori* knowledge of the magnetic field in the satellite.

References

[1] Antonucci F et al 2012, The LISA Pathfinder mission *Classical and Quantum Gravity* **29**, 124014
[2] Bortoluzzi D *et al.* 2003 Testing LISA drag-free control with the LISA technology package flight experiment *Class. Quantum Grav.* **20**, S89
[3] Heinzel G *et al.* 2005, Successful testing of the LISA Technology Package (LTP) interferometer engineering model *Classical and Quantum Gravity* **22**, S149
[4] Amaro-Seoane P *et al.* 2013, *The Gravitational Universe, Preprint* arXiv:1305.5720
[5] Lobo A *et al.* 2006 Inflight Diagnostics in LISA Pathfinder *AIP Conf. Proc.* **873**, 522
[6] Diaz-Aguiló M et al 2012, Inflight magnetic characterization of the test masses onboard LISA Pathfinder *Phys. Rev. D* **85**, 042004
[7] Hewitson M *et al.* 2009 Data analysis for the LISA Technology Package, *Class. Quantum Grav.* **26** 094003
[8] Mateos I *et al.* 2012, Temperature coefficient improvement for low noise magnetic measurements in LISA *Journal of Physics: Conference Series* **363**, 012051
[9] Diaz-Aguiló M *et al.* 2011, Neural network interpolation of the magnetic field for the LISA Pathfinder Diagnostics Subsystem *Experimental Astronomy* **30**, 1

Role of GCR positive and negative particles in charging the LISA-PF test masses in 2015

C Grimani[†,§], M Fabi[†], A J Lobo[‡,1], I Mateos[‡], D Telloni[§,¶]
[†] DiSBeF, Università degli Studi di Urbino "Carlo Bo", Urbino (PU), Italy
[§] Istituto Nazionale di Fisica Nucleare, Florence, Italy
[‡] Institut de Ciències de l'Espai (CSIC-IEEC), Barcelona, Spain
[¶] Istituto Nazionale di Astrofisica, Osservatorio Astrofisico di Torino, Pino Torinese, Italy

E-mail: catia.grimani@uniurb.it

Abstract. The LISA Pathfinder (LISA-PF) mission launch is scheduled during the second half of 2015. Galactic and solar ions with energies larger than 100 MeV/n and electrons above 10 MeV penetrate the spacecraft material and charge the gold-platinum test masses. This charging process generates spurious forces that, in some cases, may mimic the effects of genuine gravitational wave signals. A study of the test-mass charging due to galactic cosmic rays (GCRs) down to 1% in composition is reported here. The reliability of the results of this work is mainly limited by our capability to predict the energy spectra of GCRs in 2015. To this purpose, our model is applied to the expected PAMELA experiment proton data for the period January-March 2014 characterized by a positive polarity period and a level of solar modulation similar to those expected at the time of LISA-PF. The PAMELA observations will be available in the next few months. The comparison between our projections and measurements will provide valuable clues on the test-mass charging estimate uncertainty.

1. Introduction

Galactic cosmic rays (GCRs) are high energy particles reaching the Solar System from sources mostly distributed within the Milky Way. They consist approximately of 90% protons, 8% helium nuclei, 1% heavy nuclei and 1% electrons with energies ranging between 100 MeV and a few units $\times$ 10^{20} eV[1]. It is of primary importance for many experiments carried out in space to study the effects of the cosmic radiation in limiting or affecting the performance of on-board detectors.

LISA Pathfinder (LISA-PF) [2, 3] is the technology testing mission for eLISA [4], the first interferometer devoted to gravitational wave detection in space operating in the frequency interval 10^{-4}-1 Hz. LISA-PF consists of one satellite hosting inertial sensors with free-falling gold-platinum test masses. Electrodes allow for the determination of the positions of the test mass and the application of control forces where necessary. The LISA-PF spacecraft will be stationed in an orbit about the L1 Lagrange point for six months. Test-mass charging due to galactic and solar energetic particles (SEPs) [5, 6] with energies larger than 100 MeV/n represents one of the main sources of noise for the LISA-PF and eLISA missions in the lowest frequency interval [7]. The integral flux of protons and helium nuclei at energies larger than

[1] Deceased

Journal of Physics: Conference Series **610** (2015) 012025 doi:10.1088/1742-6596/610/1/012025

the nominal cut-off energy of 70 MeV/n will be monitored with particle detectors placed on LISA-PF [8, 9]. This conservative choice was made in order not to underestimate the overall particle flux charging the test-masses in case of SEP event occurrence.

The LISA-PF test-mass charging due to the most abundant species of GCRs in 2015 is estimated here. Proper projections of GCR energy spectra at the time the mission will be in space are considered. We focus on the role of electrons that decrease the net charge deposited on the test masses while increase the shot noise. Uncertainties on the prediction of the GCR proton differential flux in 2015 can be tested by applying the model we use to the expected PAMELA experiment [10] observations between January and March 2014. This interval of time was characterized by the same condition of solar polarity and similar solar modulation level with respect to those expected in 2015. We recall that the solar polarity is called positive (negative) when the Global Solar Magnetic Field (GSMF) lines exit from the Sun North (South) Pole and that the solar polarity affects cosmic-ray observations near Earth up to 40% at 100 MeV(/n) [11].

2. Galactic cosmic ray proton and helium energy spectra in 2015

The LISA-PF mission will be sent into orbit during the descending phase of the solar cycle 24. A low level of solar modulation intensity is expected during the the second half of 2015. The minimum, average, and maximum sunspot predictions during this period are 29.9-54.1-78.4, respectively [12]. In 2015 the GSMF will present a positive polarity epoch. The symmetric model in the force-field approximation by Gleeson and Axford [13] allows for the estimation of the energy spectra of cosmic rays at a distance r from the Sun, at a time t by assuming time-independent interstellar intensities. A parameter Φ is defined that can be interpreted as the energy loss experienced by cosmic-ray particles approaching the Earth from infinity. The parameter Φ for protons and helium nuclei above rigidities of 100 MV (rigidity: particle momentum per unit charge) can be assumed equal to the solar modulation parameter ϕ characterizing the intensity of the solar modulation (see [14] for details). The Gleeson and Axford model appears in very good agreement with GCR observations gathered during positive polarity periods [15]. In [16, 17, 18] we have discussed the expected values of the solar modulation parameter in 2015 ranging between 350 MV/c and 800 MV/c. The energy spectra of GCR protons and nuclei can be interpolated accordingly to the following function [1]:

$$F(E) = A\ (E+b)^{-\alpha}\ E^{-\beta} \quad \text{Particles}/(\text{m}^2\ \text{sr s GeV}). \tag{1}$$

The parameters A, b, α and β for the minimum and maximum projections of protons and helium nuclei in 2015 are reported in Table 1.

Table 1. Minimum (m) and maximum (M) projections for the energy spectra of the most abundant GCR particles in 2015.

Element	A_m	b_m	α_m	β_m	A_M	b_M	α_M	β_M
p	18000	1.54	3.67	0.88	18000	0.88	3.68	0.89
He	850	0.91	3.60	0.85	850	0.7	3.23	0.48

Proper ^{3}He and ^{4}He composition is taken into account in the following [6, 18].

In order to verify the reliability of our work, a prediction of the proton energy spectrum during the first four months of 2014 is carried out here. The PAMELA experiment collaboration is expected to publish the corresponding data in the near future. The observed average number of

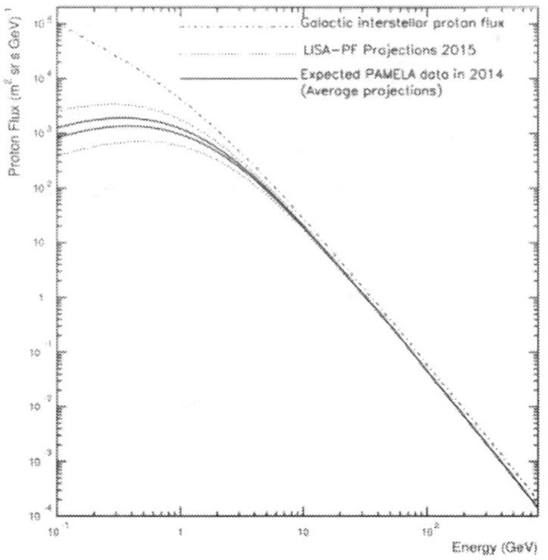

Figure 1. Minimum and maximum projections of the proton energy spectrum during the first 4 months in 2014 (continuous lines) and in 2015 (dotted lines). The dot-dashed line represents the proton spectrum at the interstellar medium [19].

sunspots between January and April 2014 was 71.2. Minimum, average and maximum sunspot projections for the same period are 42.4, 70.0, 97.6 [12], respectively, at the time of writing. Sunspot observations during the first four months of 2014 appear compatible with average sunspot projections for the same period. A similar number of sunspots was observed in 1971, 1972 and 1983, for instance. However, the solar modulation level appeared more intense for protons in 1983 during a period of negative polarity. In 1971-1972, characterized by a positive polarity epoch, the observed solar modulation parameter ranged between 500 MV/c and 600 MV/c [14]. By assuming that the solar modulation parameter varied in the same range of values during the first four months of 2014, minimum and maximum expected proton energy spectra measured by PAMELA would appear as reported in figure 1 (continuous lines). However, preliminary results for the proton differential flux in January 2014 (private communication from the PAMELA collaboration) are lower, close to our minimum projections for LISA-PF (bottom dotted curve in figure 1). We conclude that conservative (minimum and maximum rather than average) projections of GCR energy spectra should be considered even just before any mission launch.

3. LISA-PF test mass charging

We have carried out the simulation of the LISA-PF test-mass charging with the Fluka Fortran-based Monte Carlo package [20, 21]. GCR particles were propagated through the LISA-PF spacecraft by assuming an isotropic particle distribution. The same spacecraft geometry considered for a GEANT4 simulation [22] was imported in Fluka. It is worthwhile to point out that even if accurate, the geometry used in the simulation is an approximation of the actual mass distribution in the spacecraft. This approximation represents an additional source of uncertainty in the test-mass charging simulation results.

We have estimated the average test-mass net (λ_{net}) and effective (λ_{eff}) charging generated by GCR proton and helium nuclei at the time of LISA-PF data taking. The net charging rate is represented by the following equation:

$$\lambda_{net} = \sum_{j=-\infty}^{+\infty} j\lambda_j \quad \mathrm{e\ s^{-1}} \tag{2}$$

where j represents the amplitude of charging released by single events and λ_j is the rate of occurrence of these events. Positive and negative charges cancel out in the net charging computation while both contribute to the effective charging rate defined below:

$$\lambda_{eff} = \sum_{j=-\infty}^{+\infty} j^2\lambda_j \quad \mathrm{e\ s^{-1}}. \tag{3}$$

The effective charging rate represents the charging of single charges that would generate the same observed shot noise. The spectral density of the charging shot noise is expressed in terms of effective charging rate:

$$S = \sqrt{2e^2\lambda_{eff}} \quad \mathrm{e\ s^{-1}\ Hz^{-1/2}}. \tag{4}$$

The charging fluctuations at frequency f are represented by:

$$S_Q(f) = \frac{S}{2\pi f} = \frac{\sqrt{2e^2\lambda_{eff}}}{2\pi f} \quad \mathrm{e\ Hz^{-1/2}}. \tag{5}$$

In each considered case study for test-mass charging simulation, we propagated more than 2×10^6 primary particles through the LISA-PF spacecraft. This choice allows for the optimization of the computing time by limiting the uncertainty on the λ_{net} and λ_{eff} below 5% and 2%, respectively. In Table 2 we have reported the net and effective charging generated by GCRs when minimum (m) and maximum (M) projections are considered for each particle species.

Table 2. Net and effective test-mass charging of the LISA-PF test masses for the minimum (m) and maximum (M) projections of proton and helium nucleus energy spectra in 2015.

Element	λ_{net}^m e^+/s	λ_{eff}^m e/s	λ_{net}^M e^+/s	λ_{eff}^M e/s
p	14.1	168.9	32.5	295.5
^{3}He	0.22	0.92	1.9	5.6
^{4}He	0.81	1.9	3.8	10.7

4. The role of interplanetary electrons in charging the LISA-PF test masses

Electrons constitute a minor fraction in composition of the primary cosmic radiation ($e^-/p \simeq 10^{-2}$; $e^-/He \simeq 10^{-1}$). However, as low-mass, highly penetrating particles, the LISA-PF test-mass charging due to e^- must be properly taken into account.

The interplanetary absolute fluxes of electrons during minimum and maximum solar activity periods under both conditions of solar polarity were estimated in [23]. In the same work it was shown that electrons of solar origin associated with solar energetic particle events play a negligible role in charging the LISA-PF test masses. Analogous results were found for electrons produced in the magnetosphere of Jupiter. Conversely, our simulation work indicated that primary and secondary galactic electrons reduce the net charging of protons by 12% (55%) at solar minimum (maximum) during positive polarity periods. During negative polarity epochs galactic e^- were found to match 16% (65%) the positive charge released by protons at solar minimum (maximum). Finally, the effective charging due to galactic electrons was found 22%-23%(38%) of that of protons at solar minimum (maximum) during both polarity epochs.

Under the solar modulation level expected in 2015 electrons could match 50% of the absolute charge released in the test masses by protons while they may increase by about 30% the effective test-mass charging.

As a final remark, we recall that in case particle monitors with solar electron detection capabilities will be flown on eLISA, it will be possible to short-term forecast intense SEP events observed on eLISA (for a detailed discussion see [23, 24]). These improvements of particle detector performance would help in optimizing the test-mass discharging process and possibly, extending the mission lifetime.

5. Conclusions

The role of the most abundant components of GCRs in contributing to the net and effective charging of the LISA-PF test masses was considered here. In 2015 electrons are expected to match 50% of the positive charge released by protons and to contribute to approximately 30% of the proton effective charging.

5.1. Acknowledgments

This work was financially supported by the Italian Space Agency within the program "Study Exploration of the Solar System" under contract ASI-INAF I/05/07/0. IM and CG acknowledge support from the Spanish Research Ministry, contract AYA2010-15709.

References

[1] Papini P, Grimani C and Stephens A S 1996 *Nuovo Cimento* **19** 367-387
[2] Antonucci F *et al* 2011 *Class. Quantum Grav.* **28** 094001
[3] Antonucci F *et al* 2012 *Class. Quantum Grav.* **29** 124014
[4] The eLISA Consortium: Amaro Seoane P *et al Preprint* gr-qc/1305.5720
[5] Araújo H M *et al* 2005 *Astropart. Phys.* **22** 451-469
[6] Grimani C *et al* 2005 *Class. Quantum Grav.* **22** S327-S332
[7] Shaul D *et al* 2006 *AIP Conf. Proc.* **873** 172-178
[8] Cañizares P *et al* 2011 *Class. Quantum Grav.* **28** 094004
[9] Mateos I *et al* 2012 *J. Phys.: Conf. Ser.* **363** 012050
[10] Picozza P *et al* 2013 *J. Phys.: Conf. Ser.* **409** 012003
[11] Grimani C 2007 *A&A* **474** 339-343
[12] http://solarscience.msfc.nasa.gov/predict.shtml
[13] Gleeson L J and Axford W I 1968 *Ap. J.* **154** 1011-1026
[14] Usoskin I G *et al* 2005 *J. Geophys. Res.* **110** A12108
[15] Grimani C *et al* 2007 *Proc. 30th Int. Cosmic Ray Conf. (Merida)*
[16] Grimani C *et al* 2012 *J. Phys.: Conf. Ser.* **363** 012045
[17] Grimani C *et al* 2012 *Class. Quantum Grav.* **29** 105001
[18] Grimani C *et al* to be submitted to *Class. Quantum Grav.*
[19] Shikaze Y *et al* 2007 *Astrop. Phys.* **28** 154-167
[20] Battistoni G , Muraro S, Sala P R, Cerutti F, Ferrari A, Roesler S, Fassó A, Ranft J 2007 *The FLUKA code: Description and benchmarking, Proceedings of the Hadronic Shower Simulation Workshop, 2006, Fermilab 6-8 September 2006 M. Albrow, R. Raja Eds., AIP Conf. Proc.* **896** 31
[21] Ferrari A, Fassó A, Ranft J, Sala P R 2005 *FLUKA: a multi-particle transport code, CERN 2005-10*
[22] Wass P *et al* 2005 *Class. Quantum Grav.* **22** S311-S317
[23] Grimani C *et al* 2009 *Class. Quantum Grav.* **26** 215004
[24] Grimani C *et al* 2014 *Class. Quantum Grav.* **31** 045018

A Strategy to Characterize the LISA-Pathfinder Cold Gas Thruster System

M Armano[a], H Audley[b], G Auger[c], J Baird[n], P Binetruy[c], M Born[b], D Bortoluzzi[d], N Brandt[e], A Bursi[t], M Caleno[f], A Cavalleri[g], A Cesarini[g], M Cruise[h], K Danzmann[b], I Diepholz[b], R Dolesi[g], N Dunbar[i], L Ferraioli[j], V Ferroni[g], E Fitzsimons[e], M Freschi[a], J Gallegos[a], C Garcia Marirrodriga[f], R Gerndt[e], LI Gesa[k], F Gibert[k], D Giardini[j], R Giusteri[g], C Grimani[l], I Harrison[m], G Heinzel[b], M Hewitson[b], D Hollington[n], M Hueller[g], J Huesler[f], H Inchauspé[c], O Jennrich[f], P Jetzer[o], B Johlander[f], N Karnesis[k], B Kaune[b], N Korsakova[b], C Killow[p], I Lloro[k], R Maarschalkerweerd[m], S Madden[f], D Mance[j], V Martin[k], F Martin-Porqueras[a], I Mateos[k], P McNamara[f], J Mendes[m], L Mendes[a], A Moroni[t], M Nofrarias[k], S Paczkowski[b], M Perreur-Lloyd[p], A Petiteau[c], P Pivato[g], E Plagnol[c], P Prat[c], U Ragnit[f], J Ramos-Castro[qr], J Reiche[b], J A Romera Perez[f], D Robertson[p], H Rozemeijer[f], G Russano[g], P Sarra[t], A Schleicher[e], J Slutsky[s], C F Sopuerta[k], T Sumner[n], D Texier[a], J Thorpe[s], C Trenkel[i], H B Tu[g], S Vitale[g], G Wanner[b], H Ward[p], S Waschke[n], P Wass[n], D Wealthy[i], S Wen[g], W Weber[g], A Wittchen[b], C Zanoni[d], T Ziegler[e], P Zweifel[j]

[a]European Space Astronomy Centre, European Space Agency, Villanueva de la Canada, 28692 Madrid, Spain
[b]Albert-Einstein-Institut, Max-Planck-Institut für Gravitationsphysik und Universität Hannover, 30167 Hannover, Germany
[c]APC UMR7164, Université Paris Diderot, Paris, France
[d]Department of Industrial Engineering, University of Trento, via Sommarive 9, 38123 Trento, and Trento Institute for Fundamental Physics and Application / INFN
[e]Airbus Defence and Space, Claude-Dornier-Strasse, 88090 Immenstaad, Germany
[f]European Space Technology Centre, European Space Agency, Keplerlaan 1, 2200 AG Noordwijk, The Netherlands
[g]Dipartimento di Fisica, Università di Trento and Trento Institute for Fundamental Physics and Application / INFN, 38123 Povo, Trento, Italy
[h]Department of Physics and Astronomy, University of Birmingham, Birmingham, UK
[i]Airbus Defence and Space, Gunnels Wood Road, Stevenage, Hertfordshire, SG1 2AS, UK
[j]Institut für Geophysik, ETH Zürich, Sonneggstrasse 5, CH-8092, Zürich, Switzerland
[k]Institut de Ciències de l'Espai (CSIC-IEEC), Campus UAB, Facultat de Ciències, 08193 Bellaterra, Spain
[l]Istituto di Fisica, Università degli Studi di Urbino/ INFN Urbino (PU), Italy
[m]European Space Operations Centre, European Space Agency, 64293 Darmstadt, Germany
[n]The Blackett Laboratory, Imperial College London, UK
[o]Physik Institut, Universität Zürich, Winterthurerstrasse 190, CH-8057 Zürich, Switzerland
[p]SUPA, Institute for Gravitational Research, School of Physics and Astronomy, University of Glasgow, Glasgow, G12 8QQ, UK
[q]Department d'Enginyeria Electronica, Universitat Politècnica de Catalunya, 08034 Barcelona, Spain
[r]Institut d'Estudis Espacials de Catalunya (IEEC), C/ Gran Capità 2-4, 08034 Barcelona, Spain
[s]NASA Goddard Space Flight Center, 8800 Greenbelt Road, Greenbelt, MD 20771, USA
[t]CGS S.p.A, Compagnia Generale per lo Spazio, Via Gallarate, 150 - 20151 Milano, Italy

E-mail: eric.plagnol@apc.univ-paris7.fr

Journal of Physics: Conference Series **610** (2015) 012026 doi:10.1088/1742-6596/610/1/012026

Abstract. The cold gas micro-propulsion system that will be used during the LISA-Pathfinder mission will be one of the most important component used to ensure the "free-fall" of the enclosed test masses. In this paper we present a possible strategy to characterize the effective direction and amplitude gain of each of the 6 thrusters of this system.

1. Introduction

One of the aims of the ESA's LISA-Pathfinder [1] (LPF) mission is to demonstrate the feasibility of maintaining, in space, a *Test Mass* (TM) in *free fall* (i.e., following an orbit only subjected to the solar system gravitational forces) with a residual acceleration of less than $3x10^{-14}$ m/s^2 around 3 mHz. As such, LPF is a mission whose aim is to test "key technologies" that will be used for the detection of gravitational waves by LISA type [2] space missions.

In order to ensure the *free-fall* of one of its enclosed free floating Test Masses (see TM_1, figure 1), the LPF on-board system measures the position of this TM with respect to the satellite body and endeavors to maintain, via a system of micro-thrusters, the spacecraft (S/C) centered on this TM. This micro-thruster system is constituted of 6 cold gas thrusters, identical to those currently used for the Gaia [3] mission. The assumed direction cosines of the thrust provided by the thrusters is given in lines 2-4 of Table I. As the solar pressure is used to limit the number of thrusters necessary to control all 6 degrees of freedom (displacement along the X, Y and Z axis and the torques around each of those) their direction along the Z axis, which points towards the Sun, are all positive.

Because of the importance of this thruster system in ensuring the *free-fall* of TM_1, it is essential to have the possibility to characterize, in flight, the performances of each thruster. This paper presents a strategy, based on ideas suggested by W.Fichter [4], to characterize the mean direction and amplitude gain of each of the thrusters.

2. Characterizing the mean direction and amplitude gain of each thruster

The basic idea of the strategy is to apply simultaneously, in addition to the *in-loop* actuation signal, a sine force of given amplitude and frequency to each thruster. This will force a displacement of the S/C which, detected through the position error signal of TM_1, will be countered by the *in-loop* control system (DFACS : Drag Free and Attitude Control System) whose task is to maintain the S/C centered on TM_1 (and also to maintain a proper *attitude* with respect to the Sun). The quantity:

$$\vec{F}_{tot}(t)=\vec{F}^{CMD}_{DFACS}(t)-M_{S/C}\frac{\partial^2 \vec{x}_{TM1}(t)}{\partial t^2} \qquad (1)$$

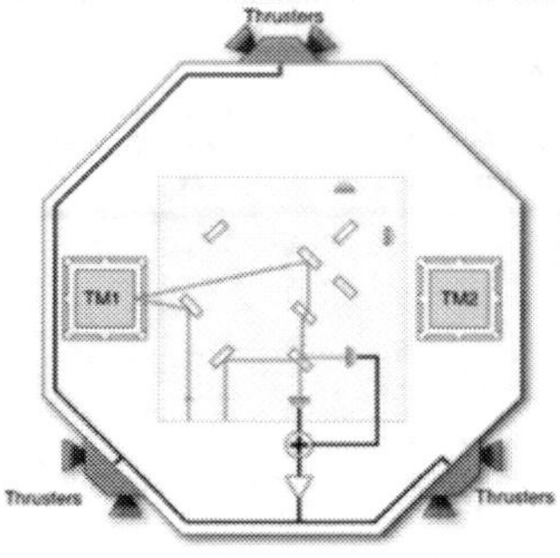

Figure 1: Schematic representation of the thruster control system.

where $M_{S/C}$ is the mass of the S/C and $\vec{x}_{TM1}(t)$ is the position vector of TM_1 with respect to the S/C, will reflect the force applied by the thruster system. The second term on the right hand side of equation 1, which accounts for the phase and gain imperfections of the control loop, is, for the actuation frequencies considered in this study (~ mHz) very small (~ 10^{-11} N) compared to the first one (~ 10^{-6} N).

It should be noted that $\vec{F}_{Tot}(t)$ will reflect the *effective* thrusts (amplitudes and directions) of the system and not the *out-of-loop* thrust and direction which is artificially applied.

Applying simultaneously different frequencies on each of the 6 thrusters and filtering $\vec{F}_{Tot}(t)$ at these corresponding frequencies allows characterization of the whole system in just one sequence.

The analysis of the direction of $\vec{F}_{Tot}(t)$ will therefore measure the direction of the thrust and its amplitude will characterize the gain of the thruster.

The data analyzed in this work comes from simulations using ESA's mission simulator SOVT for LPF which contains a complete description of the thruster system, including a non-linear dispatching algorithm, and amplitude and flutter (directional) noise.

The characterization sequence lasts of the order of 25000 seconds and figure 2 shows the three components of the DFACS requested force, $\vec{F}_{DFACS}^{CMD}(t)$. One can observe that the forces on X and Y have an average value close to zero (the solar pressure has negligible impact on these) whereas the force on the Z axis has a mean value close to 20μN, necessary to counter the solar pressure.

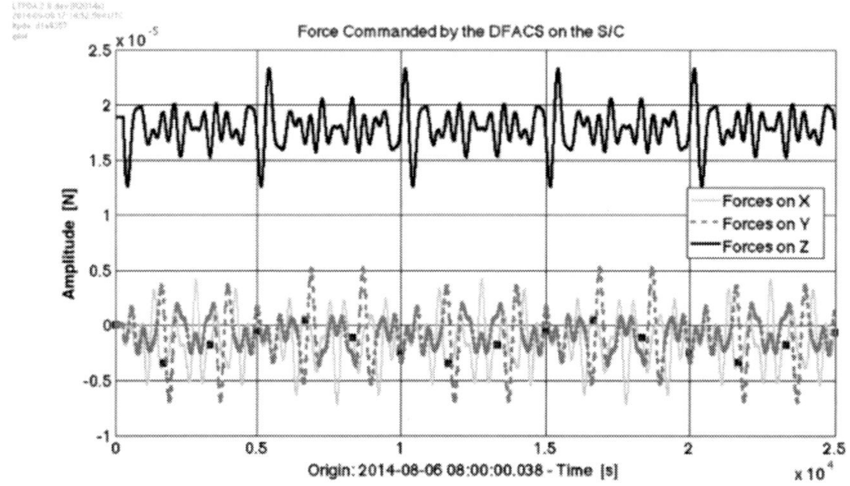

Figure 2 : Force commanded by the DFACS on the X,Y and Z (solar) axis.

Figure 3 shows, as an example, the time series obtained from $\vec{F}_{Tot}(t)$ after filtering with a band pass filter in order to isolate the S/C movement associated to the actuation of thruster N°3. The Fourier transform of this time series is used for the *PSD* method, whereas the 3D analysis of $F_X(t)$, $F_Y(t)$ and $F_Z(t)$ is at the basis of the *PCA* method (see next chapter).

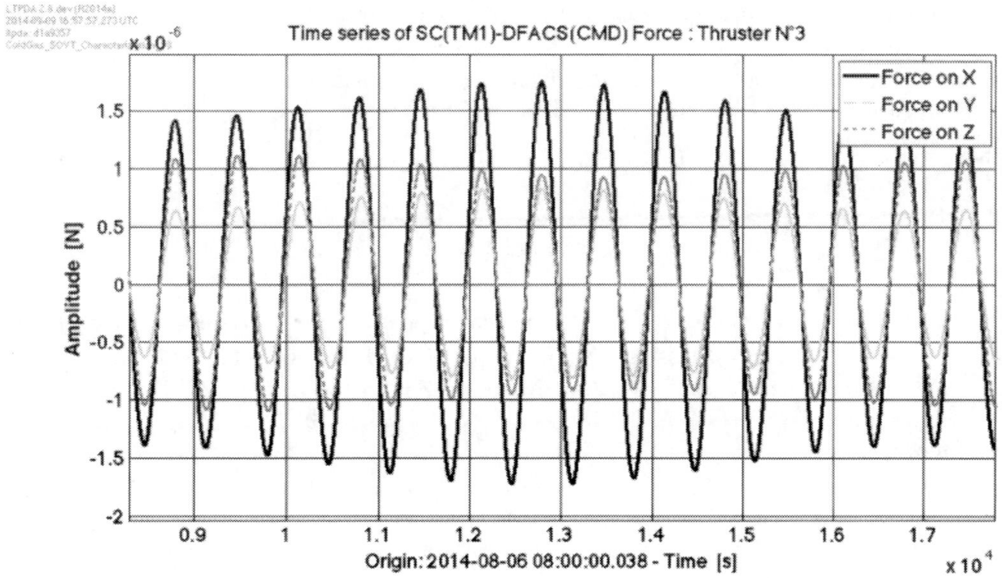

Figure 3 : A sample of the $\vec{F}_{Tot}(t)$ time series after filtering at 1.5 ± 0.12 mHz when characterizing thruster N°3.

3. Analysis Results

The mean direction and amplitude gain of the thrusters can be studied with various techniques.

After having filtered the $\vec{F}_{Tot}(t)$ time series at the frequency associated to a given thruster, one can perform a Fourier transform resulting in a three component Fourier vector ($\vec{F}_{Tot}(f)$) whose direction can be compared to the assumed direction of this thruster. This is the PSD method. In this case the applied thrust can be measured by the norm of $\vec{F}_{Tot}(f)$.

Another technique consists in considering the $\vec{F}_{Tot}(t)$ times series, regularly sampled (1 Hz), as so many points in a 3D space and perform a Principal Component Analysis (PCA) on these. Such a method yields 3 eigenvectors whose largest one will give the average thrust direction.

Table I. gives, on line 4, the difference between the assumed direction and the measured one for the two methods. Lines 2 and 3 give the frequency and thrust requested on each thruster.

Table I : Actuation frequencies, requested thrust and direction error for each thruster.

Thruster N°	1	2	3	4	5	6
Frequency (mHz)	1.10	1.30	1.50	1.70	2.10	2.70
Requested thrust (μN)	2.00	2.00	2.00	2.00	2.00	2.00
Direction using PSD (°)	0.35	0.62	0.55	0.31	0.10	0.11

4. Summary and Future Developments

The results presented show that a relatively simple strategy can, in less than 12 hours, give a precise characterization of the thruster system that LPF will have on board. The mean direction values are probably precise enough for what is needed but a more complete analysis of the two methods used (PCA and PSD) should allow for an estimate of the associated errors. The error on the direction is of particular interest as it may allow to estimate the *flutter noise* (fluctuation of the direction of a thruster) which is a difficult quantity to measure in a ground based test facility.

An extension of this strategy can also be used to measure the spectral shape of the amplitude noise of individual thrusters. This has been tested for both white and colored noise and will be published shortly.

It should also be noted that these same techniques can be applied to characterize the Colloidal Micro Newton Thruster (CMNT) system that will fly aboard LPF as part of the NASA-provided Disturbance Reduction System payload [5].

Acknowledgments

The authors would like to thank M.Hewitson (A.E.I Hannover), M.Armano (ESA), P.McNamara (ESA) and I.Thorpe (Nasa) for their help in producing the simulated data and for discussions related to this analysis.

They also would like to acknowledge the financial support of the UnivEarthS Labex program at Sorbonne Paris Cité (ANR-10-LABX-0023 and ANR-11-IDEX-0005-02) and the French space agency CNES for its financial support during these studies.

References:

[1] **The LISA Pathfinder Mission**, P.McNamara et al., 9th LISA Symposium 20, ASP Conference Series, Vol 467, page 5.

[2] **The Gravitational Universe**, The eLISA Constortium, Whitepaper submitted to ESA for the L2/L3 Cosmic Vision call. arXiv:1305.5720

[3] **The Gaia Mission** : http://www.cosmos.esa.int/web/gaia/selected-papers-articles-reports

[4] **Identification Algorithms for Micro-Propulsion System Parameters Using Drag-Free Test Masses,** T. Ziegler and W. Fichter, Space Sci Rev (2010) 151: 49–60

[5] **Colloid Microthrust Propulsion for the Space Technology 7 (ST7) and LISA Missions**, P. J. Ziemer, et al. "roceedings of the 6th International LISA Symposium, AIP Conference Proceedings, Volume 873, pp. 548-555 (2006).

Journal of Physics: Conference Series **610** (2015) 012027 doi:10.1088/1742-6596/610/1/012027

Bayesian statistics for the calibration of the LISA Pathfinder experiment

M Armano[a], H Audley[b], G Auger[c], P Binetruy[c], M Born[b], D Bortoluzzi[d], N Brandt[e], A Bursi[s], M Caleno[f], A Cavalleri[g], A Cesarini[g], M Cruise[h], K Danzmann[b], I Diepholz[b], R Dolesi[g], N Dunbar[i], L Ferraioli[j], V Ferroni[g], E Fitzsimons[e], M Freschi[a], C García Marirrodriga[f], R Gerndt[e], L Gesa[k], F Gibert[k], D Giardini[j], R Giusteri[g], C Grimani[l], I Harrison[m], G Heinzel[b], M Hewitson[b], D Hollington[n], M Hueller[g], J Huesler[f], H Inchauspé[c], O Jennrich[f], P Jetzer[o], B Johlander[f], N Karnesis[k], B Kaune[b], N Korsakova[b], C Killow[p], I Lloro[k], R Maarschalkerweerd[m], S Madden[f], D Mance[j], V Martin[k], F Martin-Porqueras[a], I Mateos[k], P McNamara[f], J Mendes[m], E Mitchell[n], A Moroni[s], M Nofrarias[k], S Paczkowski[b], M Perreur-Lloyd[p], P Pivato[g], E Plagnol[c], P Prat[c], U Ragnit[f], J Ramos-Castro[q], J Reiche[b], J A Romera Perez[f], D Robertson[p], H Rozemeijer[f], G Russano[g], P Sarra[s], A Schleicher[e], J Slutsky[r], C F Sopuerta[k], T Sumner[n], D Texier[a], J Thorpe[r], C Trenkel[i], H B Tu[g], S Vitale[g], G Wanner[b], H Ward[p], S Waschke[n], P Wass[n], D Wealthy[i], S Wen[g], W Weber[g], A Wittchen[b], C Zanoni[d], T Ziegler[e], P Zweifel[j]

[a] European Space Astronomy Centre, European Space Agency, Villanueva de la Cañada, 28692 Madrid, Spain
[b] Albert-Einstein-Institut, Max-Planck-Institut für Gravitationsphysik und Universität Hannover, 30167 Hannover, Germany
[c] APC UMR7164, Université Paris Diderot, Paris, France
[d] Department of Industrial Engineering, University of Trento, via Sommarive 9, 38123 Trento, and Trento Institute for Fundamental Physics and Application / INFN
[e] Airbus Defence and Space, Claude-Dornier-Strasse, 88090 Immenstaad, Germany
[f] European Space Technology Centre, European Space Agency, Keplerlaan 1, 2200 AG Noordwijk, The Netherlands
[g] Dipartimento di Fisica, Università di Trento and Trento Institute for Fundamental Physics and Application / INFN, 38123 Povo, Trento, Italy
[h] Department of Physics and Astronomy, University of Birmingham, Birmingham, UK
[i] Airbus Defence and Space, Gunnels Wood Road, Stevenage, Hertfordshire, SG1 2AS, UK
[j] Institut für Geophysik, ETH Zürich, Sonneggstrasse 5, CH-8092, Zürich, Switzerland
[k] ICE-CSIC/IEEC, Facultat de Ciències, E-08193 Bellaterra (Barcelona), Spain
[l] Istituto di Fisica, Università degli Studi di Urbino/ INFN Urbino (PU), Italy
[m] European Space Operations Centre, European Space Agency, 64293 Darmstadt, Germany
[n] The Blackett Laboratory, Imperial College London, UK
[o] Physik Institut, Universität Zürich, Winterthurerstrasse 190, CH-8057 Zürich, Switzerland
[p] SUPA, Institute for Gravitational Research, School of Physics and Astronomy, University of Glasgow, Glasgow, G12 8QQ, UK
[q] Universitat Politècnica de Catalunya, Enginyeria Electrònica, Jordi Girona 1-3, 08034 Barcelona, Spain

[r] NASA Goddard Space Flight Center, 8800 Greenbelt Road, Greenbelt, MD 20771, USA
[s] CGS S.p.A, Compagnia Generale per lo Spazio, Via Gallarate, 150 - 20151 Milano, Italy

E-mail: karnesis@ieec.uab.es

Abstract. The main goal of LISA Pathfinder (LPF) mission is to estimate the acceleration noise models of the overall LISA Technology Package (LTP) experiment on-board. This will be of crucial importance for the future space-based Gravitational-Wave (GW) detectors, like eLISA. Here, we present the Bayesian analysis framework to process the planned system identification experiments designed for that purpose. In particular, we focus on the analysis strategies to predict the accuracy of the parameters that describe the system in all degrees of freedom. The data sets were generated during the latest operational simulations organised by the data analysis team and this work is part of the LTPDA Matlab toolbox.

1. Introduction

The LISA Pathfinder mission [1] is a joint mission of the European Space Agency (ESA) and the National Aeronautics and Space Administration (NASA). It is aimed to test and validate key technologies for a future space-based Gravitational-Wave (GW) observatory, like the eLISA [2] mission concept. The eLISA is going to directly detect the effects of arriving GWs, via monitoring the relative motion between pairs of test masses (TMs) in free fall conditions. The LPF mission, being a demonstrator mission, consists in a single spacecraft hosting two proof masses following nominal geodesic motion, whose displacements are monitored by means of Mach-Zehnder laser interferometry. The principal requirement of the mission is expressed as

$$S_{\Delta a}^{1/2}(f) = 3\times 10^{-14}\left[1+\left(\frac{f}{3\,\mathrm{mHz}}\right)^2\right]\mathrm{ms}^{-2}\mathrm{Hz}^{-1/2}, \qquad 1\,\mathrm{mHz}\leq f\leq 30\,\mathrm{mHz}, \tag{1}$$

where the $S_{\Delta a}^{1/2}$ term, is the spectral density of the differential acceleration between the TMs. In order to reach the level of Eq. (1), the instrument must be fully characterised during the mission, and the various unknown dynamical parameters need to be estimated. For that reason, a series of system identification experiments have been proposed, each one stimulating the system in a particular manner. In the following sections we will discuss the planned experiments, together with the analysis strategies. Finally, we will present results of the algorithm performance on simulated data sets. This work is fully integrated in the standardised LTPDA toolbox [3], and is also part of the pipeline analysis to be performed during mission operations.

2. The system identification experiments

The planned system identification investigations can be divided into two main categories; those over the sensitive axis, and the cross-talk experiments [4]. The first case consists in exciting the three body system (the two TMs and the space-craft) along the x-axis, which is monitored by the most sensitive instrument, the differential interferometer (or o_{12} for short). In the second case, the system is excited along various degrees of freedom, and it aims at studying any possible cross-coupling effects that "pollute" the o_{12} optical channel. In particular, the TMs are commanded to rotate around their z-axis (ϕ angle), and along their y-axis, and the space-craft only around its z-axis (Φ angle). The commanded injection signals are, in essence, a series of sinusoids that will induce a large (in amplitude) response of the system to be used to calibrate the close-loops transfer functions, and to estimate the dynamic parameters with the desired accuracy.

3. The Bayesian data analysis framework

For the analysis of the aforementioned experiments, a Bayesian framework has been developed [5, 6], where a posterior distribution is sampled via Markov Chain Monte Carlo (MCMC) methods. The posterior distribution $\pi(\vec{\theta}|y)$ can be expressed as

$$\pi(\vec{\theta}|y) = \frac{\pi(y|\vec{\theta})p(\vec{\theta})}{\pi(y)}, \tag{2}$$

where $\vec{\theta}$ the parameter set to be estimated given the data-set y, $\pi(y|\vec{\theta})$ the likelihood function, $p(\vec{\theta})$ the prior Probability Density Functions (PDFs), and $\pi(y)$ is the so-called evidence of the model. Usually, the evidence is omitted in parameter estimation procedures, as it serves only as a normalisation constant. Then, assuming Gaussian properties of the noise and high signal-to-noise ratio (SNR), the likelihood can be written as

$$\pi(y|\vec{\theta}) = C \times e^{-\frac{1}{2}(y - h(\vec{\theta})|y - h(\vec{\theta}))} = C \times e^{-\chi^2/2}, \tag{3}$$

where C a constant, and the $(\cdot|\cdot)$ determines the natural inner product in frequency domain $(a|b) = 2\int_0^\infty \mathrm{d}f[\tilde{a}^*(f)\tilde{b}(f) + \tilde{a}(f)\tilde{b}^*(f)]/\tilde{S}_n(f)$, where $\tilde{S}_n(f)$ is the power spectral density of the noise time-series. A tilde denotes the operations in frequency domain, and an asterisk denotes the complex conjugation. The $h(\vec{\theta})$ is the so-called template, or in other words, the model of the dynamics of the system.

But for the case of the LTP, the noise time-series might be unknown, or simply not available during the system identification experiments. This can be tackled by modelling the curve of the spectral density of the noise as in [7], where the coefficients i of the $\tilde{S}_n(f)$ term are multiplied by a set j of η amplitudes to be also estimated together with the dynamics parameters

$$\tilde{S}_{n,i} \to \eta_j \tilde{S}_{n,i}, \quad i_j < i \leq i_{j+1}. \tag{4}$$

Another method has been proposed, if one can assume zero mean and Gaussian properties of the noise. Then, the noise coefficients can be marginalised out of the posterior distribution. More details can be found in [8].

4. Results from simulated data sets

We can now apply the available techniques on the experiments over the sensitive axis. A simple realisation of a differential acceleration model can be expressed as [6, 8]

$$\alpha_{12} = \left[\frac{\mathrm{d}^2}{\mathrm{d}t^2} + \omega_2^2\right] o_{12}(t-\tau) + (\omega_1^2 - \omega_2^2)o_1(t-\tau) - A_{sus}\frac{g_2(f)}{m_2}, \tag{5}$$

where ω_1^2 and ω_2^2 are the parasitic stiffnesses of the TMs, o_1 the readout of the first interferometer that measures the displacement of the first TM with respect to the space-craft, A_{sus} a control loop gain, and τ a system delay. The $g_2(f)/m_2$ is the (normalised to the mass m_2) applied force in the second TM, that may be frequency dependet. In this simple case, we have modelled a low-pass filter procedure applied to the commanded forces.

Working in the acceleration domain, has certain advantages, like no controller transfer functions appearing in the equations, but one must consider that the acceleration noise spectral density depends on the parameter set, so that $\tilde{S}_n \equiv \tilde{S}_n(\vec{\theta})$ [9]. A first approach to solve this problem is to apply the *iterative* χ^2 scheme [8]. This method is based on a loop of sequential estimations of $\vec{\theta}$ and then $\tilde{S}_n(\vec{\theta})$, then constructing the likelihood function, and maximising it

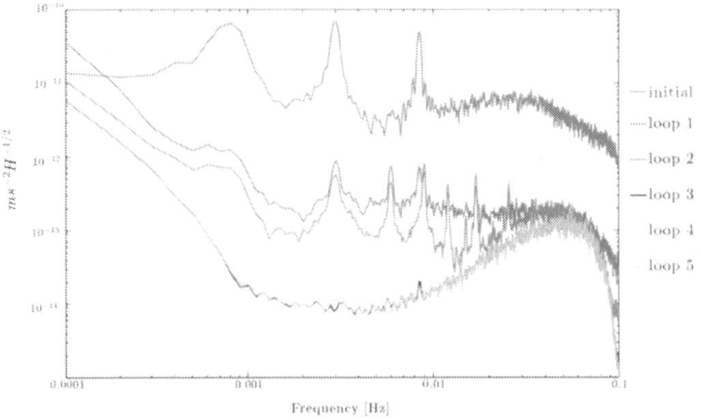

Figure 1. Example of the iterative χ^2 scheme fitting the differential acceleration α_{12}. At the end of each estimation of $\vec{\theta}_i$, with $1 \leq i \leq N_{iter}$ the likelihood function is updated taking into account the $\tilde{S}_n(\vec{\theta}_i)$ and a new estimation procedure takes place, until a convergence criterion is satisfied. These curves correspond to the fitting progress of a high dimensional model on the cross-talk experiments. See text for details.

to apply a new round of corrections on $\vec{\theta}$. An example of the iterative χ^2 scheme can be seen in Fig. 1. For practical reasons, the outer loop of estimations is performed with methods that can be faster than a MCMC run. We use the Nelder-Mead Simplex algorithm [10] until convergence is achieved, and then employ the MCMC algorithm to sample the posterior after assigning prior densities. In Fig. 2, we compare the iterative χ^2 technique to the noise modelling method of Eq. (4). As expected, there is no bias between the two methods, mostly due to the very high SNR of the particular experiments.

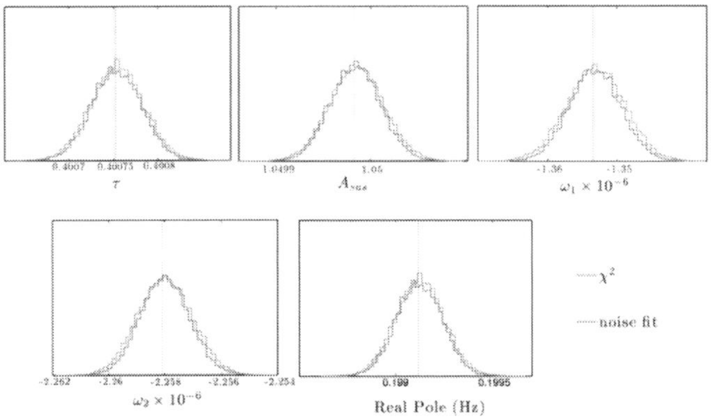

Figure 2. The PDFs of the system parameters of Eq. (5), as estimated by both techniques of the iterative χ^2 and the noise modelling of eq. (4). Since the SNR for the particular experiments is very large, we did not expect to find significant difference.

The developed MCMC methods, can be directly used for the case of the cross-talk experiments, where the SNR is smaller for certain parameters. As already mentioned, the three bodies of the system are commanded in different degrees of freedom in a consecutive manner. The first TM is commanded over ϕ and y, then the same injection signals are applied to the second TM, and finally the space-craft is commanded to rotate around its z-axis. The parameters

Journal of Physics: Conference Series **610** (2015) 012027 doi:10.1088/1742-6596/610/1/012027

to be estimated are the various cross-coupling effects that induce signal on the sensitive axis of the experiment. They can be of *(1)* geometrical origin, where the motion on other degrees of freedom is projected on the x-axis, *(2)* electrostatic origin, where distortion of the field lines from the command electrodes project undesired electrostatic forces, *(3)* gravitational origin, where a self-gravity imbalance of the space-craft could affect the TMs motion, and finally *(4)* external effects of magnetic or thermal origin.

We may choose to form an analytical model in the acceleration domain, that includes the complete span of the cross-talk experiments, yielding a total of sixteen parameters to be estimated. A long MCMC search over the parameter space produces satisfactory results for the cross-talk coefficients. In Fig. 3, the residual acceleration during the cross-talk experiment is displayed. It is evident that the estimated parameters yield a residual level identical to the noise measurement, performed in a previous day of the experiment.

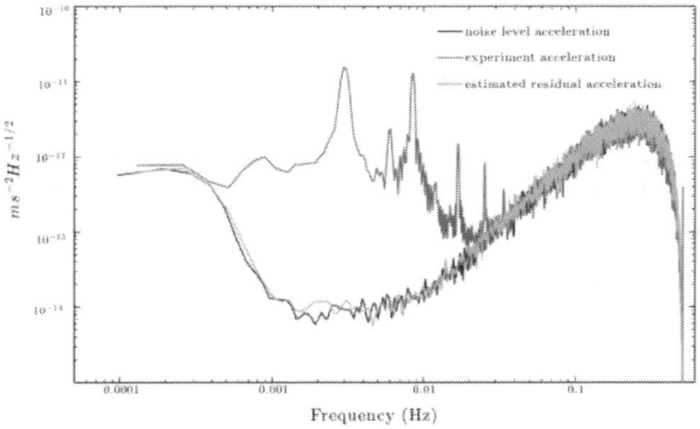

Figure 3. The residuals differential acceleration between the two TMs compared with the noise level measured in a previous day, and the total differential acceleration during the crosstalk-experiments.

The analysis of the cross-talk experiments can be more demanding, mostly due to the high dimensionality of the models, but also due to some cross-coupling coefficients contributing with a very low SNR to the overall signal measured. A suitable model is therefore required for the analysis, to quantitatively assess the physical effects that can explain more efficiently the observations. A Reversible Jump MCMC (RJMCMC) [6] algorithm has already been developed for this case, and already applied to the simulated data. The RJMCMC algorithm can be seen as a generalised MCMC method that is capable of sampling different parameter spaces simultaneously. It directly estimates the so-called Bayes factor $B_{\rm XY}$, which is defined as the ratio of the evidences of the two competing models X and Y. For the cross-talk case we compared two models, the first one (which we name X) being higher in dimension by one extra parameter, which corresponds to the cross-coupling coefficient of the θ angle to the differential acceleration. It was then proven that the new realisation of the model, which includes the new cross-talk term is more probable by $\log B_{\rm XY} = 85.0$, where Y is the model without this contribution. It is the same model that produces the residuals in Fig. 3.

5. Discussion

We have developed a Bayesian framework to be applied for the analysis of the LPF mission planned system identification experiments. Different approximations have been employed for modelling of the noise, or the treatment of an unknown level of the noise. In addition, model selection techniques have been developed in order to identify the most probable physical effects contributing to the overall differential acceleration. This is proven to be of major significance for cases where the model is complicated, like in the cross-talk experiments. The designed analysis has been successfully tested and validated in operational exercises for the preparation of the data analysis algorithms. This work is part of the pipeline analysis that is going to be run during operations.

Acknowledgments
This work was supported from AYA-2010-15709 (MICINN) and 2009-SGR-935 (AGAUR) contracts.

References

[1] McNamara P *et al* 2013, The LISA Pathfinder Mission, *ASP Conference Series* **467**
[2] eLISA Collaboration, 2013, *The Gravitational Universe*, arXiv 1305.5720
[3] Hewitson M *et al* 2009, Data analysis for the LISA Technology Package, *Class. Quantum Grav.* **26** 094003
[4] Vitale S, 2007, *Tech. Rep.* S2-UTN-TN-3045, Issue 1
[5] Nofrarias M, *et al*, Bayesian parameter estimation in the second LISA Pathfinder mock data challenge, 2010 *Phys. Rev.* D **82** 122002
[6] Karnesis N, *et al*, 2014, Bayesian model selection for LISA Pathfinder *Phys. Rev.* D **89** 062001
[7] Tyson B. Littenberg and Neil J. Cornish, 2009, Bayesian approach to the detection problem in gravitational wave astronomy, *Phys. Rev.* D **80** 063007
[8] S Vitale *et al*, 2014, Data series subtraction with unknown and unmodeled background noise, *Phys. Rev.* D **90**, 042003
[9] Congedo G, 2012, PhD thesis, *Spacetime Metrology with LISA Pathfinder*
[10] Jeffrey C. Lagarias *et al*, 1998, Convergence Properties of the Nelder–Mead Simplex Method in Low Dimensions, *SIAM Journal of Optimization*, **9**, 112-147

Magnetic field measurement using chip-scale magnetometers in eLISA

I Mateos, M. Diaz-Aguiló, L. Gesa, F. Gibert, N. Karnesis, I. Lloro, A. Lobo[1], V. Martín, M. Nofrarias, J. Ramos-Castro and C. F. Sopuerta

Institut de Ciències de l'Espai (CSIC-IEEC), 08034 Barcelona, Spain

E-mail: mateos@ice.csic.es

Abstract. Magnetic sensors are necessary devices to map the magnetic field and gradient at eLISA test masses location. Their primary goal is assessing the contribution of the magnetic effects to the acceleration noise budget. Our experience, accumulated during the magnetic diagnostics system design for LISA Pathfinder, indicates that the accuracy of the magnetic field map interpolation at the test mass is critical issue. Therefore, taking into consideration eLISA increased performance demands, an enhancement of the LISA Pathfinder magnetic subsystem is deemed necessary. A goal pursued by using alternative magnetic sensing techniques. In this study, the accuracy improvements in the magnetic field map reconstruction obtained with the currently conceived instrumental layout are demonstrated.

1. Introduction

The evolved Laser Interferometer Space Antenna (eLISA) is a dedicated space-based gravitational wave observatory proposed as a large science mission to ESA [1]. In its current design, the noise requirement in terms of free-fall accuracy is $\sqrt{2} \cdot 3\,\mathrm{fm\,s^{-2}Hz^{-1/2}}$ in a measurement bandwidth between 0.1 mHz and 1 Hz. Among the main contributions to the total noise budget are the magnetic fields and gradients present within the spacecraft, which are mostly generated by different units such as the micro-thrusters. Hence, eLISA will include a set of magnetic sensors for magnetic environment monitoring which will allow discerning the magnetic noise contribution from the overall measured acceleration noise.

The current eLISA magnetic measurement subsystem is based on LISA Pathfinder's where mature fluxgate magnetometers were selected due to their proven sensitivity and actual availability for space applications. However, their use for magnetic field estimation at the test mass (TM) position present a few drawbacks driven by the following two basic factors:

- Permalloy-core magnetometers are able to disturb their magnetic surroundings due to the employed measurement principle.
- Sensors of large size conflict with the possibility of having a sufficient number of them to properly map the magnetic field around the TMs.

In LISA Pathfinder, these factors limited the total number of tri-axial sensors to just four which, moreover, had to be placed somewhat far from the TMs (≥ 18.85 cm). As a consequence,

[1] Deceased

classical interpolation methods led to unsatisfactory magnetic field estimations which forced using alternative approaches which required magnetic environment previous knowledge [2]. In order to mitigate the aforementioned limitations and to ensure a more robust magnetic field reconstruction for eLISA purposes, a sufficient number of smaller sensor devices with lower magnetic back-action effects are required.

2. Interpolation method: Multipole expansion

The magnetic field at the TM location must be estimated only with the information given by the magnetometer located around the region of interest. Since the magnetic field surrounding the TM is considered to be mostly a *vacuum* field ($\nabla \times \mathbf{B} = \nabla \cdot \mathbf{B} = 0$), the estimated magnetic field $\mathbf{B}_e$ inferred from the magnetic field readings can be approached by a multipole expansion, where the order of the series is determined by the number of magnetometer data channels N. The magnetic field can therefore be expressed as

$$\mathbf{B}_e(\mathbf{x}) = \nabla\Psi(\mathbf{x}) = \sum_{l=1}^{L}\sum_{m=-l}^{l} M_{lm}(t)\, \nabla[r^l\, Y_{lm}(\mathbf{n})],$$

where $r \equiv |\mathbf{x}|$ and $\mathbf{n} \equiv \mathbf{x}/r$ are the spherical coordinates of the field point $\mathbf{x}$. M_{lm} and Y_{lm} are the multipole coefficients and the standard spherical harmonics of degree l and order m, respectively.

3. Magnetic sources and sensor layout

Since the magnetic environment and requirements are not yet known for eLISA, the algorithm performance verification has been based on the distribution of LISA Pathfinder known magnetic sources (modeled as magnetic dipoles). Therefore, the magnitude and location of the sources have been fixed but the dipoles orientation were randomly set (using a normal distribution) in order to evaluate the robustness of the algorithm. According to this, a series of 10^3 different magnetic scenarios have been considered for the present analysis.

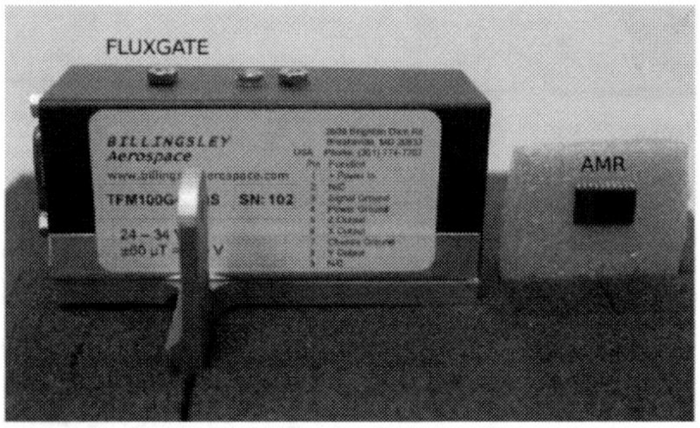

Figure 1. Fluxgate magnetometer used in LISA Pathfinder and AMR sensor proposed for eLISA.

The key aspect of the study is the use of sensors based on the anisotropic magnetoresistance (AMR) principle to replace the fluxgate magnetometers used in LISA Pathfinder shown in Figure 1. These sensors use a different physical principle to operate, but they both include a magnetic core as sensor head. However, for the purpose of maintaining the magnetic cleanliness needed for eLISA, the use of AMRs pose a clear advantage over fluxgate magnetometers due to the fact that the amount of ferromagnetic material present in their cores is much smaller. Therefore, AMR magnetometers can be placed closer to eLISA TM because of their inherently lower magnetic moment, which has been verified by measurement in a superconducting quantum interference device (SQUID) [3].

For the present analysis, the magnetometers have been symmetrically placed on the wall of the vacuum chamber, which encloses the TM in order to meet the strict vacuum requirements. Figure 2 displays the the magnetic sources distribution and the location of the eight sensors. The exact magnetic field map for a given source distribution is shown in Figure 3, where the magnetometers and the test mass positions are also represented.

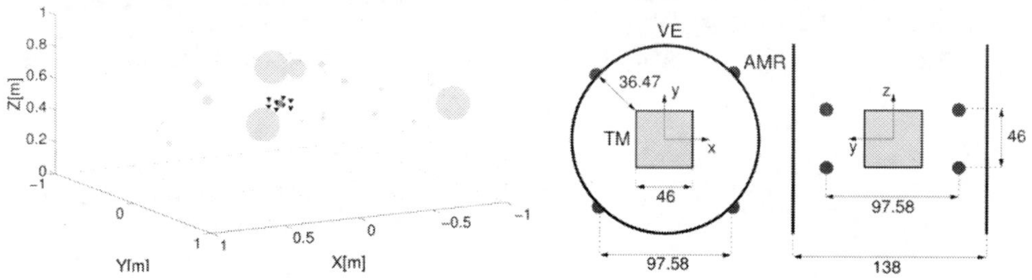

Figure 2. Left: Views of the 29 measured dipole magnetic sources in LISA Pathfinder (green dots: size is proportional to their magnetic moment), the test mass (red dot) and the 8 AMR magnetometers (black triangles). Right: Sensor array configuration on the vacuum enclosure (Units are in mm).

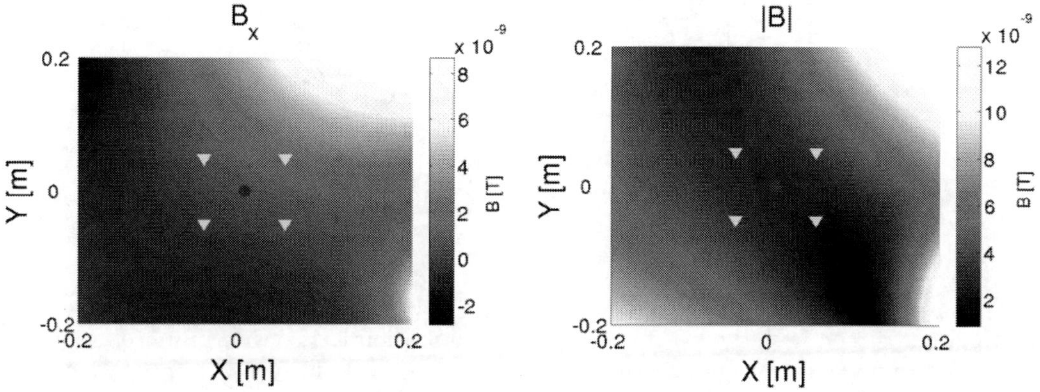

Figure 3. Magnetic field contour plot (B_x and $|B|$) of the magnetic field for a specific source configuration.

4. Results: Magnetic field reconstruction

The proposed number of sensors allows resolving the magnetic field using multipole expansion up to the hexadecapole order approximation. The resultant reconstructed field, shown in figure 4, reveals a clear resemblance to the exact one shown in the previous section.

The estimation errors at the TM location for LISA Pathfinder and for two possible configurations for eLISA (equipped with AMRs) are displayed in figure 5. The improvement with the 8-AMR configuration is encouraging and reaches mean errors lower than 1% and 2% for the magnetic field and magnetic field gradient, respectively. The reader will find detailed information on this matter in [4].

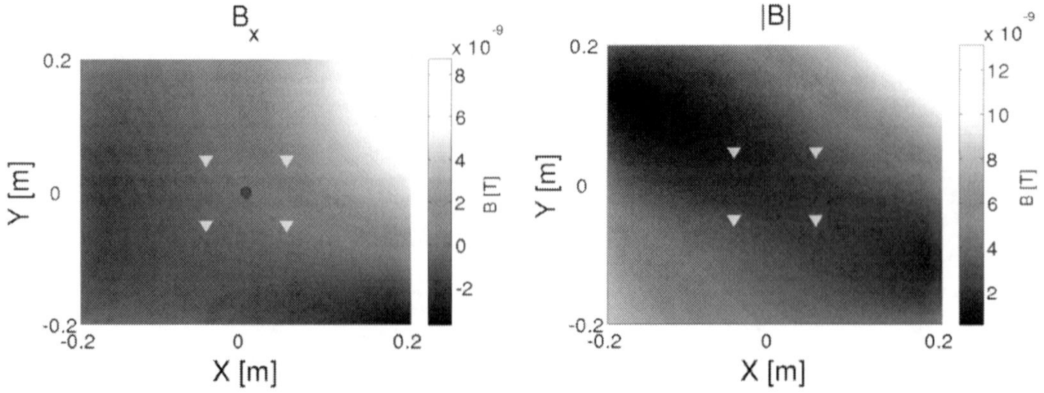

Figure 4. Contour plot of the estimated field (B_x and $|B|$) using multipole expansion with 8 magnetometers.

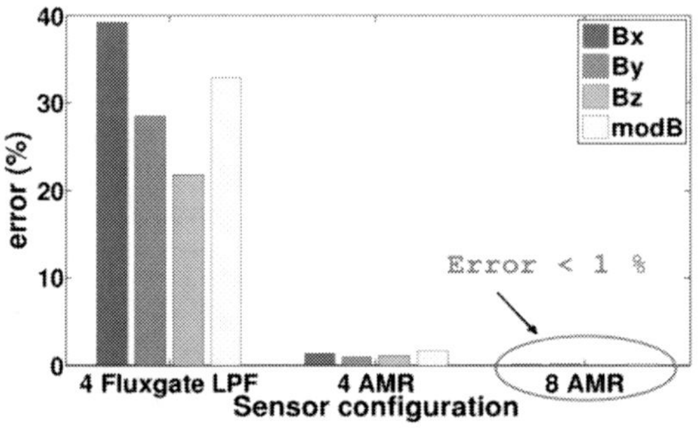

Figure 5. Relative errors of the estimated magnetic field at the position of the TM using different sensor arrangements (4 fluxgates used in LISA Pathfinder, 4 AMRs and 8 AMRs as shown in figure 2).

5. Conclusion

Multipole expansion with the suggested sensor configuration has consistently produced reliable results in estimating the magnetic field at the positions of the TM. Hence, the proposed magnetic measurement subsystem for eLISA represents a considerable improvement in the magnetic field estimation accuracy compared to LISA Pathfinder. It shall be remarked that this improvement has been achieved due to the small size and low magnetic back-action offered by AMRs which enables the possibility of not only installing more sensors but also placing them closer to the TMs. Future work will include the analysis taking into account extra details such as readout noise, unexpected offset and spatial resolution of the sensor head.

Acknowledgments

We are grateful to Airbus Defence and Space UK for providing the measured magnetic moment values for the LISA Pathfinder units. Support for this work came from Project AYA2010-15709 of *Plan Nacional del Espacio* of the Spanish Ministry of Science and Innovation (MICINN).

Journal of Physics: Conference Series **610** (2015) 012028 doi:10.1088/1742-6596/610/1/012028

References

[1] Pau Amaro *et al*, White paper on The Gravitational Universe science theme for ESA (2013).
[2] M. Diaz-Aguiló, A. Lobo and E. García-Berro, Exp. Astron., **30**, 1 (2011).
[3] I. Mateos *et al.*, ASP Conf. Ser., **467**, 341 (2013).
[4] I. Mateos *et al.*, Interpolation of the magnetic field at the test masses in eLISA (submitted).

Optical Telescope Design Study Results

J. Livas[1*] **and S. Sankar**[1,2]
[1]NASA Goddard Space Flight Center, Greenbelt, MD 20771 USA
[2]Universities Space Research Association, Greenbelt, MD 20771

E-mail: Jeffrey.Livas@nasa.gov

Abstract. We report on the results of a study conducted from Nov 2012-Apr 2013 to develop a telescope design for a space-based gravitational wave detector. The telescope is needed for efficient power delivery but since it is directly in the beam path, the design is driven by the requirements for the overall displacement sensitivity of the gravitational wave observatory. Two requirements in particular, optical pathlength stability and scattered light performance, are beyond the usual specifications for good image quality encountered in traditional telescopic systems. An important element of the study was to tap industrial expertise to develop an optimized design that can be reliably manufactured. Key engineering and design trade-offs and the sometimes surprising results will be presented.

1. Introduction

Space-based gravitational wave observatories will enable the observation of astrophysical signals in the source-rich low-frequency band from 0.0001 to 0.1 Hz. All current observatory designs require the use of an optical telescope to efficiently deliver light from one spacecraft to another. Therefore the telescope is an important technology for future gravitational-wave space missions. Our goal is to continue the development work for such a telescope, using the eLISA concept [1] as a reference mission for our particular design. The work ultimately includes fabricating and testing a prototype against the requirements for a gravitational-wave observatory.

The initial approach has been to develop a telescope design that meets a set of science-based requirements and can be manufactured on a small scale, with a modest initial target of NASA Technology Readiness Level 5 by 2018. After developing a design at Goddard Space Flight Center that met all the nominal optical specifications, we commissioned a study with an industrial partner to obtain advice from a vendor with considerable practical experience in designing and building similar optical systems.

One important choice was between different implementations of the design: on-axis vs off-axis. In the eLISA application, the telescope is to be used to transmit and receive simultaneously, and the on-axis design suffers from a significant "narcissus" reflection from the secondary mirror. The outgoing beam reflects at near normal incidence from the center of the secondary, and returns an on-axis beam that appears on the detector along with the received signal. This reflection dominates the scattered light budget and makes it extremely difficult to meet the scattered light requirement without suppressing

* To whom any correspondence should be addressed.

Journal of Physics: Conference Series **610** (2015) 012029 doi:10.1088/1742-6596/610/1/012029

that reflection. An off-axis design avoids the narcissus reflection, but was thought to be more susceptible to the presence of thermal gradients on-orbit that can introduce off-axis aberrations such as coma or astigmatism. These aberrations are not easily removed by a simple focus mechanism. An important first task for the study contractor was to independently weigh such considerations and make a recommendation regarding the most suitable design form or implementation.

Two key questions were framed for the study contractor to guide the process:
(1) Does an on-axis implementation meet the requirements, in particular for scattered light and dimensional stability?
(2) Is an off-axis implementation feasible to manufacture, while still robust enough to meet the telescope specifications in an environment of axial and transverse thermal gradients?

Once a design form was chosen, the study required a number of deliverables. These included:

- complete mechanical, thermal and optical design, including scattered light analysis
- test plan for verifying and validating requirements
- manufacturing plan for building 10 identical telescopes, including a schedule
- rough order of magnitude (ROM) cost estimate and cost breakdown for 10 telescopes

2. Key Optical Requirements and Goals of the Study

The application is a precision measurement of the relative displacement of two proof masses located in widely separated spacecraft through a pair of telescopes (not for image formation). The key optical requirements for the telescope are shown in Table 1.

Table 1: Key Telescope Optical Requirements

	Parameter	Derived From	eLISA
1	Wavelength	Nominal mission design	1064 nm
2	Net Wave front quality of as built telescope subsystem over science field of view under flight-like conditions	Pointing	$\lambda/30$ RMS
3	Field-of-Regard (Acquisition)	Acquisition	+/- 200 µrad (large aperture)
4	Field-of-Regard (Science)	Orbits	+/- 20 µrad (large aperture)
5	Field-of-View (Science)	Scattered Light	+/- 8 µrad (large aperture)
6	Science boresight accuracy	FOV, pointing	+/- 1 µrad (large aperture)
7	Telescope subsystem optical path length stability under specified environment	Path length Noise/ Pointing	$1\,pm/\sqrt{Hz} \times \sqrt{\left(1+\left(\frac{0.003}{f}\right)^4\right)}$ where $0.0001 < f < 1$ Hz
8	Afocal magnification	short arm interferometer	200/5 = 40x (+/- 0.4)
9	Mechanical length	spacecraft diameter	< 350 mm
10	Optical efficiency (throughput)	Shot noise	>0.85
11	Scattered Light	Displacement noise	$< 10^{-10}$ of transmitted power into science FOV at the detector

The telescope must be nearly diffraction limited for efficient delivery of laser power over the large separation between spacecraft. The requirements for field of view and wavefront quality are not particularly challenging. Requirements 7 and 11 are the most unusual for telescope designers and thus are considered challenging. Pathlength stability (requirement 7) is required because the telescopes are in series with the displacement measurement. The required precision of the displacement measurement is 10 pm/√Hz, and the uncertainty budget allocated to the telescope is 1 pm/√Hz over the eLISA science measurement band from 0.0001 to 0.1 Hz. The contribution from the telescope enters the measurement of the one-way displacement twice, since there is a telescope at both ends of the measurement path.

The telescope transmits and receives optical power simultaneously. The transmitted power is 1 W from the large aperture, and the expected received power is approximately 100 pW. To avoid adding broadband noise, it is desirable for the scattered light to be stable in phase relative to the local laser against which the received light is beat to form interference fringes. Our models indicate that most of the scattered light is expected to be from the mirrors, so the dimensional stability requirement should also mean that the scattered light is phase-stable. Phase stability, in turn, impacts the maximum allowable scatter. Furthermore, the scattered light must be in the same spatial and polarization mode as the local oscillator and/or received signal beam to interfere with the observatory displacement measurement. We have approximated that condition in Table 2 by specifying that the light be scattered into the science field of view.

Finally, one other requirement that is not easily captured in Table 1 is the ease with which the telescope can be manufactured. The mission will require 6 flight telescopes, plus 1-2 spares, and several telescopes for ground testing. The telescopes must be made in a relatively short time and be interchangeable to facilitate integration and ground testing. A simple and robust design is required that can be easily manufactured to minimize the need for extensive optimization during assembly and alignment.

3. Specific Trades

Table 2 shows a summary of the major issues considered as part of the study. The grey background shading indicates the highest risk issues.

Table 2: Trade Study Summary

	Material	**Silicon / Silicon Carbide**		**Glass/Graphite Composite**	
	Design Form	**On-Axis**	**Off-Axis**	**On-Axis**	**Off-Axis**
Manufacturing	Optics	No M2 beam trap/ spot implementation	Standard practice	No M2 beam trap/ spot implementation	Standard practice
Manufacturing	Structure	Contractor heritage	Contractor heritage	Contractor heritage	Contractor heritage
Manufacturing	Alignment	Standard practice	Standard practice	Standard practice	Standard practice
Environmental Test	Thermal Vacuum	Standard practice	Standard practice	Standard practice/ Outgassing concerns	Standard practice/ Outgassing concerns
Environmental Test	Launch Loads	Heritage design and strength	Heritage design and strength	Heritage design and strength	Heritage design and strength
Recurring Cost		lower for structure/ optics/testing	lower for structure /optics/testing	Higher for structure/optics/testing	Higher for structure/optics/testing
Schedule		Optic procurement drives schedule	Optic procurement drives schedule	Structure procurement/ testing drives schedule	Structure procurement/ testing drives schedule
Stability	Thermal analysis	Can meet requirements	Can meet requirements	Can meet requirements after outgassing (long term risk)	Can meet requirements after outgassing (long term risk)
Stability	Manufacturing and material variability	Use of invar in metering path	Use of invar in metering path	Use of invar in metering path; long term outgassing effects	Use of invar in metering path; long term outgassing effects
Scattered light		No M2 beam trap/ spot implementation and native performance	Standard practice	No M2 beam trap/ spot implementation and native performance	Standard practice

Each issue was evaluated for both silicon carbide and carbon fiber/glass composite materials, and both an on-axis and an off-axis implementation of the same optical design were considered.

4. Results

The study contractor recommended an off-axis silicon carbide design as the lowest risk (Table 2) and materials cost. The composite material system is known to exhibit dimensional changes consistent with water absorption and desorption [2]. The required time for the material to stabilize by water desorption in vacuum varies unpredictably and can be a month or more. This uncertainty will add to integration and testing time, thereby adding an undesirable schedule risk. Both materials have relatively low coefficients of thermal expansion (CTE), but the composite can be tailored with some effort to a level of 0.1 ppm/K, whereas silicon carbide is approximately 4 ppm/K. However, the telescope requirement for optical pathlength stability can be met with several different CTE materials, depending on the details of the telescope design form.

An on-axis implementation (using either material) was dismissed because methods for suppression of the narcissus reflection are still in the preliminary development stage. The off-axis silicon carbide design seems to possess the least technical and schedule risk.

Our industrial partner provided a cost estimate and schedule as part of a confidential bidding process so we cannot disclose the details. However, we can say generally that the recurring engineering costs are consistent with earlier NASA estimates from the LISA Project, but we also emphasize the contracted industrial partner arrived at their estimates independently and without prior knowledge of these LISA Project estimates. The cost estimate includes non-recurring engineering as well as creation of testing facilities for subsystem verification both of which were significant portions of the total cost. The schedule assessment provided by the industrial partner was somewhat longer than NASA estimates, but included time for procurement of the silicon carbide components, which is in series with the rest of the telescope manufacturing and not included in the NASA estimates. They recommended performing parallel fabrication/testing of multiple copies of the telescope, utilizing three teams of personnel with team members periodically rotating from team to team.

5. Lessons Learned

The process of working with a technically capable industrial partner has been worthwhile. The exercise forced us to work on converting the mission level specifications into lower level requirements, which could then be more easily assimilated by the contracted vendor into a design form and model. We have gained a great deal of insight by collaboration - in particular we have a better appreciation for the parts of the system that are considered difficult in the industry. We have also learned some of the capabilities of this contractor regarding both manufacture and modelling. This knowledge has allowed us the chance to address many issues from the point of view of the telescope engineer, instead of the scientist's viewpoint.

An example of this collaboration can be found in the thermal design. The thermal specifications were difficult for the vendor, and we found that an interface specification to the spacecraft was necessary, but not sufficient. We eventually provided the study contractor with a full spacecraft thermal design from our prior observatory-level work at NASA. Furthermore, we simplified the design compliance criteria to check only the lowest frequency point in the specifications to make the problem tractable for the industrial engineering team. This example of interchange and collaboration was one of many and enabled the project to proceed more smoothly.

We found that the scattered light specifications and, in particular the coupling mechanism to the observatory sensitivity, were also very unusual and thus challenging for the industrial partner. Approximations were made to facilitate more rapid optimization and communication with the study team. An example of such an approximation was the specification of a science field-of-view at the detector, instead of the full mode-overlap calculation often computed by a science team. The pathlength stability criterion is also unusual for the industry and, though the study engineering team ultimately determined that it was feasible, it does impact cost and schedule to have the industrial partner perform the verification of this requirement for the telescope subsystem. Indeed, the metrology apparatus for testing to this level of displacement sensitivity does not exist at the facility of our study contractor. Much of the quoted cost of the telescope subsystem goes towards establishing the testing facilities and purchasing the necessary equipment.

6. Summary and Outlook

Our partnership with the industrial study team was fruitful and produced several interesting results. For example, the reluctance of the engineers to recommend carbon composite materials (due mainly to technical and schedule risk) drove the choice of a light weighted silicon-carbide structure to meet the stability and thermal requirements. Integrated thermal, mechanical, and optical modelling enabled the study contractor to conclude that the off-axis design would be acceptable in the expected on-orbit environment. Another interesting conclusion was that fabrication and testing of an off-axis telescope is no more difficult than an on-axis version, especially for the higher end optical companies likely to bid on a flight-model, but the back-scattered light is expected to be a significant challenge for an on-axis design.

This study was an important first step in our communication with industry regarding the telescope subsystem. We anticipate continued collaboration with industry partners as we endeavour to push the telescope towards higher technical readiness levels. We have also begun efforts to procure and test a prototype telescope as part of these ongoing efforts at both model validation and technology maturation.

We are aware of a similar effort in Europe [3] and that approach is complementary. Both investigations have converged to an off-axis optical design, although the design approach and material choices are very different. The European study is focused on testing the telescope system level CTE at low temperatures, while this study seeks to investigate the trade-offs involved in ultimately achieving low noise contributions from back-scattered light. Our laboratory demonstration will concentrate on scattered light model validation since earlier work has indicated dimensional stability with a silicon carbide metering structure [4]. We continue to work towards the ultimate goal of a robust telescope design for space-based gravitational wave observatories.

Acknowledgements

The authors would like to acknowledge the help of the Goddard Space Flight Center eLISA Telescope team, including Petar Arsenovic, Peter Blake, John Crow, Joseph Howard, Lenward Seals, Ron Shiri, Justin Ward, and Garrett West. The work described here was supported by grant 11-SAT11-0027 awarded by NASA Strategic Astrophysics Technology program. S. Sankar acknowledges support by NASA grant NNX12AE97G.

7. References

[1] LISA assessment study report (Yellow Book), *ESA/SRE*(2011)3, URL: http://sci.esa.int/science-e/www/object/index.cfm?fobobjectid=48364.

[2] Sanjuan, J, Preston, A, Spector, A, Korytov, D, Freise, A, Dixon, G, Mueller, G, Livas, J 2011 Carbon fiber reinforced polymer dimensional stability investigations for use on the laser interferometric space antenna mission telescope *Rev. Sci. Instrum.* **82**, 124501

[3] Adrianus L. Verlaan ; Harmen Hogenhuis ; Joep Pijnenburg ; Martin Lemmen ; Stefano Lucarelli ; Dietmar Scheulen ; David Ende; LISA telescope assembly optical stability characterization for ESA. Proc. SPIE 8450, Modern Technologies in Space- and Ground-based Telescopes and Instrumentation II, 845003 (September 13, 2012); doi:10.1117/12.925112.

[4] Sanjuan, J., Preston, A, Spector, A., Korytov, D., Freise, A.,Dixon, G., Mueller, G., Livas, J. "Note: silicon carbide telescope dimensional stability for space-based gravitational wave detectors", *Rev. Sci. Instrum.* **83**(11), 116107 (2012).

Journal of Physics: Conference Series **610** (2015) 012030 doi:10.1088/1742-6596/610/1/012030

Simulations of an etched spiral axial attenuation scheme for an on-axis reflecting telescope

Aaron Spector, Guido Mueller

Department of Physics, University of Florida

Abstract

The current generation of proposed space based interferometric gravitational wave detectors all use a reflecting telescope to transfer the laser signals between the space- craft. One of the proposed telescope designs is an on-axis classical Cassegrain with the secondary mirror axially aligned to the primary mirror. Since the outgoing beam will be incident normal to the secondary, some of the light will be reflected directly back to the optical bench. Length changes between the telescope structure and the optical bench will cause this back-reflected light to introduce phase noise to the measuremen t signal. The phase noise from this process must be suppressed below$\sqrt{0}$.1 μcycles / Hz to meet the LISA requirements. We have presented simulations that demonstrate that the back-reflected power can be sufficiently attenuated by using a specifically patterned Anti-Reflective (AR) region, in the shape of a spiral, at the center of the secondary mirror.

1 Introduction

While the specific design of the first generation of space-based interferometric gravitational-wave detectors remains uncertain, all of the proposed missions use reflecting telescopes to transfer laser signals between Spacecraft (SC) separated by vast distances [1] [2]. One concern with using an on-axis telescope, like the one seen in Figure 1a, is that light will be directly back-reflected to the optical bench from the telescope secondary mirror. This light can have phase noise that will couple into the measurement signal due to the length fluctuations in the distance between the telescope and the optical bench. Using the Laser Interferometer Space Antenna (LISA) requirements on technical noise sources as a guide, our goal is to suppress the back-reflection induced phase noise below 0.1 μcycles/$\sqrt{\text{Hz}}$. To achieve this we must cast a shadow along the optical axis of the telescope and reduce the spatial overlap between the field received from the far SC and the back-reflected field. The spatial overlap (η), as defined in [3], is a measure of the interferometric contrast between these two fields evaluated over the area of the photodetector on the LISA optical bench. For length changes of 10 nm/$\sqrt{\text{Hz}}$ between the telescope structure and the optical bench η

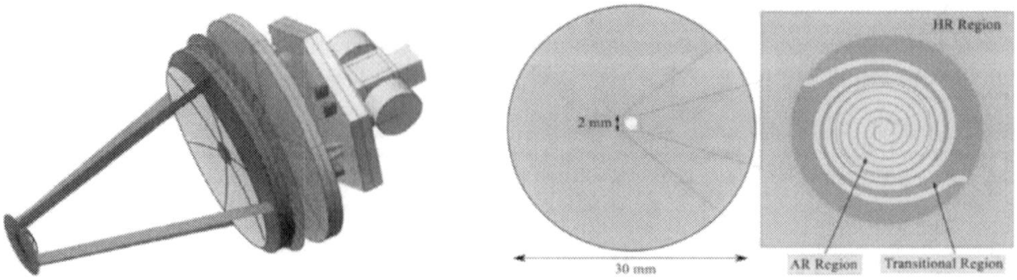

(a) On-axis telescope (b) Etched secondary and spiral pattern

Figure 1: (a) Mock design for an on-axis telescope in blue and optical bench in yellow. (b) Secondary with a gold HR coating and spiral pattern (unwound by a factor of 1000) etched into its center. White is representative of the etched region exposing the substrate.

must be less than 5×10^{-6}. If no axial attenuation scheme is present the unmitigated spatial overlap between these fields is 3×10^{-4} [3]. This means that the total back-reflected power in the spatial mode of the received field must be attenuated by 3.5 orders of magnitude.

2 Axial Attenuation with Sub-wavelength Etching

While it is possible to attenuate the axial back-reflection using a discrete Anti-Reflective (AR) region at the center of the secondary, it is difficult to reduce the spatial overlap below the LISA requirements due to light scattering from the edge of the AR region. This has motivated us to pursue coatings that use a smooth transition in the reflectivity from the Highly Reflective (HR) region to the AR region. Our previous work suggested that a sinusoidal rolloff between these regions will prevent diffraction effects from scattering light back along the optical axis of the telescope [3].

One way to obtain this transition in the reflectivity is to etch sub-wavelength features into a metallic HR coating such as gold. Doing this forms an effective index between the HR coating and the substrate that varies according to the surface density of gold at a given point on the mirror. We can then tune the surface density across the mirror to get a spatially dependent reflectivity. With a spiral etching pattern, like the one shown in Figure 1b, the surface density depends on how tightly the spiral is wound. In the center of the spiral, the AR region, the surface density of gold is set so that an equal power of light is reflected from the gold and the exposed fused silica substrate. Since the gold layer is a quarter-wave thick these two fields destructively interfere. The spiral in Figure 1b is unwound by a factor of 1000 to show its shape. In reality, the spiral will not only be much more tightly wound, but the width of the gold ridges will be on the order of hundreds of nm.

As the radius of the pattern gets larger, the spiral begins to unwind and the surface

Journal of Physics: Conference Series **610** (2015) 012030 doi:10.1088/1742-6596/610/1/012030

density of gold increases thus causing an increase in the reflectivity of that point on the mirror. This smooth rise in the reflectivity forms the transitional region between the AR and HR regions. Once the HR region is reached, the spiral ends and the surface of the substrate is entirely covered in gold. While any reflective metallic coating could be used, our theoretical analysis will focus on gold because of its high reflectivity of 98% at 1064 nm and our familiarity with etching it.

3 Spiral Design

The first step in designing the spiral was making the reflectivity of the AR region as low as possible. As mentioned earlier, this was done by adjusting the surface density and thickness of the gold coating such that the light reflected from the gold destructively interferes with the light reflected from the area where the substrate is exposed. By doing a superposition of the field reflected from fused silica with the field reflected from the gold we found the following equation for the effective reflectivity R_{e_o} in the AR region. Here d represents the thickness and S_{Au_o} is the surface density of gold. The first term, $S_{Au_o}\left(\frac{1-N_{Au}}{1+N_{Au}}\right)$, represents the coefficient of reflection for the portion of the incident field reflected from the gold. The second term is the coefficient of reflection for the portion of the field reflected by the exposed fused silica. The surface density of fused silica in the AR region, $1 - S_{Au_o}$, is multiplied by the coefficient of reflection of fused silica, with N_{Si} being the complex index of refraction of fused silica. It is then multiplied by an additional phase factor due to the thickness of the gold coating.

$$R_{e_o} = \left| S_{Au_o}\left(\frac{1-N_{Au}}{1+N_{Au}}\right) + (1 - S_{Au_o})\left(\frac{1-N_{Si}}{1+N_{Si}}\right) e^{-i\frac{2\pi}{\lambda}2d} \right|^2 \tag{1}$$

λ is the laser wavelength of 1064 nm. For gold deposited on a fused silica substrate we found that the ideal thickness is 243 nm and the ideal surface density of the gold is 16%.

Next we found the following equation for the axial angle ϕ of the spiral as a function of its radius r in the AR region. Here w is the width of the trenches in the gold coating that expose the substrate and comprise the spiral.

$$\phi_{AR}(r) = \frac{2\pi}{w}(1 - S_{Au_o})r \tag{2}$$

As the radius increases the axial angle increases linearly to maintain a constant spacing between the trenches. This keeps the surface density of gold at 16% to preserve the AR properties of this region.

With the spiral defined for the AR region the next step was to unwind the spiral to achieve the appropriate rolloff in the transitional region. As the spiral goes to larger radii the spacing between the trenches grows. This gradually increases the surface density of the metallic coating and produces the rolloff in R_e. In our case we have chosen a Hanning window function rolloff shown in the following equation due to its success in our original paper [3].

$$R_e(r) = \frac{1 - \cos\left(\pi \frac{r-a}{b-a}\right)}{2} \tag{3}$$

Using the rolloff in equation 3 we came up with the following equation for the spiral in the transitional region.

$$\phi_{trans}(r) = \frac{(1 - S_{Au_o})}{w}\left(\pi r + (b-a)\sin\left(\pi \frac{r-a}{b-a}\right)\right) \tag{4}$$

This equation continuously increases the spacing between the trenches as the spiral get larger to achieve a rolloff in the effective reflectivity shown in equation 3. Also, when $r = a$ the axial angle of the spiral is the same in equation 2 and 4 to ensure that the spiral is continuous at the interface between the AR region and the transitional region. This equation for $a < r < b$, with equation 2 for $r \leq a$, define the entire spiral.

4 Simulations

Simulations for the back-reflected distribution from the spiral were done by calculating the Huygen's integral at an object plane on the secondary for each point along a grid that defined the image plane at the input of the optical bench. Here the field calculated from the Huygen's integral was then subtracted from an expression for the propagated TEM_{00} mode. The resulting distribution was then propagated through the rest of the optical train to the location of the photodetectors using an FFT propagation script [4]. Once we had the distribution at the photodetectors we evaluated the spatial overlap between the resulting field and a simulated field received from the far SC.

5 Results

The simulated back-reflected distribution from the spiral gave a spatial overlap at the photodetector of 2×10^{-7}, demonstrating that the spiral pattern was able to reduce the mode matched power by more than 6 orders of magnitude. This comfortably exceeded the 3.5 orders of mode matched power attenuation needed to suppress the spatial overlap below the 5×10^{-6} requirements.

The resulting intensity distribution of the spiral back-reflection can be seen in Figure 2a. The unmitigated back-reflection had a relatively flat intensity distribution of roughly 20 W/m^2. The back-reflected distribution from the spiral pattern had an intensity of 6.7×10^{-5}W/m^2 on the optical axis, which was nearly 5.5 orders of axial intensity suppression. The peak intensity shown in the spiral's back-reflected distribution was 0.012 W/m^2, or 3.2 orders of intensity suppression. This bright region, however, makes up a relatively

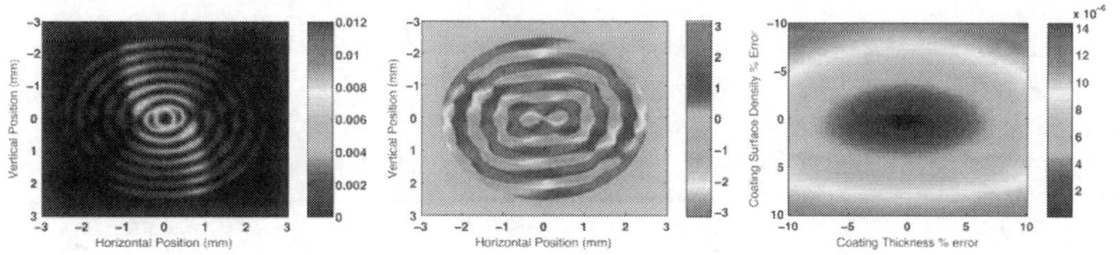

(a) Spiral back-reflection (b) Back-reflected phase (c) Coating error tolerances

Figure 2: (a) Simulated back-reflected intensity distribution from a spiral etched mirror apertured by the collimating mirror of the telescope (W/m^2). (b) Simulated phase of the back-reflection from a spiral etched mirror with the wavefront curvature removed (radians). (c) Results from a simulation on the effect of errors in the coating thickness and surface density in terms of η evaluated over the detector.

small portion of the distribution. In fact, the total power suppression was 4.2 orders of magnitude when integrating over the entire distribution seen in Figure 2a and 4.0 orders of magnitude when evaluating over the area incident on the photodetector. This would be sufficient to meet the requirements even without considering the phase of the back-reflected distribution which further attenuated the mode matched power.

Figure 2b displays the phase of the back-reflected distribution with the spherical wavefront curvature removed. From this image the spatial incoherence of the back-reflected field is quite apparent. Since the field received from the far SC has a relatively flat wavefront this will help further suppress the spatial overlap as different regions of back-reflected distribution will produce different phases that will destructively in the interference beatnote with the received field at the photodetector. This destructive interference accounted for the remaining two orders of magnitude suppression in the mode matched power.

It is important to point out that this can all be done without any additional AR coating on the substrate since the etched gold coating itself is anti-reflective. In addition to this, the tolerances on the thickness and surface density of the gold coating are realistically achievable. Figure 2c shows result of simulating the spatial overlap at the photodetector with errors in the thickness and surface density of the gold coating. Even with $\pm 4\%$ errors in the coating surface density and $\pm 7\%$ errors in the coating thickness the spatial overlap will still be suppressed below the 5×10^{-6} requirement. Our conversations with manufacturers suggest that coatings with these specifications should be attainable. The results of these simulations show that etching an AR spiral pattern into a metallic coating is a viable method for sufficiently attenuating the axial back-reflection below the LISA requirements.

Journal of Physics: Conference Series **610** (2015) 012030 doi:10.1088/1742-6596/610/1/012030

References

[1] Oliver Jennrich. "LISA technology and instrumentation". In: *Classical and Quantum Gravity* 26.15 (2009), p. 153001. URL: http://stacks.iop.org/0264-9381/26/i=15/a=153001.

[2] Jeff Livas. *SGO Mid: A LISA-Like Concept for the SGO at a Middle Cost Point*. URL: http://pcos.gsfc.nasa.gov/studies/gravwave/gravitational-wave-mission-rfis.php.

[3] Aaron Spector and Guido Mueller. "Back-reflection from a Cassegrain telescope for space-based interferometric gravitational-wave detectors". In: *Classical and Quantum Gravity* 29.20 (2012), p. 205005. URL: http://stacks.iop.org/0264-9381/29/i=20/a=205005.

[4] H Yamamoto et al. "Simulation tools for future interferometers". In: *Journal of Physics: Conference Series* 32.1 (2006), p. 398. URL: http://stacks.iop.org/1742-6596/32/i=1/a=061.

[5] Aaron Spector and Guido Mueller. "Experimental investigation of the back-reflection from an on-axis telescope for space-based interferometric gravitational-wave detectors". In: *to be published* (2014).

Initial progress with numerical modelling of scattered light in a candidate eLISA telescope

Shannon R. Sankar[1] **and Jeffrey C. Livas**[2]
[1] Universities Space Research Association
[2] National Aeronautics and Space Administration, Goddard Space Flight Center

E-mail: shannon.r.sankar@nasa.gov

Abstract. We report a numerical analysis of the scattered light from within the field of view of an eLISA-like full-duplex telescope and, by comparison to a nominal allowable scattered light specification, we place constraints on the permissible mirror surface roughness and contamination levels of the mirrors. Our analysis was performed with commercially available stray light software, typically used in non-interferometric imaging telescopes and we do not include polarization or coherent effects. This work therefore represents the early steps towards a more complete understanding of the scattered light budget.

Introduction

Designs of interferometric space-based gravitational wave observatories continue to evolve. However all designs require exchange of laser beams between vastly separated spacecraft. In order to achieve this optical link, a telescopic element is required to both expand the beam going towards the far spacecraft and efficiently collect the beam sent from that far spacecraft. Mass and volume requirements dictate that the same telescope be used for these two purposes. The nominal eLISA telescope design is operated in full-duplex configuration - transmitting a 1 W beam, while simultaneously receiving approximately 100 pW, where the power difference between the outgoing and incoming beams is due largely to diffraction over the inter-spacecraft distance. This disparity of beam powers means that backscatter from the high-power outgoing beam is the chief contributor to the telescope scattered light budget. If this scattered light possesses large spatial overlap with the local oscillator at the detector, any phase fluctuations from the scattered light can couple to the interferometric displacement measurement. Ultimately, this phase noise must be kept small, to avoid reducing the observatory sensitivity and masking the signal from a gravitational wave.

Using this argument of minimizing phase noise, Spector and Mueller [1] have placed constraints on the permissible scatter which, when rescaled to eLISA specifications, implies the permissible scattered light power seen by the detector must be below approximately 100 pW. It is very difficult to achieve this level of scatter in an on-axis telescope design due to a strong narcissus reflection from the secondary mirror. Off-axis designs have tilted optical elements, eliminating the direct narcissus. However, the non-zero surface roughness and surface contamination still couple the outgoing beam into the received mode. It is this aspect of the telescope that we have studied numerically. Our choice of scattered light modelling software is FRED by Photon Engineering LLC.

[1] To whom any correspondence should be addressed.

Journal of Physics: Conference Series **610** (2015) 012031 doi:10.1088/1742-6596/610/1/012031

Reference design

The design of the telescope plays a key role in determining the scattered light seen by the detector. For this work, we have studied a particular off-axis reflective design consisting of four curved mirrors, referred to as M1 (primary), M2 (secondary), M3 (tertiary) and M4 (quaternary). Figure 1 shows the optical layout. The details of the design formulation have been presented elsewhere [2], [3]. The telescope expands a 5 mm outgoing beam to a 200 mm beam, giving an overall afocal magnification of 40, and also possesses an internal focus located between M2 and M3.

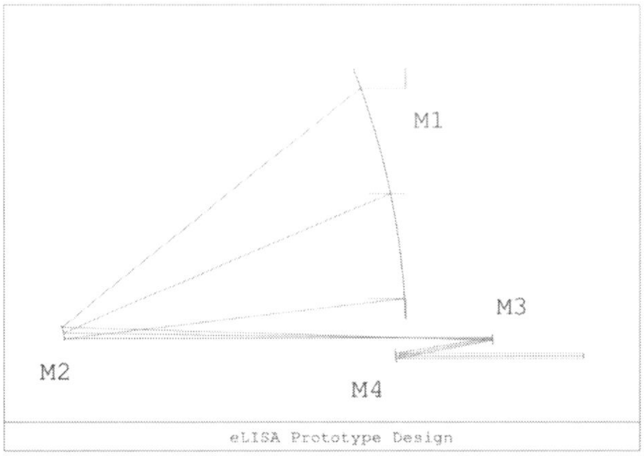

Figure 1. Scattered light was analysed using a 4-mirror, off-axis, afocal design template shown above.

Methodology for stray light calculation

In simulating the system, we chose to ignore scatter from structure or non-optical surfaces outside the field of view as separate estimates place those contributions several orders of magnitude below the scatter from the mirrors. The total stray light is then due to contributions from both particulate contamination and random residual surface errors of the mirrors. These two individual sources of scatter are addressed by the Mie scatter model and the Harvey-Shack scatter model, respectively. Both types of scatter are well implemented in most scattered light simulation software. FRED uses efficient algorithms for the calculation of the total scatter and also includes importance sampling, which greatly reduces the processing time and number of traced rays necessary for convergence of the scattered power.

Surface roughness properties ranging from 0.1 to 15 nm were applied to each mirror in turn. Similarly, particulate levels were varied on each mirror surface, simulating different cleanliness levels from CL100 to CL300 as defined by [4]. By iteratively changing these parameters on each surface and analysing the scattered power on the detector, it was possible to deduce the most relaxed parameters for which the specification of scattered power is still met.

We note that a key concept in the telescope scatted light model is the inclusion of a spatial and angular window, physically provided mainly by the overlap integral of the beams on the detector. To model this aspect of the system in a ray-tracing sense, we purposely restricted our interest to only include rays which impinge on the detector area within a small ±8 μrad circular field of view. This necessary restriction was implemented by explicitly coding an angular filter for the direction cosines of the incident rays, and rejecting scatter from outside this cone. We stress that this angular ray filter does not include coherent effects or interference, which may alter the results. More work will be done in the future to better simulate the scatter using physical optics propagation and an explicit beam overlap. Polarization information, and the effect of polarization optics, was not explicitly included in this analysis. Furthermore, the scattered power from optical elements between the detector and M4 was not included, though the methodology is scalable enough that these optics may be included in future simulations. It should be noted that such optics may contribute scattered power but are expected to produce very little phase noise in the interferometric measurement, if they are located on the ultra-stable optical bench. Finally, the coherence effects among the scattered field, the signal field and the local oscillator field were not included, except in the formulation of the permissible scatter [1] as noted in the Introduction section.

Initial Results

The initial results from the telescope scatter simulation in FRED indicate that a sizeable fraction of the scatter at the detector for this design is due to M3 and M4. The M4 contribution is zeroth-order scattering, being generated by the quaternary mirror and going directly to the simulated detector. However, for this design, the scatter contribution from M3 is largest. Our simulations imply that this property is due to a focusing effect from M4 - the quaternary mirror position, orientation and curvature are such that it efficiently reflects the low-angle scatter generated by the tertiary mirror onto the detector - an example of first order scatter. Scatter from the primary and secondary mirrors is also directed to the detector in this manner, but the overall relay efficiency of scatter from those optics to the detector is smaller than it is for M3 in this design. Table 1 presents the total scattered power from each optic, along with the surface roughness and contamination levels used in the simulation.

Table 1. Scattered light on the detector due to each mirror with optimized specifications.

Mirror	Cleanliness Level[a]	Surface Roughness (nm)	Scattered Power on Detector[b] (W)
M1	CL 300	1.5	< 5e-14
M2	CL 200	1.5	2e-12
M3	CL 200	0.5	5e-11
M4	CL 200	0.5	9e-12

[a] Contamination levels are listed in accordance with [4]. Future model refinement will include more realistic particle distributions [5].
[b] Scattered power simulations are typically accurate to within a small factor of approximately 2 or 3.

It is important to note that the modelling shows that the majority of the scatter at this level is due to the particulate contamination, not the direct surface roughness of the optical surface. Deposited film contamination and/or non-volatile residue, such as hydrocarbons or water ice, have not been modelled here.

Outlook and Future Work

Rigorous testing in an experimental setting is required to address effects which are not modelled (such as polarization and coherence) and also to verify the effectiveness of the current model in analysing other contributions. To accomplish this goal, a full scale prototype telescope is being constructed. A direct test of the stray light will be performed and we anticipate a full comparison to the scattered light model. A smaller scale testbed will be used in the meantime to verify the model as it applies to the M3-M4 pair. We expect that the model will be refined in the future as we perform experimental tests of the scattered light power at the detector.

Acknowledgments
The authors wish to acknowledge the technical assistance provided by the Goddard Space Flight Center Optics Branch, particularly Lenward Seals, Garrett West and Joseph Howard. This work was supported by NASA Strategic Astrophysics Technology award 11-SAT11-0027.

[1] Spector A, Mueller G, 2012 *Class. Quantum Grav.* 29 205005
[2] Livas J, Sankar S, in preparation, to be submitted to the *Proceedings of the 10th LISA Symposium in Gainesville FL 2014*
[3] Livas J, Arsenovic P, Crow J, Hill P, Howard J, Seals L, Shiri S, 2013 *Opt. Eng.* 52(9). 091811.
[4] Military Standard MIL-1246C Particulate contamination
[5] Peterson R, Magallanes P, Rock D, 2002 *Proc. SPIE* 4774, Optical System Contamination: Effects, Measurements, and Control VII, 79

Journal of Physics: Conference Series **610** (2015) 012032 doi:10.1088/1742-6596/610/1/012032

Sub-system mechanical design for an eLISA optical bench

Michael Perreur-Lloyd[1], Karsten Danzmann[3], Ewan D. Fitzsimons[2], Gerhard Heinzel[3], Jan-Simon Hennig[3], Christian J. Killow[1], Maike Lieser[3], David I. Robertson[1], Sönke Schuster[3], Alasdair Taylor[1], Michael Tröbs[3], Harry Ward[1], Dennis Weise[2]

[1] SUPA, Institute for Gravitational Research, School of Physics and Astronomy, University of Glasgow, Glasgow G12 8QQ, UK

[2] Airbus Defence and Space – Space Systems, D-88039 Friedrichshafen, Germany

[3] Albert Einstein Institute, Callinstrasse 38, 30167 Hannover, Germany

E-mail: michael.perreur-lloyd@glasgow.ac.uk

Abstract. We present the design and development status of the opto-mechanical sub-systems that will be used in an experimental demonstration of imaging systems for eLISA. An optical bench test bed design incorporates a Zerodur® baseplate with lenses, photodetectors, and other opto-mechanics that must be both adjustable - with an accuracy of a few micrometers - and stable over a 0 to 40°C temperature range. The alignment of a multi-lens imaging system and the characterisation of the system in multiple degrees of freedom is particularly challenging. We describe the mechanical design of the precision mechanisms, including thermally stable flexure-based optical mounts and complex multi-lens, multi-axis adjuster mechanisms, and update on the integration of the mechanisms on the optical bench.

1. Introduction

eLISA optical metrology involves interfering static optical beams with optical beams that are changing direction over time. This can lead to an unwanted coupling of beam movement to apparent longitudinal signal, so called tilt-to-piston coupling. This coupling is one of the largest sources of error in the metrology error budget for eLISA. The method proposed for minimising the effect of tilt-to-piston coupling in the eLISA baseline design is the use of imaging systems placed directly in front of the detectors.

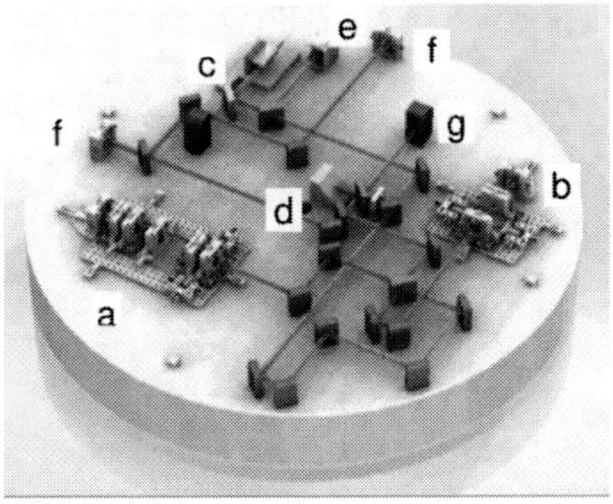

Figure 1. CAD representation of the Ø580 mm 'minimal' optical bench (OB) test bed with both classical (a) and non-classical optical imaging system (b) opto-mechanics integrated.
Amongst the various mirrors and beamsplitters on the OB, several other optical and opto-mechanical assemblies that make up the interferometer can be seen: fibre injector optical sub-assembly (c), out-of-plane telescope mirror (d), power monitor and alignment photodiodes (e, f), beam dumps (g).

To experimentally verify the performance of different optical imaging systems in a representative way a test bed has been designed and is under construction. On the test bed, the performance of the imaging systems will be tested by measuring the tilt-to-piston coupling coefficients and by investigating the alignment sensitivity of the overall imaging systems and of its sub-units. Two candidate optical designs have been developed, both of which are to be tested as part of the study. To enable the two phases of testing and investigation, precision opto-mechanical mounts for these two candidate optical designs are required. These adjustable multi-lens mechanisms must be designed to interface with the polished Ø580 x 80 mm thick Zerodur® optical bench (OB) test bed yet they must be fully removable such that the candidate designs may be interchanged.

This paper reviews the design of the opto-mechanical mechanisms and their use on the OB test bed, a representation of which can be seen in Figure 1.

2. Imaging system optical design

The motivations behind the two optical design approaches are described in [1] although the exact optical designs have been altered in line with a re-scoping of the original study. The primary performance requirement for the imaging system is to reduce the tilt-to-piston coupling to less than 25μm/rad over the entire field of view. One of the designs to be tested is a four lens optical design based upon a classical pupil relay system. This design can be seen in Figure 2. All four lenses are custom manufactured from fused silica.

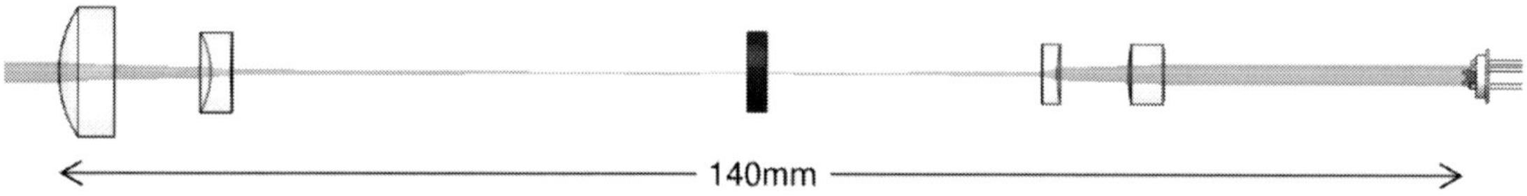

Figure 2. Four lens classical optics approach. The component in the centre of the schematic is a field stop. The component at the right hand side is a quadrant photodetector (QPD).

For the second optical design, the approach was to use Gaussian beam properties as opposed to classical ray tracing methods and to incorporate commercially available fused silica optics. This more compact design, shown in Figure 3, will also be tested as part of the investigations.

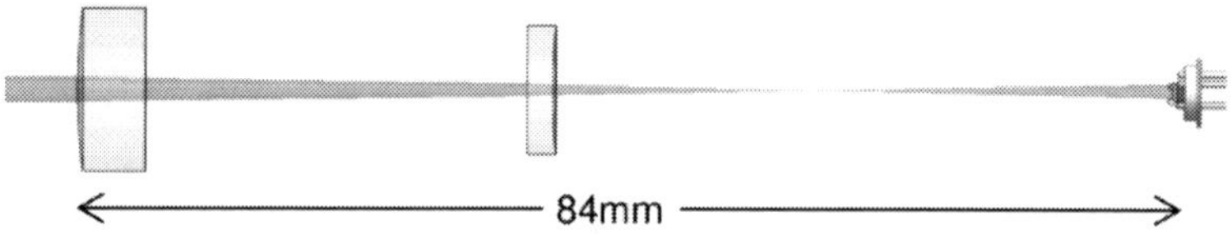

Figure 3. Two lens non-classical optics approach. The component at the right hand side is a QPD.

3. Mechanical design requirements

In each imaging system the lenses must be aligned to one another to accurately form an image on the photodetector. To do so, each lens must be able to be manipulated in five degrees of freedom (5DoF). It must also be possible to align the imaging systems, as a whole, to the photo detector on the OB. The assembled unit must then remain static and be thermally stable throughout the interferometric imaging performance testing. The alignment specifications can be seen in Table 1.

Table 1. Typical alignment adjustment specifications for each lens or lens pair.

	Specification		Tolerance
Alignment specifications	De-centre	X, Y	+/- 20 μm
	Distance tolerance to next lens	Z	+/- 50 μm
	Lens centring		+/- 3' (or ~1 mrad)
	Lens tilt	pitch/yaw	+/- 3' (or ~1 mrad)

Journal of Physics: Conference Series **610** (2015) 012032 doi:10.1088/1742-6596/610/1/012032

Thereafter, for the characterisation activity, the mounting hardware for each system must include a wider range of adjustment to investigate the effect of systematic offsets on individual mounts, lens pairs and the whole imaging system. Table 2 summarises these characterisation requirements and the resolution required by the adjustment mechanisms.

Table 2. Typical alignment characterisation specifications for each lens or lens pair.

	Specification		Tolerance
Adjustment specifications for imaging system characterisation	De-centre	X, Y	+/- 60 μm
	Distance tolerance to next lens	Z	+/- 200 μm
	Lens tilt	pitch/yaw	+/- 10' (or ~3 mrad)
Resolution of adjustment	Linear	X, Y, Z	micrometer
	Angular	pitch/yaw	<arc-minute (~100s of μrad)

4. Mechanical design detail

Both imaging system mechanical designs are made up of three equivalent modules: the super-baseplate, carrying all lens mounts; the lens pair assemblies and the QPD mount. The larger and more-complex, four-lens design, shown in Figure 4, has two of the lens pair assemblies and an additional field stop mount.

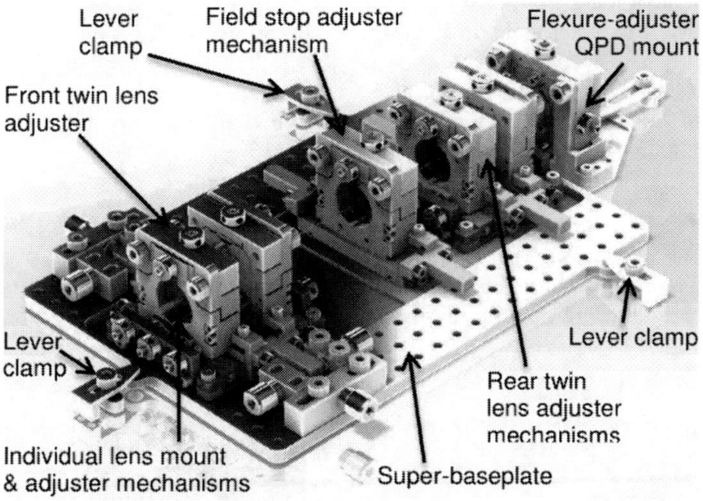

Figure 4. Four lens imaging system design. The super-baseplate is 151 x 112 mm. The red beam represents the laser light.

Techniques of ultra-precision mechanism design [2] are widely adopted throughout the imaging system design including kinematic mounting techniques for stable positioning and flexures for fine adjustment and positional control. Ultra-fine threaded screws with a thread pitch of 0.25 mm are used to provide precision adjustment of the lenses and QPD, and when employed in combination with flexures they provide the required levels of precision adjustment and clamping. The imaging system assembly and its sub-assemblies are made stable by the use of kinematic mounts: three hemispherical features on the underside of the assemblies bear against a flat surface in the interfacing part.

4.1. The super-baseplate

The super-baseplate is a precision machined titanium plate kinematically mounted to the Zerodur® baseplate by way of three underside-mounted ball-bearings. Three lever-arm clamps that attach to one of three custom titanium nuts epoxied to the Zerodur® provide the clamping. At either side of the lens mounts on the super-baseplate there are M3 mounting holes to allow the flexible attachment and removal of lateral and transverse adjusters for the twin lens and field stop assemblies.

Journal of Physics: Conference Series **610** (2015) 012032 doi:10.1088/1742-6596/610/1/012032

4.2. Lens pair assembly

The lens pair assembly, see Figure 5, consists of two individual lens holders and their 5DoF adjuster mechanisms. These assemblies are attached to a titanium baseplate which is in turn mounted to the super-baseplate via three hemispherical points on the underside. The symmetrically designed lens pair assembly is locked in place on the super-baseplate at a single point through the centre of the assembly.

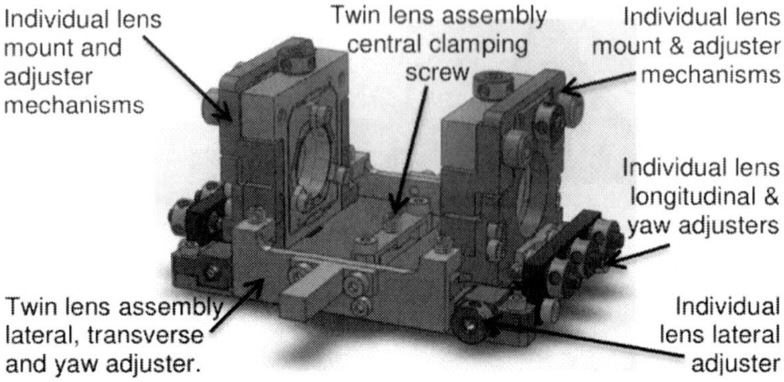

Figure 5. Isometric view of a lens pair assembly from the non-classical imaging system design.

4.3. Individual Lens Holders

The individual lens holders are constructed from an aluminium and titanium architecture. Lenses with diameters ranging from 6mm to 13mm are held in a tailored aluminium mounts that attach to a monolithic aluminium flexure mechanism which, as can be seen in Figure 6, are then mounted to a titanium frame. Thermal stability of the lens holder, and also of the overall assembly comes through careful design of the aluminium mount and the titanium frame such that they expand in opposite directions from one another results in a stable lens centre.

4.3.1. Vertical and pitch adjustment

The aluminium flexure mechanism in combination with an ultra-fine threaded screws allow for vertical adjustment of the lens in the mount. For pitch adjustment of the lens another precision adjustment screw, bearing against the aluminium lens mount but affixed to the titanium frame, utilizes a flexure hinge in the titanium frame to provide a pitch mechanism in the mount. These mechanisms can be seen in Figure 6 and Figure 7.

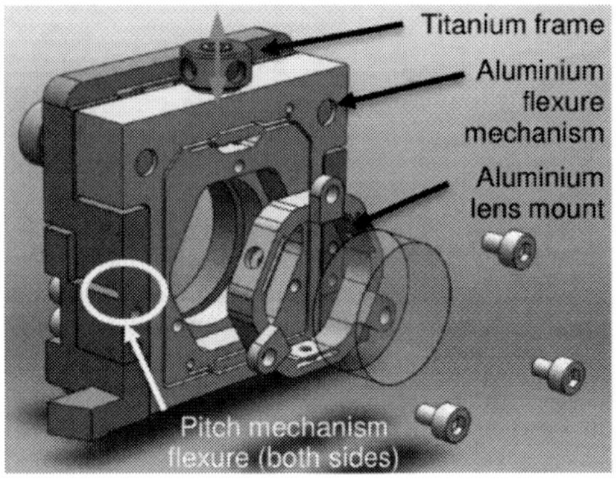

Figure 6. Exploded view of individual lens holder.

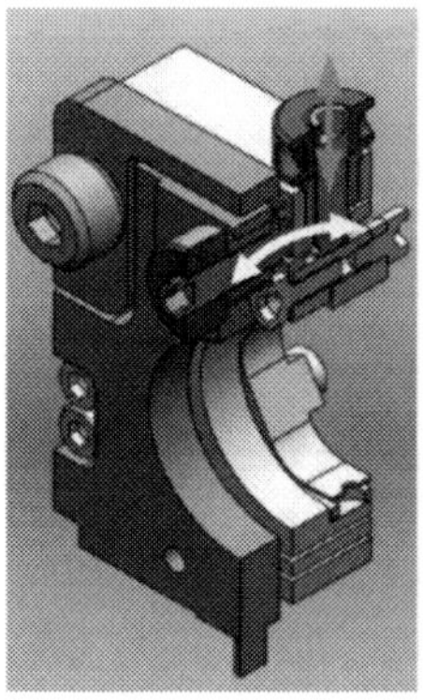

Figure 7. Cross-section view of individual lens holder showing vertical and pitch mechanism.

4.3.2. Lateral adjustment

The individual lens holder may be adjusted laterally using a titanium flexure mechanism, Figure 8 and Figure 9, that sits beneath each lens holder and is driven by fine pitch screws. The frame of the lens holder interfaces with this lateral adjuster through a 2mm pin at the base of the mount.

4.3.3. Longitudinal and yaw adjustment

Longitudinal and yaw adjustment is made using three fine pitch screws with customised screw tips as shown in Figure 10. The central screw performs the longitudinal movement but, thanks to its bevelled end, can permit rotation via the two side pusher screws.

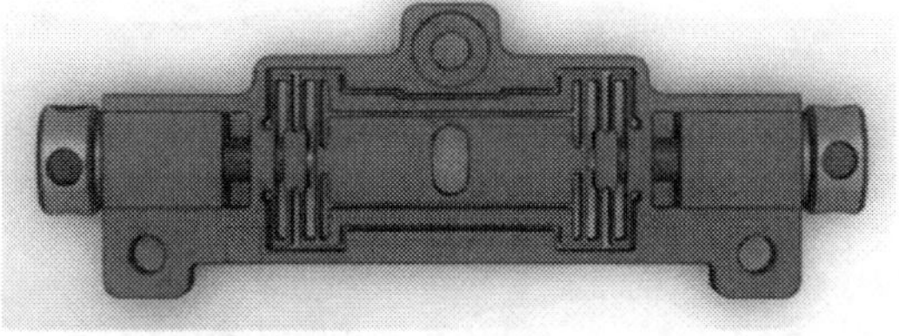

Figure 8. Top view of individual lens holder lateral adjustment mechanism with adjuster screws and locking screws attached.

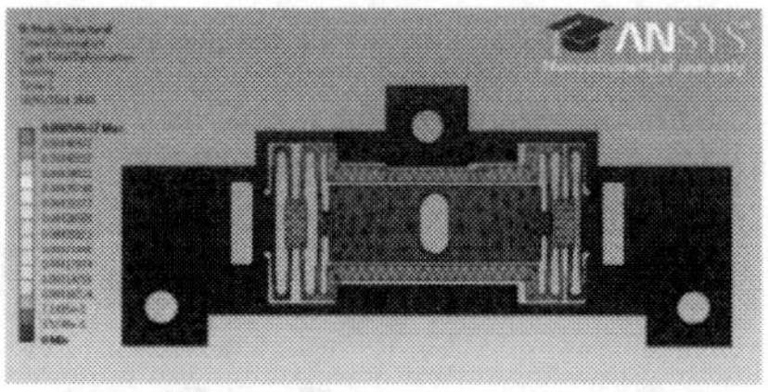

Figure 9. The individual lens holder lateral adjustment mechanism in its deformed state using ANSYS Finite Element Modeller.

4.4. QPD mount

The QPD mount, Figure 11, differs from to the aforementioned individual lens mount design in that the aluminium flexure mechanism is mounted in a rigid titanium frame with no pitch adjustment. Further, in addition to a vertical flexure mechanism, the QPD mount also incorporates a lateral flexure mechanism within the aluminium part. The QPD itself is glued to a MACOR interface collar, for electrical and thermal isolation, and this is attached to the rear of the aluminium flexure mechanism. The base of the QPD mount is titanium which, like the super-baseplates, sits on three hemispherical points and is lever-clamped to the Zerodur® OB.

Figure 10. Longitudinal and yaw adjustment mechanisms of the individual lens holder.

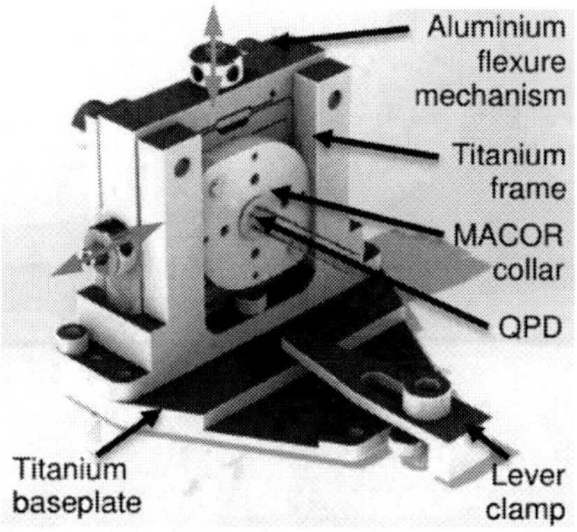

Figure 11. Rear isometric view of the QPD mount.

5. Conclusions

The design of high precision opto-mechanics to be used in the investigations of imaging systems and photodetectors for eLISA are detailed in this paper. By using an array of fine adjustment mechanisms incorporating monolithic flexures, ultra-fine precision screws and kinematic mounting techniques, the required alignment, adjustment precision, and interchangeability has been achieved in the design. These imaging system designs are currently under manufacture and shall be integrated on to an optical bench test bed for investigation. Rather than the previously planned elegant breadboard, the re-scoped study will see the construction of a 'minimal' optical bench with features necessary to the testing the

Journal of Physics: Conference Series **610** (2015) 012032 doi:10.1088/1742-6596/610/1/012032

imaging systems as a science interferometer. As part of the same re-scoped study, a telescope simulator is being developed that will generate a representative Rx beam and local oscillator beam, and include a reference interferometer to combine these beams for use on the OB.

References

[1] d'Arcio et al 2012 An Elegant Breadboard of the Optical Bench for eLISA/NGO *ICSO 2012* http://www.icsoproceedings.org/

[2] Smith, S.T., and Chetwynd, D.G. 1992 *Foundations of ultra precision mechanism design* CRC Press, Developments in nanotechnology Volume 2, Chapters 3 and 4.

Acknowledgments

This work was funded within the ESA core technology program "Optical Bench Development for LISA", ESTEC contract no. 22331/09/NL/HB. University of Glasgow acknowledges the support from the UK Space Agency (UKSA) and Scottish Universities Physics Alliance (SUPA). AEI Hannover acknowledges the support by the German Aerospace Center (DLR) (50OQ0601, 50OQ1301).

Development of a US Gravitational Wave Laser System for LISA

J Camp
NASA Goddard Space Flight Center, Greenbelt, Maryland 20771, USA

K Numata
University of Maryland, Department of Astronomy/CRESST, College Park, Maryland 20741, USA

E-mail: Jordan.B.Camp@nasa.gov

Abstract. A highly stable and robust laser system is a key component of the space-based LISA mission architecture. We will describe our plans to demonstrate a TRL 5 LISA laser system at Goddard Space Flight Center by 2016. The laser system includes a low-noise oscillator followed by a power amplifier. The oscillator is a low-mass, compact external cavity laser, consisting of a semiconductor laser coupled to an optical cavity, built by the laser vendor Redfern Integrated Optics. The amplifier is a diode-pumped Yb fiber with 2 W output, built at Goddard. We show noise and reliability data for the full laser system, and describe our plans to reach TRL 5 by 2016.

1. Introduction

The NASA's Goddard Space Flight Center (GSFC) has been producing space-borne lasers for various missions (See, for example, [1, 2]). These space instruments were based on diode-pumped solid-state Nd:YAG lasers. For future earth- and space-science missions, including the gravitational-wave missions like LISA, we are making transition to a new generation of space-based lasers. These lasers use the emerging telecom laser technology, such as fiber laser/amplifier, waveguide devices, and semiconductor lasers. These components naturally fit into the precision laser systems for the interferometric missions because of their high mechanical robustness, excellent reliability, compact form factor, and high wall-plug efficiency. We are targeting to provide TRL (technology readiness level) 5 laser for the LISA-like missions by 2016. We have invested approximately $3.5 million over the past six years on laser development for the LISA-type missions. Our research has included development of 1064 nm Planar-Waveguide external cavity laser (PW-ECL), PW-ECL reliability tests, and amplifier development, including its noise measurements and reliability studies. These are funded through multiple funding sources, such as NASA's SBIR (Small Business Innovative Research) and SAT (Strategic Astrophysics Technology) awards.

For the LISA-like missions, we are pursuing an all-fiber/waveguide space laser solution based on the MOFA (master oscillator fiber amplifier) configuration, which is a waveguide-based oscillator followed by a preamplifier and a power amplifier (Fig. 1).

Journal of Physics: Conference Series **610** (2015) 012033 doi:10.1088/1742-6596/610/1/012033

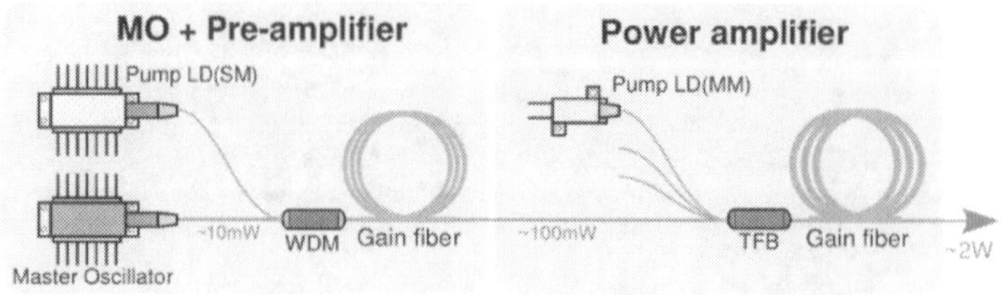

Figure 1. Concept of the MOFA configuration. Optical isolators, modulators, and redundant LDs are not shown. SM: single-mode; MM: multi-mode.

2. Master Oscillator

We have developed a fiber ring laser [3] and a fiber DBR laser for space interferometry. Although these lasers perform like the non-planar ring oscillator (NPRO) at low (<10 kHz) frequency, they have larger relaxation oscillation peak, around 1 MHz, which could affect the heterodyne interferometry typically operated near this frequency. Therefore, we shifted our focus to the development of the planar-waveguide external-cavity diode laser (PW-ECL). The Telecordia-qualified PW-ECL, built by Redfern Integrated Optics, offers advantages over solid-state lasers, including simpler design, more compact size, lower mass, and less consumption of electrical power. The narrow reflection peak of the Bragg reflector in the planar lightwave circuit (PLC) enables stable, low-noise, single-mode lasing at a selected wavelength. Originally, the PW-ECL was only available at the telecom C-band (1528~1565 nm). Based on the experience on the C-band PW-ECL, we have successfully built 1064 nm version of PW-ECL by changing the gain chip and the PLC design. The maximum output power of PW-ECL is 15 mW.

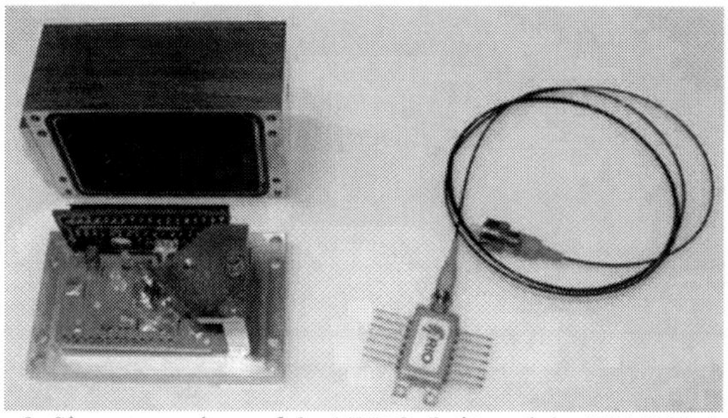

Figure 2. Size comparison of the NPRO (left) and the PW-ECL (right).

Figure 2 shows a comparison of its size with the non-planar ring oscillator (NPRO). The PW-ECL's package is much more compact than that of the NPRO, in which a strong magnet limits the size. We based our choice of the PW-ECL on an investigation of number of lasers [4]. The ECL can be locked to either a hyperfine absorption line of acetylene at 1542 nm [4], that of iodine at frequency doubled 1064 nm [5], or an optical cavity of finesse 10^5 [6] with low-residual frequency noise.

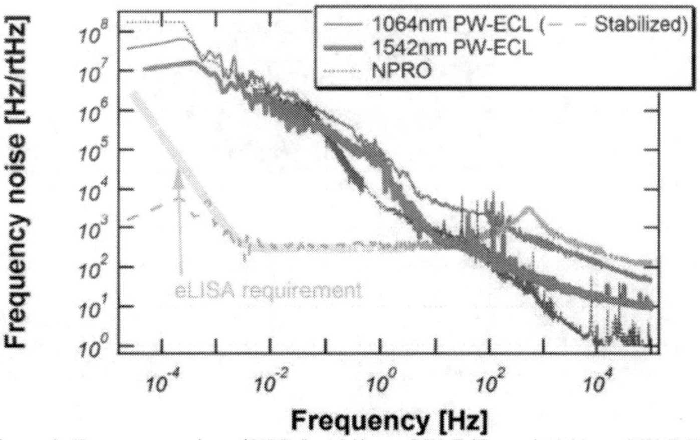

Figure 3. Frequency noise of NPRO, 1542 nm PW-ECL, and 1064 nm PW-ECL.

Figure 3 compares frequency noise of NPRO, 1542 nm PW-ECL, and 1064 nm PW-ECL. Below 0.1 Hz, the PW-ECLs make frequency noise comparable to the NPRO. Above 10 Hz, the noise drops by $1/\sqrt{f}$ and $1/f$ for the PW-ECL and NPRO, respectively; therefore their difference becomes larger at a higher frequency. At 10 kHz, the PW-ECL produces approximately 5 and 40 times higher frequency noise than the NPRO at 1542 nm and 1064 nm, respectively. When stabilized to the iodine reference, after amplified by the two stage Yb fiber amplifier and frequency-doubled by a MgO:PPLN crystal, the 1064 nm PW-ECL achieved the frequency noise requirement of eLISA [7] at low frequency. However, the high frequency noise at high Fourier frequency may still be problematic from the viewpoint of phase locking. We estimate that the RMS phase noise must be suppressed by a factor of about 100. Therefore, we are trying to improve the frequency noise performance by optimizing the gain chip and PLC designs. We are also looking into a possibility to implement a 100 MHz bandwidth intra-cavity phase modulation section within the gain chip, so that we can use it as a widely-tunable, fast frequency tuning actuator. By the time these implementations are completed, we will try to use external acousto-optic modulator in order to suppress the excess noise.

We made a detailed study of the mechanical, thermal, and radiation robustness of the PW-ECL, and found it to be qualified for use in space. Lucent Government Solutions (LGS) planned and oversaw the tests, which involved vacuum thermal cycling, hermiticity, radiation, and accelerating aging [8]. For these reasons, the 1550-nm PW-ECL was adopted as the metrology laser for the OpTIIX mission [9] on the International Space Station (ISS). We will revisit some of these reliability studies for the 1064 nm PW-ECL in FY15. We think the risk of failing the reliability test is low, since it is using the same packaging as the 1550 nm PW-ECL and the gain-chip vendor (Eagleyard) data indicates high reliability.

3. Fiber amplifiers

Figure 4. Package design of the seed laser and the pre-amplifier system.

The 15 mW maximum output power of PW-ECL may not be sufficient for a seed laser of the 2-W power amplifier in LISA-type missions, when the optical loss in phase modulator and instability caused by too high gain in the power amplifier are taken into account. Therefore, in order to have a power margin, and to have flexibility in changing mission requirements, we are developing fiber pre-amplifiers with output power of ~100 mW at 1064 nm. Figure 4 shows the design of the seed laser and the pre-amplifier system. The laser unit will house two PW-ECLs and two pump laser diodes for redundancy, as well as their control circuits.

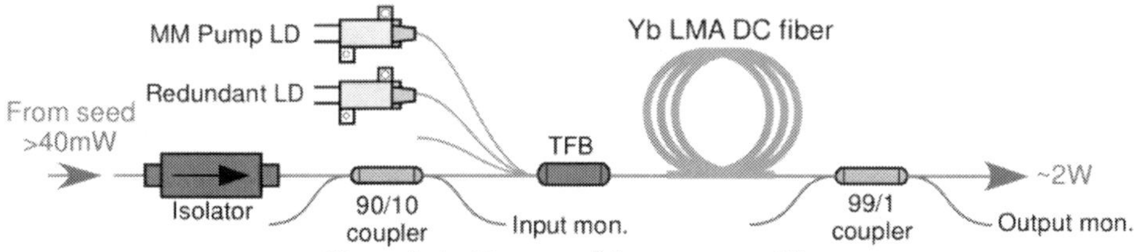

Figure 5. Architecture of the power amplifier

We worked with LGS to develop a Yb fiber power amplifier for the LISA-type missions. After trade studies and simulations, we adopted the 1064 nm amplifier architecture shown in Fig. 5. A 10-μm core, double-clad (DC), large-mode area (LMA) fiber was selected as a gain media to elevate the SBS (stimulated Brillouin scattering) threshold. It is forward, clad-pumped through a tapered fiber bundle (TFB). The forward-pumping design minimizes potential sources of feedback (e.g. from the TFB). Such feedback could result in lasing and in catastrophic damage, which we found to occur in our prototype backward-pumped amplifiers.

Using a commercial NPRO as a seed source, we evaluated the noise performance of the power amplifier, such as the frequency noise, the differential phase noise, and the intensity noise [10]. We also performed vacuum thermal cycling and gain fiber irradiation in order to look at the amplifier reliability. We have learned that the amplifier performance is not degraded by thermal cycling, but is degraded by gamma irradiation if improper gain fiber is used. The combined performance of the PW-ECL, the pre-amplifier, and the power amplifier has been in part proven in the frequency-stabilization experiment using iodine [5] mentioned above.

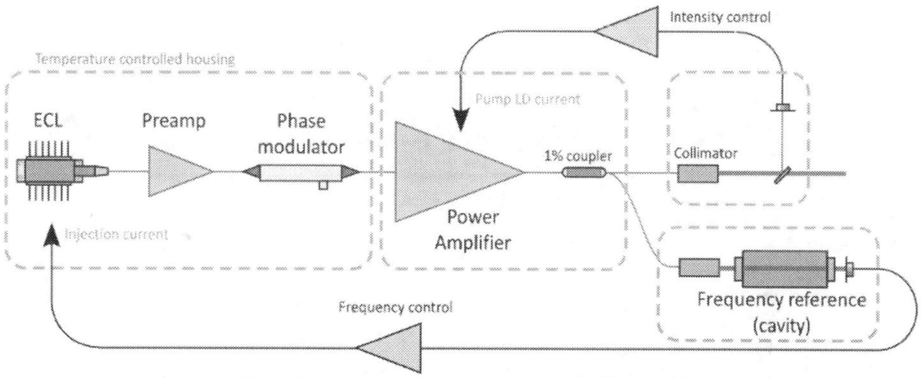

Figure 6. Planned system testing setup for FY15 and FY16.

By 2016, we will rebuild amplifiers using more realistic components, combine them with the low-noise 1064 nm PW-ECL, and perform system and environmental tests (Fig. 6). The system test will include locking the power amplifier output to an optical reference cavity, and phase-locking two laser systems, in order to show applicability of our system to the LISA-like missions.

4. Summary

NASA/GSFC has been involved in research on spaceborne lasers since the 1990s. Taking advantages of the space laser experience and the emerging telecom laser technology, we are developing laser system for the LISA-type missions. Our research has included both master laser and amplifier developments. We plan to finish to lowering PW-ECL's frequency noise and laser system testing by the end of FY15, and to perform reliability testing by the end of FY16 to achieve TRL 5.

References

[1] Afzal R S 1994 *Appl. Opt.* **33** 3184
[2] Yu A W, Novo-Gradac A M, Shaw G B, Li S X, Krebs D C, Ramos-Izquierdo L A, Unger G L and Lukemire A 2008, in Conference on Lasers and Electro-Optics/Quantum Electronics and Laser Science Conference and Photonic Applications Systems Technologies (Optical Society of America), CMQ2
[3] Numata K and Camp J 2012 *Laser Phys. Lett.* **9** 575
[4] Numata K, Camp J, Krainak M A and Stolpner L 2010 *Opt. Express* **18** 22781
[5] Numata K, Alalusi M, Stolpner L, Margaritis G, Camp J and Krainak M 2014 *Opt. Lett.* **39** 2101
[6] Clivati C, Mura A, Calonico D, Levi F, Costanzo G, Calosso C and Godone A 2011 *IEEE Transactions on Ultrasonics, Ferroelectrics and Frequency Control* **58** 2582
[7] https://www.elisascience.org/.
[8] Numata K and Camp J 2012 *J. Phys.: Conf. Ser.* **363** 012054
[9] Postman M, Sparks W B, Liu F, Ess K, Green J, Carpenter K G, Thronson H and Goullioud R, 2012 *Proc. SPIE* **8442** 84421T
[10] Numata K and Camp J 2013 *ASP Conference Series* **467** 239

Journal of Physics: Conference Series **610** (2015) 012034 doi:10.1088/1742-6596/610/1/012034

240 nm UV LEDs for LISA test mass charge control

Taiwo Olatunde, Ryan Shelley, Andrew Chilton, Paul Serra, Giacomo Ciani, Guido Mueller, John Conklin

University of Florida
E-mail: tolatunde@ufl.edu

Abstract. Test Masses inside the LISA Gravitational Reference Sensor must maintain almost pure geodesic motion for gravitational waves to be successfully detected. LISA requires residual test mass accelerations below 3 $fm/s^2/\sqrt{Hz}$ at all frequencies between 0.1 and 3 mHz. One of the well-known noise sources is associated with the charges on the test masses which couple to stray electrical potentials and external electromagnetic fields. LISA Pathfinder will use Hg-discharge lamps emitting mostly around 254 nm to discharge the test masses via photoemission in its 2015/16 flight. A future LISA mission launched around 2030 will likely replace the lamps with newer UV-LEDs. Presented here is a preliminary study of the effectiveness of charge control using latest generation UV-LEDs which produce light at 240 nm with energy above the work function of pure Au. Their lower mass, better power efficiency and small size make them an ideal replacement for Hg lamps.

1. Introduction

The Laser Interferometer Space Antenna (LISA) will act as a space based gravitational wave observatory measuring changes in separations between isolated TMs [1]. LISA requires test mass (TM) accelerations to be suppressed below 3 fm $/s^2/\sqrt{Hz}$ at all frequencies between 0.1 and 3 mHz. Each TM is enclosed in a housing that supports the sensing electrodes and protects the TM from many disturbances including solar wind and high energy particles from galactic and solar rays. However some of them can still penetrate and charge the TMs. Test mass charge couples to imbalanced DC biases on opposing conductors of the housing to produce random walk force noise [1, 2].

By measuring and controlling the motion of two free falling test masses located inside a single satellite, LISA Pathfinder (LPF) aims to test many of the complex technologies useful for LISA such as drag free control, inertial sensors and precision laser interferometry [3]. LISA Pathfinder discharge system exploits the photoelectric effect using UV light produced by Hg lamps following the method demonstrated by Gravity Probe B [4]. A proposed discharge system for LISA could follow the same principle, but replace Hg lamps with UV-LEDs. Feedthroughs would be used to direct light from UV-LEDs, operated at DC, towards either the gold coated TMs or electrode housing surfaces. Alternate charge control architectures would also be possible, such as AC charge control [2] synchronized with the 100 kHz injection voltage. This approach could utilize two UV injectors as in LPF, or perhaps only one injector, which illuminates both the TM and housing.

In theory, UV photons require energy above the work function of Au to liberate electrons. The minimum required illuminating wavelength for a pure Au surface with work function 5.1 eV is 243 nm. However, contamination generally leads to a lowering of the work function and allows higher wavelengths to be used [3]. If no bias voltages are used, then illumination of the TM would cause electron flow i_{TM} from the TM to the housing; light is reflected to the housing which gives rise to unwanted electron flow i_{EH} from the electrode housing to the TM. If on the other hand the housing is illuminated, electrons i_{EH} would flow from the housing to the TM, light is reflected to the TM and gives rise to unwanted electron flow i_{TM} from TM to electrode housing. A positive test mass discharge rate is obtained if $i_{TM} - i_{EH} > 0$. Figure 1 shows the UV LEDs evaluated and their spectral properties.

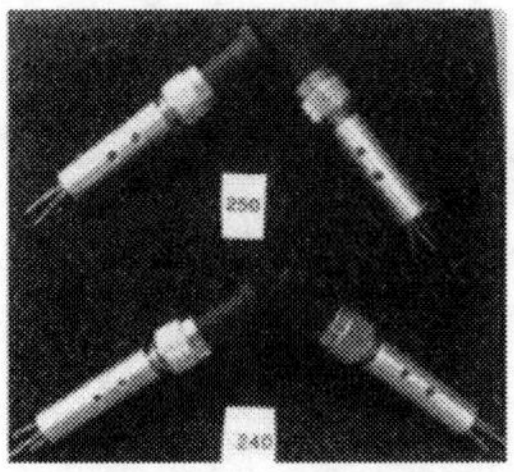

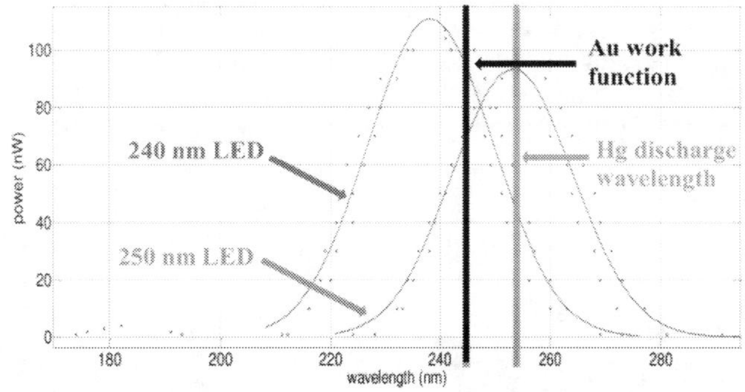

Figure 1. *Left:* 240 and 250 nm UV-LEDs from SETi. *Right:* Gaussian fit to measurements of the UV LEDs' intensity spectrum compared with the work function of Au and the Hg lamp discharge wavelength.

2. Reflectivity

In the LISA Pathfinder GRS design, both the TM and the electrode housing are gold coated. When UV light illuminates the TM directly, some of the light reflects back and is absorbed by the housing. This can create photoelectrons that flow in opposition to the intended direction. Simulations show that for ideal reflection properties, the TM absorbs around 70% of the total light when it is illuminated and about 20% is absorbed by the housing. When illuminating the housing, around 25% is absorbed by the housing and about 15% by the TM [5]. The remaining percentage is lost in gaps in the electrode housing or surfaces that are not useful for charge control. Reflectivity measurements for different wavelength UV LEDs, Au coating techniques, and coating ages are made to understand and bound the effects of reflection on the performance of a UV LED based charge management system.

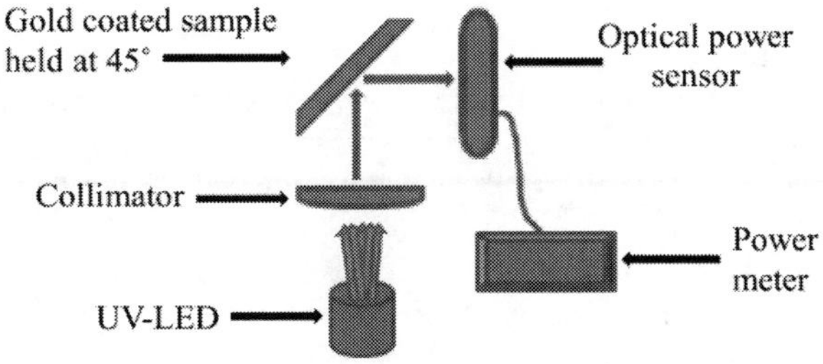

LED wavelength	Reflectivity (%)
240 nm	24.1 ± 0.2
250 nm	25.1 ± 0.1

Figure 2. *left:* Set-up for reflectivity measurements. *Right:* Results of Reflectivity Measurements

The measurement of the reflectivity of the gold coating on the TM and electrode housing is performed by shining collimated light from a single 240 nm or 250 nm UV LED at a gold coated aluminium sample held at 45 deg. The reflected light is detected by using Thorlabs PM100D power meter with S120VC detector head. A bi-convex lens with a focal length of 20 mm is used to collimate the light. Reflectivity is computed as the power measured after reflection from the gold sample, divided by the power previously measured by placing the sensor directly after the collimator. Figure 2 shows the measurement setup and measured reflectivity, which varies only slightly between the 240 nm and 250 nm UV LEDs.

3. Quantum Efficiency (QE) measurements

Quantum efficiency is the number of photoelectrons emitted from the surface per incident UV photon. It is therefore one of the critical properties governing the performance of a charge management system. A dedicated experiment for measuring QE has been assembled and is shown in Figure 3. An Au coated sample is isolated inside a 90 mm diameter hollow integrating sphere using Ultem, which has a resistivity of 10^{17} Ω-m [2]. An ultra-high vacuum UV fiber optic cable with a 600 μm core is used to direct the light from the UV-LED to the sample. The sphere is biased to +9 V and the sample to −9 V. The UV-LED current is then varied over a range of 10-24 mA, using an ILX Lightwave LDX-3200 precision current source. This range of current corresponds to 4-10 μW of UV optical power, and at their maximum current draw of 25 mA, both the 240 nm and 250 nm UV LEDs consume roughly 180 mW of electrical power. The resulting photocurrent is measured using a Keithley 6485 Picoammeter and then converted to number of photoelectrons. A zero degree incident angle is used for the QE measurements at a vacuum pressure of 10^{-5} Torr.

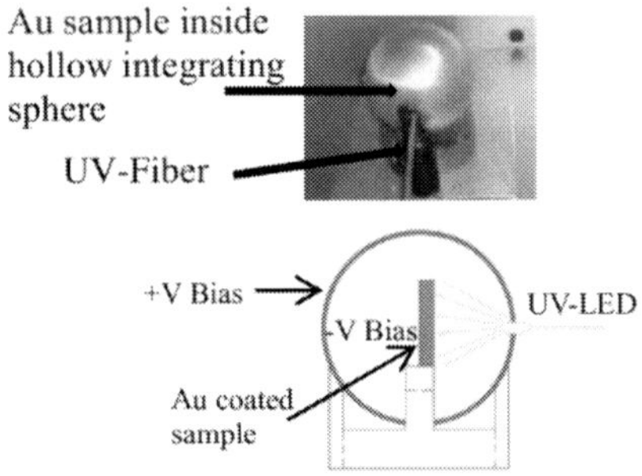

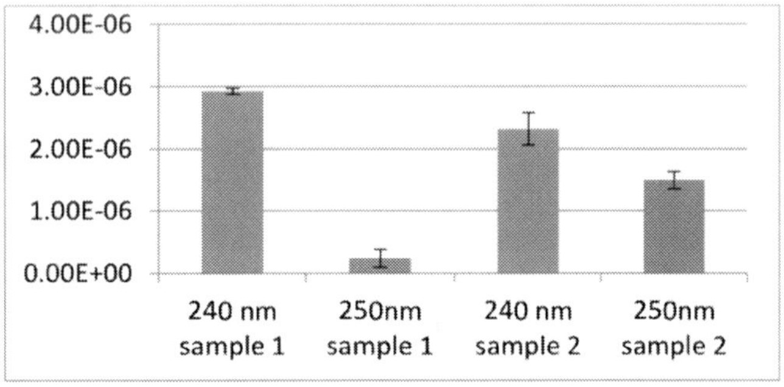

Figure 3. *Left:* QE experimental set-up. *Right*: QE measurement results

The QEs of two different samples were measured. Sample 1 with a gold layer 200 nm thick was coated in March 2013, and Sample 2 with a gold layer 500 nm thick was coated in September 2013 (6 months later). The 240 nm and 250 nm UV-LEDs are similar in terms of power consumption, reflectivity and cost, but as expected and shown in Figure 3, the 240 nm LEDs demonstrate higher QE than the 250 nm LEDs.

4. UV LED fiber coupler and temperature controller

The lifetime of the 240 nm and 250 nm UV-LEDs are reported by SETi to be 1,000 and 1,650 hours respectively, when driven at 20 mA, near their maximum power, corresponding to roughly 10 μW of optical power. Although, during charge control only ~100 nW of UV power is required. The maximum operating temperature is 30 °C at the case of the package and every additional 10 °C cuts total operating hours in half. Proper heat management can dramatically improve the lifetime of these devices. A UV LED fiber optic coupler with built in temperature control is being developed for space applications of the UV LED. The design of the coupler is depicted in Figure 5.

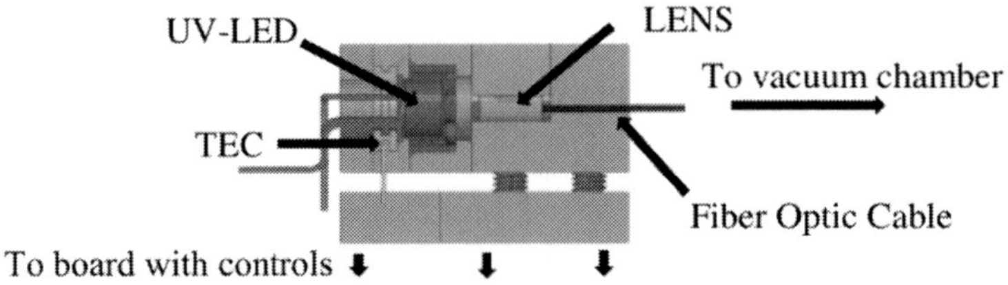

Figure 4. Proposed heat dissipation management design

Journal of Physics: Conference Series **610** (2015) 012034 doi:10.1088/1742-6596/610/1/012034

Light from the UV-LED is coupled to a fiber optic cable by means of a fused silica bi-convex lens with a focal length of 10 mm. A fused silica ferrule and a borosilicate glass sleeve are used to hold the fiber against the lens. The whole assembly is glued using a UV-curable low outgassing optical epoxy. The holder securing the fiber/lens in front of the UV-LED includes a Thermo-Electric Cooler (TEC). A set of driving electronics controls the TEC and applies current to the UV LED. Heat from the TEC due to contact with the UV-LED dissipates through the body of the assembly to a supporting electronics board.

5. Initial TM charging of the UF torsion pendulum.

The UF torsion pendulum is a technology development facility aimed at testing the performance of new inertial sensors like the LISA GRS and related technologies [6]. Both GRS mock-ups in the pendulum, shown in Figure 5, have three UV ports. One port is directed at the TM, the second toward the housing, and the third illuminates both. As an initial test of the effectiveness of the UV LED charge control system, a 240 nm UV LED was used to charge up the TM using the UV port directed at the TM. Figure 5 shows a plot of the pendulum angle as a function of time. At roughly t = 5000 s, the UV LED is turned on and within 5 minutes the pendulum becomes unstable due to positive charge on the TM relative to housing.

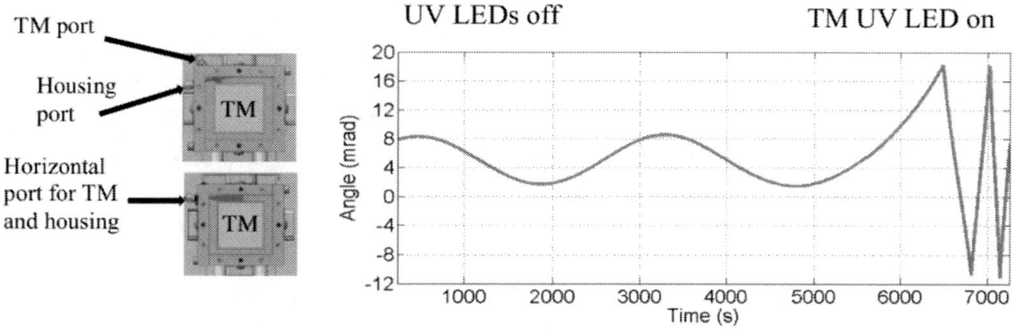

Figure 5. Plot of UV LED building up charge on TM

6. Future Work

UV LEDs are an attractive alternative to Hg lamps as UV sources for the LISA charge management system. They offer relatively high optical power (10 μW) at wavelengths near 240 nm and reduced volume, mass and power with respect to Hg lamps. The 240 nm UV LEDs consistently exhibit higher quantum efficiencies with Au surfaces compared with 250 nm UV LEDs, with little change in power consumption and reflectivity. The 240 nm LEDs have also been used to demonstrate movement of the TM charge using the UF Torsion Pendulum. In the near future, we plan to construct custom UV LED fiber couplers with built in temperature control and perform thermal, vibration and shock tests to verify their compatibility with the space environment. Finally, both DC and AC charge control using various UV light injection directions will be demonstrated using our torsion pendulum facility.

References

[1] Shaul DNA, et al 2008 *Int. J. Mod. Phys. D* **17** 993-1003
[2] K Balakrishnan, et al arXiv:12020585v1
[3] T Ziegler, et al 2009 *J. Phys.: Conf. Ser.154 012009.*
[4] JW Conklin 2008 J. Phys.: Conf. Ser. 140(1) 114002.
[5] Daniel Hollington 2011 Ph.D. Thesis Imperial College London
[6] Andrew Chilton, et al 2014 *J. Mod Phys. this series.*

Journal of Physics: Conference Series **610** (2015) 012035 doi:10.1088/1742-6596/610/1/012035

An electro-optical simulator for eLISA LOT: Lisa On Table

P. Gruning, H. Halloin, J. Brossard, P. Prat, S. Baron, C. Buy, P. Jimenez

Laboratoire APC
10, rue Alice Domon et Léonie Duquet
75205 Paris Cedex 13

E-mail: pierre.gruning@apc.univ-paris7.fr

Abstract. This paper describes the progress on a hardware simulator of eLISA developed at the APC laboratory in Paris, France. It is designed to simulate realistic noise and the appropriate delays resulting from the huge distances the laser has to travel between the eLISA spacecrafts. We present the experimental setup consisting of an optical and electric interferometer. Also, the time delay interferometry noise reduction method and its performance on simulated signals in the simplest transponder case will be described.

1. Introduction

eLISA's capability to detect the tiny variations in space-time due to gravitational waves relies on many parameters such as accurate processing algorithms (TDI), precise feedback loops (arm-locking) and very low noise, extremely high performance instruments (phasemeters). Hardware simulators are necessary to characterize the detection devices, test the numerical models and study the influence of the hardware on the detection algorithms. The LOT (LISA On Table) experiment has been designed to adress this issue. It is an electro-optical simulator composed of both, an optical and an electric interferometer.

The LOT is expected to be a very complete hardware simulator representative of eLISA, offering the additional flexibility to define ad-hoc fluctuations in the frequency, amplitude or phase of the laser and propagation delays. The control electronics also allows to accelerate the simulation, so that one year of data streams could be generated within a few days. The DDS channels are splitted in order to generate both, the electric beat notes and the optical ones using acousto-optical modulators (AOM) to shift the laser frequency and also to inject the frequency noise with the appropriate delays. Finally, the phasemeter recovers the signals from both interferometers.

Furthermore, in it's final version, besides the fact of being able to perform measurements of realistic eLISA signals, the LOT will be extremely useful as a hardware in the loop simulator during eLISA's assembly, integration, testing and verification (AIT/AIV) phases where the interactions between different hardware devices will be studied.

Journal of Physics: Conference Series **610** (2015) 012035 doi:10.1088/1742-6596/610/1/012035

2. Experimental setup

This part describes the main aspects of the LOT, on one hand the electric interferometer including the command-control devices, the signal generation and the phasemeter and on the other hand the optical interferometer.

2.1. Electronic LOT

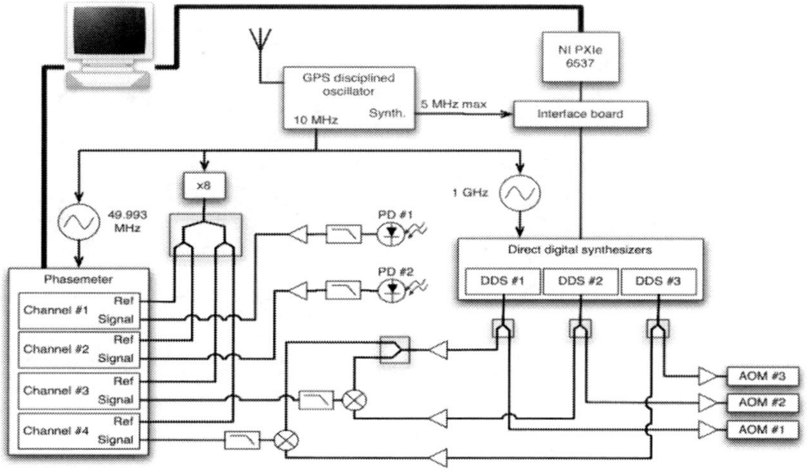

Figure 1: Overwiew of the electric part

Computer controlled DDS (Direct Digitial Synthesizer, model Agilent AD9912) are used to generate the analog signals from the data streams generated by NI PXIe driven by a labview program. The DDS are able to generate RF signal up to 400 MHz with an accuracy of 3.6 mHz and the possibility to adjust the phase of the signal with a precision of 0.38 mrad. The electronic clocks are derived from a 10 MHz high stability, GPS disciplined oscillator to reduce the differential jitter noise.

As for eLISA, channel 1 (i.e. the local arm) is mixed respectively with channel 2 and 3 (i.e. the distant arms) in order to simulate the two local signals. These two data streams are recorded by the phasemeter on channel 3 and 4, the other channels being used for the optical signals recovered by the photodiodes (PD). The phasemeter is a prototype developed by the Albert Einstein Institut in Hanover, Germany [1] which transmits the output data at a rate of about 23.8 Hz. Both, the DDS and the phasemeter are controlled by a labview program with predefined mathematics models (computed by a separate C library for maximum efficiency) used to generate the RF frequencies and the simulated noise, including user-defined delays. These delays are held fixed in the present work, but will be representative of the eLISA orbits once the command program is combined with LISACode [8, 9].

2.2. Optical LOT

The optical part of the LOT is mainly based on a Mach-Zehnder interferometer combined with three AOMs (acousto-optic modulator) which are driven by the DDS to shift the laser frequency on the arms in order to obtain heterodyne interferences. The beatnotes are measured by photodiodes and sent to the phasemeter.

The optical interferometer is shown on Fig.2. which represents one module of the simulator. The LOT is presently composed of two of those modules, each one represents one satellite of

Journal of Physics: Conference Series **610** (2015) 012035 doi:10.1088/1742-6596/610/1/012035

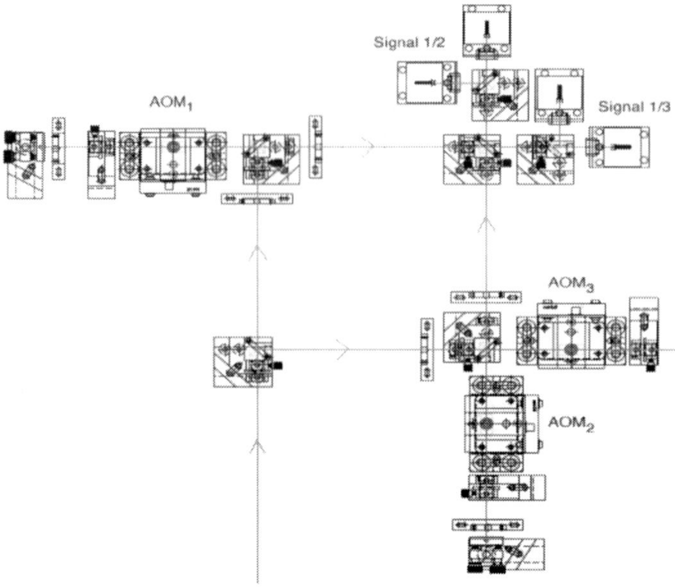

Figure 2: The optical interferometer

the eLISA configuration. For this study only one module has been used, representative of the mother spacecraft. As for eLISA, the module representing one satellite has three interfering beams, one local and two distant. A single laser source at 1064 nm (Innolight Mephisto 500) is used to produce these beams.

The local arm's frequency can be shifted with AOM 1 (driven by DDS channel 1). In the same way, AOM 2 and 3 (driven by DDS channel 2 and 3) induce the frequency shifts for the two distant arms. Free space A&A MT110-B50 TeO2 AOMs have been used for the LOT. The maximum diffraction efficiency ($\geq 80\%$) is reached at 110 MHz, with a bandwidth of ± 15 MHz with an efficiency greater than 65%. This broad modulation bandwidth is particularly useful for simulations of the Doppler effect. Each of these distant arms interferes with the local arm to produce a heterodyne signal detected by 48 MHz bandwith photodiodes with power noise below 8 pW/$\sqrt{\text{Hz}}$. Signal 1/2 is the resulting beatnote from the interfering beams shifted by AOM 1 and 2 and in the same way, signal 1/3 results from the interference between the beams shifted by AOM 1 and AOM 3 separated using orthogonal polarizations.

Also, each AOM produces two beams, one frequency shifted to produce the beat note for the heterodyne interferometer (order 1 beam), and one unaffected beam (order 0) which will be used for optical path length compensation using a dark fringe stabilization scheme.

All the experiment is performed in a clean room. The optical table is placed on an air cushion to reduce high frequency noise and all optical devices are placed under a box in order to reduce noises induced by air flow and temperature fluctuations. Also, a heat device is fixed on the top to implement temperature layers so that air perturbation induced by eventual warm spots on the table are quickly absorbed.

3. Time Delay Interferometry tests

This section explains briefly the TDI principle in the simplest case of a transponder type configuration. First results of noise reduction in this case will be shown in the second part on simulated delayed signals, both on the electric and optical LOT.

3.1. *TDI principle*

The dominant noise in eLISA is, by many orders of magnitudes, the laser phase fluctuations. These noises are however transported from one spacecraft to another and appear as differences between the local laser and the distant, delayed one. Their contribution can be canceled by properly delaying and combining the signals: this method is known as time delay interferometry (TDI) (see e.g. [2,3,10–15]). TDI however requires a precise knowledge of the light time between spacecraft (at the meter level over 10^6 km). Moreover, these equations assumed perfect clocks on each spacecraft, with no relative drifts nor jitter. Ranging and clock noise transfer can be performed by adding auxiliary modulations on the laser links [4]. After TDI and clock noise corrections [7], the residual phase noise should be due to the test masses (residual acceleration noise at low frequency) and the optical noise (including shot noise and path length fluctuations on the optical benches).

For the first experimental measurements performed on the LOT, the simplified, 'transponder' equation (which require only two simulated data streams), has been used with the additional constraint of a static constellation (i.e the delay operators $D_{2'1}D_{12'}$ and $D_{31'}D_{1'3}$ are constant and commute). In that case, the simplest TDI combination (X, first generation) leads to:

$$X_{1^{st}} = (1 - D_{31'}D_{1'3})s_{TT;1} - (1 - D_{2'1}D_{1'2})s_{TT;1'} \quad (1)$$

$s_{TT;1}$ and $s_{TT;1'}$ being the two simulated data streams. This combination synthesizes an equivalent Michelson interferometer with equal arms.

3.2. *TDI test on delayed signals*

To perform a first test of TDI with the LOT in a realistic configuration, a white noise (amplitude 280 Hz/$\sqrt{\text{Hz}}$) was simulated and applied on the local arm, as well as on the distant (delayed) arms 2 and 3. This setup is representative of a static 'transponder' eLISA used by equation 1. The signal sent to the local arm is not delayed, whereas the signal on arm 2 is delayed by 6.502081 s and arm 3 by 6.502081 s (equal arm lengths configuration).

A second test is performed with the same configuration as above but with different delays on the distant arms. The signal on arm 2 is delayed by 6.502081 s and arm 3 by 6.711826 s (unequal arm lengths configuration). These delays corresponds to the round-trip time of the laser between the eLISA spacecrafts. The resulting amplitude spectral densities are shown on figure 3 and 4.

For theses tests, frequencies of 108 MHz, 112.5 MHz and 112.7 MHz have been applied respectively on the local arm, on the distant arm 2 and 3 resulting in beat notes of 4.5 MHz and 4.7 MHz for the electric interferometer and twice that for the optical one due to a double pass of the laser through the AOMs.

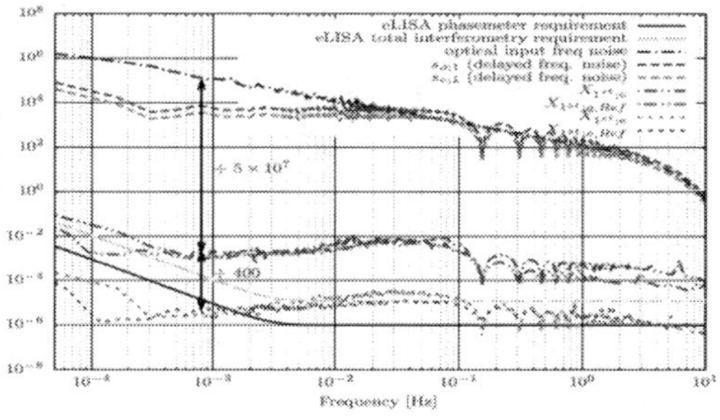

Figure 3: TDI on equal delayed signals (6.502081 s)

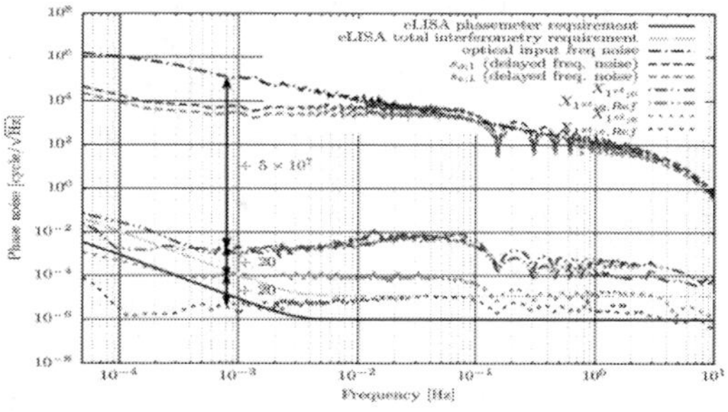

Figure 4: TDI on unequal delayed signals (6.502081 s and 6.711826 s)

Journal of Physics: Conference Series **610** (2015) 012035 doi:10.1088/1742-6596/610/1/012035

$s_{o,1}$ and $s_{e,1}$ are the measured signals , 'o' stands for optic and 'e' for electric. $X_{1^{st};o}$ and $X_{1^{st};e}$ are the signals after TDI has been applied. The $X_{1^{st},Ref}$ shows the reference level after application of TDI on signals without WN modulation.

For equal arm length we get a noise reduction of 5×10^7 on the optical signals at 1 mHz and 2×10^{10} for the electric one. The lower performance for the optical LOT is expected due to the optical path length fluctuations. For the unequal arm case the optical noise reduction is unchanged but limited to 10^9 for the electric one. This degradation could be due to phase jitter between the synthesized clocks driving the noise injection (1 GHz for the DDS operations) and the timestamps of the recorded phases (49.993 MHz for the phasemeter synchronization).

4. Comparison between the LOT and eLISA

This section is intended to make a brief comparison between the LOT and eLISA regarding the main parameters adjustable on our simulator and also some performances.

	LOT (current version)	LOT (final version)	eLISA
Laser frequency noise	560 $\frac{Hz}{\sqrt{Hz}}$ (simulated and adjustable)	560 $\frac{Hz}{\sqrt{Hz}}$ (simulated and adjustable)	280 $\frac{Hz}{\sqrt{Hz}}$
Laser power	Up to 0.5 W (3mW on photodiodes on standard operation mode)	Up to 0.5 W (3mW on photodiodes on standard operation mode)	2W (a few pW on the photodiodes)
TDI performance	5×10^7	At least 10^8	10^8
Delays	6.5s and 6.7s (adjustable but constant)	variable eLISA-like delays	Between 6.47s and 6.87s
Doppler shift	none	eLISA-like Doppler shift	amplitude up to 4MHz with frequency less than $10^{-7} Hz$

Table 1: Comparative table between the current version of the LOT, the final version and eLISA

In our case we simulated a laser frequency noise with twice the value of the expected eLISA laser frequency noise, also, regarding the laser power, a certain level had to be maintained in order to keep the dark fringe stabilization scheme working (see 5. Next steps). With this stabilization scheme implemented, eLISA-like TDI performances are expected. Concerning the delays, the command/control software allows only constant delays but in the future, pre-calculated delay variations (using realistic orbital models) will be used. These same models will also allow Doppler shift simulation. Currently, only a "pseudo-realistic" Doppler shift simulation is possible by adding a simple frequency modulation on the signals.

5. Next steps

The results of the first measurements, as described above, demonstrate the validity of the concept but also clearly show evidence of the major sources of noise: jitter in the simulated delays and optical path-length noise. The next steps and future works will address these two points and also increase the representativity of the simulation.

Journal of Physics: Conference Series **610** (2015) 012035 doi:10.1088/1742-6596/610/1/012035

- An FPGA (Field Programmable Gate Array) board will be inserted right after the NI PXIe 6537 communication board to take charge of buffering, delaying and synchronizing the command frames to the AOMs. It will also allow to add frequency corrections, taken from digital or analog error signals to mimic the phase lock of the laser sources on the distant spacecraft as well as the implementation of the arm-locking (see [5]) stabilization scheme. Additionally, a ranging pilot tone (frequency modulation around 1 Hz) will be implemented to monitor the effective delay of the frames (this technique has already been successfully used in [6]).
- The optical bench phase noise will be actively compensated using the interference of 'direct' (i.e. order 0) beams. Actually, these beams are unaffected by the RF signals on the AOMs and can be used to form an homodyne interferometer. The LOT interferometer arm lengths can therefore be stabilized using mirror dithering (above 1 kHz, i.e. outside the frequency band of eLISA) and dark fringe stabilization scheme. The compensation is expected to approach the interferometry requirement of eLISA.
- Another effort is currently being made to couple the present command-control system for the LOT, to the LISACode simulation software [8,9] also developed in our laboratory. Once achieved, this work will allow the simulations of realistic propagation delays, taking into account Sagnac effect, variable delays and Doppler shifts. It will also be possible to directly compare the 'numerical' results of TDI (as given by LISACode) to the same algorithm applied on optically or electronically simulated beat notes.
- Some space has been saved on the optical bench to insert electro-optical modulators. These modulators are planned to simulate the implementation of clock noise transfer and ranging (see [16]). However, the current architecture of LOT only allows clock noise transfer between the two daughter and the mother spacecraft and not between the two daughter spacecrafts.

6. Conclusion

This paper reported the current status of the LISA On Table experiment. The principle has been validated, with beatnotes recorded both on optical and electronic interferometers. The limited performance due to time jitter will be resolved soon, the FPGA card being under development. Also, the active compensation is about to be installed to improve the noise reduction of 5×10^7 for the optical frequency noise.

Acknowledgment

This work has been funded by the French Space Agency (CNES), under grants R-S07/SU-0001-012 and R-S08/SU-0001-012. The authors wish to thank G. Heinzel and I. Bykov (Albert Einstein Institut, Hanover, Germany), who provided the phasemeter and adapted it to our needs.

References

[1] Bykov, I., Delgado, J.J.E., Marin, A.F.G., Heinzel, G., Danzmann, K.: Lisa phasemeter development: Advanced prototyping. Journal of Physics: Conference Series **154**(1), 012,017 (2009)

[2] Dhurandhar, S., Ni, W.T., Wang, G.: Numerical simulation of time delay interferometry for a lisa-like mission with the simplification of having only one interferometer. Advances in Space Research **51**(1), 198 – 206 (2013)

[3] Dhurandhar, S.V.: Time-delay interferometry and the relativistic treatment of lisa optical links. Journal of Physics: Conference Series **154**(1), 012,047 (2009)

[4] Heinzel, G., Esteban, J.J.J., Barke, S., Otto, M., Wang, Y., Garcia, A.F., Danzmann, K.: Auxiliary functions of the lisa laser link: ranging, clock noise transfer and data communication. Classical and Quantum Gravity **28**, 4008 (2011)

[5] McKenzie, K., Spero, R.E., Shaddock, D.A.: Performance of arm locking in lisa. Phys. Rev. D **80**, 102,003 (2009)

[6] Mitryk, S.J., Mueller, G., Sanjuan, J.: Hardware-based demonstration of time-delay interferometry and tdi-ranging with spacecraft motion effects. Phys. Rev. D **86**, 122,006 (2012)

[7] Otto, M., Heinzel, G., Danzmann, K.: Tdi and clock noise removal for the split interferometry configuration of lisa. Classical and Quantum Gravity **29**(20), 205,003 (2012)

[8] Petiteau, A.: De la simulation de lisa a l'analyse des donnees. Ph.D. thesis, Université Paris Diderot - Paris 7 (2008)

[9] Petiteau, A., Auger, G., Halloin, H., Jeannin, O., Plagnol, E., Pireaux, S., Regimbau, T., Vinet, J.Y.: Lisacode: A scientific simulator of lisa. Physical Review D **77**, 23,002 (2008)

[10] Shaddock, D.A., Tinto, M., Estabrook, F.B., Armstrong, J.W.: Data combinations accounting for lisa spacecraft motion. Phys. Rev. D **68**, 061,303 (2003)

[11] Shaddock, D.A., Ware, B., Spero, R.E., Vallisneri, M.: Postprocessed time-delay interferometry for lisa. Phys. Rev. D **70**, 081,101 (2004)

[12] Tinto, M., Dhurandhar, S.V.: Time-delay interferometry. Living Reviews in Relativity **8**(4) (2005)

[13] Tinto, M., Shaddock, D.A., Sylvestre, J., Armstrong, J.W.: Implementation of time-delay interferometry for lisa. Phys. Rev. D **67**, 122,003 (2003)

[14] Vallisneri, M.: Geometric time delay interferometry. Physical Review D **72**(4), 042,003 (2005)

[15] Vallisneri, M.: Erratum: Geometric time delay interferometry [phys. rev. d 72, 042003 (2005)]. Physical Review D **76**(10), 109,903 (2007)

[16] Sweeney, D., Mueller, G.: Experimental verification of clock noise transfer and components for space based gravitational wave detectors. Opt. Express **20**(23), 25,603–25,612 (2012)

Journal of Physics: Conference Series **610** (2015) 012036 doi:10.1088/1742-6596/610/1/012036

A noise simulator for eLISA: Migrating LISA Pathfinder knowledge to the eLISA mission

M.Armano[a], H.Audley[b], G.Auger[c], J.Baird[n], P.Binetruy[c], M.Born[b], D.Bortoluzzi[d], N.Brandt[e], A.Bursi[t], M.Caleno[f], A.Cavalleri[g], A.Cesarini[g], M.Cruise[h], K.Danzmann[b], I.Diepholz[b], R.Dolesi[g], N.Dunbar[i], L.Ferraioli[j], V.Ferroni[g], E.Fitzsimons[e], M.Freschi[a], J.Gallegos[a], C.García Marirrodriga[f], R.Gerndt[e], LI.Gesa[k] F.Gibert[k], D.Giardini[j], R.Giusteri[g], C.Grimani[l], I.Harrison[m], G.Heinzel[b], M.Hewitson[b], D.Hollington[n], M.Hueller[g], J.Huesler[f], H.Inchauspé[c], O.Jennrich[f], P.Jetzer[o], B.Johlander[f], N.Karnesis[k], B.Kaune[b], N.Korsakova[b], C.Killow[p], I.Lloro[k], R.Maarschalkerweerd[m], S.Madden[f], D.Mance[j], V.Martín[k], F.Martin-Porqueras[a], I.Mateos[k], P.McNamara[f], J.Mendes[m], L.Mendes[a], A.Moroni[t], M.Nofrarias[k], S.Paczkowski[b], M.Perreur-Lloyd[p], A.Petiteau[c], P.Pivato[g], E.Plagnol[c], P.Prat[c], U.Ragnit[f], J.Ramos-Castro[qr], J.Reiche[b], J.A.Romera Perez[f], D.Robertson[p], H.Rozemeijer[f], G.Russano[g], P.Sarra[t], A.Schleicher[e], J.Slutsky[s], C.F.Sopuerta[k], T.Sumner[n], D.Texier[a], J.Thorpe[s], C.Trenkel[i], H.B.Tu[g], D.Vetrugno[g], S.Vitale[g], G.Wanner[b], H.Ward[p], S.Waschke[n], P.Wass[n], D.Wealthy[i], S.Wen[g], W.Weber[g], A.Wittchen[b], C.Zanoni[d], T.Ziegler[e], P.Zweifel[j]

[a] European Space Astronomy Centre, European Space Agency, Villanueva de la Cañada, 28692 Madrid, Spain
[b] Albert-Einstein-Institut, Max-Planck-Institut für Gravitationsphysik und Universität Hannover, 30167 Hannover, Germany
[c] APC UMR7164, Université Paris Diderot, Paris, France
[d] Department of Industrial Engineering, University of Trento, via Sommarive 9, 38123 Trento, and Trento Institute for Fundamental Physics and Application / INFN
[e] Airbus Defence and Space, Claude-Dornier-Strasse, 88090 Immenstaad, Germany
[f] European Space Technology Centre, European Space Agency, Keplerlaan 1, 2200 AG Noordwijk, The Netherlands
[g] Dipartimento di Fisica, Università di Trento and Trento Institute for Fundamental Physics and Application / INFN, 38123 Povo, Trento, Italy
[h] Department of Physics and Astronomy, University of Birmingham, Birmingham, UK
[i] Airbus Defence and Space, Gunnels Wood Road, Stevenage, Hertfordshire, SG1 2AS, UK
[j] Institut für Geophysik, ETH Zürich, Sonneggstrasse 5, CH-8092, Zürich, Switzerland
[k] Institut de Ciències de l'Espai (CSIC-IEEC), Campus UAB, Facultat de Ciències, 08193 Bellaterra, Spain
[l] Istituto di Fisica, Università degli Studi di Urbino/ INFN Urbino (PU), Italy
[m] European Space Operations Centre, European Space Agency, 64293 Darmstadt, Germany
[n] The Blackett Laboratory, Imperial College London, UK
[o] Physik Institut, Universität Zürich, Winterthurerstrasse 190, CH-8057 Zürich, Switzerland
[p] SUPA, Institute for Gravitational Research, School of Physics and Astronomy, University of Glasgow, Glasgow, G12 8QQ, UK
[q] Department d'Enginyeria Electrònica, Universitat Politècnica de Catalunya, 08034 Barcelona, Spain
[r] Institut d'Estudis Espacials de Catalunya (IEEC), C/ Gran Capità 2-4, 08034 Barcelona, Spain
[s] NASA Goddard Space Flight Center, 8800 Greenbelt Road, Greenbelt, MD 20771, USA
[t] CGS S.p.A, Compagnia Generale per lo Spazio, Via Gallarate, 150 - 20151 Milano, Italy

E-mail : henri.inchauspe@apc.univ-paris7.fr

Journal of Physics: Conference Series **610** (2015) 012036 doi:10.1088/1742-6596/610/1/012036

Abstract. We present a new technical simulator for the eLISA mission, based on state space modeling techniques and developed in MATLAB. This simulator computes the coordinate and velocity over time of each body involved in the constellation, i.e. the spacecraft and its test masses, taking into account the different disturbances and actuations. This allows studying the contribution of instrumental noises and system imperfections on the residual acceleration applied on the TMs, the latter reflecting the performance of the achieved free-fall along the sensitive axis. A preliminary version of the results is presented.

1. Introduction

The LISA Pathfinder [1] mission will be launched next year and will provide valuable information on the technology and performances that will be used for the future eLISA mission [2]. It is thus of a great importance to be able to migrate the knowledge acquired from LISA Pathfinder to this future mission, thus giving the physics community a tool to understand its performances and to suggest possible improvements.

The LISA Technology Package (LTP), which will be tested during the LISA Pathfinder mission, involves many subsystems as, for example, the micro-propulsion system, the interferometer readout or the controllers. Each of these subsystems might show some imperfections and could impact the final noise budget. These features of the system could affect the motion of the TMs, preventing them from achieving perfect free-fall, impacting the performances of eLISA, particularly at frequencies around and below 1 mHz. Evaluating the level of the residual acceleration is therefore of particular importance since its measurement will not be easily measurable in flight.

A plan for an end-to-end simulation is in preparation within the community and the simulator presented here will be one of the important building blocks of this effort.

This paper will first describe the mathematical approach used to model the system, in particular the LTI approximation and its justification. Then an overview of the simulator structure, showing which physical phenomena are considered and how they are modeled will be presented. Finally, preliminary results of this simulator, especially the influence of individual instrumental noises on the residual acceleration applied on the TMs along the sensitive axis will be presented.

2. A description of the simulator and of the relevant reference frames

2.1. The eLISA constellation

The eLISA design consists of 3 S/C orbiting around the Sun, forming a constellation in a quasi-equilateral triangular shape. Each S/C contains actually several bodies, i.e. the S/C itself and one or two TMs. In the eLISA configuration, the "Mother" S/C contains 2 TMs (as well as 2 optical benches and 2 telescopes) and the two other "Daughters" contain only one TM. The arm length of this space interferometer is planned to be 1 million km.

To fully determine the dynamics of the system, one has to know the evolution of the 3x6 coordinates and their 3x6 corresponding derivatives per S/C over time. This includes the position of the center of mass (CoM) of the S/C and those of the TMs, their attitude, their linear and angular velocities. At the initial point of the orbits, the 3 bodies are assumed to be rigidly linked, i.e. the TMs are caged but released immediately. This implies that the bodies are in their nominal position, i.e. the S/C follows the programmed orbit, the TMs are centered inside their housing and the velocities of the 3 bodies are the same. Because the CoM positions of the 3 bodies are different, they will not naturally follow identical orbits: one of the tasks of the system on-board controllers will be to prevent any potential drift between the S/C and the TMs.

Journal of Physics: Conference Series **610** (2015) 012036 doi:10.1088/1742-6596/610/1/012036

Great care has been taken in defining the reference frames w.r.t. which the different coordinates are expressed. The S/C position and attitude are expressed in a proper inertial reference frame, the *O-frame*, that will be introduced later. The TM coordinates are expressed in the housing frames. These frames are represented by an orthonormal basis whose origin is the center of the housing and whose unit vectors are perpendicular to the faces of the housing. They are fixed w.r.t. the S/C, i.e. their orientation are fixed in the reference frame B attached to the S/C. The set of reference frames used in this simulator is represented in the figure 1.

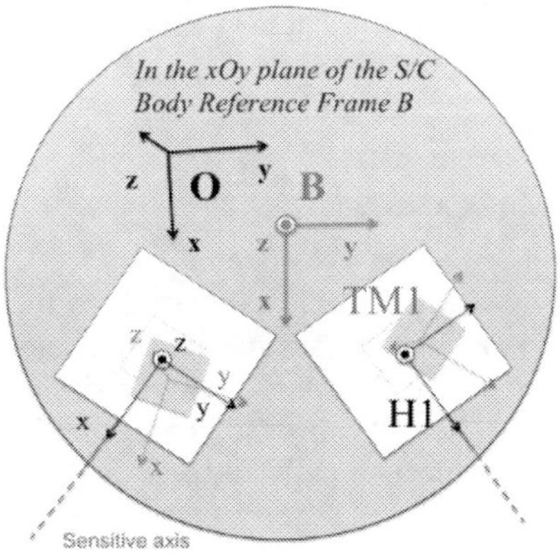

Figure 1. Reference frames involved in the simulator. The B frame is attached to the CoM of the S/C and the unit vectors, fixed w.r.t. the S/C, are used to describe the S/C orientation. H1 and H2 frames are attached to the housings and stay fixed w.r.t. the S/C. Their unit vectors are perpendicular to the faces of the housings. TM1 and TM2 are attached to the TMs CoM. Orientation of TM1 and TM2 w.r.t. H1 and H2 give the attitude of the TMs inside the S/C.

2.2. *Equations of motion*

The dynamics of the three bodies are constrained by 2 equations:

- Newton equations for translational motion. The corresponding solutions are the position vector of the center of mass of the bodies and their linear velocities.
- Euler equations for rotational motions. The solutions are the set of Euler angles able to describe the orientation of the bodies and their angular velocities.

These equations require that the coordinates be expressed in Galilean reference frames. If this is not the case, inertial terms have to be introduced to reflect the apparent inertia forces viewed from the point of view of the actual non-inertial reference frame. Such is the case for the position of the TMs expressed in the housing frame, which is a rotating frame w.r.t. the Galilean reference frame J attached to the Sun. As new rotating reference frame, called the orbital frame *O*, will be introduced, additional inertia terms will also appear in the dynamics of the S/C.

2.3. *A Linear Time Invariant (LTI) State Space Model (SSM)*

A convenient way to parameterize the dynamics is the so-called space state representation, see for example M.Weyrich [4]. It consists in expressing a set of equations in a matrix form whose solution is a vector (the State Vector) containing all the solutions of the individual equations. This vector contains all the coordinates involved in the dynamics.

A particular case occurs when the modeled system is linear and time-invariant (LTI model). This implies that the SSM matrices are constant and independent of the states. With this property, solving, assembling and converting (e.g. in frequency domain) the model reduces to handling this set of SSM matrices. Working in the frequency domain also requires this time-invariance.

The linearization process is simplified by the presence of system controllers. They allow the system to reach a stable equilibrium state. For instance, the controllers will force the TMs to be centered in their housing at all times. Thus, this equilibrium point becomes the natural point around which the linearization is made. The dynamics of the S/C presents a specific problem. If the attitude of the S/C is expressed in the *EclipticJ2000-frame*[1] an equilibrium point cannot be defined. Note that the position of the S/C does not need to be linearized, only the attitude has to be. Expressing the attitude dynamics of The S/C w.r.t. the *EclipticJ2000-frame* will therefore not allow for a long-term linearization. To correct for this, a new reference frame is introduced in the next section.

2.4. *The orbital reference frame O*

The strategy is to find a reference frame in which a linearization around an equilibrium point of the S/C dynamics can be performed. One can note that:

- The S/C follow, to a certain degree the geodesic determined by their initial (t=0) position and velocity. Instead of calculating their exact (non-geodesic!) trajectory as a function of time, one calculates the deviations from this initial geodesic.
- The controllers will maintain the proper attitude of the S/C such that their receiving telescopes and emitted lasers point to the opposite S/C.

For each S/C, the *O-frame* is then constructed in the following way:

- The origin of the *O-frame* follows the initial geodesic of the associated S/C.
- The orientation of the *O-frame* evolves in such as way that its X-axis points to the center of the line joining the two opposite S/C. This quantity is known, with sufficient precision, from the a-priori knowledge of the initial geodesic of the 3 S/C. Note that during the mission, the orientation of the S/C is controlled by sensing the waveform of the incoming lasers and that in the simulator the geometry of the emitting/receiving telescopes are mechanically fixed to the S/C body.

2.5. *The residual acceleration estimation*

One of the main aims of the simulator is to calculate the residual acceleration along the sensitive axis (i.e. the laser link), thus testing the efficiency of the *drag-free* strategy. The simulator calculates, at all times, the position of the S/C and of the TMs with respect to the *O-frame*. Computing the acceleration of the TMs w.r.t. *O-frame* will provide the residual acceleration, with one caveat however. As the TMs do not follow the *O-frame* geodesic, a residual gravitational pull needs to be precisely calculated and accounted for. The final acceleration and its projection on the sensitive axis are then calculated. This quantity is therefore representative, to within a $\sqrt{2}$ factor[2], of the residual acceleration between two opposite TMs.

3. Control strategy, Noises and imperfections

Many subsystems are involved in LTP to in order to allow the TMs to be in free-fall. Each of them has some degree of imperfections that may have some repercussions over the quality of the free-fall. Because the system requires a complex control strategy, every noise and imperfection may interact through the control loop, making the analysis of the noise contributions possibly counter-intuitive.

3.1. *The Control strategy*

The goal of the control process is:

- To allow the telescopes of a given S/C to point towards the distant S/C, thus assuring that the laser link is maintained between them.
- To keep the TMs well centered and well oriented in their housing.

There are 15 coordinates to control, 3 for the S/C (attitude), and 6 for each TM (position and attitude). To ensure this, three types of control are defined:

- The first one is the *Drag-Free* control and corresponds to the actuation of the micro-propulsion system that positions the S/C w.r.t the TMs, assuring a control of the position of the TMs without

---

[1] An inertial reference frame centred on the Sun whose main axis (x-axis) is on the ecliptic plane and is pointing towards the Earth position at the standard vernal equinox epoch (J2000). The z-axis is normal to the ecliptic plane and the y-axis completes the basis.

[2] We suppose here that the residual acceleration noise of distant TMs are gaussian and uncorrelated. Multiplying one of them by a $\sqrt{2}$ factor is statistically equivalent to adding them.

any force applied on it. This *Drag-Free* control will be used to constrain the x-coordinates of the TMs in their housing, i.e. along the sensitive axis, because we want to avoid as much as possible any applied force along the sensitive axis. An additional degree of freedom allows controlling the z-coordinate of TM1.

- The second one is the *Attitude* control which orients the S/C in order for it to point correctly towards the opposite S/C.
- The last one is the *Suspension* control. This corresponds to the actuation of the capacitive system that can exert electrostatic force on the TMs to correct their orientation and position on the Y-Z direction (except for the Z position of TM1).

One can emphasize here that in this context, the TM will not, strictly speaking, follow pure geodesics. However, on the sensitive axis, they will be as undisturbed as possible.

3.2. Noises and imperfections

In the eLISA system control loop, many subsystems come into play, as much for the actuation as for the measurements. In the simulator, all the subsystems are represented as symbolic blocks, all connected with links that represent inputs and outputs, see figure 2. Each block has a state space representation that models the subsystems properties, as for instance cross-talk matrices or time delays. A corresponding noise block is added to all actuations and measurement signals. Each noise is described by an amplitude and a frequency dependence by means of a transfer function.

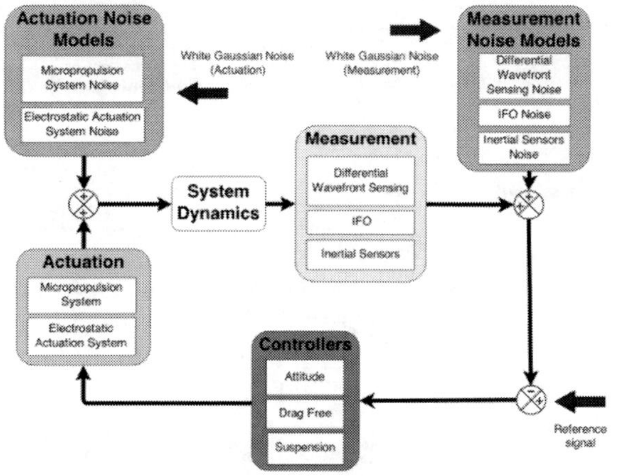

Figure 2. Block diagram of the simulator

To associate these blocks with their proper parameter, information from the LISAPathfinder SSM has been used. Relevant information has also been taken from F.Cirillo [5] and from valuable information provided by industrial support to ESA, in particular concerning the noise transfer functions. However, this information lies on specifications rather than on rigorous experimental characterizations. Since LPF will test this very kind of technology in space, these approximations will be replaced by the realistic values as measured during the mission.

4. Preliminary results

The simulator has been run for a mission duration of about a couple of days (a few tens of minutes of CPU time), enough to evaluate, over the eLISA measurement band (10^{-5}-10^{-1} Hz), the behavior of the system. At the end, the simulator provides the evolution of the 18 coordinates of the system (and their derivatives) from which the residual acceleration can be computed and their associated Power Spectral Density (PSD). If necessary, calculations over the whole duration of the mission can also be performed in a very reasonable time.

Several runs have been made, activating only one source of noise at a time. A final calculation is done with the presence of all noises. Figure 3 shows, as a function of frequency, the total noise budget superimposed on the contribution of each component.

These results are very preliminary and will evolve as each noise contribution is actualized, taking into account the best present knowledge. They should not be taken as representative of the performance level of the future eLISA mission.

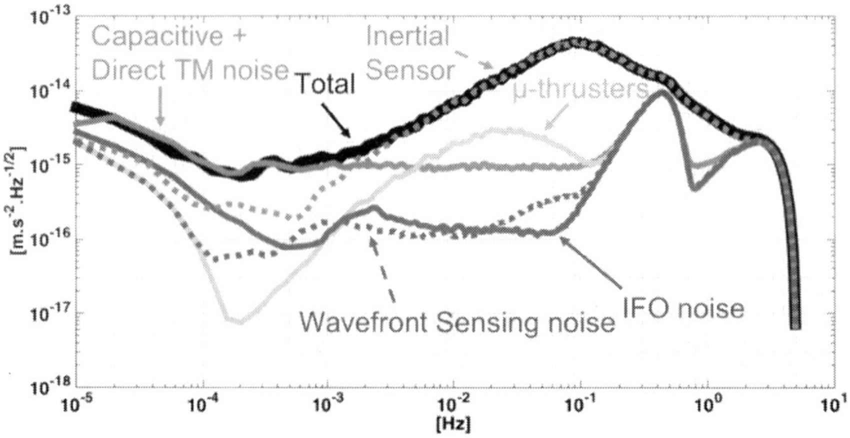

Figure 3. Decomposition of the contributions of individual noises to the PSD of the residual acceleration

5. Conclusion

This work presents a first version of an eLISA simulator aimed at estimating the residual acceleration noise on the sensitive axis. It provides an efficient and very complete framework to evaluate and analyze the performances of this future mission. Because of its modularity, the impact of every system component (sensors, actuations, thrusters, controllers…) can be studied and adapted to the best present knowledge. The results of the LISAPathfinder mission will be essential to provide the simulator with realistic values and will show which subsystem has to be improved, and by what amount, in order for eLISA to satisfy its scientific requirements.

Without reaching the complexity level of an industrial simulator, it is complete enough to study its main components. Its ease of use should provide the scientific community with a tool to optimize the scientific performances of the mission.

The simulator is also constructed in such a way that the configuration of the system can easily evolve. For example, testing configurations with one test mass per S/C, spherical test masses, improved sensors and actuators can be studied. A quantitative comparison between a two-arm configuration (eLISA) and three-arm one (LISA) is also easy to implement.

Once the impact of each subsystem has been adapted to the level of the best knowledge available, the residual acceleration achieved for each system configuration can be transferred to data analysis programs, such as LISACode [3], in order to quantify the scientific performances of the future mission.

Acknowledgments

E.P. and H.I would like to acknowledge the financial support of the UnivEarthS Labex program at Sorbonne Paris Cité (ANR-10-LABX-0023 and ANR-11-IDEX-0005-02) and the French space agency CNES for its financial support during these studies. The authors would like to thank the LISA and LISAPathfinder teams at AEI and Prof. Stefano Vitale for some very fruitful discussions.

References

[1] Antonucci F, et al. 2011 Classical Quantum Gravity **28** 094001
[2] eLISA Consortium 2013 arXiv:1305.5720
[3] Petiteau A, et al 2008 Phys. Rev. D **77** 023002
[4] Mario Weyrich thesis, University of Stuttgart, 2008
[5] Francesca Cirillo thesis, Università di Pisa, 2007

10th International LISA Symposium (LISAX) IOP Publishing
Journal of Physics: Conference Series **610** (2015) 012037 doi:10.1088/1742-6596/610/1/012037

Development of a Micro-Thruster Test Facility which fulfils the LISA requirements

Franz Georg Hey[1,2], A. Keller[1], U. Johann[1], C. Braxmaier[3,4], M. Tajmar[2], E. Fitzsimons[1], D. Weise[1]

[1] AirbusDS, Space Systems, Friedrichshafen, Germany
[2] Technische Universität Dresden, Germany
[3] Universität Bremen, Germany
[4] DLR Bremen, Institute of Space Systems, Germany

E-mail: `franz.hey@astrium.eads.net`

Abstract. In the context of investigations for a sufficient attitude control thruster for LISA, we have developed a thruster test facility which consists of a highly precise thrust balance coupled with plasma diagnostics. In parallel to the test facility development, investigations to downscale a High Efficiency Multistage Plasma Thruster (HEMP-T) are also being carried out. The thruster has been used to demonstrate the measurement capabilities of the facility. The setup allows a parallel operation of all instruments and can also be used for other types of µN propulsion systems including cold gas thrusters. The thrust balance consists of two pendulums. As read out a heterodyne laser interferometer is used. Differential wave front sensing (DWS) enables the measurement of the pendulum tilt which, via suitable calibration using an electrostatic comb, can be converted to a thrust. The whole setup is a symmetric configuration enabling a common-mode rejection of the dominant noise sources (e.g. seismic noise etc.). The thrust balance has a demonstrated precision of 0.1 µN. Based on our unique design, this precision can be attained down to 10^{-3} Hz. Thus, the measurement setup is especially suitable for characterising the thrust noise of potential eLISA propulsion candidates. We give an overview of the design, the present performance and the future plans.

1. Introduction

In 2009 the Laboratory for Enabling Technologies (LET) at AirbusDS Friedrichshafen started a research and development project to investigate possible micro-Newton thrusters for highly precise attitude control of satellites for Laser Interferometer Space Antenna (LISA) like missions. Within the framework of the project, a feasibility study to downscale a High Efficiency Multi Stage Plasma Thruster (HEMP-T) to micro-Newton thrust is performed. A schematic thruster assembly is given in figure 1 a). It depicts a cut through the cylindrical HEMP-T. The dielectric discharge chamber (in yellow) is surrounded by a system of periodically poled permanent magnets. This typical magnet stack forms a special magnetic field topology which separates the discharge chamber into several magnetic cells. Furthermore, the field confines the plasma on the rotation axis of the thruster. Therefore, the plasma has ideally almost no interaction with the dielectric walls surrounding the discharge chamber. On the left side of the scheme (upstream end of the discharge chamber) the anode is mounted. The anode operates as the gas inlet as well. On the right side of the scheme (downstream) a neutraliser cathode is placed. Between

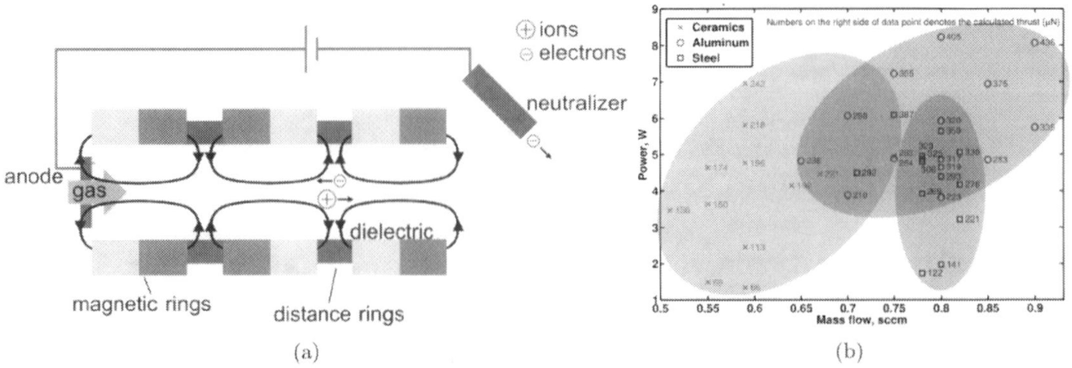

Figure 1. Schematic view of the HEMP-T concept is presented in a). As an example, an arrangement with a cylindrical discharge channel and three magnet rings is shown, such that 3 cusp zones with high radial magnetic field are formed in the discharge channel [1]. In b) the operation space of the different micro-HEMP-Ts is presented.

the anode and the cathode the potential difference ranges from 300 V to 2000 V. The HEMP-T works as an electrostatic thruster i.e. the thruster relies on coulomb forces to accelerate a propellant of charged particles. HEMP-Ts use xenon as propellant, therefore, the plasma consists of positive charged xenon ions. The ionisation of the neutral xenon gas to the plasma is realised by bombarding the plasma with electrons (electron bombardment thruster).

For the development of micro newton thrusters and for the basic research of micro-Newton ion propulsion, a sufficient test facility is required. Such facilities are available e.g. at the University of Giessen [7]. The standard set of diagnostic hardware consists of a Faraday Cup, or Faraday Array, and a Retarding Potential Analyser (RPA). The tools are used to characterise the basic parameter such as ion current density, or ion energy of a thruster plume, which allows an indirect thrust measurement. But also direct thrust measurement systems are in use [8] [2] [9]. Especially the challenging requirements of future space telescope missions like LISA require a detailed characterisation of the potential thruster candidates. In particular, the direct measurement and the characterisation of the thrust noise is imported to get reliable informations of the Attitude and Orbit Control System (AOCS) jitter, but as far as we know the existing thrust balances are not long term stable enough to characterise micro-Newton thruster in the complete LISA measurement bandwidth. Therefore, we developed a highly precise and highly stable micro-Newton thrust balance which fulfils the LISA requirements, especially in point of thrust noise.

2. Measurement Setup

The developed thrust balance is a double pendulum balance. The applied force (or thrust) of the thruster under test, deflects the measurement pendulum. The detected pendulum deflection can be converted into the applied thrust by multiplication with the pendulum spring rate. In general, the performed measurements are a differential measurement between the two fully symmetric pendulums. This allows the suppression of seismic and thermal mechanical noise which leads to the unique stability of the presented setup. The balance, in back view, is shown in figure 2 c). The balance consist of the pendulum arm (d), the major balance structure and the bearing (b) and two dielectric mirrors (c). Each pendulum is damped by an eddy current brake (a). A plastic tube (h) is used as gas supply for the investigated thruster. The thruster is not shown in figure 2 c). An electro static comb (f) [5] is used to calibrate the balance,. A heterodyne

(a) (b) (c)

Figure 2. Picture of the whole micro-Newton thruster test facility, Image a) shows the facility from outside, in b) the inner space of the vacuum tank is shown. Picture c) shows the micro-Newton thrust balance in back view.

interferometer is used as translation sensor. Due to the use of quadrant photo diodes, differential wave front sensing allows an independent measurement of the pendulum tilt in addition to the normal translation measurement. The thrust balance can be used in two different operation modes, open loop and closed loop. In open loop operation the pendulum translation is directly measured, the difference between the equilibrium of the pendulum and the measured translation can be converted to the applied thrust. In closed loop operation each pendulum is actively controlled to neutral position by a voice coil. As controller input the DWS signal is used.

The balance is placed in a 1500 litre vacuum chamber. Two turbo molecular and one cryo pumps with together 11400 $^{\text{litre}}/_{\text{s}}$ pumpage are used to evacuate the tank. Therefore, the basis pressure is $4e^{-7}$ mbar and the working pressure, with 0.5 sccm gas ballast, is $1e^{-6}$ mbar. The vacuum chamber is shown in figure 2 a). To decouple the facility from the ground, the vacuum tank is mounted on an ITEM structure which is supported by four optical isolators. In addition to the thrust balance, a basic set plasma diagnostics is also part of the test facility. It consists of 15 Faraday Cups and one RPA. The instruments are mounted on a rotatable jib arm, which can rotate 180 ° around the thruster. Figure 2 b) presents the internal space of the vacuum tank, with the thrust balance and the plasma diagnostic. The facility design allows a parallel operation of the thrust balance and the plasma diagnostic. Thus, a comparison of the direct and of the indirect thrust measurement can be performed.

In parallel to the thrust balance development we continued the downscaling of the HEMP-T principle to micro-Newton Thrust. The characterisation of the different micro-HEMP-T designs were performend in the presented facility, as well as in the thruster test facility at the University of Giessen [7].

3. Results

As result of the parametric experimentally downscaling campaign a thrust of 66 µN by an ISP of 200 s were achieved, but on higher thrust levels (around 400 µN) the thruster has an ISP larger then 1500 s [6]. The operation space of the micro-HEMP-T is presented in figure 1 b).

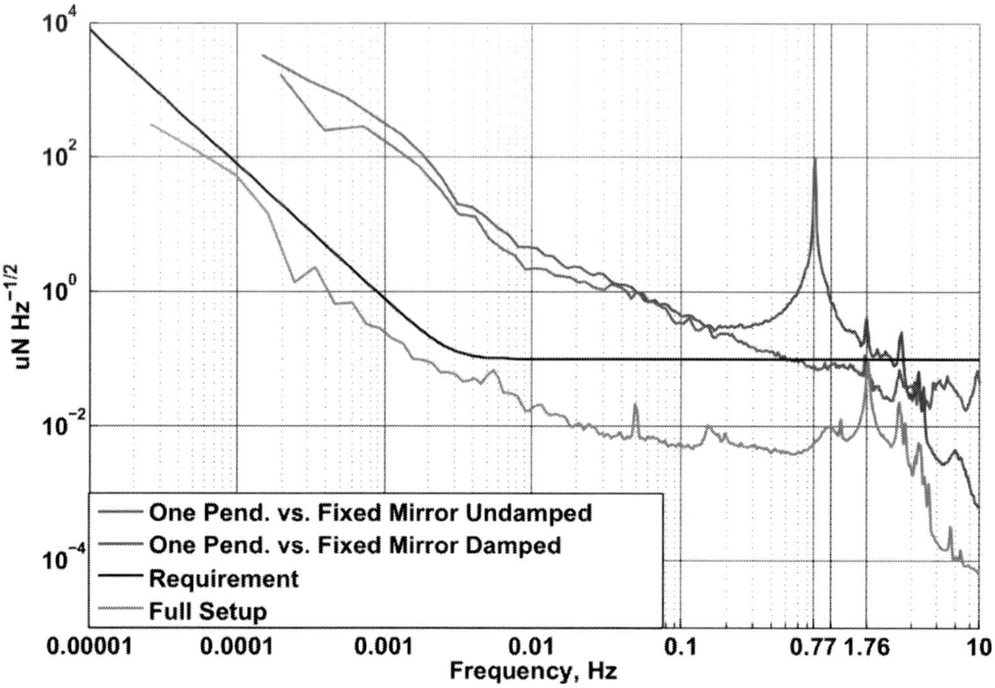

Figure 3. Summary of the balance performance. The blue plot presents the performance of one undamped pendulum. The purple plot shows the performance of a damped pendulum. The Black curve is the requirement. The red plot shows the performance of the full balance setup with active common mode rejection.

To characterise the thrust balance and the whole test facility, we performed a set of different tests with and without active thruster. Figure 3 summarises the thrust balance performance and the effect of different parts of the thrust balance assembly, in point of thrust noise. The presented measurements were performed with the fully integrated setup. A deactivated thruster was mounted on the balance. The cryo pump had to be deactivated during the measurements because of the still insufficient noise shielding between cryo pump and vacuum chamber. The Power Spectral Density (PSD) presents the thrust in µN, normalised by the frequency in $\sqrt{\text{Hz}}$ versus the frequency. The black plot presents the targeted balance requirement, which is the LISA thrust noise requirement [4]. The blue curve presents the noise measurement of a single undamped pendulum versus a fixed mirror, hence no common mode rejection occurs. The observed eigenfrequency of the pendulum is 0.77 Hz, this is close to the estimated eigenfrequency of 0.8 Hz. The amplitude of the free swinging pendulum is the dominant noise term of the measurement resolution. Down to lower frequencies pink noise occurs and limits the resolution in point of long term stability. The purple plot presents the performance of one damped pendulum versus a fixed mirror (no common mode rejection). The damper suppresses the eigenfrequency. Therefore, at higher frequencies the noise level decreases on 1.5 order of the magnitudes, compared to the blue plot. At lower frequencies pink noise dominates the PSD.

The noise measurement of the whole setup, which means two damped pendulums, is shown as the red plot. Therefore, the noise level decreases of one order of the magnitude. Down to lower frequencies the rise of the noise level is reduced. The plot illustrates that the assembly is stable enough to fulfil the LISA requirement.

4. Conclusions

The micro-Newton thruster test facility at AirbusDS in Friedrichshafen, which consist of a highly precise thrust Balance and a plasma diagnostic system, is operational. The balance fulfils the LISA requirement in point of thrust noise and thrust resolution. This allows the characterisation of potential micro-Newton thruster candidates for LISA. First tests with an active thruster were performed [3]. Moreover, the results of our HEMP-T downscaling were given. Especially on higher thrust levels the micro-HEMP-T is able to operate with an ISP over 1500 s, which is promising. Therefore, we will continue the downscaling of the HEMP-T principle. In the next months, an advanced noise shielding of the cryo pump will be installed, this will lead to thrust measurements with permanent running cyro pump. It is planned to perform the characterisation of different micro-Newton thrusters in our micro-Newton thruster test facility.

References

[1] Tim Brandt, Thomas Trottenberg, Rodion Groll, Franz Georg Hey Frank Jansen, Ulrich Johann, Holger Kersten, and Claus Braxmaier. Simulation of a down-scaled hemp-thruster. *RGCEP-2014*, 2014.

[2] Manuel Gamero-Castaom, Vlad Hruby, and Manuel Martnez-Snchez. A torsional balance that resolves sub-micro-newton forces. *IEPC-01-235*, 2001.

[3] Franz Georg Hey, Andreas Keller, Dirk Papendorf, Claus Braxmaier, Martin Tajmar, Dennis Weise, and Ulrich Johann. Development of a highly precise micro-newton thrust balance. *IEPC-2013-275*, 2013.

[4] Oliver Jennrich et al. *LISA, unveiling a hidden universe*. European Space Agency, 2011.

[5] W.A. Johnson and L.K. Warne. Electrophysics of micromechanical comb actuators. *Microelectromechanical Systems, Journal of*, 4(1):49 –59, mar 1995.

[6] Andreas Keller, Franz Georg Hey, Peter Koehler, Marcel Berger, Claus Braxmaier, Davar Feili, Dennis Weise, and Ulrich Johann. Parametric study of hemp-thruster, downscaling to n thrust levels. *IEPC-2013-269*, 2013.

[7] Andreas Keller, Peter Khler, Waldemar Grtner, Benjamin Lotz, Davar Feili, Philipp Dold, Marcel Berger, Claus Braxmaier, Dennis Weise, and Ulrich Johann. Feasibility of a down-scaled hemp-thruster. *IEPC-2011-138*, 2011.

[8] S. M. Merkowitz, P. G. Maghami, Sharma, W. D. Willis, and C. M. Zakrzwski. A micro-newton thrust-stand for lisa. *Classical and Quantum Gravity*, 19(7):1745, 2002.

[9] Denis Packa, Jean Bonnet, and Simone Rocca. Thrust measurements with the onera micronewton balance. *IEPC-2007-118*, 2007.

The UF Torsion Pendulum, a LISA Technology Testbed: Sensing System and Initial Results

Andrew Chilton, Ryan Shelley, Taiwo Olatunde, Giacomo Ciani, John W. Conklin, Guido Mueller

Abstract. The upcoming LISA Pathfinder mission will test the Gravitational Reference Sensor and the Disturbance Reduction System for a future LISA-like space mission. While LISA Pathfinder is expected to show that the technology for LISA exists and meets the LISA requirements, it is likely that LISA Pathfinder will also reveal areas where future improvements can be made and might be necessary. Some of these are already well known (such as the discharging system). After all, the technology for LISA Pathfinder was frozen about 10 years ago or about 30 years before a LISA-like mission will be launched. The case for continued testing and development of the technology is clear. The University of Florida is currently building a torsion pendulum-based test facility to explore new techniques and also to develop a base in the United States for state-of-the-art Gravitational Reference Sensor technologies.

1. Introduction

LISA [1][2] is the most mature concept for detecting gravitational waves from space and has the potential to both open up a new realm of astronomy and provide a probe of fundamental physics. To do this it must measure gravitational wave strains of $\approx 10^{-23}$ over a ten day integration time, and thus must be sensitive to length strains on the order of $\frac{\text{pm}}{\sqrt{\text{Hz}}}$ in the measurement band of 0.1 mHz to 100 mHz. One obstacle to this ambitious goal is residual acceleration noise on the test masses, which limits the sensitivity of LISA-like missions below a few mHz. LISA-like missions typically use the following requirement for the residual acceleration noise on each test mass:

$$s_a^{1/2} < 3 \times 10^{-15} \left(1 + \left(\frac{f}{3\ \text{mHz}}\right)^2\right)^{1/2} \frac{\text{m}}{\text{s}^2\sqrt{\text{Hz}}} \quad (1)$$

The UF Torsion Pendulum is a LISA technology development facility which will build on the work done at the University of Trento and is aimed at testing the performance of new technologies for the LISA Gravitational Reference Sensor (GRS) that enable this requirement to be met.

2. Concept and Motivation

The LISA spacecraft (SC) ensure the purity of free fall of their test masses using the Drag-Free Attitude Control System (DFACS). The GRS continually senses the position and orientation of the TMs. These signals are used by the DFACS to maneuver the SC around the TMs with μN thrusters. Conceptually, the GRS consists of a set of electrodes which surround the TM. These electrodes are mounted inside the electrode housing (EH) and are connected to the GRS electronic. Two of the electrodes are used as injection electrodes where an AC signal is injected to polarize the TM. This polarization is measured with the sensing electrodes. Furthermore, the electrodes are also used for electrostatic actuation of the TMs. The GRS completely encloses the TM and is ideally the only device which interacts with it. Most spurious forces on the

TM will depend on these interactions and need to be minimized to meet the acceleration noise requirements in Eq. 1.

Ground-testing is critical to develop noise models for these interactions and to investigate different designs and operational modes of the GRS. The University of Trento has previously constructed a torsion pendulum [3] to decouple form the Earth's gravitational field. Their experimental results still set the standard in this field and our design is similar. We attached hollow mock-ups of LISA TMs to the ends of the arms of a cross-shaped structure which serves as the pendulum's inertial member. This entire structure is suspended from a torsion fiber. The rotational or torsional degree of freedom of the pendulum is then decoupled from Earth's gravitational field and in 'free fall' above its mHz resonance frequency. The lever arm of the pendulum allows us to treat its rotation as a simple translation of the TM inside the GRS. Hollow TMs are used to reduce the mass of the pendulum. The reduced mass allows to reduce the thickness of the suspension fiber which reduces the thermal noise. The reduced mass also increases the accelerations and displacements caused by all surface forces which enables us to study these least understood forces in greater detail [4] while accelerations caused by bulk forces are typically independent of the mass. Two of the four TMs are then encased in separate GRSs to study their interactions with the TMs.

3. Mechanical Design

3.1. Torsion Pendulum

The inertial member structure is comprised of four cubic test masses which are mounted on four 22 cm long struts. Each 46 mm, 480 g test mass is hollow and fabricated from aluminum to minimize mass. The aluminum surfaces are gold-coated to minimize stray surface potentials. Each test mass is constructed from five individual parts that are bonded together with EPO-TEK H20E conductive epoxy. Two of the TMs are surrounded by simplified electrode housings (EHs). These two TMs are fixed to the main pendulum structure via quartz rings to electrically isolate them. The inertial member is shown in Figure 1, and the full system in Figure 2.

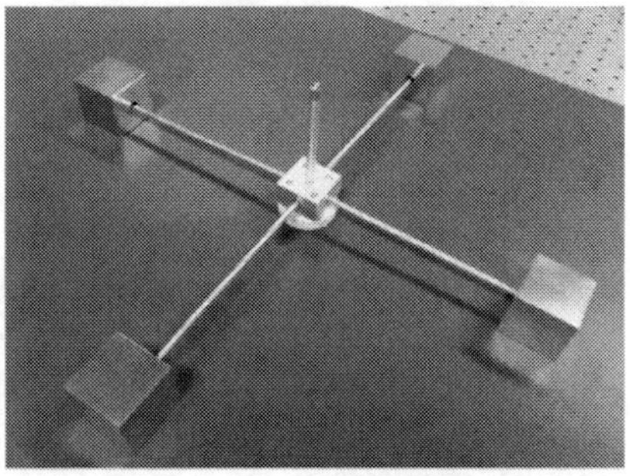

Figure 1: Inertial Member. The hollow TM mockups are mounted at the ends of four rods, each in turn attached to a central assembly.

The suspension fiber is fixed to a shaft that elevates the attachment point to a position 10 cm above the center of mass of the pendulum. This maximizes the natural frequency of the pendulum in the tilting mode so that it is well above the measurement band and the rotational mode frequency. The suspension fiber itself is a 1 m long tungsten fiber with a diameter of 50 μm. The fiber thickness was minimized for the given weight of 480 g (with a safety factor of roughly 2) to minimize fiber thermal noise. This fiber is glued to a custom fiber attachment cap at each end such that a standard screw can be attached. A magnetic damper is located above the main fiber to damp out the swinging mode of the pendulum. The damper consists of two sets of neodymium magnets fixed to the vacuum chamber. An aluminum disc-shaped conductor is fixed to the suspension fiber within their magnetic field so that any swing in the pendulum is damped by eddy currents formed in the conductor. The magnetic damper is shown in Figure 3b. A thicker (250 μm diameter) tungsten fiber is located above the magnetic damper. This fiber preserves all degrees of freedom of the pendulum, while experiencing a negligible amount of deformation during torsional oscillation. Using similar custom caps, the thick fiber is fixed to the top of the vacuum chamber.

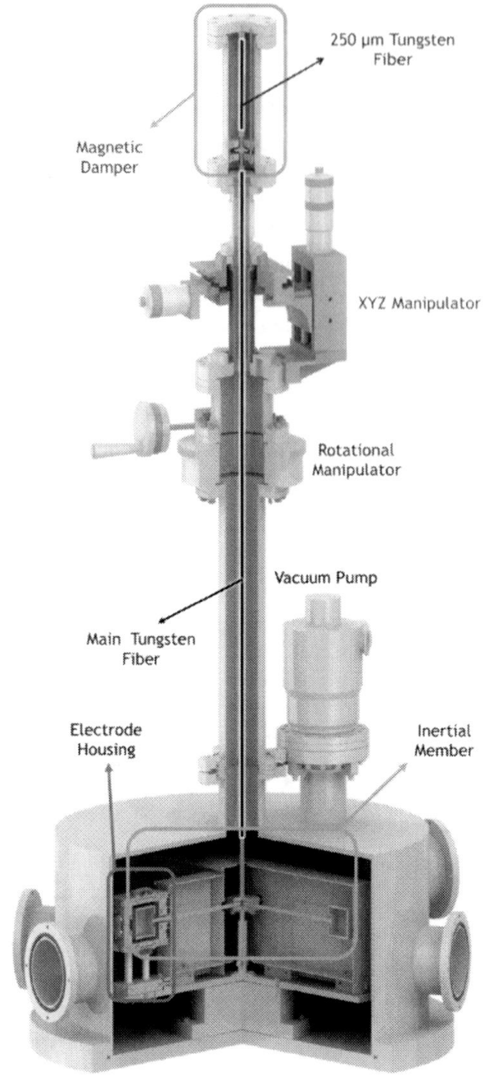

Figure 2: Complete UF Torsion Pendulum assembly. Various features are pointed out.

3.2. Simplified Electrode Housing

The current electrode housing is a simplified version of the LISA Pathfinder GRS, and includes a total of 6 electrodes: 1 located on each of the interior faces. In the future, we will replace this EH with potential flight models that include electrodes to sense all degrees of freedom. The inertial member of the pendulum hangs from the suspension fiber such that two of the test masses remain enclosed in electrode housings for capacitive readout. The structure of the housing is also gold-coated aluminum. The electrodes are electrically isolated from the structure of the housing with ceramic bushings. In the sensitive direction (in which the pendulum rotates about its axis), the gaps between the electrodes and test mass surface are 8 mm. This gap size was chosen (in contrast to the 4 mm gap size in the LPF GRS) to minimize interactions between the GRS and the TM while we improve our facility as a whole.

UV optical fibers connect directly to the structure of the electrode housing so that UV light can be used to control the charge of the test mass. There are 3 available optical fiber ports located on each electrode housing: one that illuminates the test mass, one that illuminates the housing, and one that illuminates both. The electrode housing is shown in Figure 3a.

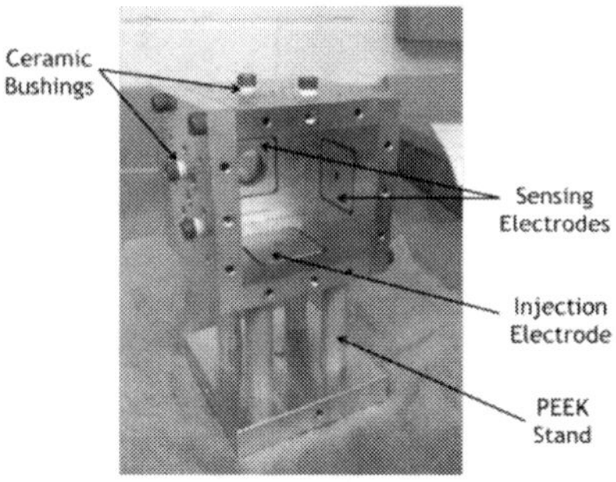

(a)

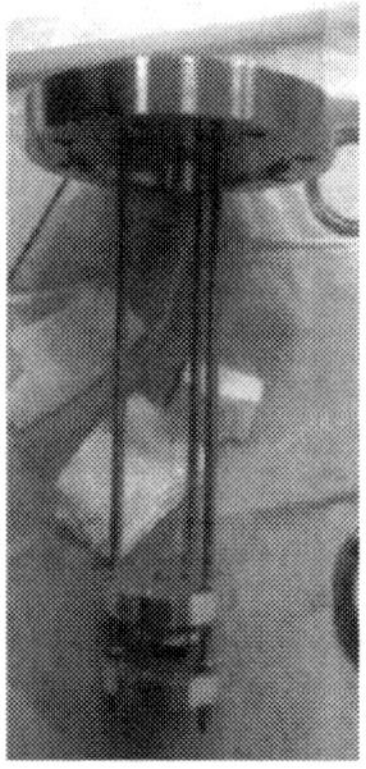

(b)

Figure 3: Photographs of the (a) electrode housing and (b) eddy current damper, both prior to integration with the pendulum.

All of the abovementioned components are located inside a vacuum chamber. The vacuum chamber is 60 cm in diameter. It includes a tall

Mode	Rotating	Swinging	Tilting	Bouncing
Parameters	$\Gamma = \frac{\pi r^4}{2L}\left(E_r + \frac{mg}{\pi r^2}\right)$ $E_r = 161$ GPa $I_r = 1.65 \times 10^{-2}$ $\mathrm{kg} \cdot \mathrm{m}^2$	$L = 1\mathrm{m}$ $g = 9.81\ \mathrm{m/s^2}$ $m = 477$ g	$I_t = 8.37 \times 10^{-3}$ $kg \cdot m^2$ $h = 98.8$ mm	$k = \frac{E_b \pi r^2}{L}$ $r = 25\ \mu\mathrm{m}$ $E_b = 411$ GPa
Equation	$f_r = \frac{1}{2\pi}\sqrt{\frac{\Gamma}{I_r}}$	$f_s = \frac{1}{2\pi}\sqrt{\frac{g}{L}}$	$f_t = \frac{1}{2\pi}\sqrt{\frac{mgh}{mh^2+I_t}}$	$f_b = \frac{1}{2\pi}\sqrt{\frac{k}{m}}$
Predicted Value	$f_r = 393\ \mu$Hz	$f_s = 455$ mHz	$f_t = 948$ mHz	$f_b = 6.55$ Hz
Measured Value	$f_r = 310 \pm 8\ \mu$Hz	$f_s = 434.2$ ± 0.1 mHz	$f_t = 1.27$ ± 0.03 Hz	$f_b = 6.08$ ± 0.04 Hz

Table 1: Natural frequencies of pendulum, both as computed from theory and as measured (see Fig.7). Damping terms have been left out as they only negligibly influence the frequencies. The measured values are the average of three separate overnight measurements.

tube that encloses the suspension fiber. At the top of the tube, the fiber is attached to both a rotational manipulator to allow adjustment of the orientation of the pendulum, and a translational XYZ manipulator to allow for calibration and centering, as shown in Fig. 2.

Finally, we include a summary of the different parameters used to predict the natural frequencies of the pendulum, shown in Table 1. The rotational and bouncing mode frequencies were constrained by minimizing the fiber radius for the inertial member design, and all modes were constrained by practical length considerations. For the swing, tilt, and bounce mode, the discrepancies between predicted and measured values are not critical and can be attributed to machining and assembly tolerances as well as to some oversimplifications in the model. The difference in values for f_r is due to the presence of charge on our pendulum and is explained more thoroughly below.

4. Electronic Design

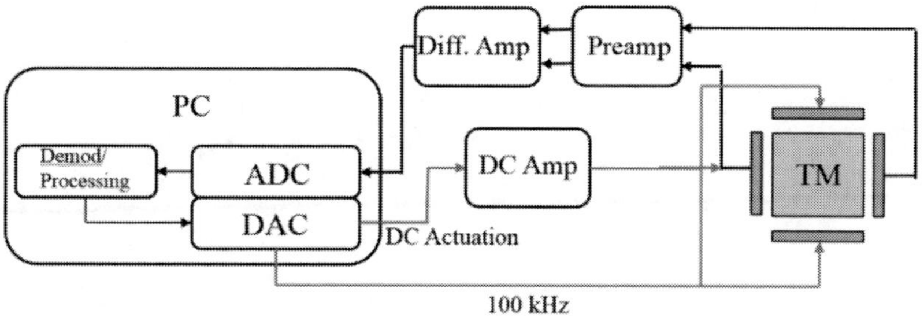

Figure 4: Block diagram of signal flow in the UF Torsion Pendulum readout scheme.

4.1. Analog Electronics

The GRS measures the position of the TM by measuring the position-dependent capacitance of a set of electrodes towards the TM. Our electronics function by polarizing the TM at 100 kHz, converting the current flowing through the TM/electrode capacitors to a voltage with a transimpedence amplifier, and digitizing the result. The transimpedence amplifiers hold the sensing electrodes at virtual ground with respect to the TM, which guarantees that the voltage drop between the TM and electrodes is only due (discounting noise sources) to the injection voltage. This is important not only to maintain purity of the 100 kHz signal, but also to avoid unintentional forces on the TM.

In more detail, the injection voltage V_{inj} is applied directly to a pair of electrodes on a given GRS prototype as shown in Fig. 3a, so the resultant TM voltage at 100 kHz (the frequency to be assumed throughout the text) is

$$V_{TM} = \frac{2C_{inj}}{C_{tot}} V_{inj} \equiv \alpha V_{inj}, \tag{2}$$

where C_{inj} is the capacitance of one injection electrode towards the TM, and C_{tot} is the total capacitance of the TM towards the EH and electrodes (all held at ground). Further consider the position dependence of both capacitances in the "soft" DOF in the parallel plate model:

$$C_{\pm}(x) = \frac{\epsilon_0 A_{elec}}{d \mp x} \tag{3}$$

$$\approx \frac{\epsilon_0 A_{elec}}{d} \left(1 \mp \frac{x}{d} + \left(\frac{x}{d}\right)^2 \pm O\left(\left[\frac{x}{d}\right]^3\right)\right) \tag{4}$$

$$\to \Delta C \equiv C_+ - C_- = \frac{-2\epsilon_0 A_{elec}}{d}\frac{x}{d} + O\left(\left[\frac{x}{d}\right]^3\right) \tag{5}$$

d is the gap between the TM and electrode when the TM is centered, and A_{elec} is the area of that electrode. For small displacements ($\frac{x}{d} \ll 1$) the difference between two opposing capacitances is close to linear in the TM displacement. Hence, a measurement of the differential capacitance is a measurement of the position of the TM with respect to the center of the housing. The electronics shown in Fig. 5 shows the sensor circuit. The 100 kHz signal passes through the initial high-pass filter to the transimpedence amplifiers and finally to the differential amplifier. The total transfer function is

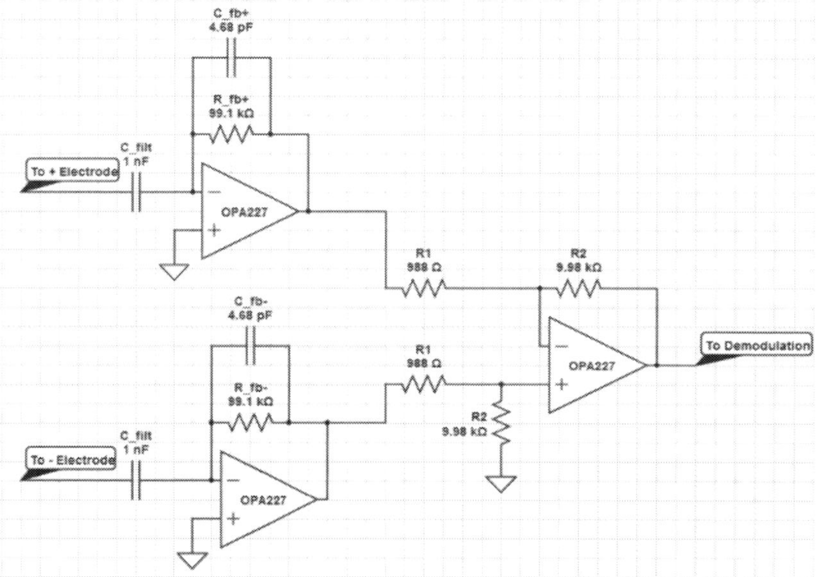

Figure 5: Circuit diagram of the UF Torsion Pendulum analog readout electronics.

$$V_{out} = -\frac{R_2}{R_1}\frac{\Delta C}{C_{fb}}\alpha V_{inj} = -\frac{R_2}{R_1}\frac{\alpha V_{inj}}{C_{fb}}\frac{2\epsilon_0 A_{elec}}{d}\frac{x}{d} \tag{6}$$

Therefore, the amplitude of the readout electronics output at 100 kHz, which is the desired readout signal, is linear in the TM displacement. The expected limiting noise source is shot noise in R_{fb} at an equivalent displacement noise level of $10\text{nm}/\sqrt{\text{Hz}}$.

4.2. Digital Electronics

The signal is a 100 kHz signal with an amplitude which is proportional to the displacement and which changes phase by 180° at the lock point. This signal is then demodulation digitally on board the DAQ card. The 100 kHz signal from the readout electronics is passed through a single-pole antialiasing filter with a corner frequency of 350 kHz before it is digitized with a 700 kHz sampling rate. The digitized signal is then multiplied with both the 100 kHz signal which was used to polarize the TM and that same signal phaseshifted by 90°. We use cascaded integrator-comb (CIC) filters to integrate and downsample these products by a factor of 8192, for a final data rate of 85.6 Hz. These demodulated signals are passed to the host computer for storage. By observing the In-phase/Quadrature vector swept out by the pendulum's motion, we can rotate the demodulation phase such that all of the information is in one quadrature to maximize signal-to-noise ratio. The dataflow for this scheme is shown in Fig. 6.

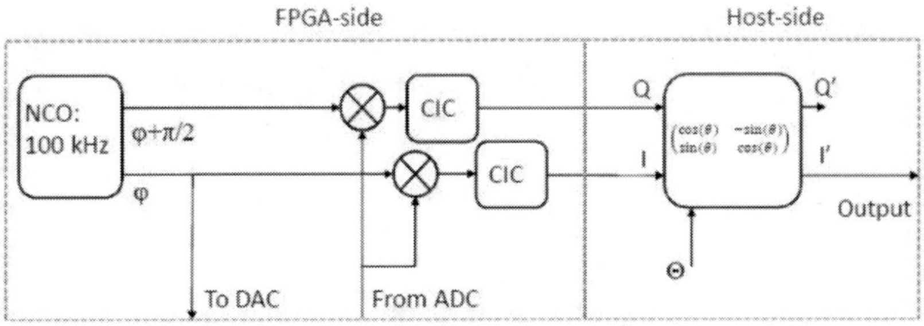

Figure 6: Block diagram of the digital side of the readout electronics. Here ϕ is the Numerically-Controlled Oscillator (NCO) angle and Θ is the demodulation angle.

The DAQ is also used to apply actuation signals to the pendulum. The force between two capacitor plates (in an idealized infinite-parallel-plate model) is given by

$$F_{cap}(x) = -\frac{1}{2}\frac{\partial C}{\partial x}(V_1 - V_2)^2 \tag{7}$$

where V_1 and V_2 are the voltages on the capacitor plates. Therefore, we can apply a force on the TM by applying a voltage on the appropriate electrode and pull the TM in that direction. We currently have monopolar DC amplifiers (which we will upgrade to bipolar AC amplifiers shortly) which can amplify up to 1000 V for intially stabilizing the pendulum. For precision actuation, we use the 10 V output of the DAQ directly. These signals are first sent through a low-pass filter to avoid interference with the readout, and thereafter directly applied to the electrodes.

5. Current Noise Performance

We present here a measurement of the noise performance of the facility at its current state of development and a plot of the theoretical limit on the performance of the pendulum. After allowing the chamber to pump down to its base pressure of 2×10^{-5} torr, both sensors (referred to as "Gamma" and "Delta") were calibrated using the micrometer translation stage shown in Fig. 2. The equilibrium point of the system was moved to the center of the sensor using the rotation stage, and the pendulum's rotational mode was damped using the electrostatic actuation. The actuation was then shut off and the pendulum was allowed to freely oscillate over a weekend for nearly 59 hours. During this period, the positions of the TMs in both the Gamma and Delta sensors were recorded. The motion is dominated by the swinging motion which is correlated in both degrees of freedom. We are more interested in the differential motion which is directly associated with the torsional rotation of the pendulum. Fig. 7 shows the linear spectral density (LSD) from the Gamma sensor signal; the LSD of the Delta sensor signal looks identical, and the differential signal ϕ_{LSD}. The signal has been scaled by the lever arm of 22 cm while the

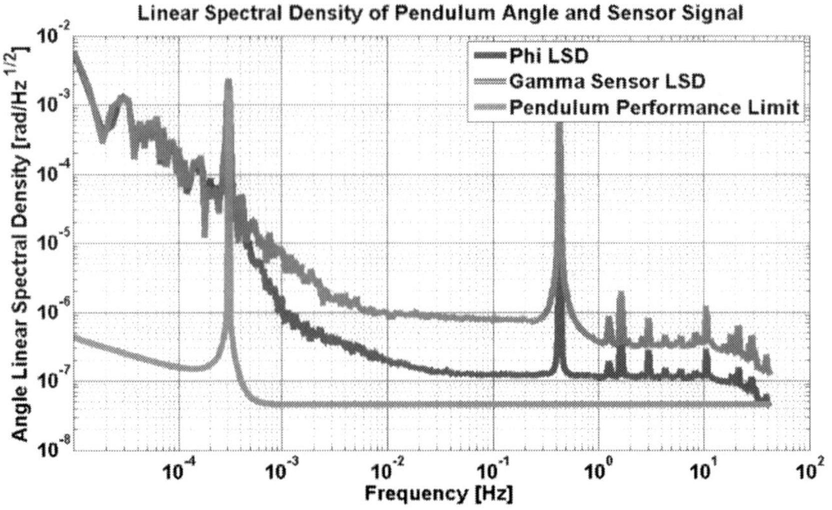

Figure 7: Linear Spectral Density of TM angle and position, with the theoretical pendulum performance limit in red.

differential signal is scaled by the distance between both test masses to express them as angles. The LISA requirement and the breakdown of the pendulum performance limit is shown in Fig. 8.

Note firstly that the rotational mode peak is somewhat less than what it should be from theory (see Table 1). This reduction in resonant frequency is due to the presence of charge on one of the TMs, which we measure to be on the order of 10^{-10} C. Our charge control system (see[ ]) has thus far been unable to remove it completely. As mentioned above, charge on the TMs adds (to first order) a negative stiffness which counteracts the natural stiffness of the fiber and lowers the resonant frequency. All other resonances were close to theory.

A large amount of common-mode noise between the sensors in the 1 to 100 mHz band is taken out in the Phi LSD; this implies that it resides in the swing mode. Our ability to remove this noise in the Phi spectrum depends strongly on the accuracy of our relative calibration between the sensors, so investigating this flat, non-electronic noise between 10 mHz and the swing resonance is a prime concern. We are also limited around 1 mHz by an unknown noise source that goes as 1/f, here partly obscured by the rotational resonance. At the lowest frequencies (below the resonance), the cross-coherence between Phi and temperature (measured but not shown here) is nearly 1; while this is currently below our measurement band, it will likely become important as we improve the

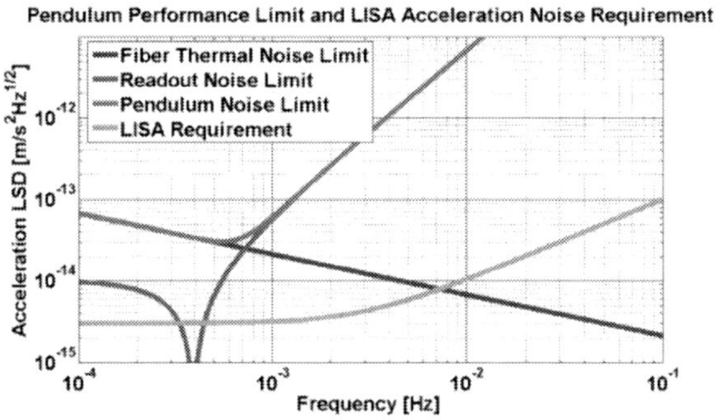

Figure 8: UF Torsion Pendulum theoretical noise limit and LISA sensitivity requirement. Here the acceleration noise is an equivalent noise on a LISA TM, accounting for the difference in mass between the LISA TMs and our pendulum. The fiber thermal noise term assumes a fiber Q=3000, and the readout noise uses the 10 nm/$\sqrt{\text{Hz}}$ theoretical value from the readout electronics. The "Pendulum Noise Limit" assumes a 2 kg LISA TM.

performance of the pendulum at higher frequencies, and as such we will build a thermal room around the pendulum to stabilize the thermal environment.

Finally, analysis of the prototype readout electronics suggests the noise floor at higher frequencies (above the swing mode resonance) should be around 10 $\frac{\text{nm}}{\sqrt{\text{Hz}}}$, but in reality it is several times higher than this; the noise is definitely electronic in origin as it appears at the same level when the GRS is replaced by two static capacitors. Since there is actually a strong cross-coherence between the two sensors above 100 mHz, we suspect this higher noise level may be due to either insufficient shielding or a common ground loop. We will continue to investigate the cause of this excess and improve the design of the readout electronics to bring them in line with the LISA GRS performance requirement.

6. Future Work

The ongoing mission of the UF Torsion Pendulum group is to measure the performance of various GRS designs and test new technologies with which we can improve that performance. Practically speaking, this means we will be measuring the transfer function of various physical parameters related to the GRS into external torque. With appropriate models of the LISA environment, this will allow us to assert that a given GRS model would perform at a certain level during the actual mission. For exmaple, we will integrate heating elements to vary the temperature of different parts of the GRS (both net and differentially) or the pressure inside the chamber. Investigations into these transfer functions will begin as soon as we finish the commissioning phase for the facility and produce a first prototype of a LISA-like GRS, both of which are currently in progress. We will also use the facility to test alternative hardware and techniques for the Charge Management System. Another broader goal is to reduce sources of noise in the pendulum as much as possible - not only to be able to make more sensitive transfer function measurements, but also to be able to make statements about the level of excess noise from unmodeled sources present in the system. As a simple example of this procedure, insulating styrofoam will be added around the vacuum chamber to improve thermal stability. To further improve thermal stability, the entire vacuum chamber will be placed in a two-meter-deep concrete pit, which should provide some isolation against building vibrations as well.

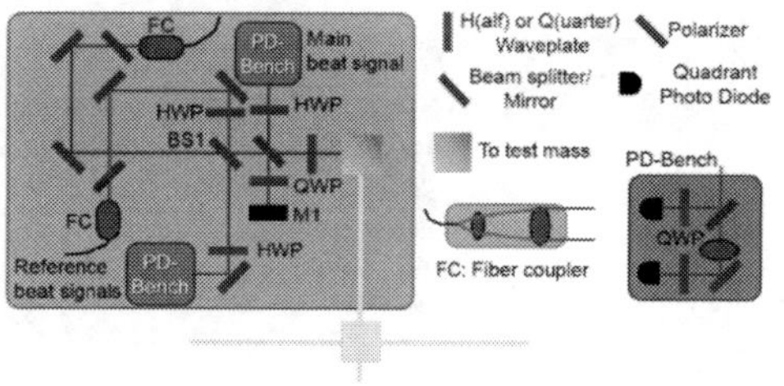

Figure 9: Interferometric readout scheme for the UF Torsion Pendulum.

Furthermore, we will implement an interferometric readout for the pendulum. A potential design is shown in Fig. 9. It is a polarization-multiplexed heterodyne interferometer, wherein (in Fig. 9) the red and blue lasers are orthogonally polarized, and offset by the heterodyning frequency. A reference beat signal between the two lasers is taken on the reference PD-bench before BS1. The blue laser then gathers phase information from the test mass, while the red laser gathers a constant amount of phase from a reference mirror (M1). The beat between these signals is then measured on the main PD-bench. The advantage of this scheme is a minimization of the space on the whole bench which needs to be ultrastable. Finally, we will be able to do wavefront sensing with the Guoy telescope on each PD-bench to also measure the tilt of the TM.

7. References

[1] Stebbins R *et al.* (2009). Astro2010 RFI Response.
[2] Amaro-Seoane P *et al.*, (2012), arXiv:1201.3621
[3] Carbone L *et al.* (2007) Phys. Rev. D 75, 042001
[4] Ciani G (2008) Ph.D. Thesis, University of Trento.
[5] Hewitson M *et al.* (2009) *Class. Quantum Grav.* **26** 094003
[6] Olatunde T *et al.* (2014) *LISA Symposium X Proceedings* This Issue.

Journal of Physics: Conference Series **610** (2015) 012039 doi:10.1088/1742-6596/610/1/012039

Adhesive Bonding for Optical Metrology Systems in Space Applications

Martin Gohlke[1], Thilo Schuldt[1,2], Klaus Döringshoff[5], Achim Peters[5], Ulrich Johann[3], Dennis Weise[3], Claus Braxmaier[1,4]

[1] German Aerospace Center (DLR), Robert Hooke Straße 7, 28359 Bremen, Germany
[2] Hochschule für Technik, Wirtschaft & Gestaltung, Brauneggerstr. 55, 78462 Konstanz, Germany
[3] Airbus D&S, Claude-Dornier-Straße, 88039 Friedrichshafen, Germany
[4] Center of Applied Space Technology and Microgravity (ZARM), University Bremen, Am Fallturm 1, 28359 Bremen, Germany
[5] Humboldt University Berlin, Newtonstrasse 15, 12489 Berlin, Germany

E-mail: `martin.gohlke@dlr.de`

Abstract. Laser based metrology systems become more and more attractive for space applications and are the core elements of planned missions such as LISA (NGO, eLISA) or NGGM where laser interferometry is used for distance measurements between satellites. The GRACE-FO mission will for the first time demonstrate a Laser Ranging Instrument (LRI) in space, starting 2017. Laser based metrology also includes optical clocks/references, either as ultra-stable light source for high sensitivity interferometry or as scientific payload e.g. proposed in fundamental physics missions such as mSTAR (mini SpaceTime Asymmetry Research), a mission dedicated to perform a Kennedy-Thorndike experiment on a satellite in a low-Earth orbit. To enable the use of existing optical laboratory setups, optimization with respect to power consumption, weight and dimensions is necessary. At the same time the thermal and structural stability must be increased. Over the last few years we investigated adhesive bonding of optical components to thermally highly stable glass ceramics as an easy-to-handle assembly integration technology. Several setups were implemented and tested for potential later use in space applications. We realized a heterodyne LISA related interferometer with demonstrated noise levels in the pm-range for translation measurement and nano-radiant-range for tilt measurements and two iodine frequency references on Elegant Breadboard (EBB) and Engineering Model (EM) level with frequency stabilities in the 10^{-15} range for longer integration times. The EM setup was thermally cycled and vibration tested.

Adhesive bonding of optical components is a well-known technique. The advantages w.r.t. standard opto-mechanical setups are the mechanical and thermal stability; a disadvantage is that components once bonded can not be adjusted anymore. However, there are many application where a later adjustment is not needed or required. Beside adhesive bonding, hydroxide-catalysis (HC [1]) bonding and optical contacting are well-known methods to attach optical components. The obtained long-term thermal and mechanical stabilities are impressive. However, the techniques have higher requirements on the components (surface properties) and a cleanroom environment during integration is needed. The presented technique of adhesive bonding works with a very liquid adhesive to achieve thin layer thicknesses. It can be carried out with less effort than HC bonding. For many optical setups adhesive bonding might be a simple alternative to HC bonding and optical contacting.

Figure 1. Left: Photograph of the testboard developed for comparing adhesive bonding and hydroxide-catalysis bonding. Right: Heterodyne interferometer developed in the LISA context, realized using adhesive bonding technology for integration of the optical components.

1. Assembly Integration Technology

In a first activity, different adhesives were tested. The space-qualified Hysol 9313 two-component epoxy is very liquid, has a strong adhesive strength (28.9 MPa) and is thus well suited for the application described here. In tests, layer thicknesses of a few μm were demonstrated (compared to HC bonding layer thicknesses in the range of several dozen nm [2]). A thin adhesive layer is important for the later instrument performance: the thinner the film thickness, the smaller the film thickness differences, which potentially lead to errors and tilts of the bonded component. A second effect is the smaller influence of thermal expansion of the adhesive.

After these preliminary tests a simple test board was designed, cf. figure 2, left. Here, the integration steps for positioning the optical components were tested. Six mirrors ($20 \times 15 \times 7\,\mathrm{mm}^3$) were attached to a $10 \times 10 \times 4\,\mathrm{cm}^3$ base-plate made of Zerodur. The surface of the base was specified with $\lambda/10$, this is similar to the requirements of HC bonding and guarantees the parallelism of the laser beam w.r.t. the surface of the base-plate. A pitch of the optical component cannot be adjusted by tilting. Therefore, the mirrors have specific requirements where the angle between the base and the optical surface has be within $90° \pm 2$ arcseconds. Two of the six mirrors were HC bonded to the base-plate, they act as reference mirrors for the later performance tests. The other four mirrors were attached using adhesive bonding. Therefore, a specifically designed jig was used in order to bring the substrate in the right position and fix them for 24 hours during the curing process of the adhesive [3]. The surfaces of the base of the mirrors were polished for the HC bonding and matt-grinded for the adhesive bonding.

2. Optical Setups

2.1. Heterodyne Interferometer

In a first application, we integrated a heterodyne interferometer, developed in the LISA context. 17 optical components (beam splitters and mirrors made of fused silica) were integrated to a $20 \times 20 \times 4\,\mathrm{cm}^3$ base-plate made of Zerodur where a contrast of 90 % of the superimposed beams after the recombination BS was achieved. This is an indication of the small tilting of the bonded optical component w.r.t the surface of the base-plate. The noise performance of the interferometer is in the pm range for translation measurements and in the nrad range for tilting (both in the LISA frequency band) and it was not limited by the mechanical design.

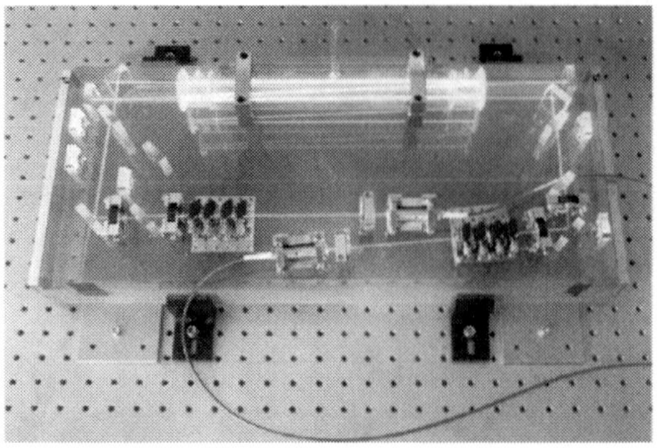

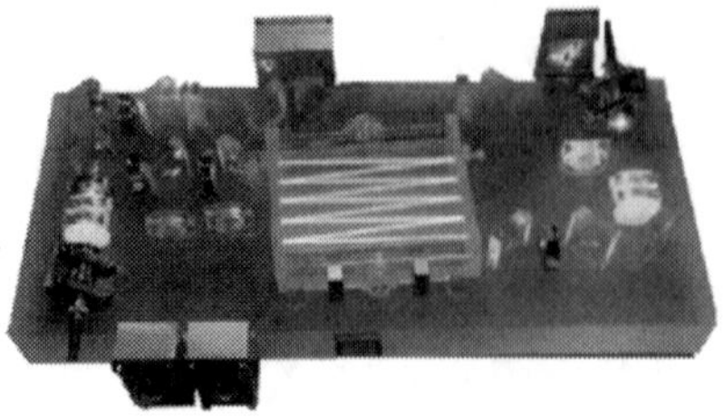

Figure 2. Left: Photograph of the fully integrated iodine standard on elegant breadboard level. Right: The engineering model of the iodine standard with the nine-pass iodione cell.

2.2. Iodine-Based Frequency References

Two setups on elegant breadboard (EBB) and engineering model (EM) level were developed, based on a state-of-the-art laboratory setup realized at the Humboldt-University Berlin [4]. The laboratory setup employs modulation transfer spectroscopy using an 80 cm long iodine cell. A frequency stability in the range of 3×10^{-15} for integration times between 100 s and 10,000 s was demonstrated. Based on this setup, an EBB and later an EM was developed, which should be smaller and lighter than the existing setup. They should also have a better thermal and mechanical stability in order to survive environmental tests such as thermal cycling and vibrational testing, both without spoiling the reached performance [5].

With respect to its later application in space, a fiber-coupled spectroscopy setup on elegant breadboard (EBB) level was realized using adhesive bonding technology as described above. In order to reduce the size of the setup the beam path through the iodine cell was folded. In this triple pass configuration, the length of the cell can be reduced to 30 cm while the effective length is still 90 cm. The optical setup can now be integrated on a smaller $55 \times 25 \times 5\,\text{cm}^3$ base-plate. The base-plate is made of OHARA Clearceram-Z HS, a thermally and mechanically highly stable glass ceramics with a low CTE of $2 \cdot 10^{-8}\,\text{K}^{-1}$. All optical components which possibly influence the beam pointing stability (beamsplitter, mirrors etc.) are integrated using adhesive bonding. All other components (polarizer, wave plates, etc.) which do not influence the optical path were held in special holders made of Invar (CTE of $\sim 10^{-6}\,\text{K}^{-1}$). Additional plates made of Invar were also attached to the sides of the base-plates, they are used for integration of detectors and cell cooling.

Based on the experience with the EBB setup, a more compact and ruggedized setup on engineering model (EM) level was designed and realized. The main issue for a compact setup is the realization of a compact multipass cell. The cell uses a $10 \times 10 \times 3\,\text{cm}^3$ fused silica spacer with wedged windows. The cell of the EM is designed for a nine-pass configuration (corresponding to 90 cm interaction length as used before). All optical components are made of fused silica. They are integrated on a $38 \times 18 \times 10\,\text{cm}^3$ base-plate also made of fused silica. Performance tests at the Humboldt University Berlin show similar frequency stability of about $1 \cdot 10^{-14}$ at an integration time of 1 s and below $5 \cdot 10^{-15}$ at integration times between 10 s and 100 s for the EM and EBB setups, comparable to the laboratory setup. The EM setup was also environmentally tested including thermal cycling (-20°C to +60°C) and vibrational testing (sine vibration up to 30 g; random vibration up to 25.1 g). The frequency stability was measured before and after the tests where no degradation was observed.

Journal of Physics: Conference Series **610** (2015) 012039 doi:10.1088/1742-6596/610/1/012039

3. Conclusion

Adhesive bonding as discussed in this article is seen as an easy-to-handle technique for realizing space-compatible optical systems. A high sensitivity heterodyne interferometer was realized showing noise levels in the pm range for translations and nrad range for tilting. Two compact iodine-based frequency references were developed, showing a frequency stability in the 10^{-15} range at integration times between 100 s and 1.000 s. These iodine references are an attractive alternative to the baseline cavity-based frequency references and fulfill the requirements for LISA and NGGM. Another advantage is the absolute frequency stability of an iodine setup. The offset frequencies are in the range of only a few kHz and below which would save time in the lock acquisition between several spacecraft.

References

[1] E. J. Elliffe, J. Bogenstahl, A. Deshpande, J. Hough, C. Killow, S. Reid, D. Robertson, S. Rowan, H. Ward and G. Cagnoli: Hydroxide-catalysis bonding for stable optical systems for space, Classical and Quantum Gravity, vol. 22 (2005), no. 10, p. 257

[2] N. L. Beveridge: Characterisation of silicon-silicon hydroxide catalysis bonds for future gravitational wave detectors, PhD Thesis, University of Glasgow, 2012

[3] S. Ressel, M. Gohlke, D. Rauen, T. Schuldt, W. Kronast, U. Mescheder, U. Johann, D. Weise, and C. Braxmaier: Ultrastable assembly and integration technology for ground- and space-based optical systems, Appl. Opt., vol. 49 (2010), no. 22, pp. 42964303.

[4] K. Döringshoff, M. Reggentin, M. Nagel, E. Kovalchuk, A. Keetman, T. Schuldt, C. Braxmaier, and A. Peters: Iodine based optical frequency reference with 10^{-15} stability, Proceedings of the 26th European Frequency and Time Forum, EFTF 2012, 2012.

[5] T. Schuldt, J. Stühler, K. Döringshoff, J. Pahl, E. V. Kovalchuk, M. Franz, U. Johann, A. Peters, C. Braxmaier: An Iodine-Based Ultra-Stable Optical Frequency Reference and its Application in Fundamental Physics Space Missions, Proceedings of the 27th European Frequency and Time Forum, EFTF 2014, 2014.

Journal of Physics: Conference Series **610** (2015) 012040 doi:10.1088/1742-6596/610/1/012040

Residual Gas Noise in the Test-mass Module for DECIGO Pathfinder

K Okutomi[1], T Akutsu[2], M Ando[3], M Nikaido[4], N Tanaka[2], Y Torii[2], S Sato[5], K Izumi[6], D Chen[7], DPF Working Group

[1] raduate University for Advanced Study, Shonan illage, ayama, anagawa 24 - 1 3, a an

[2] ational Astronomical bservatory of a an, sawa 2-21-1, o yo 181-8588, a an

[3] e artment of hysics, he University of o yo, o yo 113- 33, a an

[4] chanomi u University, 2-1-1, o yo 112- 12, a an

[5] Faculty of Engineering, osei University, a inocho, o yo 184-8584, a an

[6] LI anford bservatory, 127124 Rte 1 , Richland, WA 354, USA

[7] e artment of Astronomy, he University of o yo, o yo 113- 33, a an

E-mail koki.okutomi@nao.ac.jp

Abstract. ECI athfinder is the first milestone mission for ECI , a future gravitational wave antenna. In F, residual gas noise acting on the test mass is estimated to increase and e ceed the re uirement for force noise of 1×1 $^{-15}$ $/\sqrt{\ }$ due to geometry of the test-mass module. We erformed a Monte Carlo simulation to calculate the residual gas noise and found that the engineering model of the test-mass module cannnot satisfy the re uirement. o reduce the gas noise, we resent revised geometry of the test-mass module using comb-li e electrodes.

1. Introduction

ECI is a future s ace gravitational-wave detector which aims to detect gravitational waves around .1 1 . ECI has Fabry- erot cavities between s ace crafts with a baseline of 1, m. As the first ste toward ECI , a small satellite called ECI athfinder F is lanned to verify some of the ey technologies re uired for ECI 2 . F has a 3 -cm Fabry- erot interferometer with free-falling test masses.

In these s ace missions, force noise acting on the test masses is estimated to be the largest noise source to limit the sensitivity of the detector. o achieve the strain noise of less than 2×1 $^{-24}$ $^{-1/2}$ around .1 , ECI has stringent re uirement for force noise of 4×1 $^{-17}$ $/\sqrt{\ }$ in deci-hert band, which is 1 5 of LISA s re uirement. In the F mission, the force noise is re uired to be below 1×1 $^{-15}$ $/\sqrt{\ }$. owever, force noise from residual gas molecules is found to increase and e ceed the re uirement for F. his increase is caused by narrow ga s between the test masses and surrounding modules.

he F test masses are laced in each test-mass modules, housings with electrodes and mechanical systems for recise control of a osition of the test mass. he electrodes ma e units of ca acitive sensor and electrostatic actuator mounted on the inner surfaces of the housing, develo ed with reference to the LISA athfinder ravity Reference Sensor. For an engineering model of the test-mass module, each inner surface faces on a surface of the test mass with a

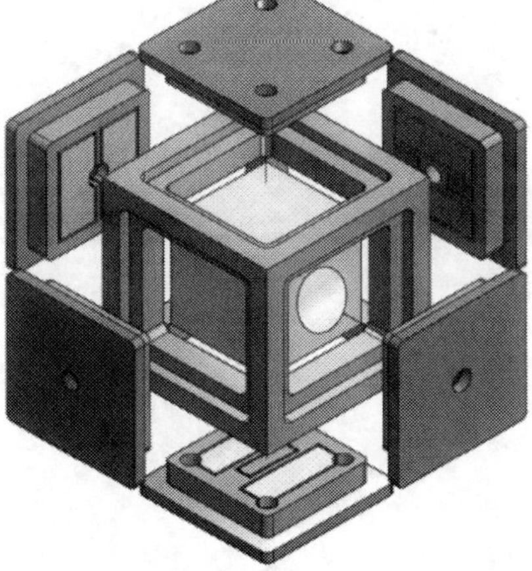

Figure 1. An assembly diagram of the test-mass housing for the engineering model of F. he test mass is laced inside a frame vivid green , each surface is faceed by sensor actuator electrodes yellow , in ection electrodes small light-brown rectangles and guard rings emerald green . Surrounding covers gray are also figured.

distance of 3 mm. his ga is smaller than 1 1 of a scale of the cubic test mass 5 mm on a side. Such a narrow ga causes increasing of force noise from molecular im acts, as investigated in the case of LISA athfinder 3 and ground detectors 4 .

o reduce the residual gas noise, we studied geometry of the test-mass module with a Monte-Carlo simulation. In this a er, a test-mass module with comb-li e electrodes is resented to mitigate increasing gas noise. Involving the change of the geometry, we also discussed actuation force with di erent sha es of the electrodes.

2. Residual Gas Noise in the Test-mass Module

Residual gas is one of force noise sources acting directly on the test mass. he test mass is disturbed by random molecular collisions and behaves li e Brownian motion. If we regard the collisions as inde endent im ulses, force noise from gas molecules has a fre uency inde endent s ectrum.

In F, gas force noise in the test-mass module is considered to be larger than the re uirement of 1×1^{-15} $/\sqrt{\ }$. his roblem comes from an increasing e ect of uctuation force with limited s ace around the test mass.

he cubic test masses are located in each housing called test-mass module. his module rovides a local control system of the test mass. Figure 1 shows an assemble diagram of the housing. Electrodes mounted on the inner surfaces of the housing ma e ca acitive sensors and electrostatic actuators which can monitor control a osition of the test mass without any mechanical contact to the test mass. When the test mass is oating at the center of the housing, the ga between the surfaces of the test mass and the inner walls of the housing is 3 mm.

he geometry with such narrow ga s around the test-mass raises additional gas force noise. We can attribute the increasing of gas noise to correlated collision im ulses from molecules in the narrow ga s. A molecure which entered into the ga re eats collisions with the test mass and walls until it esca es from the ga . here is an interaction time τ, the ty ical time during which a molecule stays in the ga and generates correlated collisions. Force noise on the test mass is nown to increases ro ortionally with the interaction time τ 4 .

We erformed a Monte Carlo simulation to estimate the residual gas noise with the engineering model of the housing. his simulation traces the tra ectory of gas molecules in a 2 ×2 ×2 cm cubic volume containing the test mass and the housing the electrodes and the guard rings shown in Figure 1. he si e of the volume is chosen as a ty ical scale of a s ace around the test-mass module in the satellite. he electrodes and the guard rings enclose the test

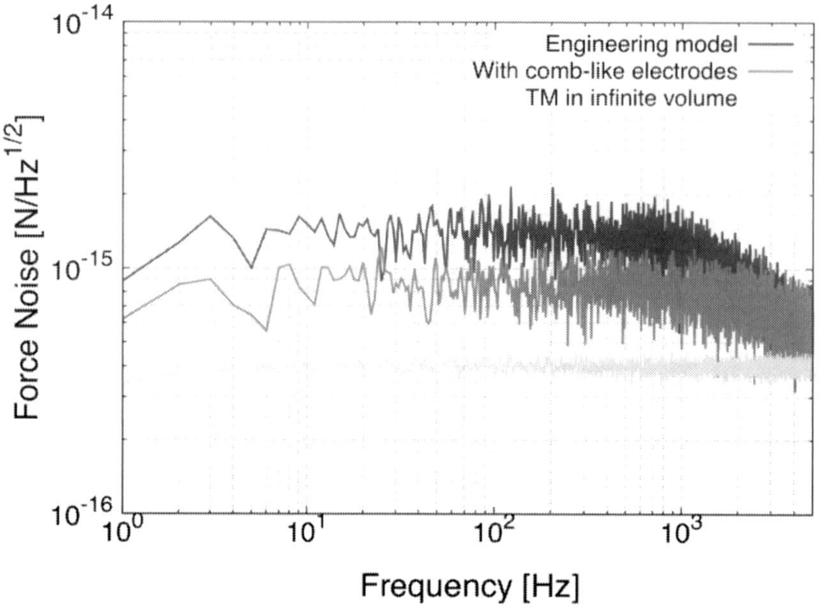

Figure 2. Simulated force noise s ectrum from residual gas molecules for the housing with the engineering model blue and the housing with comb-li e electordes red , calculated for 1 $^{-6}$ a of $_2$ at room tem erature. A force noise s ectrum with a test mass M in infinite gas volume green is also lotted for com arison.

mass at a distance of 3 mm. his calculation model doesn t include the frame and the covers see Figure 1 .

In this simulation, molecules are distributed uniformly and randomly in the volume with random velocities following a Ma well-Bolt mann distribution. Collisions of the molecules with the walls are com letely inelastic once molecules are absorbed, then subse uently reemitted following a cosine angular distribution, inde endent of the incoming velocity. At each collision with the test mass, time and transfered momentum vector are recorded. We have run 1, molecules simultaneous simulation for 1 seconds. After that, we converted obtained ower s ectrum density of a lied force noise on the test mass into that in the ressure of 1 $^{-6}$ a using the e uation of state. ere we assumed the tem erature as a room tem erature and s ecies of molecule as $_2$. We ignored molecular stic ing time on the surface and collisions of molecules with each other. his simulation methods are similar to those of the numerical study in 4 .

he calculated ower s ectrum density of the gas force noise in the housing for the engineering model is described as a blue s ectrum in Figure 2. We have confirmed that the used code is valid with com aring the result of the infinite gas volume case and an analytical calculation using e uation 11 in 5 . Com aring the green and the blue s ectrum, there is about increase in the result of the bread board model below a cut-o fre uency $f \approx 1$. his cut-o fre uency corres onds to a ty ical interaction time $\tau \approx \omega^{-1}$, a time scale that molecules re eat correlated collision im ulses in a narrow ga 4 . Assuming that the gas noise has a at s ectrum at lower fre uencies, calculated acceleration noise is about by a factor of 2 larger than the re uirement.

here are several strategies to mitigate the increased gas noise. ne is to lower residual gas ressure because the force brought from gas molecules to the test mass is ro ortional to the

Journal of Physics: Conference Series **610** (2015) 012040 doi:10.1088/1742-6596/610/1/012040

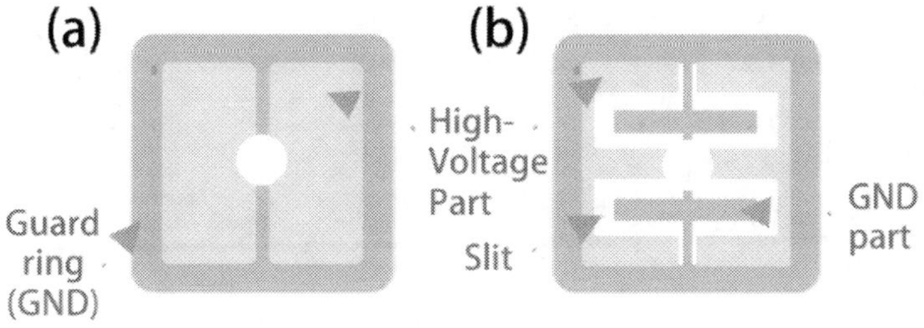

Figure 3. attern configurations of electrodes for the test-mass module. a Electrodes in the bread board model. b Comb-li e electrode, which has slits white art through the surface.

ressure, however it s di cult. Another is to e and the ga between the test mass and the housing. In this time, we have to consider a trade o between reducing the gas noise and the erformance of the electrostatic instruments of the test-mass module. In articular, a wider ga between the test mass and the electrodes ma es electrostatic actuators less owerful so that it cannot a ly enough force to control the test mass. In the ne t section, we will resent an alternative way to reduce the residual gas noise.

3. Mitigation of Gas Noise with Comb-like Electrodes

We designed comb-li e electrodes for the housing to reduce the gas force noise. Figure 3 shows the configuration of the comb-li e electordes. his configuration is designed to reduce the gas noise by e anding the s ace around the test mass. he comb-li e electrodes have slits through which gas molecules can es ace from the ga . Because the gas-noise increase is ro ortional to the ty ical interaction time τ, the slits can reduce the gas noise by shortening the time τ.

he result of residual gas noise simulation with comb-li e electrodes is lotted as a red curve in Figure 2. It shows that comb-li e electrodes has about half as large noise as the engineering model. A higher cut-o fre uency and less noise level are su osed to re ects the e ects of the slits. hus the model with comb-li e electrodes satisfies the force noise re uirement of $1 \times 1\ ^{-15}\ /\sqrt{\ }$. ne concern is that other heavier s ecies of molecule than $_2$ we assumed here are highly li ely to a ear in the orbit, e.g. $_2$ and $_2$. Because the force from gas molecules is ro ortional to their mass, the force noise will be larger by a factor of a s uare root of the molecular mass ratio. Moreover, other concern is deterioration in electrostatic shielding of the test mass by the electrodes. More slits would increase the sensitivity to any stray fields outside the housing and conse uently s oil the force noise to some e tent. herefore, further o timi ation of the comb configuration may be necessary.

In addition, we have to ta e the trade o of loss of the area of electrodes into account. Ma ing slits cannot avoid to decrease the area of the electrodes. Because electrostatic force a lied by an electrode is ro ortional to its area, we estimated how much loss we can allow to the area of the electrodes by finite element analysis. In this analysis, we modeled the test mass housing e ce t for the frame and the covers using C MS L. As the first ste of the study, we used an in ection voltage of C 3 V to estimate the actuation force reduction. his voltage was a lied to the two electorodes in direction to calculate the resulting electrostatic force. hen the electrostatic force a lied on the test mass is calculated. We assumed that other electrodes, guard rings and a lane at infinite distance are grounded.

able 1 shows estimated ma imum actuation force. With comb-li e electrodes, there can be a dro in actuation force involved to decreasing their area. owever the force er area is increased from the sim le electrodes. his im lies a ossibility of satisfying the re uired actuation e ciency while reducing the gas noise by the use of comb-li e electrodes. Com arison of this result to a case with sim le electrodes and e anded ga will be done in future study. Based on this result, we will roceed to erform more thorough calculations with AC voltage

Table 1. Ma imum actuation force with di erent sha es of electrode simulated with C MS L.

	et actuation force μ	Force area μ cm^2
Sim le electrode	45.	6.8
Comb-li e electrode	22.5	8.3

in ections, which will be used in the actual satellite.

4. Acknowledgement

he authers are grateful for su ort from the rant-in-Aid for Strategic evelo ment by A A, rant-in-Aid for Fundamental evelo ment for on-Board Instruments by A A, and Financial Su ort rogram for International Conference resentations by the Course-by-Course Education rogram in S E AI.

References

1 awamura S *et al* 2 11 *Class. Quantum Grav.* **28** 4 11
2 Ando M *et al* 2 1 *Class. Quantum Grav.* **27** 84 1
3 Cavalleri A *et al* 2 *Phys. Rev. Lett.* **103** 14 6 1
4 olesi R *et al* 2 11 *Phys. Rev. D* **84** 63 7
5 Cavalleri A *et al* 2 1 *Phys. Lett. A* **374**, 34 3365 336
6 rasso S, Altucci C, Barone F, Rago ino , Solimeno S, ham- u, inet , Abbate R 1 8 *Phys. Lett. A* **244** 36 37

10th International LISA Symposium (LISAX) IOP Publishing
Journal of Physics: Conference Series **610** (2015) 012041 doi:10.1088/1742-6596/610/1/012041

Transportable setup for amplifier phase fidelity measurements

M Tröbs[1], C Bogan[1], S Barke[1], G Kühn[1], J Reiche[1], G Heinzel[1] and K Danzmann[1]

[1] AEI Hannover, (MPI für Gravitationsphysik und Leibniz Universität Hannover), Callinstr. 38, 30167 Hannover, Germany

E-mail: `michael.troebs@aei.mpg.de`

Abstract. One possible laser source for the Laser Interferometer Space Antenna (LISA) consists of an Ytterbium-doped fiber amplifier originally developed for inter-satellite communication, seeded by the laser used for the technology demonstrator mission LISA Pathfinder. LISA needs to transmit clock information between its three spacecraft to correct for phase noise between the clocks on the individual spacecraft. For this purpose phase modulation sidebands at GHz frequencies will be imprinted on the laser beams between spacecraft. Differential phase noise between the carrier and a sideband introduced within the optical chain must be very low. We report on a transportable setup to measure the phase fidelity of optical amplifiers.

1. Introduction

The LISA mission [1, 2] aims to detect gravitional waves by measureing pathlength variations between satellites.

Although the laser concept ten years ago was a single NPRO emitting 1 W of laser power[3], now the baseline is a master-oscillator power-amplifier system with an electro-optical modulator (EOM) between oscillator and amplifier for clock noise transfer [4], ranging, and data communication [5, 6].

On each spacecraft, clock information will be modulated on the laser sent to the distant spacecraft as phase modulation sidebands at GHz frequencies. The phase of the sideband-sideband signals will be measured and used to correct for differential clock noise between the satellites. This requires that the phase of the sidebands is not disturbed by other elements in the transmission chain. This work focuses on the possible effects of the fiber amplifier behind the EOM.

The effect of differential phase noise within the transmission chain $\Delta\phi_{\mathrm{PRO}}$ should affect the phase read out of the carrier-carrier signal by not more than $6\,\mu\mathrm{rad}/\sqrt{\mathrm{Hz}}$,

$$\Delta\phi_{\mathrm{PRO}}(f) \leq 6\,\mu\mathrm{rad}/\sqrt{\mathrm{Hz}} \cdot u_{\mathrm{PL}}(f) \tag{1}$$

where the shape factor $u_{\mathrm{PL}}(f)$ is given by

$$u_{\mathrm{PL}}(f) = \sqrt{1 + \left(\frac{2.8\,\mathrm{mHz}}{f}\right)^4}. \tag{2}$$

The ancillary modulation error $\Delta\phi_{\mathrm{AME}}$ between a sideband at frequency f_{EOM} and the carrier is given by

$$\Delta\phi_{\mathrm{AME}}(f) \leq \Delta\phi_{\mathrm{PRO}} \cdot \frac{f_{\mathrm{EOM}}}{f_{\mathrm{het}}} \tag{3}$$

where f_{het} is the maximum heterodyne frequency. With f_{EOM}=2.4 GHz and f_{het}=24 MHz we obtain

$$\Delta\phi_{\mathrm{AME}}(f) \leq 0.6\,\mathrm{mrad}/\sqrt{\mathrm{Hz}} \cdot \sqrt{1+\left(\frac{2.8\,\mathrm{mHz}}{f}\right)^4}. \tag{4}$$

Possible noise sources between carrier and sideband are optical pathlength changes in the amplifier, stimulated Brillouin scattering (SBS), and nonlinear dispersion in combination with seed laser frequency noise. These effects have been discussed previously [7].

2. Flight representative optical amplifier

One option for the LISA laser system is to use the laser developed for the technology demonstrator mission LISA Pathfinder as seed laser in combination with an EOM and optical amplifier originally developed for laser communication terminals [8, 9].

The output power of the amplifier in the desired polarization was measured with an integrating sphere and photo diode and is shown in Fig. 1. The amplifier was seeded by a LISA-like signal (narrowband carrier + two sidebands containing 10% of the total power each). The nominal pump diodes were operated. The redundant pump diodes were not used.

The onset of SBS leads to increased power noise in the amplifier output in a millisecond time scale. The absence of SBS was verified by monitoring the output powers of the individual amplifier stages while slightly tapping on the amplifier using an Allen key. Figure 2 shows relative

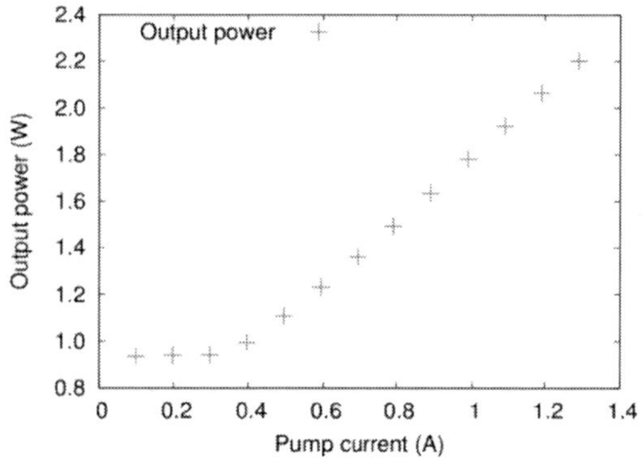

Figure 1. Measured amplifier output power.

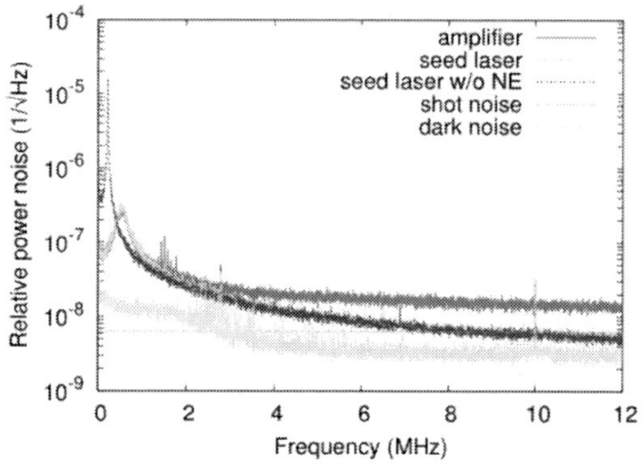

Figure 2. Relative power noise.

power noise (RPN). It is compatible with LISA requirements [10]. It was measured using an in-house photo detector by Tesat-Spacecom GmbH & Co. KG. Dark noise consists of electronic noise of the photo detector, preamplifier and spectrum analyzer. The relative power noise of the seed laser was measured without and with active noise eater. The amplifier was seeded only with active noise eater. It was operated at 2.2 W output power. The trace "shot noise" was plotted as consistency check. Approx. 8 mA photocurrent were detected. The measured noise of the seed laser shows only the same noise as the shot noise of the photocurrent at high frequencies. The finite bandwidth of the detector causes the slight reduction of the measured noise below the shot noise.

3. Test setup

Figure 3 shows the simplified setup to measure differential phase noise of a fiber amplifier. Light from two lasers is interfered and used to seed the amplifier under test. One of the lasers acts as carrier, the other mimicks a single phase modulation sideband. A fraction of the light is detected by photo diodes before and behind the amplifier. A control loop is used to offset-phase lock the difference frequency of both lasers to 2.4 GHz. The signals from both photo diodes are digitized, mixed with a slightly different frequency (2.4 GHz + 1.6 kHz), and low-pass filtered. A phasemeter implemented in software [11] was used to determine the phase difference between both signals. Differential phase noise between carrier and phase modulation sideband is visible as phase noise in this difference signal.

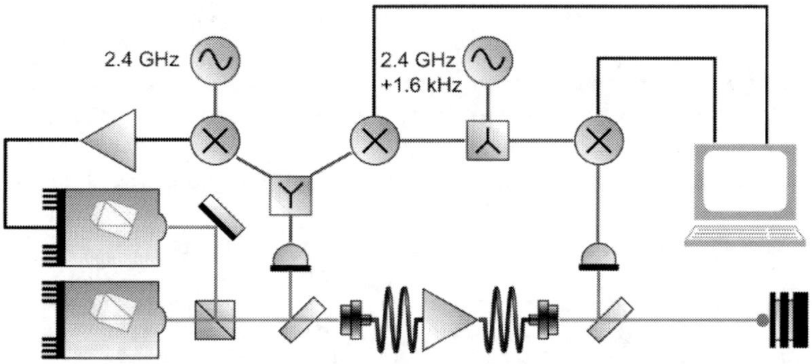

Figure 3. Simplified setup to measure fiber amplifier differential phase noise.

Figure 4 shows the planned setup for differential phase noise measurements. It differs in some details from the simplified setup shown in Fig. 3. The planned setup is implemented using three breadboards to ensure transportability. The first breadboard houses the two lasers, the second the actual interferometric measurement setup, and the third is the interface to the fiber amplifier under test with a mode-matching telescope and a power-attenuation stage consisting of two polarising beam splitters and a half-wave plate in between.

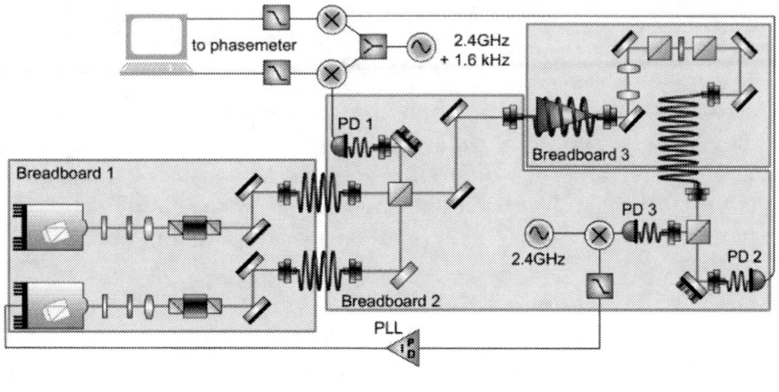

Figure 4. Planned setup for differential phase noise measurements.

A third photo diode (labelled PD3) is used for the offset-phase lock of the lasers. With only two photo diodes, the local oscillator of the offset-phase lock coupled into the respective phasemeter channel leading to a parasitic signal that is present even in the absence of laser light.

Journal of Physics: Conference Series **610** (2015) 012041 doi:10.1088/1742-6596/610/1/012041

This parasitic signal can disturb the phase measurement. Similar parasitic signals have been observed and investigated previously [12].

The kilohertz signal produced from photo diode 2 is split and the two nominally identical signals are digitized and processed by two phasemeter channels. The difference between these two signals is used to measure the phasemeter sensitivity.

The mixers that convert the GHz signals to kilohertz are known to show a signal amplitude-dependent phase shift [7]. Hence, the amplitudes of the phasemeter input signals could be stabilized by control loops that acted on piezo-driven tilt mirrors in front of fiber couplings before the (fiber-coupled) photo diodes.

Half-wave plates and polarizing beam splitters are used before and after fiber couplings for polarization control. Room temperature fluctuations are suppressed by passive thermal isolations.

4. Sensitivity

Figure 5 shows the current status of the transportable setup. Most electronics are integrated in the rack. Breadboards one and two from Figure 4 are placed on top of the rack and a simple passive fiber is used instead of the amplifer under test and breadboard three. Photo diode three for the laser offset phase lock and the mixer are placed in the smaller of the boxed on-top of breadboard one. The larger box contains the photo diodes and mixers that feed the phasemeter. Figure 6 shows the sensitivity of the setup, measured with a passive polarization-

Figure 5. Current status of the transportable setup.

maintaining single-mode fiber instead of breadboard 3 and the fiber amplifier under test. The trace labelled "phasemeter" shows the difference of two nominally identical phasemeter channels. The trace labelled "fiber" shows the difference between the signals before and behind the fiber. Although a sufficient sensivity has been shown, it is not yet reliably reproducible. Currently, noise investigations are in progress to ensure a constant and reliable performance.

5. Conclusions

A laser system candidate for LISA with space-qualified hardware is available. It consists of the seed laser for the LISA Technology Package as well as an electro-optical modulator and Ytterbium-doped fiber amplifier originally developed for optical communication. A transportable test setup to measure the phase fidelity of the amplifier has shown sufficient sensitivity. Noise hunting is in progress to ensure a constant noise level of the setup.

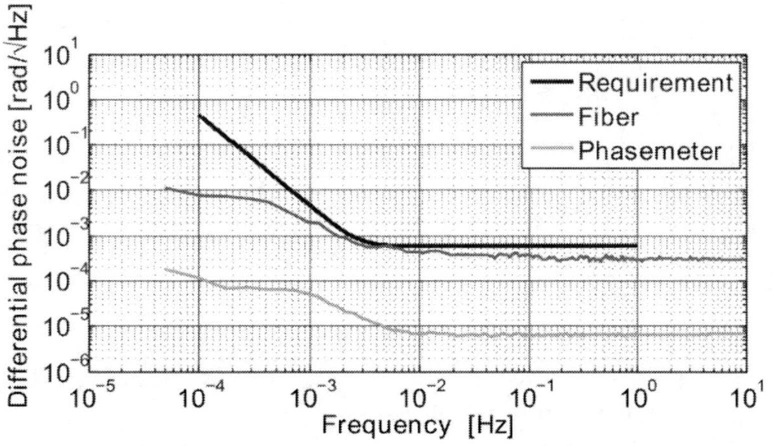

Figure 6. Sensitivity of the setup, measured with a passive polarization-maintaining single-mode fiber instead of breadboard 3 and fiber amplifier under test.

6. Acknowledgements

We acknowledge support by the German Aerospace Center (DLR) (50OQ0601, 50OQ1301) and thank the German Research Foundation for funding the Cluster of Excellence QUEST (Centre for Quantum Engineering and Space-Time Research). We thank Tesat-Spacecom GmbH & Co. KG for the possibility to perform measurements on the flight-representative hardware and acknowledge Stefan Seitz for operating the amplifier.

References

[1] Danzmann K and Rüdiger A 2003 *Class. Quantum Grav.* **20** S1–S9
[2] Jennrich O 2009 *Class. Quantum Grav.* **26** ISSN 0264-9381
[3] Tröbs M 2005 *Laser development and stabilization for the spaceborne interferometric gravitational wave detector LISA* Ph.D. thesis University of Hannover
[4] Barke S, Tröbs M, Sheard B, Heinzel G and Danzmann K 2010 *Appl. Phys. B* **98** 33–39
[5] Esteban J J, García A F, Barke S, Peinado A M, Cervantes F G, Bykov I, Heinzel G and Danzmann K 2011 *Opt. Express* **19** 15937–15946 URL `http://www.opticsexpress.org/abstract.cfm?URI=oe-19-17-15937`
[6] Heinzel G, Esteban J J, Barke S, Otto M, Wang Y, Garcia A F and Danzmann K 2011 *Classical and Quantum Gravity* **28** 094008 URL `http://stacks.iop.org/0264-9381/28/i=9/a=094008`
[7] Tröbs M, Barke S, Theeg T, Kracht D, Heinzel G and Danzmann K 2010 *Opt. Lett.* **35** 435–437
[8] Smutny B, Kaempfner H, Muehlnikel G, Sterr U, Wandernoth B, Heine F, Hildebrand U, Dallmann D, Reinhardt M, Freier A, Lange R, Boehmer K, Feldhaus T, Mueller J, Weichert A, Greulich P, Seel S, Meyer R and Czichy R 2009 5.6 Gbps optical intersatellite communication link URL `http://dx.doi.org/10.1117/12.812209`
[9] Muehlnikel G, Kämpfner H, Heine F, Zech H and Troendle D 2012 The Alphasat GEO Laser Communication Terminal Flight Acceptance Tests *Proc. of the International Conference on Space Optical Systems and Applications (ICSOS)* URL `http://icsos2012.nict.go.jp/pdf/1569586689.pdf`
[10] Tröbs M, d'Arcio L, Barke S, Bogenstahl J, Bykov I, Dehne M, Diekmann C, Fitzsimons E D, Fleddermann R, Gerberding O, Hennig J S, Hey F G, Hogenhuis H, Killow C J, Kochkina E, Kullmann J, Lieser M, Lucarelli S, Perreur-Lloyd M, Pijnenburg J, Robertson D I, Shoda A, Sohmer A, Taylor A, Wanner G, Ward H, Weise D, Heinzel G and Danzmann K 2012 Testing the LISA optical bench *Proc. of the International Conference on Space Optics* URL `http://congrex.nl/icso/2012/papers/FP_ICSO-110.pdf`
[11] Heinzel G, Wand V, Garcia A, Jennrich O, Braxmaier C, Robertson D, Middleton K, Hoyland D, Rüdiger A, Schilling R, Johann U and Danzmann K 2004 *Class. Quantum Grav.* **21** S581–S587
[12] Wand V, Bogenstahl J, Braxmaier C, Danzmann K, Garca A, Guzmn F, Heinzel G, Hough J, Jennrich O, Killow C, Robertson D, Sodnik Z, Steier F and Ward H 2006 *Classical and Quantum Gravity* **23** S159 URL `http://stacks.iop.org/0264-9381/23/i=8/a=S21`

Journal of Physics: Conference Series **610** (2015) 012042 doi:10.1088/1742-6596/610/1/012042

Towards a FPGA-controlled deep phase modulation interferometer

M Terán[1], V Martín [1], Ll Gesa[1], I Mateos[1], F. Gibert[1], N. Karnesis[1], J Ramos-Castro[2], T S Schwarze[3], O Gerberding[3], G Heinzel[3], F Guzmán[4] and M Nofrarias[1]

[1] Institut de Ciències de l'Espai, (CSIC-IEEC), Bellaterra, Spain
[2] Universitat Politècnica de Catalunya, Barcelona, Spain
[3] Albert-Einstein-Institut, Max-Planck-Institut für Gravitationsphysik, Hannover, Germany
[4] National Institute of Standards and Technology, Gaithersburg, MD 20899, USA

E-mail: nofrarias@ice.cat

Abstract. Deep phase modulation interferometry was proposed as a method to enhance homodyne interferometers to work over many fringes. In this scheme, a sinusoidal phase modulation is applied in one arm while the demodulation takes place as a post-processing step. In this contribution we report on the development to implement this scheme in a fiber coupled interferometer controlled by means of a FPGA, which includes a LEON3 soft-core processor. The latter acts as a CPU and executes a custom made application to communicate with a host PC. In contrast to usual FPGA-based designs, this implementation allows a real-time fine tuning of the parameters involved in the setup, from the control to the post-processing parameters.

1. Introduction

Deep phase modulation interferometry [1] was proposed as a method to enhance homodyne interferometers to work over many fringes, allowing for instance continuous real-time tracking of free falling test mass, as required for space based gravitational wave detectors [2].

The advantage of the proposed deep phase modulation scheme is that simplifies the required optical setup, driving and modulation electronics when compared with heterodyne based detection experiments [4]. The technique was proposed as an extension of the so called $J_1 \ldots J_4$ [3] methods, which involve a sinusoidal phase modulation at a fixed frequency with modulation depths up to 5 rad in one arm of the interferometer. These methods extract the encoded phase by means of analytical formulae to solve for the unknowns, among them the phase.

The deep phase modulation scheme uses however higher order harmonics ($n \geq 10$) to extract the phase information from the modulated output. For a phase modulated homodyne interferometer, this can be expressed as

$$V_{PD}(t) = V_{DC}(\phi) + \sum_{n=1}^{\infty} a_n(m, \phi) \cos(n\,(\omega_m\, t + \Psi)) \qquad (1)$$

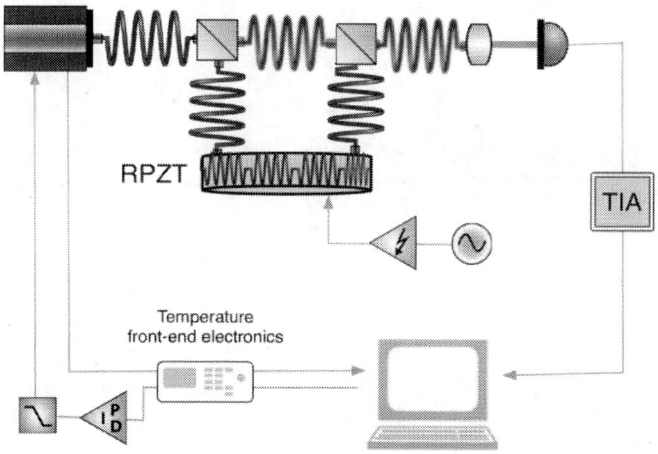

Figure 1. Deep phase modulation is implemented through an all-fiber Mach-Zehnder interferometer that uses a piezo tube with optical fiber coiled around it as a phase modulator. The measured output voltage is post processed in an external PC.

with

$$a_n(m, \phi) = k\, J_n(m) \cos\left(\phi + n\,\frac{\pi}{2}\right) \tag{2}$$

$$V_{DC}(\phi) = A(1 + C\, J_0(m)\, \cos\phi) \tag{3}$$

where $J_n(m)$ are Bessel functions, ϕ is the interferometer phase, m is the modulation depth, ω_m is the modulation frequency, Ψ is the modulation phase, C is the contrast, and A combines nominally constant factors such as light powers and photodiode efficiencies. Hence, the proposed scheme requires an implementation that allows high index phase modulation and precise phase extraction, as we describe in the following.

2. Experiment description and results

Our implementation of the deep phase interferometry is an all-fiber Mach-Zehnder interferometer operating at 1064 nm which uses a piezo tube with 5 m of optical fiber wrap around it to increase the pathlength in one of the interferometer arms. In our current setup, displayed in Fig.1, we have modified the LISA Pathfinder temperature front-end electronics [5] to control the laser temperature. Data acquisition and post-processing is performed in a PC.

In order to obtain the interferometer phase, the Fourier coefficients $a_n(m, \phi)$ in Eq. (2) are first obtained through a Fast-Fourier Transform of a segment of data. Then, the coefficients $\{k, m, \phi, \Psi\}$ are obtained by minimisation of

$$\chi^2 = \sum_{n=1}^{10} |\widetilde{V}_{PD}(n) - a_n(m, \phi) e^{i\,n\,\Psi}|^2 \tag{4}$$

where $\widetilde{V}_{PD}(n)$ is the n-th harmonic of the measured voltage at the output of the photodiode. A Levenberg-Marquardt algorithm is used to process the measured output and obtain the set

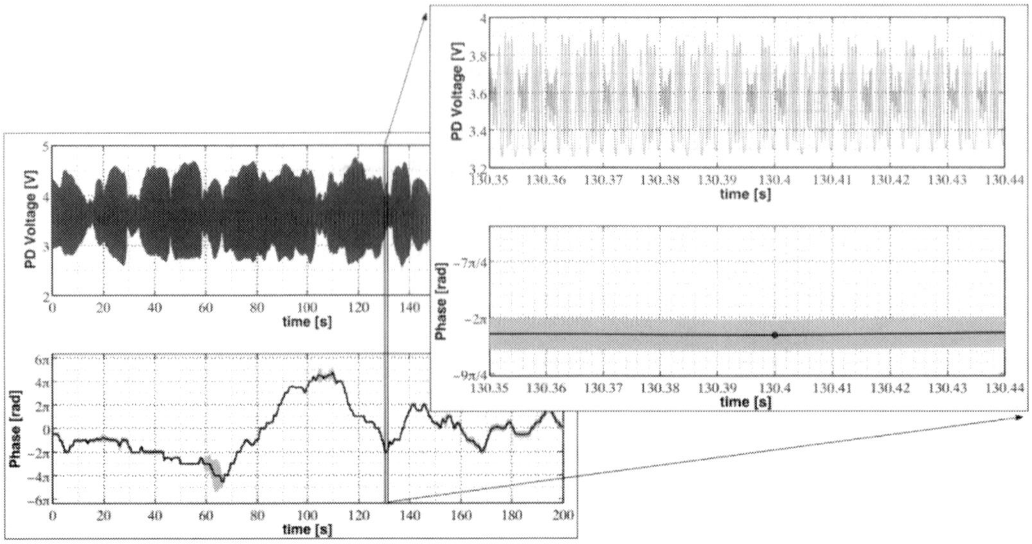

Figure 2. *Top:* Output voltage measured at the photodiode. *Bottom:* Measured phase after the post-processing. The shadowed area shows 95 % confidence intervals due to fit errors. The inset shows a zoom on a 54 ms period where the phase modulation on the measured voltage can be clearly appreciated.

of coefficients. In our proof-of-principle implementation we have applied a 200 Hz modulation signal to the piezo with a modulation depth $m \simeq 9$. Our current low frequency sensitivity with a table-top experiment on air is $10\,\mu\mathrm{m}/\sqrt{\mathrm{Hz}}$ at 10 mHz. Figure 2 shows the photodiode output sampled at 10 kHz while the bottom panel is the associated phase after post-processing in windows of 4000 samples, yielding an effective phase sampling of 2.5 Hz. The shaded areas show 95 % confidence intervals due to fit errors.

3. The System On Chip approach

While the original deep phase modulation interferometer was implemented in a scheme similar to the one shown in Fig. 1, recent developments have improved the design to shift the generation of signal modulation and the phase extraction to a FPGA [6].

In the same line, we have developed the software infrastructure that will allow a FPGA-based phasemeter configurable in real-time thanks to the System On Chip (SoC) approach. As shown in Fig 3, the following components have been synthesized in a Xilinx© FPGA: A Gaisler© LEON 3 Soft-Core CPU, a 4DSP© FMC116 ADC wrapped in a custom made component that communicates directly with the CPU using AMBA technology bus, and a custom embedded RTEMS Application running on SoC, that is in charge of acquiring, processing and transmitting data to Host PC Application trough ethernet TCP/IP, system monitoring and configuration managing. In parallel, the Host PC Application manages the user interface to customize the system and data persistence.

4. Conclusions

We have shown an implementation of deep phase interferometer based on an all-fiber Mach-Zehnder interferometer with a piezo tube with fiber coiled around it acting as a phase modulator.

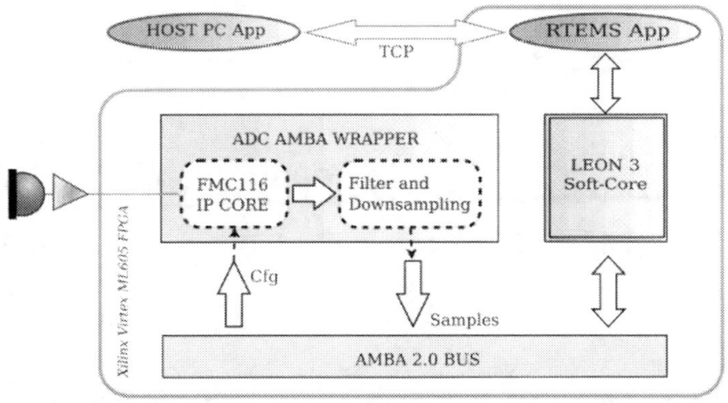

Figure 3. A schematic of the System On Chip (SoC) infrastucture in our current design.

Our proof-of-concept experiment achieves a low frequency performance of $10\,\mu\text{m}/\sqrt{\text{Hz}}$ at 10 mHz.

We have also implemented the required software infrastucture to run the required post processing tasks in a dedicated FPGA. Our effort has been focused so far in achieving a data acquisition system based on the SoC approach where we can build more complex data processing tasks at later stages. The LEON 3 Soft-Core CPU embedded in the FPGA will allow a more flexible control of the experiment once it is fully integrated in the optical setup.

Future developments of our setup include the already mentioned integration of the FPGA in the optical experiment, the implementation of the digital analysis, modulation and post-precessing in the FPGA and the integration of the metrology experiment in vacuum conditions.

References

[1] Heinzel G*et al.* 2010, *Deep phase modulation interferometry*, Opt. Exp., Vol. 18, No. 18, 19076

[2] Amaro-Seoane P *et al.* 2013, *The Gravitational Universe, Preprint* arXiv:1305.5720

[3] Sudarshanam V S, Claus R O 1993, *Generic $J_1 \ldots J_4$ method of optical-phase detection: Accuracy and range enhancement* , J. Mod. Opt. 40, 483-492

[4] Heinzel G *et al.* 2005, *Successful testing of the LISA Technology Package (LTP) interferometer engineering model*, Class. Quantum Gravity 22, S149S154.

[5] Sanjuán J *et al.* 2007, *Thermal diagnostics front-end electronics for LISA Pathfinder* Rev. Sci. Instrum. 78, 104904

[6] Schwarze T S *et al.* 2014, *Advanced phasemeter for deep phase modulation interferometry* Optics Express, Vol. 22, Issue 15, pp. 18214-18223

A brief comparison of optical pathlength difference and various definitions for the interferometric phase

Gudrun Wanner[1], **Sönke Schuster**[1], **Michael Tröbs**[1] **and Gerhard Heinzel**[1]

[1] Max-Planck-Institute for Gravitational Physics (Albert-Einstein-Institute) , Callinstr. 38, D-30167 Hannover, Germany
and
Leibniz University of Hannover, Hannover, Germany

E-mail: gudrun.wanner@aei.mpg.de

Abstract. In this paper we discuss that the phase readout in low noise laser interferometers can significantly deviate from the underlying optical pathlength difference (OPD). The cross coupling of beam tilt to the interferometric phase readout is compared to the OPD. For such a system it is shown that the amount of tilt to phase readout coupling depends strongly on the involved beams and their parameters, as well as on the detector properties and the precise definition of the phase. The unique single element photodiode phase is therefore compared to three common phase definitions for quadrant diodes. It is shown that neither phase definition globally shows the least amount of cross coupling of angular jitter.

1. Introduction

A frequent use of laser interferometers is to sense distance variations. Assume a very simple and perfect Mach-Zehnder interferometer where a mirror in one beam path is rotated around the beam reflection point. For simulations and interpretation the interferometer can then be reduced to the setup shown in Fig. 1. Here, the rays labeled z_{MB} and z_p indicate the propagation axis of measurement and reference beam respectively. In this paper, we discuss this very simple setup and its resulting phase readout. We highlight the fact that the resulting phase readout can deviate significantly from what might be expected from the change in the optical path, even in absence of standard noise sources such as laser frequency noise, electronic readout noise or thermal noise. This significant deviation originates from the phase- and intensity profiles of laser beams which cannot be sufficiently represented by plane waves or rays and secondly from incomplete beam detection and beam clipping on finite photodiodes as well as on the actual definition of the interferometric phase in the case of quadrant diodes.

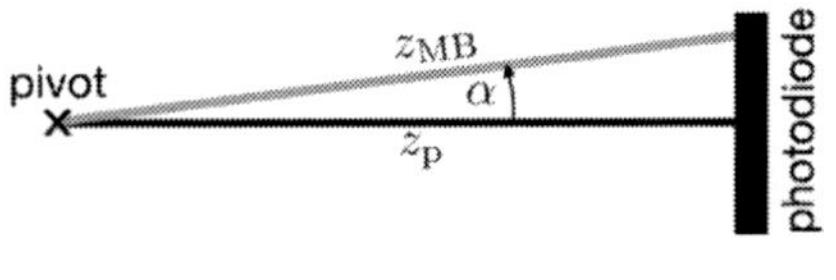

Figure 1: Effective beam paths in a simple laser interferometer, if a mirror in the measurement beam path rotates around the beam reflection point (thus coinciding with the pivot) by $\alpha/2$. For $\alpha = 0$ both beams impinge orthogonally on the center of the photodiode.

Journal of Physics: Conference Series **610** (2015) 012043 doi:10.1088/1742-6596/610/1/012043

2. Interferometric phase vs optical pathlength difference (OPD)

The difference between OPD and interferometric phase can be seen for instance by looking at the tilt to length (TTL) coupling in the given simplified setup.

With no medium involved, the OPD can be computed geometrically: it is the difference between the distances z_{MB} and z_p the measurement and reference beam propagate between pivot and diode (see Fig. 1):

$$\mathrm{OPD} = z_{\mathrm{MB}} - z_p = \left(\frac{1}{\cos(\alpha)} - 1\right) z_p \approx \frac{\alpha^2}{2} z_p \,. \tag{1}$$

Assuming rotation angles in the order of 100 µrad and a distance z_p of about 20 cm, the OPD is in the order of nanometers. For picometer interferometers such as LISA and LISA Pathfinder, it can be seen from this very simple and general example, that the coupling of tilt (i.e. angular jitter) to the OPD can be a significant effect. However, the OPD is not what is measured in these missions - or generally in interferometers, but the interferometric phase. We define here the interferometric phase as the phase that can be extracted from the photocurrent resulting from the two interfering beams on a photodiode (see [1] for a mathematical description of the interferometric phase).

To show that the interferometric phase can deviate significantly from the OPD, we will show two special cases.

Example case 1: We assume that both beams are circular fundamental Gaussian beams. Let the waist of both beams be located halfway between pivot and photodiode and the distance z_p between pivot and photodiode is chosen to be twice the Rayleigh range of the beams. The wavefronts of both beams on the detector surface can then be approximated by a section of a sphere which has its center in the pivot of the rotation. This means, that a mirror tilt maps the wavefront onto itself. For small rotations we therefore expect negligible cross coupling from tilt to the interferometric phase. We have shown previously [1] with a numerical simulation, that for this special case the coupling is well below pm-level.

Example case 2: Assume the same case as before, but let the center of rotation be an arbitrary point. If the detector is a large single element detector which does not clip either beam, and if the parameters of the interfering beams are matched, the coupling of tilt into the interferometric phase is negligible, as we will show below.

We have shown before [2, eq (34)] that the resulting phase is then $\phi \approx \alpha^2 z/(4z_0)$, where z_0 is the Rayleigh range of both beams and z the distance from their waist. In order to compare the interferometric phase to the OPD, it needs to be converted to a length, which we call the longitudinal pathlength signal $\mathrm{LPS} = \phi/k$. For typical wavenumbers k in the order of $2\pi/(1\,\mu\mathrm{m}) \approx 10^7/\mathrm{m}$, tilt angles in the order up to a few hundred µrad, and assuming that Rayleigh range z_0 and distance z from the waist are of comparable order, we see that the tilt to length coupling is negligible:

$$\mathrm{LPS} \approx \frac{\alpha^2}{k}\frac{z}{4z_0} \approx \frac{(10^{-4})^2}{10^7}\,\mathrm{m} \ll 1\,\mathrm{pm}\,. \tag{2}$$

In both examples the OPD lies in the order of nanometers while the phase readout is negligible. In example case 1 this is due to the very specific choice of beam parameters. This shows that the nature of the beams (e.g. circular fundamental Gaussian beams) needs to be accounted for when predicting the interferometric phase. In example case 2 the vanishing cross coupling results from the cancellation of two effects [3], which holds for arbitrary but matched fundamental Gaussian beams that are detected without clipping by a single element photodiode.

If either beam is clipped for instance by the slits of a quadrant photodiode (QPD) or because the diode is insufficiently large ($\lesssim 3$ times spot size on the detector), a significant coupling can occur (see Fig. 2(a)). This shows that clipping effects and therefore the detector shape have significant impact on the phase readout.

Interferometric phase of a quadrant diode Quadrant photodiodes (QPDs) are often used for phase detection in interferometers, because their use allows to locate the centroid as well as the relative angle of the incident beams[4, 5]. Any quadrant diode naturally provides not one but four photocurrents, which need to be combined in order to generate one phase signal. Consequently, there are several options to define a phase signal for quadrant diodes. One possibility to describe these signals is via the complex amplitude C of the photocurrent, which we define as the integral over the detector surface of the product of the electric field of the measurement beam and complex conjugated electric field of the reference beam and which is given by:

$$C := \int \mathrm{d}S E_m E_r^* = \bar{P}c \exp(i\phi) =: A \exp(i\phi) \, . \tag{3}$$

Here, $\bar{P}$ is the time averaged detected power and c the measured contrast of the interference. The complex amplitude C can be determined experimentally for instance by a single bin discrete Fourier transform, as done in LISA Pathfinder [6]. The interferometric phase ϕ is then simply the argument of the complex amplitude C. If a large single element detector is assumed, S can be set to the detector plane. If a quadrant detector is assumed, there are four finite integrals to be performed, resulting in four complex amplitudes C_n and consequently four interferometric phases ϕ_n, and four contrasts c_n (n=1,..,4). The sum of the complex amplitudes plus the hypothetical complex amplitudes of the slits would exactly be equal to the complex amplitude of a single element detector of identical diameter. Therefore, the definition of a QPD phase closest to the phase of a single element photodiode is:

$$\phi_{\mathrm{LPF}} = \arg\left(\sum_{n=1}^{4} C_n\right) = \arg\left(\sum_{n=1}^{4} A_n \exp(i\phi_n)\right) \, . \tag{4}$$

This phase definition was chosen in LISA Pathfinder (LPF). Alternatively the arithmetic mean of the quadrant phases ϕ_n could be computed:

$$\phi_{\mathrm{AP}} = \frac{1}{4}\sum_{n=1}^{4} \phi_n = \frac{1}{4}\arg\left(\prod_{n=1}^{4} C_n\right) \, . \tag{5}$$

Here, the product notation on the right hand side is a better choice for implementation, since it is more robust to phase jumps than the arithmetic mean on the left hand side.
While ϕ_{AP} might be an intuitive choice, it has clear disadvantages compared with the LPF definition: in the case of poorly centered beams the phases of all quadrants contribute equally to ϕ_{AP}, even though the phase of one or several quadrants might originate from a very low number of photons, while the phases of other quadrants might originate from strong signals. This case is naturally compensated in ϕ_{LPF}, because the summation of the complex amplitudes is effectively a vector addition.
The power imbalance problem in the averaged phase can be accounted for, by weighing the quadrant phases for instance by the amplitudes $A_n := \bar{P}_n c_n$:

$$\phi_{\mathrm{WAP}} = \frac{A_1\phi_1 + A_2\phi_2 + A_3\phi_3 + A_4\phi_4}{A_1 + A_2 + A_3 + A_4} \, . \tag{6}$$

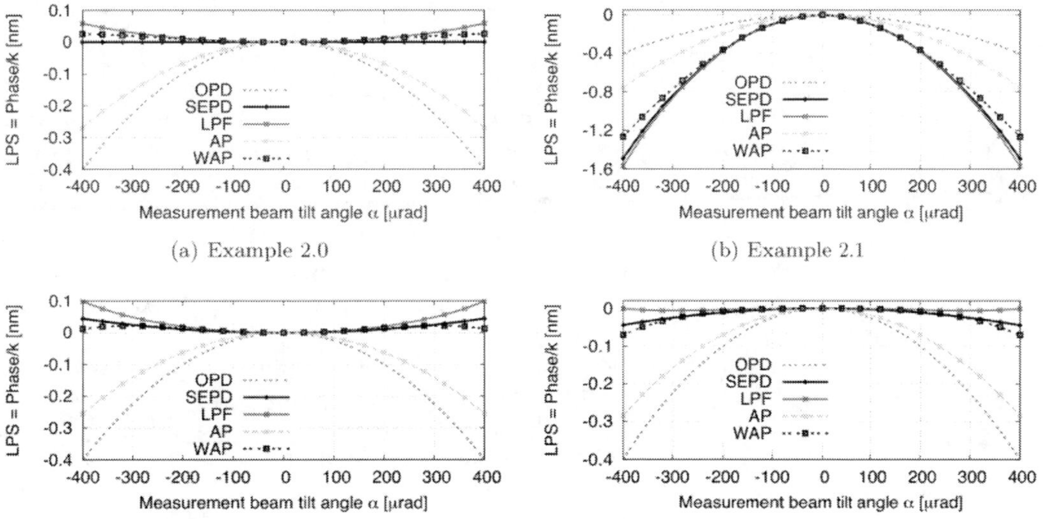

(a) Example 2.0 (b) Example 2.1

(c) Example 2.2 (d) Example 2.3

Figure 2: Longitudinal pathlength signal (LPS) for the setup of example case 2 with the settings listed in Tab. 1. These graphs were generated with the numerical software tool IfoCAD[7]. Subfig. (a) is the matched beam parameter case described in example case 2. Subfig.(b)-(d) show the same setup for different choices of mismatched beam parameters.

Parameter description	value	Parameter description	value	
diameter of circular QPD	10 mm	measurement beam power	1.2 mW	
photodiode gap width	70 μm	reference beam power	0.7 mW	
λ: wavelength laser beams	1064 nm	z_p: distance from pivot to photodiode	5 mm	
	Ex 2.0	Ex 2.1	Ex 2.2	Ex 2.3
distance RB from waist to QPD [m]	-0.4	-0.4	0	0
distance MB from waist to QPD [m]	-0.4	-0.6	0	0
RB Rayleigh range [m]	0.5	3.137	2.5	2.0
MB Rayleigh range [m]	0.5	4.124	2.0	2.5

Table 1: Parameter settings for Figure 2. RB: reference beam, MB: measurement beam

The ϕ_{WAP} is a first order approximation of ϕ_{LPF}, since the first series expansion of the complex amplitude is $C \approx A + iA\phi$:

$$\phi_{\mathrm{LPF}} \approx \arg\left(\sum_{n=1}^{4}(A_n + iA_n\phi_n)\right) \approx \arctan\left(\frac{A_1\phi_1 + A_2\phi_2 + A_3\phi_3 + A_4\phi_4}{A_1 + A_2 + A_3 + A_4}\right) \tag{7}$$

$$\approx \frac{A_1\phi_1 + A_2\phi_2 + A_3\phi_3 + A_4\phi_4}{A_1 + A_2 + A_3 + A_4} = \phi_{\mathrm{WAP}}\ . \tag{8}$$

There are of course numerous further options to define the phase readout for quadrant detectors. The definitions $\phi_{\mathrm{LPF}}, \phi_{\mathrm{AP}}$ and ϕ_{WAP} are those we regard as most intuitive and physically closest to the phase readout of a single element detector. A direct comparison of these signals is shown in Fig. 2 for variations of example case 2. These numerical simulations were generated with the software tool IfoCAD[7], the given phase definitions and the settings listed in Tab. 1. More detailed information on the IfoCAD algorithms to compute interferometric phases are given in [1, 8]. Subfig. 2(a)) shows the case of matched beam parameters. As discussed above, the LPS of a single element diode (SEPD) does not sense the measurement beam tilt, provided that the detector is sufficiently large that the beams are fully detected - which is the case in this example. However, all here defined phase signals of a quadrant diode show a significant amount of TTL cross coupling. Since we assume identically shaped SEPDs and QPDs, ϕ_{LPF} equals ϕ_{SEPD} in the hypothetical limit that the gap width goes to zero. Therefore one can say that any difference occurring between ϕ_{SEPD} and ϕ_{LPF} results fully from the clipping at the insensitive gaps between the quadrants of the QPD. Therefore, Subfig. 2(a) shows the vanishing TTL coupling on SEPDs as described above as well as the TTL cross coupling on QPDs originating from incomplete beam detection.

For mismatched beam parameters, the amount of TTL cross coupling of each phase signal depends on the exact setup and the defined beam parameters, as shown in Fig 2(b)-2(d). It can be seen that either of the three QPD LPS signals can show the least amount of TTL cross coupling. In other words, neither signal is globally an optimal definition for a phase readout. The beam parameters in the examples were chosen to highlight this fact, but are otherwise arbitrary. A detailed comparison of the phase signals is work in progress and will be discussed in a separate paper.

3. Conclusions

For high performance interferometers, such as LISA Pathfinder, and the Laser Ranging Instrument (LRI) of GRACE Follow-On, it is necessary to distinguish carefully the optical pathlength difference from the interferometric phase, since these parameters could deviate significantly, depending on beam parameters and setup. For future high performance interferometers, such as in space based gravitational wave detectors like LISA and DECIGO, and in future interferometric geodesy missions beyond GRACE Follow-On, the fact that the cross coupling of angular jitter to phase noise depends on the phase definition could be used to reduce the resulting phase noise. For this however, dedicated analyses need to be performed which also include effects such as the robustness of the phase signal to phase jumps, coupling of other optical noise types or the robustness to quadrant failure.

Acknowledgments

We gratefully acknowledge support by Deutsches Zentrum für Luft- und Raumfahrt (DLR) with funding of the Bundesministerium für Wirtschaft und Technologie with a decision of the Deutschen Bundestag (DLR project reference No 50 OQ 1301).

4. References

[1] Wanner G *et al* 2012 *Optics Communications* **285** 4831–4839
[2] Wanner G, Heinzel G 2014 *Applied Optics* **53**, No. 14
[3] Schuster S, Wanner G, Tröbs M, Heinzel G: "Vanishing tilt-to-length coupling for a singular case in two-beam laser interferometers with Gaussian beams" *submitted to Applied Optics*
[4] Morrison E *et al* 1994 *Automatic alignment of optical interferometers Appl. Opt.* **33**, No. 22
[5] Morrison E *et al* 1994 *Experimental demonstration of an automatic alignment system for optical interferometers Appl. Opt.* **33**, No. 22
[6] Heinzel G *et al Class. Quantum Grav.* **21**, (2004) 581-587
[7] `http://www.lisa.aei-hannover.de/ifocad/`
[8] Kochkina E *et al ASP Conference Series*, **467**

Journal of Physics: Conference Series **610** (2015) 012044 doi:10.1088/1742-6596/610/1/012044

Environmental Effects for Gravitational-wave Astrophysics

Enrico Barausse[1,2], **Vitor Cardoso**[3,4] **and Paolo Pani**[3,5]

[1]CNRS, UMR 7095, Institut d'Astrophysique de Paris, 98bis Bd Arago, 75014 Paris, France
[2]Sorbonne Universités, UPMC Univ Paris 06, UMR 7095, 98bis Bd Arago, 75014 Paris, France
[3]CENTRA, Departamento de Física, Instituto Superior Técnico, Universidade de Lisboa, Avenida Rovisco Pais 1, 1049 Lisboa, Portugal.
[4]Perimeter Institute for Theoretical Physics, Waterloo, Ontario N2L 2Y5, Canada.
[5]Dipartimento di Fisica, "Sapienza" Università di Roma, P.A. Moro 5, 00185, Roma, Italy

Abstract. The upcoming detection of gravitational waves by terrestrial interferometers will usher in the era of gravitational-wave astronomy. This will be particularly true when space-based detectors will come of age and measure the mass and spin of massive black holes with exquisite precision and up to very high redshifts, thus allowing for better understanding of the symbiotic evolution of black holes with galaxies, and for high-precision tests of General Relativity in strong-field, highly dynamical regimes. Such ambitious goals require that astrophysical environmental pollution of gravitational-wave signals be constrained to negligible levels, so that neither detection nor estimation of the source parameters are significantly affected. Here, we consider the main sources for space-based detectors – the inspiral, merger and ringdown of massive black-hole binaries and extreme mass-ratio inspirals – and account for various effects on their gravitational waveforms, including electromagnetic fields, cosmological evolution, accretion disks, dark matter, "firewalls" and possible deviations from General Relativity. We discover that the black-hole quasinormal modes are sharply different in the presence of matter, but the ringdown signal observed by interferometers is typically unaffected. The effect of accretion disks and dark matter depends critically on their geometry and density profile, but is negligible for most sources, except for few special extreme mass-ratio inspirals. Electromagnetic fields and cosmological effects are always negligible. We finally explore the implications of our findings for proposed tests of General Relativity with gravitational waves, and conclude that environmental effects will not prevent the development of precision gravitational-wave astronomy.

1. Introduction

Today, we have convincing indirect evidence from binary pulsars [1] for the existence of gravitational waves (GWs), which are a generic prediction of General Relativity (GR) and other relativistic theories of gravity. The ground-based GW detectors LIGO and Virgo are currently being updated to advanced configurations [2, 3] expected to achieve sensitivities sufficient for detecting signals from binaries of stellar-mass black holes (BHs) and/or neutron stars within the end of this decade. Detection of these signals will allow measuring the masses and spins of the binary components with accuracies comparable to current X-ray probes [4]. On the same timescale, pulsar-timing arrays will target signals from widely separated massive-BH binaries [5], and on a longer timescale spaced-based detectors will detect these systems at smaller separations, including the binary's merger and ringdown phases. Also, ESA has recently selected GWs as the science theme for its L3 mission with launch slot 2034. One possible mission that would explore this science theme is given by the space-based detector eLISA [6], whose "Pathfinder" mission will be launched in 2015 [7]. Detectors such as eLISA will estimate the source parameters, and in particular the masses and spins of massive BHs, to within fractions of a percent and up to $z \sim 10 - 15$ [6], which will permit testing models for the symbiotic coevolution of massive BHs and their host galaxies [see e.g. Refs. [8, 9, 10, 11, 12]]. Furthermore, GW detectors will allow for precision tests of GR in the currently unexplored highly dynamical, strong-field regime [13, 14, 15].

Estimates of the accuracy of GW detectors in measuring the source parameters usually do not account for the realistic astrophysical environments surrounding the sources – such as electromagnetic fields, accretion disks and dark matter (DM) – based on the expectation/hope that their effect will be negligible. However, a careful examination is needed to assess the environment's impact on GW observables, so as to determine whether precision GW physics is possible at all: unmodeled deviations (due to environmental effects) from the pure "vacuum" gravitational waveforms predicted by GR may degrade the signal-to-noise ratio and the parameter estimation accuracy, potentially jeopardizing tests of gravity theories and astrophysical models. On the other hand, if these effects are non-negligible and can be modeled, they may provide important information about the environments of GW sources.

This article examines the impact of environmental effects on the most powerful source of GWs, namely the inspiral, merger and ringdown of BH binaries, both with comparable and extreme mass-ratios. Our analysis follows that of Ref. [16], but we focus here in particular on the implications for GW astrophysics with an eLISA-like mission.

2. Matter effects in compact-object binaries

The best understood environmental effect in compact-object binaries (including BH binaries) is the possible presence of a gaseous accretion disk. During the binary's inspiral, the disk affects the orbital evolution in various ways. The masses and spins of the compact objects change under *accretion* of gas [17, 18]. Furthermore, matter exerts a *gravitational pull* on the binary's components, modifying their trajectories [19, 18]. Finally, the gravitational interaction of the compact objects with their own wake in a gaseous medium produces *dynamical friction* [17, 20, 18] and can give rise to *planetary migration* [21, 22]. All these effects impact the binary's orbital evolution, and therefore the gravitational waveforms.

Other effects potentially affecting a binary's motion and GW emission are the cosmological expansion and strong electromagnetic fields. The former effects are expected to be small and comparable to those of the cosmological constant, $\Lambda \sim 10^{-52}\text{m}^{-2}$. As for the latter, astrophysical BHs are believed to be almost neutral because of quantum discharge effects [23], electron-positron pair production [24, 25, 26], and charge neutralization by astrophysical plasmas. Nevertheless, scenarios where BHs acquire an electric charge via various mechanisms, typically related to the presence of a magnetic field, have been put forward to explain some BH-driven high-energy phenomena [27]. Classical induction by an external magnetic field B can produce a charge Q satisfying [28]

$$q \equiv \frac{Q}{\sqrt{G}M} \lesssim 2 \times 10^{-6} \frac{M}{10^6 M_\odot} \frac{B}{10^8\text{Gauss}} . \tag{1}$$

The magnetic field of the massive BHs powering Active Galactic Nuclei (AGNs) is believed to be $\lesssim 10^3 - 10^5$ Gauss [29, 30, 31]. In the following, to account for the uncertainties in the magnetic field measurements, we assume *very* conservative upper limits, $B = 10^8$ Gauss and $q = 10^{-3}$.

DM might also affect the motion of BH binaries and their GW signal. Reference [32] showed that if a massive BH grows adiabatically, the DM density in its vicinity also increases and gives rise to a "spike". However, these spikes were later found to be efficiently destroyed and diluted into "cores" by binaries of massive BHs with comparable masses [33], thus making them less relevant for binaries living in galaxies that have recently experienced a major merger (i.e. most

Milky-way type galaxies). A similar flattening of these spikes is induced by DM scattering off stars [34, 35] and by the possible off-center formation of the BH seeds at high redshift [36]. The combination of these effects is expected to produce DM profiles with a shallow slope (rather than a spike) near BH binaries, with reference densities $\rho_{\rm DM} \sim 10^2 - 10^3 M_\odot/{\rm pc}^3$ for binaries with mass ratio 1 : 1 and 1 : 10, respectively [33].

In satellite galaxies that have never undergone mergers the situation might be different. Reference [37] found that if massive BHs grow from "light" seeds with mass $M_{\rm seed} \sim 10^2 M_\odot$ at $z \sim 20$, they may still be surrounded by DM spikes at $z \sim 0$. However, these spikes are unlikely to pose a problem for precision GW astronomy, because the BHs residing at their center would have mass $M_{\rm bh} \sim M_{\rm seed}$, and would thus emit outside the frequency band of space-based detectors (such as eLISA). On the other hand, if massive BHs form from "heavy" seeds with mass $M_{\rm seed} \sim 10^5 M_\odot$ at $z \sim 15$, Ref. [37] find that ~ 100 satellite galaxies containing BHs with mass $M_{\rm bh} \sim M_{\rm seed}$ surrounded by spikes would be present in a Milky-Way type halo. These BHs might possibly produce GWs in eLISA's band, e.g. if they were to capture a stellar-mass BH and form an extreme mass-ratio inspiral (EMRI), see also Ref. [38].

Because of the various scenarios discussed above, we take our reference value for the DM density to be

$$\rho_{\rm DM} = 10^3 M_\odot/{\rm pc}^3 = 4 \times 10^4 {\rm GeV/cm}^3\,, \tag{2}$$

for massive BH binaries with comparable masses and for EMRIs in Milky-way type galaxies, but we also entertain the extreme possibility of higher DM densities (i.e. spikes) of $\sim 10^{10} - 10^{12} M_\odot/{\rm pc}^3$ in satellite galaxies.

3. Environmental effects in the inspiral

Using the conservative reference values discussed above, we studied how realistic astrophysical environments affect the inspiral phase of compact binaries, during which the binary's separation shrinks adiabatically under GW emission. More specifically, we consider the gravitational pull of the matter surrounding the binary, the effect of gas accretion onto the binary's components, dynamical friction and planetary migration, BH charges induced by external magnetic fields and cosmological expansion/acceleration effects. We do not consider gravitational interactions with stars, which are known to be *(i)* unimportant for massive BH binaries in the eLISA band [see however [39] for massive BH binaries detectable with pulsar-timing arrays] and *(ii)* only important for a few percent of the EMRIs detectable with eLISA [40], *if* stars have a cuspy distribution *very close* to massive BHs [which is uncertain observationally even for our Galaxy, c.f. Ref. [41]]. Similarly, gravitational interaction of an EMRI with a second nearby massive BH (to within few tenths of a pc from the EMRI) may also produce a detactable effect on eLISA waveforms [42]. The fraction of EMRIs in such a situation is wildly uncertain, but probably on the order of a few percent [see discussion in Ref. [42]].

Our main conclusions are:

(i) Environmental effects can be safely neglected for most compact-object binaries detectable with eLISA. Table 1 shows the corrections to the periastron shift and GW phase in a variety of environments, in the case of EMRIs, which are the most affected sources because they spend $\sim 10^5 - 10^6$ inspiral cycles in the frequency band of eLISA-like detectors (as opposed to $\sim 10^2$ cycles for massive-BH binaries with comparable masses). Even with our rather extreme reference values, matter corrections are at most marginally detectable and usually completely negligible. As an approximate criterion, for an (unmodeled) effect to have a measurable effect on eLISA waveforms, it should introduce a dephasing $\delta_\varphi \gtrsim 0.1$ rads during the mission's expected lifetime of 1 year.

(ii) An exception is given by EMRIs in geometrically-thin, radiatively-efficient accretion disks, where environmental effects are sufficiently large to affect estimates of the source

Journal of Physics: Conference Series **610** (2015) 012044 doi:10.1088/1742-6596/610/1/012044

Table 1. Maximum environmental corrections to a typical EMRI's periastron shift ($\delta_{\rm per}$) and GW phase (δ_φ) over a typical eLISA's mission duration of one year. $\delta_{\rm per}$ and δ_φ are respectively the relative and absolute corrections to vacuum values. Dissipative effects such as GW radiation reaction, dynamical friction and hydrodynamic drag from accretion give negligible $\delta_{\rm per}$ and are thus not shown. We consider two BHs with masses ($10M_\odot$, $M = 10^6 M_\odot$), on a quasicircular inspiral ending at the innermost stable circular orbit (ISCO) $r = 6GM/c^2$. The periastron shift is computed at $r = 10GM/c^2$. Conservative environmental reference values are $q = 10^{-3}$, $B = 10^8$ Gauss, $\rho_3^{\rm DM} = \rho_{\rm DM}/(10^3 M_\odot/{\rm pc}^3)$. We assume a Shakura-Sunyaev disk model with viscosity parameter $\alpha = 0.1$ and Eddington ratio $f_{\rm Edd} = 10^{-4}$ ($f_{\rm Edd} = 1$) for thick and thin disks, respectively. The scaling with the parameters can be found in Ref. [16], and in Ref. [21] for planetary migration.

	Correction	$\lvert\delta_{\rm per}\rvert$	$\lvert\delta_\varphi\rvert$[rads]
thin disks	planetary migration	—	10^4
	dyn. friction/accretion	—	10^2
	gravitational pull	10^{-8}	10^{-3}
	magnetic field	10^{-8}	10^{-4}
	electric charge	10^{-7}	10^{-2}
	gas accretion	10^{-8}	10^{-2}
	cosmological effects	10^{-31}	10^{-26}
thick disks	dyn. friction/accretion	—	10^{-9}
	gravitational pull	10^{-16}	10^{-11}
DM	accretion	—	$10^{-8}\rho_3^{\rm DM}$
	dynamical friction	—	$10^{-14}\rho_3^{\rm DM}$
	gravitational pull	$10^{-21}\rho_3^{\rm DM}$	$10^{-16}\rho_3^{\rm DM}$

parameters, prevent accurate tests of GR, and possibly even jeopardize detection. We found indeed that accretion, dynamical friction and planetary migration due to a thin disk are typically more important than GW emission at separations $\gtrsim\sim 20-40$ gravitational radii, and are stronger than second-order gravitational self-force effects [43] at any separation [16]. The disk's gravitational pull is instead weaker. This confirms and extends results by Refs. [17, 21, 22, 18]. However, EMRIs will be detectable with eLISA only at $z \lesssim 1$ [6], where most galactic nuclei are quiescent rather than active. Because radiatively-efficient thin disks are mainly expected in AGNs, Ref. [16] estimates that only a few percent of the EMRIs detected by eLISA will be significantly affected. BHs in quiescent nuclei are expected to be surrounded by thick radiatively-inefficient disks, whose effect is typically sub-dominant even relative to second-order self-force corrections. Overall, our results thus confirm that EMRI detection and parameter estimation with eLISA should only be marginally affected by the environment.

(iii) The early dynamics of massive BH binaries with comparable masses in gas-rich environments (such as a dual AGN) might be dominated by accretion and/or dynamical friction rather than by GW emission, from the time they enter the eLISA band down to $\sim 60-70$ gravitational radii. At smaller separations (where most of the signal is produced) the dynamics is instead driven by GW radiation reaction.

(iv) DM accretion and dynamical friction produce effects on BH binaries that are typically larger than the gravitational pull from the DM halo. Nevertheless, all these effects are negligible for both detection and parameter estimation in the case of eLISA binaries, except (possibly) for EMRIs in satellite galaxies. As discussed in section 2, DM spikes with densities $\rho \sim 10^9 - 10^{12} M_\odot/{\rm pc}^3$ may survive in such systems, in which case DM environmental effects on EMRIs (especially accretion onto the central massive BH) may be marginally comparable to GW emission.

4. Environmental effects in BH ringdown

In vacuum, the late-time signal after a massive BH merger is governed by the final BH's quasinormal modes (QNMs). This "ringdown" can be used to infer the remnant's mass and spin,

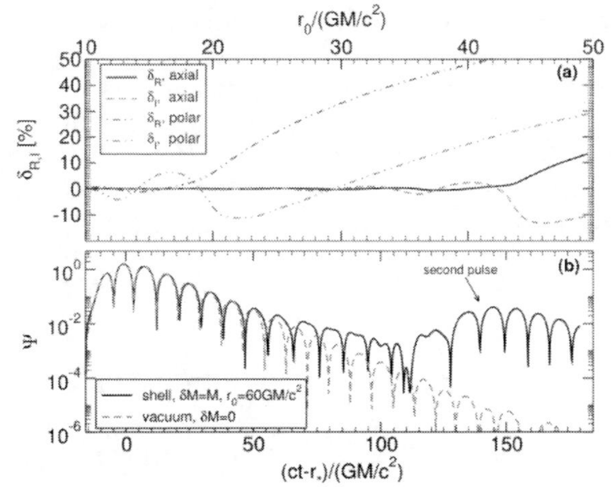

Figure 1. (a) Deviations of the real and imaginary parts of the fundamental $l = 2$ QNM frequencies for a BH with mass M surrounded by a thin shell of mass $\delta M = 0.01M$ located at r_0. In the large-r_0 limit, δ_R and δ_I grow linearly. (b) Gravitational-axial waveform for the scattering of a gaussian profile off a thin shell - BH configuration with $r_0 = 60GM/c^2$ and $\delta M = M$. The gaussian is initially placed close to the ISCO, with width $\sigma = 4GM/c^2$. The effect of matter appears in the time domain as a second pulse at late times, but the isolated BH QNMs dominate most of the response.

and to test the no-hair theorem of GR with space-based detectors [13]. We have performed a detailed study of the ringdown emission of "dirty BHs", i.e. BH geometries not exactly described by the Schwarzschild metric because containing matter [44]. Our results can be summarized as follows [see Ref. [16] for derivations and more details]:

(i) The mode spectrum, as defined by the poles of the relevant Green's function in a *frequency-domain analysis*, is extremely rich. For each mode of an isolated Schwarzschild BH we find an *infinite* set of matter-driven modes, plus one mode that parametrically corrects the isolated-BH mode (i.e., that reduces to it for vanishing matter stress-energy).

(ii) For matter configurations localized far from the BH or very close to its horizon, the deviations from the isolated BH QNMs are arbitrarily large, even for very small matter densities. This surprising behavior is illustrated in the upper panel of Fig. 1, for an infinitely thin shell of matter. While a thin-shell geometry is astrophysically unrealistic, it is useful as a proof of concept [45]. Our results show that this behavior is generic and appears for arbitrary composite BH-matter distributions [16].

(iii) An extensive search shows that although the spectrum of dirty BHs drastically differs from isolated ones, the modes of the *isolated* BH play a dominant role in a *time-domain evolution* of the composite system. In other words, although the QNMs of isolated BHs are no longer poles of the Green's function, they nevertheless dominate the ringdown waveform at intermediate time scales. This is illustrated in the bottom panel of Fig. 1, which shows the response of a thin shell – BH system to a Gaussian perturbation. After the Gaussian wavepacket "hits" the circular photon orbit, where isolated BH QNMs are known to reside [46], the *composite* system rings down in its lowest *isolated* BH QNMs. The modes of the composite system get excited only at late times (when the wavepacket reaches the shell).

(iv) Our analysis confirms that the lowest QNMs of isolated BHs *can* be used to estimate the mass of massive BHs to 0.1% accuracy [13], in spite of our ignorance on the details of their astrophysical environments.

(v) The QNMs of dirty BHs are typically localized farther away from the object than in isolated geometries. During a BH merger, these modes are excited to low amplitudes and at very late times. Therefore, they typically play a subdominant role in the merger waveforms, but will likely dominate over Price's power-law tails [47]. In principle, these modes might be excited to large amplitudes during a binary's inspiral, thus providing important information on the matter distribution, possibly even close to the horizon.

Consider for instance Planck-density "firewalls" [one possible consequence of quantum effects near the horizon [48], see also Ref. [49] for a similar earlier proposal], which we model by a spherical shell of mass δM located at $r_0 \sim 2M + \ell_P$, where $\ell_P \sim 1.6 \times 10^{-35}$m is the Planck length. We estimate the (relative) changes in the fundamental QNM frequency to be given by

$$\delta_R \sim -2\left[1 - 0.01\log\left(\frac{10^6 M_\odot}{M}\right)\right]\frac{\delta M}{10^{-2}M}\,. \tag{3}$$

This suggests that a firewall of mass as low as $\delta M \sim 10^{-4}M$ introduces a correction of the order of a few percent in the BH ringdown frequencies. Such corrections might be observable with Advanced LIGO/Virgo or eLISA, if the modes are excited to appreciable amplitudes. We stress, however, that the correction to the QNM frequencies depends (linearly) on the total mass of the firewall, which is still subject of debate [50, 51].

In summary, matter drastically changes the frequency-domain QNM spectrum, but does not prevent GW observatories from detecting BH ringdowns in realistic environments using isolated-BH templates. In very optimistic scenarios, detections could even be used to extract the mass distribution parameters and investigate accretion disks, DM halos or even firewalls [16]. These statements are supported by Table 2, which shows an upper limit to the QNM corrections in a variety of astrophysical scenarios.

5. Intrinsic limits to tests of gravity theories

Quantifying environmental effects around compact binaries is also important for testing GR. So far, GR has been tested in situations involving weak gravitational fields and mildly-relativistic velocities [52, 53]. GW observations of relativistic compact binaries will provide the possibility of testing GR in the unexplored strong-field, highly dynamical regime, i.e. in systems with characteristic velocities $v \sim c$ and gravitational potentials/curvatures orders of magnitude larger than those probed so far [54, 14, 15]. For these tests it is essential to understand the effect of matter, in order to avoid mistaking environmental effects for deviations from GR. More

Table 2. Upper limits on the environmental corrections to the BH QNMs. We define $\delta_{R,I} = 1 - \omega_{R,I}/\omega^{(0)}_{R,I}$, where $\omega_{R,I}$ is the real (imaginary) part of the ringdown frequency in the presence of environmental effects, whereas $\omega^{(0)}_{R,I}$ is the same for an isolated BH with the same total mass. Environment reference values are as in Table 1. The spherical and ring-like matter distributions have mass $\delta M \sim 10^{-3}M$. The scaling with the parameters is shown in Ref. [16].

Correction	$\lvert\delta_R\rvert$[%]	$\lvert\delta_I\rvert$[%]
spherical near-horizon distribution	0.05	0.03
ring at ISCO	0.01	0.01
electric charge	10^{-5}	10^{-6}
magnetic field	10^{-8}	10^{-7}
gas accretion	10^{-11}	10^{-11}
DM halos	$10^{-21}\rho_3^{\rm DM}$	$10^{-21}\rho_3^{\rm DM}$
cosmological effects	10^{-32}	10^{-32}

specifically, matter effects will provide an intrinsic limit to the precision of GR tests. If the gravity-theory modifications introduce effects smaller than the environmental ones, the former will be extremely hard to detect, unless the environment is precisely modeled. For instance, from the analysis of the previous section it is clear that testing GR will be extremely difficult with EMRIs in thin-disk environments, where planetary migration, dynamical friction and accretion are as or more important than GW emission.

In this section we therefore focus on systems and environmental effects that are *smaller* than GW emission, and compare them to possible deviations from GR, considering both the "conservative effects" due to the modified spacetime geometry, and the "dissipative effects" caused by the modified GW emission. We perform this analysis for a generic gravity theory,

$$\begin{aligned} S &= \frac{c^4}{16\pi\mathcal{G}}\int dx^4\sqrt{-g}\left[R+\partial^2\boldsymbol{\Psi}+\sum_i a_i U_i(\boldsymbol{\Psi},\boldsymbol{g},\partial\boldsymbol{\Psi},\partial\boldsymbol{g},...)\right] \\ &+ S_m^{(0)}[\boldsymbol{\Psi}_m, g_{\mu\nu}] + \sum_i b_i S_{m,i}^{(1)}[\boldsymbol{\Psi}_m,\boldsymbol{\Psi},\boldsymbol{g},...]\,, \end{aligned} \tag{4}$$

where $\boldsymbol{\Psi}$ is a generic field encoding the corrections to GR, $\mathcal{G}$ is an effective gravitational coupling (not necessarily matching the value of G measured by a Cavendish-type experiment), $\boldsymbol{\Psi}_m$ schematically represents the matter fields, U_i are nonminimal interaction terms, and we have linearized the matter action S_m. The couplings a_i and b_i parametrize deviations from GR (i.e. the theory reduces to GR for $a_i = b_i = 0$) and have different physical dimensions, depending on the theory. All corrections to GR will be proportional to some power of a_i and b_i. Although approximate, our analysis is largely theory-independent, as the action (4) includes most modifications of GR proposed in the literature [e.g. scalar-tensor theories, Chern-Simons gravity, khronometric gravity, Einstein-æther theory, etc; see Ref. [16] for details].

In Table 3 we present the intrinsic lower constraints that can be placed by measuring the orbital decay rate of a binary inspiral in some widely studied modified gravity theories. These particular cases can be mapped to the generic parametrization (4) by identifying each theory's coupling constants to suitable powers of a_i or b_i. As can be seen, these intrinsic limits are much less stringent than current observational bounds.

Table 3. Intrinsic lower constraints on some modified gravity theories due to environmental effects in the orbital decay rate of a binary inspiral. We consider Brans-Dicke (BD) theory, Einstein-Dilaton-Gauss-Bonnet (EDGB) gravity, Dynamical Chern-Simons (DCS) gravity, Einstein-æther theory (Æ) and Hořava gravity. We define $v_3 = v/(10^{-3}c)$, $\rho_3^{\rm DM} = \rho_0/(10^3 M_\odot/{\rm pc}^3)$, $\rho_2^{\rm disk} = \rho_0/(10^2{\rm kg/m}^3)$, $M_{10} = M_T/(10M_\odot)$, $B_8 = B/(10^8{\rm Gauss})$, $q_3 = Q/(10^{-3}M)$, $R_{10} = R/(10GM/c^2)$, $R_{\rm DM} = R/(7\times10^6 GM/c^2)$, where M_T is the total mass of the binary, ν is the symmetric mass ratio, v is the orbital velocity, R and ρ_0 are the characteristic location and density of the matter distributions. The dimensionless coupling constants $\omega_{\rm BD}$, ζ_3, ζ_4, $\mathcal{F}$ and the parameters $\gamma_{\hat\alpha}$, S, δ_m and $\beta_{\rm dCS}$ are defined as in Ref. [16] and Ref. [15]. For accretion disks we use a spherical distribution $\rho\sim\rho_0(R/r)^{\hat\alpha}$, which can approximate either thick disks ($\rho_0\sim10^{-10}{\rm kg/m}^3$, $\hat\alpha\sim3/2$) or thin disks ($\rho_0\sim10^2{\rm kg/m}^3$, $\hat\alpha\sim15/8$) . The normalization coefficients $\mathcal{P}$ and $\mathcal{T}$ are defined in the last row and last column.

	Intrinsic lower bound				
Theory	magnetic fields	Pull of DM profile $\rho\sim\rho_0(R/r)^{3/2}$	Pull of disk profile $\rho\sim\rho_0(R/r)^{\hat\alpha}$	electric charge	coefficient $\mathcal{T}$
BD	$\omega_{\rm BD}^{-1}\gtrsim10^{-6}\mathcal{PT}$	$\omega_{\rm BD}^{-1}\gtrsim10^{-19}\mathcal{PT}$	$\omega_{\rm BD}^{-1}\gtrsim10^{-1-5\hat\alpha}\mathcal{PT}$	$\omega_{\rm BD}^{-1}\gtrsim10^{-15}\mathcal{PT}$	$\left[\frac{0.1}{S}\right]^2$
EDGB	$\zeta_3\gtrsim10^{-12}\mathcal{PT}$	$\zeta_3\gtrsim10^{-25}\mathcal{PT}$	$\zeta_3\gtrsim10^{-7-5\hat\alpha}\mathcal{PT}$	$\zeta_3\gtrsim10^{-21}\mathcal{PT}$	$\left[\frac{\nu}{0.1}\right]^4\left[\frac{1}{\delta_m}\right]^2$
DCS	$\zeta_4\gtrsim10^{6}\mathcal{PT}$	$\zeta_4\gtrsim10^{-7}\mathcal{PT}$	$\zeta_4\gtrsim10^{-7-5\hat\alpha}\mathcal{PT}$	$\zeta_4\gtrsim10^{-3}\mathcal{PT}$	$\left[\frac{\nu}{0.1}\right]^2 v_3^{-6}\left[\frac{1}{\beta_{\rm dCS}}\right]$
Æ/Hořava	$\mathcal{F}\gtrsim10^{-9}\mathcal{PT}$	$\mathcal{F}\gtrsim10^{-22}\mathcal{PT}$	$\mathcal{F}\gtrsim10^{-4-5\hat\alpha}\mathcal{PT}$	$\mathcal{F}\gtrsim10^{-18}\mathcal{PT}$	1
coefficient $\mathcal{P}$	$B_8^2 M_{10}^2 v_3^{-4}$	$\rho_3^{\rm DM} M_{10}^2 v_3^{-1} R_{\rm DM}^{3/2}$	$\gamma_{\hat\alpha}\rho_2^{\rm disk} M_{10}^2 v_3^{2\hat\alpha-4} R_{10}^{\hat\alpha}$	$q_3^2 v_3^4$	

6. Conclusions

We have quantified the impact of realistic astrophysical environments on GW signals from BH binaries, including the effect of electromagnetic fields, cosmological evolution, accretion disks and DM. Our analysis shows that GW astronomy has the potential to become a precision discipline, because environmental effects are typically too small to affect the detection of GW signals and the estimation of the source's parameters. The few and rather extreme cases in which environmental effects might leave a detectable imprint should be seen as an opportunity, i.e. given a sufficiently sensitive detector and adequate modeling of these effects, GW astrophysics can be used to obtain information about the density and velocity of the matter surrounding GW sources.

For example, the DM density profile in galactic nuclei is still poorly understood, e.g. DM spikes may grow around the massive BHs dwelling at the center of galaxies [32]. Our analysis shows that if these DM spikes survive till $z \sim 0$ (as may be the case in satellite galaxies [37]), their effect would be detectable by eLISA in EMRIs, and GW observations could be used to constrain DM profiles near massive BHs (c.f. also Ref. [38]). Likewise, the corrections due to accretion disks depend on their geometry and on the accretion rate. EMRIs with exceptionally large SNR could be used to constrain models of accretion onto massive BHs. These intriguing possibilities requires more sophisticated modeling, which is beyond the scope of our analysis. (See however Ref. [17, 18, 19, 20, 21, 22] for some work in this direction.)

A quantitative analysis of environmental effects is also vital to allow tests of GR to be performed with GW observations, without mistaking the effect of astrophysical matter for a breakdown of GR. Our results therefore yield intrinsic lower bounds on the magnitude of the deviations from GR that can be tested with space-based GW detectors. Although admittedly approximate, ours is a largely model- and theory-independent analysis. While sufficient for our purposes, more sophisticated modeling (e.g. including the effect of BH spins) would be required to estimate environmental effects from real GW data when they become available.

Finally, our analysis has considered enviromental effects on a single source. If the effect under consideration is universal (as would be the case for a modification of gravity), one may enhance eLISA's sensitivity to it by combining different sources. One might try to apply a similar technique to matter effects such as those due to accretion disks, DM, etc. However, in that case the effects may be completely different in different sources, and it is not guaranteed that correlating several sources will help detect them.

Acknowledgments

We thank Pau Amaro-Seoane, Leor Barack, Emanuele Berti, Marc Casals, Luciano Rezzolla, Joe Silk and Nico Yunes for useful comments. This work was supported by the European Union through grants GALFORMBHS PCIG11-GA-321608, NRHEP 295189-IRSES, DyBHo–256667 ERC Starting Grant, aStronGR-IEF-298297, AstroGRAphy-2013-623439 and by FCT-Portugal through projects CERN/FP/123593/2011, IF/00293/2013. Research at Perimeter Institute is supported by the Government of Canada through Industry Canada and by the Province of Ontario through the Ministry of Economic Development & Innovation. Computations were performed on the "Baltasar Sete-Sois" cluster at IST, XSEDE clusters SDSC Trestles and NICS Kraken through NSF Grant No. PHY-090003.

References

[1] R. A. Hulse and J. H. Taylor, Astrophys. J. Lett. **195**, L51 (1975)
[2] G. M. Harry (LIGO Scientific Collaboration), Class. Quant. Grav. **27**, 084006 (2010)
[3] F. Acernese *et al.* (Virgo Collaboration), `https://tds.ego-gw.it/ql/?c=8940`(2012)
[4] S. Vitale, R. Lynch, J. Veitch, V. Raymond, and R. Sturani (2014), arXiv:1403.0129 [gr-qc]
[5] G. Hobbs, *et al.*, Class. Quant. Grav. **27**, 084013 (2010), arXiv:0911.5206 [astro-ph.SR]
[6] P. A. Seoane *et al.* (eLISA Collaboration)(2013), arXiv:1305.5720 [astro-ph.CO]

[7] F. Antonucci, *et al.*, Class. Quant. Grav. **28**, 094002 (2011), arXiv:1012.5968 [gr-qc]
[8] A. Sesana, J. Gair, E. Berti, and M. Volonteri, Phys. Rev. D **83**, 044036 (2011), arXiv:1011.5893 [astro-ph.CO]
[9] J. R. Gair, A. Sesana, E. Berti, and M. Volonteri, Class. Quant. Grav. **28**, 094018 (2011), arXiv:1009.6172 [gr-qc]
[10] E. Berti and M. Volonteri, Astrophys. J. **684**, 822 (2008), arXiv:0802.0025
[11] E. Barausse, Mon. Not. Roy. Astron. Soc. **423**, 2533 (2012), arXiv:1201.5888 [astro-ph.CO]
[12] A. Sesana, E. Barausse, M. Dotti, and E. M. Rossi(2014), arXiv:1402.7088 [astro-ph.CO]
[13] E. Berti, V. Cardoso, and A. O. Starinets, Class. Quant. Grav. **26**, 163001 (2009), arXiv:0905.2975 [gr-qc]
[14] J. R. Gair, M. Vallisneri, S. L. Larson, and J. G. Baker, Living Rev. Rel. **16**, 7 (2013), arXiv:1212.5575 [gr-qc]
[15] N. Yunes and X. Siemens, Living Rev. Rel. **16**, 9 (2013), arXiv:1304.3473 [gr-qc]
[16] E. Barausse, V. Cardoso, and P. Pani, Phys. Rev. D **89**, 104059 (2014), arXiv:1404.7149 [gr-qc]
[17] E. Barausse and L. Rezzolla, Phys. Rev. D **77**, 104027 (2008), arXiv:0711.4558 [gr-qc]
[18] C. F. Macedo, P. Pani, V. Cardoso, and L. C. Crispino, Astrophys. J. **774**, 48 (2013), arXiv:1302.2646
[19] E. Barausse, L. Rezzolla, D. Petroff, and M. Ansorg, Phys. Rev. D **75**, 064026 (2007), arXiv:gr-qc/0612123 [gr-qc]
[20] E. Barausse, Mon. Not. Roy. Astron. Soc. **382**, 826 (2007), arXiv:0709.0211 [astro-ph]
[21] N. Yunes, B. Kocsis, A. Loeb, and Z. Haiman, Phys. Rev. Lett. **107**, 171103 (2011), arXiv:1103.4609 [astro-ph.CO]
[22] B. Kocsis, N. Yunes, and A. Loeb, Phys. Rev. D **84**, 024032 (2011), arXiv:1104.2322 [astro-ph.GA]
[23] G. Gibbons, Commun. Math. Phys. **44**, 245 (1975)
[24] P. Goldreich and W. H. Julian, Astrophys. J. **157**, 869 (1969)
[25] M. A. Ruderman and P. G. Sutherland, Astrophys. J. **196**, 51 (1975)
[26] R. Blandford and R. Znajek, Mon. Not. Roy. Astron. Soc. **179**, 433 (1977)
[27] B. Punsly, Astrophys. J. **498**, 640 (1998)
[28] R. M. Wald, Phys. Rev. D **10**, 1680 (1974)
[29] W.-M. Zhang, Y. Lu, and S.-N. Zhang, Chinese J. Astron. and Astrophys. Suppl. **5**, 347 (2005), astro-ph/0501365
[30] N. A. Silant'ev, M. Y. Piotrovich, Y. N. Gnedin, and T. M. Natsvlishvili, A&A **507**, 171 (2009), arXiv:0909.1207 [astro-ph.CO]
[31] N. A. Silant'ev, Y. N. Gnedin, S. D. Buliga, M. Y. Piotrovich, and T. M. Natsvlishvili, Astrophysical Bulletin **68**, 14 (2013), 1203.2763 [astro-ph.HE]
[32] P. Gondolo and J. Silk, Phys. Rev. Lett. **83**, 1719 (1999), astro-ph/9906391
[33] D. Merritt, M. Milosavljevic, L. Verde, and R. Jimenez, Phys. Rev. Lett. **88**, 191301 (2002), arXiv:astro-ph/0201376 [astro-ph]
[34] G. Bertone and D. Merritt, Phys. Rev. D **72**, 103502 (2005), arXiv:astro-ph/0501555 [astro-ph]
[35] D. Merritt, Phys. Rev. Lett. **92**, 201304 (2004), astro-ph/0311594
[36] P. Ullio, H. Zhao, and M. Kamionkowski, Phys. Rev. D **64**, 043504 (2001), arXiv:astro-ph/0101481 [astro-ph]
[37] G. Bertone, A. R. Zentner, and J. Silk, Phys. Rev. D **72**, 103517 (2005), arXiv:astro-ph/0509565 [astro-ph]
[38] K. Eda, Y. Itoh, S. Kuroyanagi, and J. Silk, Phys. Rev. Lett. **110**, 221101 (2013), arXiv:1301.5971 [gr-qc]
[39] V. Ravi, J. Wyithe, R. Shannon, G. Hobbs, and R. Manchester(2014), arXiv:1404.5183 [astro-ph.CO]
[40] P. Amaro-Seoane, P. Brem, J. Cuadra, and P. J. Armitage, Astrophys. J. Lett. **744**, L20 (2012), arXiv:1108.5174 [astro-ph.CO]
[41] R. Schödel, D. Merritt, and A. Eckart, Astronomy & Astrophysics **502**, 91 (2009), arXiv:0902.3892 [astro-ph.GA]
[42] N. Yunes, M. Coleman Miller, and J. Thornburg, Phys. Rev. D **83**, 044030 (2011),
[43] A. Pound(2014), arXiv:1404.1543 [gr-qc]
[44] M. Visser, Phys. Rev. D **46**, 2445 (1992), arXiv:hep-th/9203057 [hep-th]
[45] P. Leung, Y. Liu, W. Suen, C. Tam, and K. Young, Phys. Rev. Lett. **78**, 2894 (1997), arXiv:gr-qc/9903031 [gr-qc]
[46] V. Cardoso, A. S. Miranda, E. Berti, H. Witek, and V. T. Zanchin, Phys. Rev. D **79**, 064016 (2009), arXiv:0812.1806 [hep-th]
[47] R. H. Price, Phys. Rev. D **5**, 2419 (1972)
[48] A. Almheiri, D. Marolf, J. Polchinski, and J. Sully, JHEP **1302**, 062 (2013), arXiv:1207.3123 [hep-th]
[49] S. L. Braunstein, S. Pirandola, and K. Zyczkowski, Phys. Rev. Lett. **110**, 101301 (2013), arXiv:0907.1190 [quant-ph]
[50] M. Abramowicz, W. Kluz'niak, and J. P. Lasota, Phys. Rev. Lett. **112**, 091301 (2014), arXiv:1311.0239 [gr-qc]
[51] W. Israel(2014), arXiv:1403.7470 [gr-qc]

[52] C. M. Will, *Theory and Experiment in Gravitational Physics* (Cambridge University Press, Cambridge, 1993)
[53] C. M. Will, Living Rev. Rel. **9**, 3 (2006), arXiv:gr-qc/0510072 [gr-qc]
[54] D. Psaltis, Living Rev. Rel. **11**, 9 (2008)

Journal of Physics: Conference Series **610** (2015) 012045 doi:10.1088/1742-6596/610/1/012045

Gravitational radiation from compact binaries in scalar-tensor gravity

R N Lang

Department of Physics, University of Florida, Gainesville, Florida 32611, USA

E-mail: lang@phys.ufl.edu

Abstract. General relativity (GR) has been extensively tested in the solar system and in binary pulsars, but never in the strong-field, dynamical regime. Soon, gravitational-wave (GW) detectors like Advanced LIGO and eLISA will be able to probe this regime by measuring GWs from inspiraling and merging compact binaries. One particularly interesting alternative to GR is scalar-tensor gravity. We present progress in the calculation of second post-Newtonian (2PN) gravitational waveforms for inspiraling compact binaries in a general class of scalar-tensor theories. The waveforms are constructed using a standard GR method known as "direct integration of the relaxed Einstein equations," appropriately adapted to the scalar-tensor case. We find that differences from general relativity can be characterized by a reasonably small number of parameters. Among the differences are new hereditary terms which depend on the past history of the source. In one special case, binary black hole systems, we find that the waveform is indistinguishable from that of general relativity. In another, mixed black hole-neutron star systems, all differences from GR can be characterized by only a single parameter.

1. Introduction

2015 marks the centennial of Einstein's theory of general relativity (GR). In the last 100 years, GR has withstood every test thrown at it. Nevertheless, we still expect GR to fail at some energy scale, since we have no consistent quantum theory of gravity. Modifications to gravity may also explain the phenomena of dark matter and dark energy. It is therefore important that we keep testing Einstein's theory as it moves into its second century. Notably, general relativity has not yet been tested in the strong-field, dynamical regime that exists in coalescing compact binaries made of neutron stars and/or black holes. These systems are the primary targets of current and future gravitational-wave (GW) detectors, such as Advanced LIGO and Advanced Virgo on the ground and eLISA in space. By carefully measuring the gravitational waveform of a compact binary with these detectors, and comparing it to templates from different theories, we can set limits on deviations from general relativity or possibly discover new physics.

A particularly important alternative to GR is the collection of scalar-tensor theories of gravity [1]. These theories have a long history, dating back over 50 years, and while they are perhaps an overly simple modification to Einstein's theory (with just the addition of some scalar fields to GR's metric tensor $g_{\mu\nu}$), they remain well-motivated. For instance, many so-called $f(R)$ theories, which may explain the acceleration of the universe without dark energy, can be expressed in the form of a scalar-tensor theory [2]. Scalar-tensor theories are also potential low-energy limits of string theory [1].

Journal of Physics: Conference Series **610** (2015) 012045 doi:10.1088/1742-6596/610/1/012045

Most analyses of gravitational waveforms in alternative theories of gravity include only the lowest-order effects in the waveform model. However, real templates used by gravitational-wave detectors must be calculated to high post-Newtonian (PN) order, an expansion in powers of the source velocity v/c. In this paper, we report on progress in calculating gravitational waveforms in a subclass of scalar-tensor theories to second post-Newtonian (2PN) order. The details of this work can be found in papers by Mirshekari and Will [3] (hereafter MW) and Lang [4] (hereafter L14). Throughout the paper, we use units in which $c = 1$. Greek indices run from 0-3, while Latin indices run from 1-3. Repeated indices are summed over.

We are specifically interested in theories with a single, massless scalar field ϕ. In the simplest and most familiar scalar-tensor theory, known as Brans-Dicke theory, the coupling function $\omega(\phi)$ is a constant; here, however, we allow it to be any function with a Taylor expansion. We work in the "Jordan frame," in which the scalar field does not couple directly to the matter (though, of course, ϕ will indirectly impact the matter through its role in determining the metric). With all of these assumptions, the scalar-tensor field equations are given by

$$G_{\mu\nu} = \frac{8\pi}{\phi}T_{\mu\nu} + \frac{\omega(\phi)}{\phi^2}\left(\phi_{,\mu}\phi_{,\nu} - \frac{1}{2}g_{\mu\nu}\phi_{,\lambda}\phi^{,\lambda}\right) + \frac{1}{\phi}\left(\phi_{;\mu\nu} - g_{\mu\nu}\Box_g\phi\right), \tag{1a}$$

$$\Box_g\phi = \frac{1}{3+2\omega(\phi)}\left(8\pi T - 16\pi\phi\frac{\partial T}{\partial\phi} - \frac{d\omega}{d\phi}\phi_{,\lambda}\phi^{,\lambda}\right), \tag{1b}$$

where $G_{\mu\nu}$ is the Einstein tensor, $T_{\mu\nu}$ is the stress energy of matter and nongravitational fields, and $T \equiv g^{\alpha\beta}T_{\alpha\beta}$ is its trace. We use commas to denote ordinary derivatives. Semicolons denote covariant derivatives (taken using $g_{\mu\nu}$ in the usual way), and $\Box_g \equiv g^{\alpha\beta}\partial_\alpha\partial_\beta$.

Since scalar-tensor theories do not obey the strong equivalence principle, the motion and gravitational-wave emission of a binary depend on the internal composition of its constituent bodies. To handle this complication, we have adopted the approach of Eardley [5]. We treat the matter stress-energy tensor as a sum of delta functions located at the position of each compact object. However, instead of assigning each body a constant mass, we let the mass be a function of the scalar field, $M_A = M_A(\phi)$. This gives the matter an indirect dependence on ϕ [hence the term $\partial T/\partial\phi$ in (1b)], even though we still work in the Jordan frame. In the final results, this dependence will appear as the "sensitivity" of the mass to variations in the scalar field,

$$s_A \equiv \left(\frac{d\ln M_A(\phi)}{d\ln\phi}\right)_0, \tag{2}$$

as well as derivatives of this quantity. (The subscript 0 means that the derivative should be evaluated using the asymptotic value of the scalar field, ϕ_0.) In the weak-field limit, the sensitivity is proportional to the Newtonian self-gravitational energy per unit mass of the body. For neutron stars, the sensitivity depends on the mass and equation of state of the star, with typical values 0.1–0.3 [6]. For black holes, $s = 0.5$, and all derivatives vanish.

2. Solving the field equations

To solve the field equations, we use a method known as direct integration of the relaxed Einstein equations (DIRE). While DIRE was originally developed for GR by Will, Wiseman, and Pati [7–10], it can be easily adapted to scalar-tensor theory if the Einstein equations are replaced by the field equations (1a)-(1b).

The first step is to introduce new variables which will help us place the field equations in a "relaxed" form. We assume that far away from the sources, the metric reduces to the Minkowski metric, $\eta_{\mu\nu}$, and the scalar field tends to a constant ϕ_0. We introduce a rescaled scalar field,

$$\varphi \equiv \frac{\phi}{\phi_0}\,. \tag{3}$$

Next, we define a conformally transformed metric, $\tilde{g}_{\mu\nu} \equiv \varphi g_{\mu\nu}$, with determinant $\tilde{g}$, and a "gravitational field,"

$$\tilde{h}^{\mu\nu} \equiv \eta^{\mu\nu} - \sqrt{-\tilde{g}}\tilde{g}^{\mu\nu}\,. \tag{4}$$

We use a tilde to differentiate this field from the gravitational field defined in general relativity, $h^{\mu\nu}$, which has the same definition but with the normal metric $g_{\mu\nu}$ in place of the conformally transformed metric $\tilde{g}_{\mu\nu}$. [See, for instance, (2.2) in [8].] We impose the Lorenz gauge condition

$$\tilde{h}^{\mu\nu}{}_{,\nu} = 0\,. \tag{5}$$

Then the field equation (1a) reduces to

$$\Box_\eta \tilde{h}^{\mu\nu} = -16\pi\tau^{\mu\nu}\,, \tag{6}$$

where $\Box_\eta \equiv \eta^{\alpha\beta}\partial_\alpha\partial_\beta$ is the flat-spacetime wave operator and the source is

$$\tau^{\mu\nu} \equiv (-g)\frac{\varphi}{\phi_0}T^{\mu\nu} + \frac{1}{16\pi}(\Lambda^{\mu\nu} + \Lambda_s^{\mu\nu})\,. \tag{7}$$

Here g is the determinant of $g_{\mu\nu}$. The first term of (7) represents the *compact* piece of the source. The quantity $\Lambda^{\mu\nu}$ represents the gravitational-field contribution to the stress energy. It is constructed so that it has the same functional form as in GR (except with components of $\tilde{h}^{\mu\nu}$ taking the place of components of $h^{\mu\nu}$). The quantity $\Lambda_s^{\mu\nu}$ is an entirely new contribution to the stress energy which depends on φ. Both $\Lambda^{\mu\nu}$ and $\Lambda_s^{\mu\nu}$ are given explicitly in L14, Eqs. (2.9)-(2.11). The scalar field equation (1b) can also be written as a flat-spacetime wave equation,

$$\Box_\eta \varphi = -8\pi\tau_s\,, \tag{8}$$

with source

$$\tau_s \equiv -\frac{1}{3+2\omega}\sqrt{-g}\frac{\varphi}{\phi_0}\left(T - 2\varphi\frac{\partial T}{\partial\varphi}\right) - \frac{1}{8\pi}\tilde{h}^{\alpha\beta}\varphi_{,\alpha\beta} + \frac{1}{16\pi}\frac{d}{d\varphi}\left[\ln\left(\frac{3+2\omega}{\varphi^2}\right)\right]\varphi_{,\alpha}\varphi_{,\beta}\sqrt{-\tilde{g}}\tilde{g}^{\alpha\beta}\,. \tag{9}$$

The wave equations (6) and (8) can be solved formally in all spacetime by using a retarded Green's function,

$$\tilde{h}^{\mu\nu}(t,\mathbf{x}) = 4\int \frac{\tau^{\mu\nu}(t',\mathbf{x}')\delta(t'-t+|\mathbf{x}-\mathbf{x}'|)}{|\mathbf{x}-\mathbf{x}'|}d^4x'\,, \tag{10a}$$

$$\varphi(t,\mathbf{x}) = 2\int \frac{\tau_s(t',\mathbf{x}')\delta(t'-t+|\mathbf{x}-\mathbf{x}'|)}{|\mathbf{x}-\mathbf{x}'|}d^4x'\,. \tag{10b}$$

The delta function in both these equations restricts the integration to being over the past flat-spacetime null cone emanating from the field point $(t,\mathbf{x})$. To obtain explicit solutions, we divide the spacetime into two regions. The *near zone* is defined as the area with $|\mathbf{x}| = R < \mathcal{R}$, where $\mathcal{R}$ is the characteristic wavelength of gravitational radiation from the system. Everything outside the near zone ($R > \mathcal{R}$) is the *radiation zone*.

We carry out the integrals (10a) and (10b) differently depending on the location of the field point $\mathbf{x}$ of interest (near zone or radiation zone). Once the field point is chosen, we then evaluate each integral in two separate pieces: one integral over source points $\mathbf{x}'$ in the near zone and another over source points in the radiation zone. This makes for a total of four different integration techniques, described in detail in [8] and L14. It has been shown (in GR) that all terms dependent on the boundary radius $\mathcal{R}$ in one-half of the integral are exactly canceled by pieces in the other half of the integral, leaving the final answer, as expected, independent of this arbitrary parameter [8, 9]. In our work, we simply assume this property and ignore any terms which depend on $\mathcal{R}$.

3. Equations of motion

The first step in finding the gravitational waveform is to calculate the equations of motion for the compact binary. To do so, we must evaluate $\tilde{h}^{\mu\nu}$ and φ at field points in the near zone, where they influence the bodies. As stated above, this usually involves an integration over source points in both zones. It turns out, however, that only near-zone points contribute at the order to which we work.

The calculation of the near-zone fields is an iterative one. We begin by finding the lowest order pieces of $\tilde{h}^{00}$ and φ, which depend only on the compact piece of the source. These fields are then plugged back into (7) and (9) to generate the next-highest-order source. The process continues, alternating between calculating fields and plugging them back into the source expressions, until the fields are known to the desired post-Newtonian order. The details and results of this procedure can be found in MW.

Once we have the fields, we can use (3) and (4) to solve for the metric $g_{\mu\nu}$. From the metric, we calculate Christoffel symbols just as in GR. The final equations of motion are a bit trickier, however, due to the violation of the strong equivalence principle. Using the Bianchi identity, we can write the equations of motion as

$$T^{\mu\nu}{}_{;\nu} = \frac{\partial T}{\partial \phi}\phi^{,\mu}\,. \tag{11}$$

A more useful form, involving the Christoffel symbols, is given in MW, Eq. (2.5). We plug in for the symbols, integrate over the extent of each body, and then convert to relative coordinates. The final 2.5PN equations of motion can be written as

$$\begin{aligned} a^i = &-\frac{G\alpha m}{r^2}\hat{n}^i + \frac{G\alpha m}{r^2}(A_{\rm PN}\hat{n}^i + B_{\rm PN}\dot{r}v^i) + \frac{8}{5}\eta\frac{(G\alpha m)^2}{r^3}(A_{1.5\rm PN}\dot{r}\hat{n}^i - B_{1.5\rm PN}v^i) \\ &+ \frac{G\alpha m}{r^2}(A_{2\rm PN}\hat{n}^i + B_{2\rm PN}\dot{r}v^i)\,. \end{aligned} \tag{12}$$

Here $\hat{\mathbf{n}}$ is a unit vector pointing from body 2 to body 1, r is the orbital separation, $\mathbf{v}$ is the relative velocity, $\mathbf{a}$ is the relative acceleration, $m \equiv m_1 + m_2$, and $\eta \equiv m_1 m_2/m^2$. The A and B coefficients contain all the details; they are given in MW. We have also defined

$$G \equiv \frac{1}{\phi_0}\frac{4 + 2\omega_0}{3 + 2\omega_0}\,, \tag{13}$$

where $\omega_0 \equiv \omega(\phi_0)$. This definition is chosen so that $g_{00} = -1 + 2GU$ matches the expression in GR for a perfect fluid (i.e., with all sensitivities equal to zero). However, with this definition, the coupling in the equations of motion is not simply G, but rather $G\alpha$, with $\alpha \equiv 1 - \zeta + \zeta(1 - 2s_1)(1 - 2s_2)$ and $\zeta \equiv 1/(4 + 2\omega_0)$.

The even orders in the equations of motion (0PN, 1PN, and 2PN) correspond to conservative parts of the motion. These equations can be derived from a Lagrangian and admit conserved energy and momentum quantities; for details, see MW. The odd terms (1.5PN and 2.5PN) represent radiation-reaction terms which dissipate energy and cause the binary to coalesce. In

GR, radiation reaction begins at 2.5PN order with quadrupole radiation; the 1.5PN component in our results is due to the well-known phenomenon of dipole radiation.

4. Calculating the gravitational waves

To date, we have calculated the tensor gravitational waves $\tilde{h}^{ij}$ to 2PN order. The calculation requires integrating (10a) for field points in the radiation zone. To simplify matters, we need only be concerned with the results in a subset of the radiation zone, the *far-away zone*, where $R >> \mathcal{R}$. Unlike for the equations of motion, we must consider source points in both the near and radiation zones.

Taking the limit of large R, the near-zone contribution to the tensor GWs, $\tilde{h}^{ij}_{\mathcal{N}}$, is given by

$$\tilde{h}^{ij}_{\mathcal{N}}(t, \mathbf{x}) = \frac{2}{R}\frac{d^2}{dt^2}\sum_{m=0}^{\infty} \hat{N}^{k_1}\cdots\hat{N}^{k_m} I_{\rm EW}^{ijk_1\cdots k_m}(\tau)\,, \tag{14}$$

where $\hat{\mathbf{N}} \equiv \mathbf{x}/R$ is the direction from the source to the detector and $\tau \equiv t - R$ is the retarded time. The quantities $I_{\rm EW}^{M+2}$ are known as "Epstein-Wagoner" (EW) moments [11]. They make use of the near-zone fields calculated in MW and involve a variety of different integrals. It is useful to consider the post-Newtonian orders of the various EW moments. The lowest order moment, $I_{\rm EW}^{ij}$, has a quadrupole character. As in general relativity, we define its lowest order piece to be 0PN order. It also has higher-order contributions, specifically at 1PN, 1.5PN, and 2PN orders. When substituted into (14), it generates GWs at all of these orders.

The three-index (octupole) moment $I_{\rm EW}^{ijk}$ begins at 0.5PN order but also comprises pieces at 1.5PN and 2PN orders. When plugged into (14), it generates GWs at all of these orders. Each successive EW moment begins one-half post-Newtonian order higher than the previous moment. The final moment we need is the six-index moment $I_{\rm EW}^{ijklmn}$, which consists only of a 2PN contribution (and generates 2PN gravitational waves).

Once they have been calculated to the desired order for an arbitrary number of bodies, the EW moments are simplified to the two-body case of interest and rewritten in relative coordinates. Finally, they are substituted back into (14), which involves taking time derivatives. Whenever an acceleration appears in one of these derivatives, we substitute in the equations of motion (12).

The tensor waves also include contributions from radiation-zone source points. The first step to finding them is to derive an expression for the source τ^{ij} in the radiation zone. Since there are no compact sources at $R > \mathcal{R}$, τ^{ij} is composed purely of field terms. The details of this derivation, and of the subsequent calculation of the GWs, are found in L14.

It turns out that the radiation-zone integrals produce GWs at 1.5PN and 2PN orders. For the sake of clarity, we separate GWs generated by Λ^{ij} from those generated by Λ_s^{ij}. The former produces terms which differ from the GR versions only by a scaling factor. For instance, at 1.5PN order, we find the contribution

$$\tilde{h}^{ij} = \frac{4G(1-\zeta)m}{R}\left[\frac{11}{12}\dddot{\mathcal{I}}^{ij} + \int_0^{\infty} ds \overset{(4)}{\mathcal{I}}{}^{ij}(\tau - s)\ln\frac{s}{2R+s}\right]. \tag{15}$$

Similar terms exist at 2PN order. The moment $\mathcal{I}^{ij}$ is defined in L14, Eq. (6.3a). The first term in (15) is *instantaneous*, depending only on the state of the binary at the current (retarded) time. All near-zone contributions to the waveform are also instantaneous. The second term in (15), however, is *hereditary*, depending on the entire past history of the source. Specifically, the presence of the logarithm means that this is a "tail" term, arising physically from the scattering of the waves off the background spacetime. This phenomenon is present in GR at this order; scalar-tensor theory only modifies the coefficient.

Journal of Physics: Conference Series **610** (2015) 012045 doi:10.1088/1742-6596/610/1/012045

The field terms Λ_s^{ij} produce new pieces of the gravitational waveform not seen in GR. For instance, at 1.5PN order, we have

$$\tilde{h}^{ij} = \frac{4G(1-\zeta)m_s}{R}\left(-\frac{1}{12}\dddot{\mathcal{I}}_s^{ij}\right) + \frac{4}{R}\frac{1-\zeta}{\zeta}\left(\frac{1}{6}\int_{-\infty}^{\tau}\ddot{\mathcal{I}}_s^i(\tau')\ddot{\mathcal{I}}_s^j(\tau')d\tau' - \frac{1}{6}\dot{\mathcal{I}}_s^{(i}\ddot{\mathcal{I}}_s^{j)} - \frac{1}{18}\mathcal{I}_s^{(i}\dddot{\mathcal{I}}_s^{j)}\right), \tag{16}$$

where m_s, $\mathcal{I}_s^i$, and $\mathcal{I}_s^{ij}$ are defined in L14, Eqs. (6.10), (6.11a), and (6.11b). Similar terms exist at 2PN order. Eq. (16) contains a hereditary contribution which is *not* a tail. Terms like these—hereditary but without logarithms—do appear in GR, but not until 2.5PN order. Their appearance beginning at 1.5PN order in scalar-tensor theory results from the nonvanishing of the scalar dipole moment, $\mathcal{I}_s^i$, the same effect which causes radiation reaction to appear at 1.5PN order in the equations of motion. The 1.5PN hereditary integral can be described as a "gravitational-wave memory" term. It contains a zero-frequency, or DC, component, in addition to the usual oscillatory terms. (The new 2PN hereditary integral, not shown, does not.)

5. Discussion

To find the final tensor waveform, we add the contributions from the near and radiation zones. It can be written as a post-Newtonian expansion,

$$\tilde{h}^{ij} = \frac{2G(1-\zeta)\mu}{R}[\tilde{Q}^{ij} + P^{1/2}Q^{ij} + PQ^{ij} + P^{3/2}Q^{ij} + P^2Q^{ij} + O(\epsilon^{5/2})]_{\rm TT}, \tag{17}$$

where the superscripts on P denote the PN order of each term. "TT" denotes the transverse-traceless projection; to save time, we throw away pieces at any stage of the calculation which will not survive this final projection. Each term in the expansion can be found in L14; note that there the 1.5PN and 2PN terms have been divided into the contributions from the near zone and the contributions from the radiation zone [e.g., (15) and (16)].

The expressions are significantly more complicated than the GR equivalents, Eq. (6.11) of [8]. Nevertheless, all deviations can be characterized by a small number of parameters, including G, ζ, α, and several others ($\bar{\gamma}$, $\bar{\beta}_i$, $\bar{\delta}_i$, $\bar{\chi}_i$, $\mathcal{S}_+$, and $\mathcal{S}_-$), given in or near MW Table I. All of these parameters, as expected, depend on the fundamental parameters of the theory (Taylor coefficients of the coupling function $\omega(\phi)$ and the asymptotic scalar field ϕ_0) and/or the sensitivities and sensitivity derivatives of the sources. The same set of parameters characterizes the deviations of the equations of motion from the GR version.

In general, the tensor waves in scalar-tensor theory have the same general functional form as the GR waves, with only the addition of these parameters to the coefficients. In some cases, however, the existence of a nonvanishing scalar dipole moment generates entirely new terms. One example is the 1.5PN contribution from the quadrupole moment; in GR, this moment only contributes at 0PN, 1PN, and 2PN orders. Others include the new hereditary integrals.

For binary black holes, the expressions simplify considerably. The 2.5PN equations of motion and 2PN tensor waves turn out to have the exact same forms as in general relativity if the masses are rescaled by a factor $1/(1-\zeta)$. Since the masses of the bodies are defined by their Keplerian motion, this rescaling is unmeasurable. Therefore, to these orders, the motion and tensor GW emission of two black holes in scalar-tensor theory are indistinguishable from the motion and emission in GR. This result is not surprising. Hawking showed that stationary, asymptotically flat black holes in vacuum are identical in both theories [12], leading to conjectures that the same might be true for black hole binaries. The result is supported by other analytic studies (to all PN orders, but only leading order in the mass ratio) [13] and numerical studies [14]. Of course, it depends strongly on the assumptions we made (no potential for the scalar field, a time-independent ϕ_0). One unfortunate consequence of this result is that eLISA will be unlikely to provide information about these particular scalar-tensor theories; ground-based detectors, which will detect many inspirals involving neutron stars, are more suited to the task.

For a system containing one neutron star (say, body 1) and one black hole, the expressions also simplify somewhat. In this case, the motion and waves are identical to those in GR through 1PN order (after a mass rescaling). At 1.5PN order, deviations start to occur. However, the deviations are always parametrized by the single parameter $Q = \zeta(1-\zeta)^{-1}(1-2s_1)^2$. Because Q contains no information on the derivatives of the coupling function $\omega(\phi)$, we cannot, at this order, formally distinguish the waveform produced in the Brans-Dicke theory from that produced in a general scalar-tensor theory of the type we consider. One could imagine doing so by measuring masses and sensitivities of multiple neutron stars.

Work is underway on the calculation of the scalar gravitational waves. Some of the low-order pieces can be found in [15], Eqs. (101) and (102). The process is similar to that for the tensor waves; however, there are some unique complications. For instance, the near-zone calculation also relies on the computation of moments (like the EW moments). In the scalar case, however, these include monopole and dipole moments. Since we defined the quadrupole moment (and the waves it generated) as 0PN order, the monopole and dipole moments turn out to begin at -1PN and -0.5PN orders, respectively. These moments must then be calculated to relative 3PN and 2.5PN orders in order to find the complete 2PN gravitational waves. The number of integrals required to find the 3PN-accurate monopole moment is several times that needed for any of the tensor EW moments; furthermore, several of the integrals are extremely lengthy to compute. Things become even more difficult when considering the energy loss from the system, ultimately needed to determine the gravitational-wave phasing. While the 2PN energy loss from tensor waves can be found by knowing up to the 2PN form of the waves, the 2PN energy loss from scalar waves requires knowledge of certain 2.5PN pieces of the waveform. It also requires the 3PN equations of motion, one-half PN order beyond that calculated in MW. There is a still a great deal of work to do in calculating even the simplest scalar-tensor gravitational waveforms.

Acknowledgments

We thank the organizers of the 10th International LISA Symposium for an excellent conference. We are grateful to Clifford Will and Saeed Mirshekari for their hard work on the first phase of this calculation and their useful insights on the rest. This work was supported in part by the National Science Foundation, Grants No. PHY-0965133, No. PHY-1260995, and No. PHY-1306069.

References

[1] Fujii Y and Maeda K I 2003 *The Scalar-Tensor Theory of Gravitation* (Cambridge, UK: Cambridge University Press)
[2] de Felice A and Tsujikawa S 2010 *Living Rev. Relativity* **13** 3
[3] Mirshekari S and Will C M 2013 *Phys. Rev.* D **87** 084070
[4] Lang R N 2014 *Phys. Rev.* D **89** 084014
[5] Eardley D M 1975 *Astrophys. J.* **196** L59–62
[6] Will C M and Zaglauer H W 1989 *Astrophys. J.* **346** 366–77
[7] Wiseman A G 1992 *Phys. Rev.* D **46** 1517–39
[8] Will C M and Wiseman A G 1996 *Phys. Rev.* D **54** 4813–48
[9] Pati M E and Will C M 2000 *Phys. Rev.* D **62** 124015
[10] Pati M E and Will C M 2002 *Phys. Rev.* D **65** 104008
[11] Epstein R and Wagoner R V 1975 *Astrophys. J.* **197** 717–23
[12] Hawking S W 1972 *Commun. Math. Phys.* **25** 167–71
[13] Yunes N, Pani P and Cardoso V 2012 *Phys. Rev.* D **85** 102003
[14] Healy J, Bode T, Haas R, Pazos E, Laguna P, Shoemaker D M and Yunes N 2012 *Classical Quantum Gravity* **29** 232002
[15] Will C M 2014 *Living Rev. Relativity* **17** 4

A Time Domain Waveform for Testing General Relativity

Cédric Huwyler[1], **Edward K. Porter**[2], **Philippe Jetzer**[1]

[1] Physik-Institut, Universität Zürich, Winterthurerstrasse 190, 8057 Zürich, Switzerland
[2] François Arago Centre, APC, Université Paris Diderot, CNRS/IN2P3, CEA/Irfu, Observatoire de Paris, Sorbonne Paris Cité, 10 rue A. Domon et L. Duquet, 75205 Paris Cedex 13, France

E-mail: chuwyler@physik.uzh.ch

Abstract. Gravitational-wave parameter estimation is only as good as the theory the waveform generation models are based upon. It is therefore crucial to test General Relativity (GR) once data becomes available. Many previous works, such as studies connected with the ppE framework by Yunes and Pretorius, rely on the stationary phase approximation (SPA) to model deviations from GR in the frequency domain. As Fast Fourier Transform algorithms have become considerably faster and in order to circumvent possible problems with the SPA, we test GR with corrected time domain waveforms instead of SPA waveforms. Since a considerable amount of work has been done already in the field using SPA waveforms, we establish a connection between leading-order-corrected waveforms in time and frequency domain, concentrating on phase-only corrected terms. In a Markov Chain Monte Carlo study, whose results are preliminary and will only be available later, we will assess the ability of the eLISA detector to measure deviations from GR for signals coming from supermassive black hole inspirals using these corrected waveforms.

1. Introduction

This decade will witness the advent of direct gravitational wave (GW) detection: advanced ground-based detectors such as LIGO and Virgo are scheduled to come online as early as 2015 and should hopefully provide numerous detections of stellar mass systems. Within the same time frame, pulsar timing array analysis is expected to advance to a point where the first detection of supermassive black hole binaries (SMBHBs) in the nHz range will be possible. By 2034, the European Space Agency plans to launch a space-based LISA-like GW detector called eLISA; the theme of "The Gravitational Universe" was recently chosen for the ESA Cosmic Vision L3 mission in order to nourish the development of a space-based GW mission.

The vast majority of algorithms in GW astronomy are based on the concept of matched filtering. The detection of signals and the process of parameter estimation are conducted by assuming a particular underlying gravitational theory that is, a priori, considered to be true; in our case this is the theory of general relativity (GR). That a waveform based on this theory is able to detect the signals present in the data, however, does not mean that it is the theory providing the best fit to the data. Alternative theories of gravity, deviating only slightly from GR, could also provide waveforms that also detect the signal just as well, albeit with different best-fit model parameters. Thus, if we use the wrong waveform for parameter estimation, the inferred parameters will have a fundamental bias [1] and will not fully reflect reality. This raises the question of which particular theory of gravity should be preferred in the light of the measured data. By observing SMBHBs, which are among the strongest sources of GWs, alternative theories can be compared to GR via a Bayesian inference, where the odds ratio provides a useful tool in measuring the betting odds for a certain alternative theory against GR. The space of available model hypotheses can then be populated with *specific* alternative theories (that are usually based on a modified Einstein-Hilbert action, such as scalar-tensor theories, Chern-Simons theory, etc.) or *generic* features that are not present in GR (such as a non-zero graviton mass, Lorentz violation, variable $G(t)$, etc.). One could then test all these (classes of) theories against GR. It is noteworthy that such approaches all have the common feature that they modify either the phase or the amplitude of the GW modelled by GR [1]. Interested in the mere fact of whether there exists a theory that describes the data better than GR, instead of testing for specific theories or generic features, one could thus just impose a set of theories that modify certain post-Newtonian (PN) amplitude or phase coefficients of GR, without considering a particular action or a particular physical feature. One would then of course have to make sure that deviations originate from fundamental physics and not from other effects, such as astrophysical pollution or pathologies in the chosen PN approximation scheme.

A considerable amount of work has been done in the aforementioned field. Arun, Iyer, Qusailah and Sathyprakash [2] turned the question of testing GR into the exact measurement of the PN coefficients Ψ_i of the stationary phase approximation (SPA) instead of measuring individual binary parameters, while Yunes and Pretorius [1] added an amplitude correction to introduce their parameterized post-Einsteinian (ppE) scheme as

$$\tilde{h}(f) = \tilde{h}_{\rm GR}(f)\,(1+\alpha\, u^a)\, e^{i\beta\, u^b}, \tag{1}$$

where tilde denotes a Fourier transform, $u = G\mathcal{M}\pi f/c^3$ is the reduced frequency, G is the gravitational constant, c is the speed of light and $\mathcal{M} = M\eta^{3/5}$ is the chirp mass with total mass $M = m_1 + m_2$ and symmetric mass ratio $\eta = m_1 m_2/M^2$. In this scheme, the ppE parameters (a, α, b, β) can in general be real numbers. There are several studies regarding the performance of the ppE scheme in model selection [3, 4, 5]. Sampson et al. have shown in [4] that leading order corrections as applied to GR in Eq. (1) are already enough to distinguish GR from any competing theory. There is also the TIGER pipeline devised for ground-based detector data analysis [6].

The previously mentioned approaches are constructed directly in the frequency domain and use the SPA to motivate the corrections they introduce to GR. The SPA was initially introduced into the field of GW astronomy as it saved on computational cost by making a Fast Fourier Transform (FFT) unnecessary, without much loss in accuracy. We chose not to rely on the SPA for the following reasons: i) FFTs have nowadays become considerably faster and use only 3-8% of the time needed for waveform generation, ii) several studies have shown issues with the SPA, especially in the high-mass regime; there modifications such as the introduction of unphysical mass ratios or pseudo-4PN terms are of some help [7, 8], iii) even if the time-domain PN approximation breaks down by producing non-monotonic phase and frequency evolutions below a certain orbital separation, the SPA waveforms still look fine at this point and do not indicate their apparent failure, and iv) as direct time-domain waveforms have been more commonly used to assess the parameter estimation capabilities of LISA-like detectors.

In the following sections we will quickly review the waveform models in the time domain and the SPA, create a modified time domain waveform and compare it to a phase-only ppE scheme.

2. Waveform Models

2.1. Time Domain

In the time domain, the gravitational waveform is governed by the evolution of orbital frequency and phase [9]:

$$\begin{aligned}\omega(\Theta) &= \frac{c^3}{8GM}\left[\Theta^{-3/8} + \left(\frac{743}{2688} + \frac{11}{32}\eta\right)\Theta^{-5/8} - \frac{3\pi}{10}\Theta^{-6/8}\right.\\ &\quad \left. + \left(\frac{1855099}{14450688} + \frac{56975}{258048}\eta + \frac{371}{2048}\eta^2\right)\Theta^{-7/8}\right], \qquad (2)\\ \Phi(\Theta) &= \Phi_C - \frac{1}{\eta}\left[\Theta^{5/8} + \left(\frac{3715}{8064} + \frac{55}{96}\eta\right)\Theta^{3/8} - \frac{3\pi}{4}\Theta^{2/8}\right.\\ &\quad \left. + \left(\frac{9275495}{14450688} + \frac{284875}{258048}\eta + \frac{1855}{2048}\eta^2\right)\Theta^{1/8}\right], \qquad (3)\end{aligned}$$

respectively. Here, $\Theta(t) = \frac{\eta c^3}{5GM}(t_c - t)$ is the dimensionless time parameter, where t_c is the formal time at coalescence. The GW polarisations can then be written as

$$h_{+,\times} = \frac{2GM\eta}{c^2 D_L}x\left[H^{(0)}_{+,\times} + x^{1/2}H^{(1/2)}_{+,\times} + xH^{(1)}_{+,\times} + x^{3/2}H^{(3/2)}_{+,\times} + x^2H^{(2)}_{+,\times}\right], \qquad (4)$$

where the post-Newtonian parameter $x = (GM\omega/c^3)^{2/3}$ is a function of the orbital frequency ω, D_L is the luminosity distance parameter and the $H^{(n)}_{+,\times}$ denote the harmonics of the phase which are proportional to sines and cosines of integer multiples of Φ and can be found in [9]. In the low-frequency approximation, which is the case for the vast majority of eLISA sources, the individual polarisations can be projected on the detector arm by using the antenna pattern functions $F_k^{+,\times}$: $h_k(t) = F_k^+(t)\,h_+(t) + F^\times(t)\,h_\times(t)$ for the kth channel of the detector.

2.2. Frequency Domain

Taking only the dominant harmonic into account, the Fourier transform of the time domain waveform is in the SPA expressed as

$$\tilde{h}_k(f) = \mathcal{A}(f)\,e^{i\Psi(f)}, \qquad (5)$$

where $\mathcal{A}(f) \propto f^{-7/6}$ and the SPA phase is defined to be

$$\Psi(f) = 2\pi f t(f) - 2\Phi[t(f)] - \frac{\pi}{4}, \qquad (6)$$

or in terms of a 2PN expansion,

$$\begin{aligned}\Psi(f) = 2\frac{t_c c^3}{GM}x^{3/2} - 2\Phi_c - \frac{\pi}{4} + \frac{3x^{-5/2}}{128\eta}\left[1 + \left(\frac{3715}{756} + \frac{55\eta}{9}\right)x - 16\pi x^{3/2}\right.\\ \left. + \left(\frac{15293365}{508032} + \frac{27145\eta}{504} + \frac{3085\eta^2}{72}\right)x^2\right]. \qquad (7)\end{aligned}$$

3. Modified Waveforms

In contrast to [1], we chose to take only modifications to the phase into account and neglect any corrections to the amplitude (i.e. $a = \alpha = 0$). Eq. (1) amounts to a modification of the SPA phase in terms of

$$\Psi_{\rm NGR}(b, \beta; u) = \Psi_{\rm GR}(u) + \beta\, u^b. \tag{8}$$

where 'NGR' stands for 'non-GR'. In a similar way, however slightly more adapted to Eq. (1), we chose to modify the time domain phase

$$\Phi_{\rm NGR}(i, \kappa; \Theta) = \Phi_{\rm GR}(\Theta) - \frac{1}{\eta}\kappa\,\Theta^{\frac{5-2i}{8}}, \tag{9}$$

with $\kappa \in \mathbb{R}$. If leading order corrections are enough to detect a deviation from GR in the frequency domain in Eq. (8), then in principle they should also be sufficient in Eq. (9). For the sake of a comparison between corrected time domain and SPA waveforms, however, we include higher orders to Eq. (9) such that

$$\Phi_{\rm NGR}(i, \kappa; \Theta) = \Phi_{\rm GR}(\Theta) - \frac{1}{\eta}\sum_i \kappa_i \Theta^{\frac{5-2i}{8}}. \tag{10}$$

Our aim is now to fix the coefficients κ_i in such a way that the Fourier transform of a waveform with an orbital phase given by Eq. (10) is as close as possible to an SPA waveform with a phase described by Eq. (8). To this end, let us construct a relation which allows us to compare the coefficients of Eqs. (8) and (10) by taking the derivative of the definition of the SPA phase. Noting that $\frac{d\Phi}{du}(u) = \frac{dt}{du}\frac{c^3}{G\mathcal{M}}u$, we arrive at

$$\frac{d\Psi}{du} = 2\frac{c^3}{G\mathcal{M}}\,t(u). \tag{11}$$

This enables us to write the time-of-frequency and the frequency-of-time functions in the simple form

$$t(u) = \frac{1}{2}\frac{G\mathcal{M}}{c^3}\frac{d\Psi}{du}, \qquad u(t) = \frac{G\mathcal{M}}{c^3}\frac{d\Phi}{dt}. \tag{12}$$

Since one requires $u[t(u)] = u$, the coefficients of $\Psi(u)$ given the coefficients of $\Phi(\Theta)$ can be computed by evaluating

$$u[\Theta(u)]_{\rm 2PN} = u\left(1 + \sum_{k=0}^{4} u^{k/3}\mathcal{A}_k\right) = u, \tag{13}$$

Table 1. An overview of how the corrections $\kappa_i(b, \beta)$ have to be fixed in order to approximately represent a leading order phase-only ppE waveform after a Fourier transform.

b	-5/3	-4/3	-1	-2/3	-1/3
κ_0	16β	0	0	0	0
$\kappa_{1/2}$	0	$8\beta\eta^{1/5}$	0	0	0
κ_1	$-16\beta\Phi_1$	0	$4\beta\eta^{2/5}$	0	0
$\kappa_{3/2}$	$-\frac{32}{3}\beta\Phi_{3/2}$	$-\frac{32}{5}\beta\Phi_1\eta^{1/5}$	0	$2\beta\eta^{3/5}$	0
κ_2	$16\beta\left(\frac{4}{5}\Phi_1^2 - \frac{1}{3}\Phi_2\right)$	$-\frac{64}{15}\beta\Phi_{3/2}\eta^{1/5}$	$-\frac{12}{5}\beta\Phi_1\eta^{2/5}$	0	$\beta\eta^{4/5}$

Journal of Physics: Conference Series **610** (2015) 012046 doi:10.1088/1742-6596/610/1/012046

expanded up to 2PN order in u. Assuming that κ_i and β are small enough such that all terms can be expanded at linear order, and by setting $\mathcal{A}_k$ to zero results in a linear system that can be solved for $\kappa_i(b,\beta)$. We chose to fix the correction orders between 0PN and 2PN, in particular, $b \in \{-5/3, -4/3, -1, -2/3, -1/3\}$ and $i \in \{0, 1/2, 1, 3/2, 2\}$; this corresponds to 0PN, 0.5PN, 1PN, 1.5PN and 2PN corrections to the phase/frequency, respectively. For simplicity, we chose not to consider 'negative' PN terms due to dipole radiation because we plan to apply the waveforms to SMBH inspirals where, at least in scalar-tensor theories, there appears to be no dipole radiation present [10]. Subsequently, we fix one particular value of b at a time and allow the full sum of corrections proportional to κ_i in the time domain phase. We have solved the linear system and have computed $\kappa_i(b,\beta)$ for each value of b. The results are listed in table 1. In the table, some patterns are clearly visible: Not surprisingly, the κ_i are always proportional to β. For a frequency domain phase correction entering at nPN order, $n = (3b+5)/2$, $\kappa_{i<n} = 0$ and the κ_i are proportional to $\eta^{\frac{2n}{5}}$. Furthermore, the lowest order correction κ_i has the numeric pre-factor 2^{4-2i}. In the first off-diagonal element, the coefficients are proportional to $\Phi_0 = 1$, in the secondary off-diagonal to $\Phi_{1/2} = 0$, in the third to Φ_1 and in the $4^{\rm th}$ to $\Phi_{3/2}$. In the $5^{\rm th}$ off-diagonal, terms proportional both to Φ_1^2 and Φ_2 can appear.

As previously mentioned, we chose to consider only the leading order corrections and are thus able to write the corrected time domain phase as (using $i = (3b+5)/2$)

$$\Phi^{(\pm)}_{\rm NGR}(b,\beta;\Theta) = \Phi_{\rm GR}(\Theta) \pm 2^{-1-3b}\beta\,\eta^{3b/5}\Theta^{-3b/8}, \tag{14}$$

with a modified orbital frequency given by

$$\omega^{(\pm)}_{\rm NGR}(b,\beta;\Theta) = \omega_{\rm GR}(\Theta) \pm 2^{-4-3b}\frac{3\beta}{5}\frac{c^3}{GM}\,\eta^{3b/5+1}\Theta^{-3b/8-1}. \tag{15}$$

Equations (14) and (15) can then be used in the GW polarisations described by Eqs. (4) to generate a modified waveform that approximately uses the same β parameter as a leading order phase-only ppE scheme. We assume that our coupling constant β is manifestly positive as we would like to distinguish between possible positive and negative corrections to the GR waveform.

4. Summary

We have introduced alternative theory corrections to the time domain phase and orbital frequency for a GW generated by a compact binary inspiral in GR. In order to compare these waveforms to leading order phase-corrected frequency domain waveforms, we have established a relation between the magnitude parameters κ and β. We have implemented the corrected time domain waveforms into a Markov Chain Monte Carlo code that assesses the ability of the eLISA detector to discriminate between GR and alternatives theories in the context of SMBHB inspirals. Preliminary results are available, but will be published in [11].

References

[1] Yunes N and Pretorius F 2009 *Phys. Rev. D* **80** 122003
[2] Arun K G, Iyer B R, Qusailah M S S and Sathyprakash B S 2006 *Phys. Rev. D* **74** 024006
[3] Cornish N, Sampson L, Yunes N and Pretorius F 2011 *Phys. Rev. D* **84** 062003
[4] Sampson L, Yunes N and Cornish N 2013 *Phys. Rev. D* **87** 102001
[5] Sampson L, Cornish N and Yunes N *Phys. Rev. D* **89** 064037
[6] Li T G F, Del Pozzo W, Vitale S, Van Den Broeck C, Agathos M, Veitch J, Grover K, Sidery T, Sturani R and Vecchio A 2012 *Phys. Rev. D* **85** 082003
[7] Pan Y, Buonanno A, Baker J G, Centrella J, Kelly B J, McWilliams S T, Pretorius F and van Meter J R 2008 *Phys. Rev. D* **77** 024014
[8] Damour T, Iyer B R and Sathyprakash B S 2000 *Phys. Rev. D* **62** 084036
[9] Blanchet L 2014 *Living Reviews in Relativity* **17** 2
[10] Lang R N 2014 *Phys. Rev. D* **89** 084014
[11] Huwyler C, Porter E and Jetzer P 2014 *in preparation*

Journal of Physics: Conference Series **610** (2015) 012047 doi:10.1088/1742-6596/610/1/012047

Tidal polarizability effects in neutron star mergers

S Bernuzzi[1]**, A Nagar**[2]**, S Balmelli**[3]**, T Dietrich**[1] **and M Ujevic**[4]

[1]Theoretical Physics Institute, University of Jena, 07743 Jena, Germany
[2]Institut des Hautes Etudes Scientifiques, 91440 Bures-sur-Yvette, France
[3]Physik-Institut, Universität Zürich, 8057 Zürich, Switzerland
[4]Centro de Ciências Naturais e Humanas, Universidade Federal do ABC, 09210-170, Santo André, São Paulo, Brazil

Abstract. Using the analytical effective-one-body model and nonlinear 3+1 numerical relativity simulations, we investigate binary neutron star mergers. It is found that, for nonspinning binaries, both the mass-rescaled gravitational wave frequency at merger and the specific binding energy at merger almost uniquely depend on the tidal coupling constants κ_2^T, which are functions of the stars' Love numbers, compactnesses and mass ratio. These relations are quasiuniversal in the sense that there is an additional dependence on the spins, which is linear for realistic spins values $\chi \lesssim 0.1$. In the effective-one-body model, the quasiuniversality is a direct consequence of the conservative dynamics of tidally interacting bodies. In the context of gravitational wave astronomy, our findings may be used to constrain the neutron stars' equation of state using waveforms that accurately model the merger.

1. Introduction

Coalescing neutron stars are among the most promising sources for the currently operating, ground-based interferometric gravitational waves (GW) detectors [1]. In particular, starting from the next few years, the advanced configurations of the LIGO/Virgo network are expected to detect $\sim 0.4 - 400$ events per year [2]. The late part of the coalescing process is likely to provide, via measurement of the tidal polarizability, crucial information about the neutron stars' internal structure and equation of state (EOS) [3, 4, 5, 6].

At present time, an accurate modelling of binary neutron star (BNS) mergers is only possible with numerical relativity (NR). Many improvements have been done in recent years, which led, e.g., to phasing analysis for multiorbit simulations [7, 8, 9] or to studies of the gauge-invariant relation between energy and angular momentum [10, 11].

Nevertheless, some important aspects still need to be clarified. For instance, there is insufficient knowledge about the role of finite mass ratio and finite mass effects at merger. BNS-related studies (see e.g. [6]) typically neglect these effects, insofar as they estimate the end of the quasiadiabatic BNS inspiral and the amount of GW energy emitted during the coalescence by means of the simple Schwarzschild quantities $2M\Omega_{\rm LSO}^{\rm Schw} \approx 0.13608$ and $E_{b\,\rm LSO}^{\rm Schw} \approx -0.0572$ taken in correspondence of the Schwarzschild last stable orbit (LSO).

We use new multi-orbit NR data and the analytical effective-one-body (EOB) approach to further investigate neutron stars mergers. We find that the GW frequency and binding energy at the moment of merger are only characterized, besides of the spins, by certain dimensionless tidal coupling constants (a fact also empirically observed in [5] for the frequency). For more details, see Ref. [12].

2. Methods

2.1. LSO and merger within the EOB

The EOB approach maps the general relativistic 2-body problem into the dynamics of a particle with mass $\mu = M_A M_B/M$ moving in an effective metric generated by a central mass $M = M_A + M_B$ [13]. It consists of three building blocks (see e.g. Ref. [14]): (i) a Hamiltonian $H_{\rm EOB}$; (ii) a factorized gravitational waveform; and (iii) a radiation reaction force $\mathcal{F}_\varphi$. The EOB Hamiltonian $H_{\rm EOB} = M\sqrt{1+2\nu(\hat{H}_{\rm eff}-1)}$ is constructed upon an effective Hamiltonian, that in the nonspinning case reads $\hat{H}_{\rm eff}(u,p_{r_*},p_\varphi) = \sqrt{A(u;\nu)\left(1+p_\varphi^2u^2+2\nu(4-3\nu)u^2p_{r*}^4\right)+p_{r*}^2}$. Here, $\nu = \mu/M$, $u \equiv 1/r$ (r is a dimensionless separation radius), while p_φ and $p_{r*} \equiv \sqrt{A/B}\,p_r$ are dimensionless (orbital and radial) momenta. PN results are included, in a resummed way, into the potentials $A(u;\nu)$ and $B(u;\nu)$. We implement the spin following Ref. [15], and taking the spin-orbit coupling at NNLO [16]. Finite size effects can be reproduced by adding to the point-mass contribution $A^0(u;\nu)$ of the radial potential a suitable tidal term $A^T(u;\nu)$. In $A^0(u;\nu)$, we exploit the full analytical information we posses so far, including terms up to 4PN [17], and use a (1,4)-Padé resummation. Tidal terms formally start at 5PN and are included up to NNLO order. They have the structure [18]

$$A^T(u) = -\sum_{\ell=2}^{4}\kappa_\ell^T u^{2\ell+2}(1+\bar{\alpha}_1^{(\ell)}u+\bar{\alpha}_2^{(\ell)}u^2), \tag{1}$$

with only $\bar{\alpha}_{1,2}^{(2),(3)}$ known analytically [19]. The full EOS information is encoded into the tidal coupling constants ($\ell \geq 2$)

$$\kappa_\ell^T \equiv 2\left[\frac{1}{q}\left(\frac{X_A}{C_A}\right)^{2\ell+1}k_\ell^A + q\left(\frac{X_B}{C_B}\right)^{2\ell+1}k_\ell^B\right], \tag{2}$$

where $q = M_A/M_B \geq 1$, $X_A \equiv M_A/M = q/(1+q)$, $X_B \equiv M_B/M = 1/(1+q)$, while $k_\ell^{A,B}$ and $C_{A,B}$ are the dimensionless Love numbers and compactness of star A and B. We stress that, in this work, we use a fully analytic EOB model, that does not contain any parameter calibrated against NR.

For a fixed angular momentum p_φ, circular orbits satisfy $\partial_u H_{\rm eff}(u,p_\varphi,p_r\equiv 0)=0$. Within the adiabatic ($\mathcal{F}_\varphi = 0$) EOB model, the LSO ($u_{\rm LSO}, p_{\varphi\rm LSO}$) is defined as the inflection point of the radial effective potential $\hat{H}_{\rm eff}(u)$, and thus must satisfy the additional equation $\partial_u^2 H_{\rm eff}(u,p_\varphi,p_r\equiv 0)=0$. Below $u_{\rm LSO}^{-1}$, no stable circular orbit exists, while below $p_{\varphi\rm LSO}$ no circular orbit is possible at all. The gravitational wave frequency and the binding energy at the LSO are simply calculated as $M\Omega_{\rm LSO} = \mu^{-1}\partial_{p_\varphi}H_{\rm EOB}(u_{\rm LSO},p_\varphi,p_r\equiv 0)|_{p_{\varphi\rm LSO}}$ and $E_{b\,\rm LSO} = (H_{\rm EOB}(u_{\rm LSO},p_{\varphi\rm LSO},p_r\equiv 0)-M)/\mu$.

A more complete approach is provided by the nonadiabatic EOB ($\mathcal{F}_\varphi \neq 0$). In this case the dynamics is continued after the LSO crossing, where it shows, as also NR simulations do, a characteristic peak in the GW amplitude. We define (for both EOB and NR) the peak of the $l=m=2$ mode as the moment of merger.

2.2. NR simulations

Our NR simulations employ the BAM code and the method described in [20, 21], but with some different features: (i) we use the Z4c formulation of Einstein's equations [22]; (ii) GWs are extracted from an extended wavezone [23]. The binaries are equal-mass, irrotational configurations with different EOS. A $\Gamma = 2$ polytropic EOS model is employed to simulate different compactnesses. The evolutions cover about ten orbits up to merger. These are among

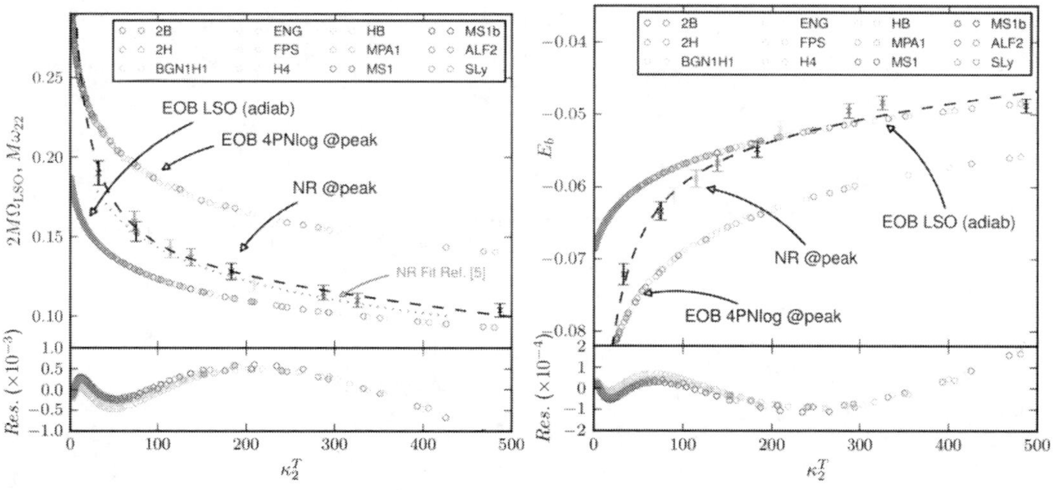

Figure 1. Top panels: GW frequency (left) and binding energy (right) at the LSO (adiabatic EOB) and at merger (nonadiabatic EOB and NR) as a function of κ_2^T for different EOS. NR data are fitted with a function of type $f(0)(1+n_1\kappa+n_2\kappa^2)/(1+d_1\kappa)$ (black dashed line). Bottom panels: residuals of the LSO data. The picture is taken from Ref. [12].

the longest BNS simulations ever performed, and some of the few where an error analysis is available [8, 10]. For each NR data set, we compute the binding energy per reduced mass, $E_b^{\rm NR}$, subtracting the GW energy loss from the initial ADM mass, following [24, 10, 11]. Here, differently from previous works, all the multipoles are included. GW frequency and binding energy are extracted at the moment of merger. We estimate error bars due to truncation errors and waveform finite extraction uncertainties from resolution tests for fewer configurations.

3. Quasiuniversal κ_ℓ^T relations

We investigated the dependence of the GW frequency $2\Omega_{\rm LSO}$ and of the specific binding energy $E_{b\,\rm LSO}$ at merger on the tidal coupling constant κ_2^T for different EOS. Within the EOB approach, we considered 12 realistic EOS and varied the masses from $1.3M_\odot$ up to the maximum allowed mass. Instead, for NR simulations, we fixed the isolation mass $M = 2\times 1.35M_\odot$ and considered the set of compactnesses $C_A = C_B = (0.12, 0.14, 0.16, 0.18)$ for the EOS MS1, MS1b, H4, ALF2, MPA1, ENG and SLy. We found that both $2\Omega_{\rm LSO}$ and $E_{b\,\rm LSO}$, once parametrized by the tidal coupling constants, are essentially independent of the EOS. From Fig. 1, which shows the plots for equal mass ratio and no spin, one sees that deviations from universality are below 0.2% in the EOB-generated curves. Because of the structure given by Eq. (1), this behavior is actually not surprising. It is a consequence of the fact that, at the LSO, the term $\kappa_2^T u^6$ dominates over the higher multipoles $\kappa_\ell^T u^{2\ell+2}$, $\ell \geq 3$. As it can be clearly seen, NR results qualitatively confirm this prediction. At the quantitative level, however, there is a gap between EOB and NR that is around $20-30\%$ for $2\Omega_{\rm LSO}$ and $10-20\%$ for $E_{b\,\rm LSO}$. The reason for this disagreement is due to the fact that our EOB model does not include neither nonlinear tidal interactions, nor hydrodynamic effects. Despite being incomplete, the EOB is able to catch deep physical connections in such a complex process as the merger.

We then used the adiabatic EOB to further investigate the merger, adding small spins $\chi = \pm 0.1$ and considering the additional mass ratio $q = 2$. The curves turn out to be almost independent of the mass ratio (see Fig. 2), despite the fact that the interval of κ_2^T gets narrower when q deviates from 1 (which is simply due to the smaller range of possibilities of picking, in

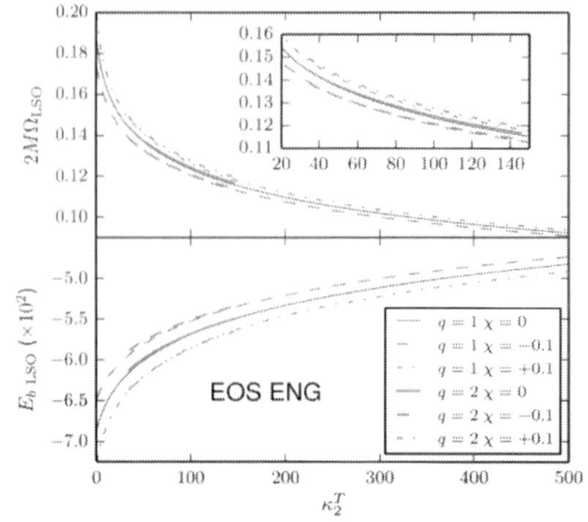

Figure 2. GW frequency (top) and binding energy (bottom) according to the adiabatic EOB model at the LSO, for different spins and mass ratios. The picture is taken from Ref. [12].

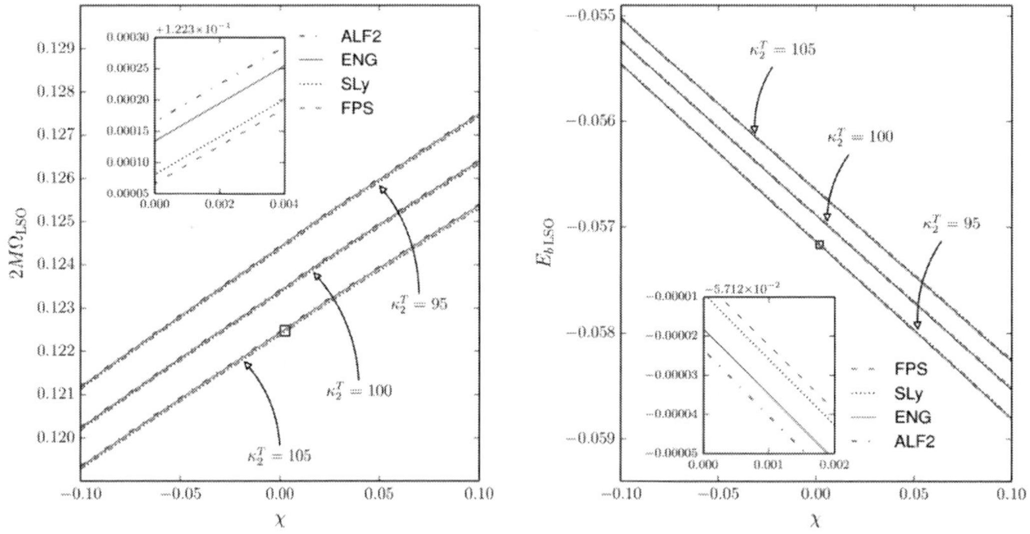

Figure 3. Linear dependence of GW frequency (left) and binding energy (right) on small spins $\chi_1 = \chi_2 \equiv \chi$ ranging from -0.1 to 0.1, for four different EOS. Both quantities are taken at the LSO of the adiabatic EOB model, with the tidal coupling contants fixed at the values $\kappa_2^T = 95, 100, 105$. The small boxes indicate the scale at which deviation from universality shows up.

the allowed interval, two masses with the wished mass ratio). The difference between $q = 1$ and $q = 2$ is $\leq 0.5\%$. By contrast, as it can be seen in Figs. 2-3, the curves are sensitive to the spin, showing an almost linear dependence. Since this implies a certain degeneracy between spin and tidal coupling constant, we denote the relations as *quasi*-universal.

Journal of Physics: Conference Series **610** (2015) 012047 doi:10.1088/1742-6596/610/1/012047

4. Conclusions

We have identified the κ_ℓ^T as fundamental coupling constants of the binary tidal interactions, together with κ^T-universal relations and their physical origin. This provides, for instance, a better insight into the empirical relation found in Ref. [5].

Our findings have implications for GW astronomy. A single measurement of the GW frequency at merger allows to extract the value of κ_ℓ^T. If both masses can be determined from the earlier template, it would be consequently possible to strongly constrain the EOS. In view of this, the κ_ℓ^T relations exhibit certain analogies with the merger properties found in Ref. [25] and with the so called $I-\lambda-Q$ relations [26, 27], see also [28]. Of course, a detailed study about the possibility of constraining the EOS deserves a study on its own.

Finally, the EOB could be used to provide a simple criterion, more precise than the Schwarzschild LSO, for stopping the waveform template.

Acknowledgments

We thank B. Brügmann, T. Damour, C. Van den Broeck for useful comments. This work was supported in part by DFG grant SFB/Transregio 7 "Gravitational Wave Astronomy" and the Graduiertenakademie Jena. S.B. and Si.B. thank IHES for hospitality during the development of part of this work. Si.B. is supported by the Swiss National Science Foundation. M.U. is supported by CAPES under BEX 10208/12-7, and thanks TPI Jena for hospitality during the development of this work. Simulations were performed on the LRZ cluster in Münich.

References

[1] Abadie J *et al.* (LIGO Scientific Collaboration, Virgo Collaboration) 2010 *Class.Quant.Grav.* **27** 173001
[2] Aasi J *et al.* (LIGO Scientific Collaboration, Virgo Collaboration) 2013
[3] Read J S, Markakis C, Shibata M, Uryu K, Creighton J D *et al.* 2009 *Phys.Rev.* **D79** 124033
[4] Damour T, Nagar A and Villain L 2012 *Phys.Rev.* **D85** 123007
[5] Read J S, Baiotti L, Creighton J D E, Friedman J L, Giacomazzo B *et al.* 2013 *Phys.Rev.* **D88** 044042
[6] Del Pozzo W, Li T G F, Agathos M, Broeck C V D and Vitale S 2013 *Phys. Rev. Lett.* **111** 071101
[7] Baiotti L, Damour T, Giacomazzo B, Nagar A and Rezzolla L 2010 *Phys. Rev. Lett.* **105** 261101
[8] Bernuzzi S, Thierfelder M and Brügmann B 2012 *Phys.Rev.* **D85** 104030
[9] Hotokezaka K, Kyutoku K and Shibata M 2013 *Phys.Rev.* **D87** 044001
[10] Bernuzzi S, Nagar A, Thierfelder M and Brügmann B 2012 *Phys.Rev.* **D86** 044030
[11] Bernuzzi S, Dietrich T, Tichy W and Bruegmann B 2014 *Phys.Rev.* **D89** 104021
[12] Bernuzzi S, Nagar A, Balmelli S, Dietrich T and Ujevic M 2014 *Phys.Rev.Lett.* **112** 201101
[13] Damour T and Nagar A 2009 The Effective One Body description of the Two-Body problem (*Preprint* `gr-qc/0906.1769`)
[14] Damour T, Nagar A and Bernuzzi S 2013 *Phys.Rev.* **D87** 084035
[15] Damour T 2001 *Phys. Rev.* **D64** 124013
[16] Nagar A 2011 *Phys.Rev.* **D84** 084028
[17] Bini D and Damour T 2013 *Phys.Rev.* **D87** 121501
[18] Damour T and Nagar A 2010 *Phys. Rev.* **D81** 084016
[19] Bini D, Damour T and Faye G 2012 *Phys.Rev.* **D85** 124034
[20] Brügmann B, Gonzalez J A, Hannam M, Husa S, Sperhake U *et al.* 2008 *Phys.Rev.* **D77** 024027
[21] Thierfelder M, Bernuzzi S and Brügmann B 2011 *Phys.Rev.* **D84** 044012
[22] Bernuzzi S and Hilditch D 2010 *Phys. Rev.* **D81** 084003
[23] Hilditch D, Bernuzzi S, Thierfelder M, Cao Z, Tichy W *et al.* 2013 *Phys. Rev.* **D88** 084057
[24] Damour T, Nagar A, Pollney D and Reisswig C 2012 *Phys.Rev.Lett.* **108** 131101
[25] Bauswein A and Janka H T 2012 *Phys.Rev.Lett.* **108** 011101
[26] Yagi K and Yunes N 2013 *Science* **341** 365–368
[27] Yagi K, Stein L C, Pappas G, Yunes N and Apostolatos T A 2014 *Phys. Rev. D* **90**(6) 063010
[28] Maselli A, Cardoso V, Ferrari V, Gualtieri L and Pani P 2013 *Phys. Rev. D* **88**(2) 023007

The remains of a spinning, hyperbolic encounter

L De Vittori[1], A Gopakumar[2], A Gupta[2], P Jetzer[1]

[1] Physik-Institut, Universität Zürich, 8057 Zürich, Switzerland

[2] Department of Astrophysics, Tata Institute of Fundamental Research, Mumbai, India

Abstract. We review a recently proposed approach to construct gravitational wave (GW) polarization states of unbound spinning compact binaries. Through this rather simple method, we are able to include corrections due to the dominant order spin-orbit interactions, in the quadrupolar approximation and in a semi-analytic way. We invoke the 1.5 post-Newtonian (PN) accurate quasi-Keplerian parametrization for the radial part of the dynamics and impose its temporal evolution in the PN accurate polarization states equations. Further, we compute 1PN accurate amplitude corrections for the polarization states of non-spinning compact binaries on hyperbolic orbits. As an interesting application, we perform comparisons with previously available results for both the GW signals in the case of non-spinning binaries and the theoretical prediction for the amplitude of the memory effect on the metric after the hyperbolic passage.

1. Introduction

In preparation for first direct observations of GWs with ground-based detectors, expected in the next few years, it is desirable to have GW templates for every possible initial configuration of the two-body problem, and that as precise as possible. Our recent work belongs to that program, investigating in particular the gravitational radiation from hyperbolic encounters and its effects on the metric.

Previous work on the subject includes the Newtonian treatment of the gravitational radiation field associated with non-spinning objects on unbound orbits, done by Turner [1], its first extension to the first PN order in the dynamics done by Turner and Will [2] and completed by Junker & Schäfer [3], the full 1PN corrected waveforms, including PN accurate polarization states h_+ and $h_\times$, available in [4], where a generalized true anomaly parametrization was employed.

No treatment of spinning binaries on hyperbolic orbits can be found in the literature, apart from the work we review here, by De Vittori et al. [5]. More details about GW radiation were described for the non-spinning case only, e.g. the quadrupolar order energy and momentum loss during the interaction [6] and its 1PN extension [7], or the analytic formula for the power spectrum [8], which generalizes the high eccentricity limit [1] and the parabolic limit [9].

In this summary we review our recent work [5], where we obtained temporally evolving GW polarization states for spinning compact binaries in PN accurate hyperbolic orbits. We included leading order spin-orbit interactions effects, and the conservative non-spinning orbital dynamics is 1PN accurate. Therefore, we were able to describe spinning compact binaries along unbound orbits up to 1.5 PN order accuracy. Additionally, we incorporated 2.5 PN radiation reaction effects while computing the quadrupolar order polarization states h_+ and $h_\times$.

Journal of Physics: Conference Series **610** (2015) 012048 doi:10.1088/1742-6596/610/1/012048

1.1. The Memory Effect

The plots for GW signals at PN order reveal a non-vanishing difference between the wave amplitudes at $t = \infty$ and $t = -\infty$, which is called the linear *memory effect*. In the following, we will review the basic concepts of this memory, as well as some crude estimates of its amplitude at Newtonian order, available in Refs. [10, 11, 12] and briefly summarized in [13].

Consider an ideal GW detector in which truly free-falling test masses are only sensitive to gravitational forces, like e.g. a ring of test masses in space, a Pulsar Timing Array (PTA), or a LISA-like detector. A GW without memory causes the detector to return to its initial configuration. Instead, after the passage of a GW with memory the displacement state of the detector is different than long before the wave had passed. This net change in the amplitude of the metric induced by a GW is what we define as the memory effect.

Ground based detectors are not ideal detectors and this kind of memory cannot be stored in their signal, since internal forces, e.g. the wires on the pendulum, bring the test masses back to their original positions. However, for new generation detectors and especially for PTA this signature should be taken seriously into account since it appears in the polarization states already at Newtonian order.

The quadrupolar order GW polarization states we require for a system of two orbiting masses m_1 and m_2 at a distance R, can be expressed through the transverse-traceless part of the radiation field h_{ij}^{TT}, where the quadrupolar order contribution, as is well-known, reads

$$h_{ij}^{\mathrm{TT}}\big|_{\mathrm{Q}} = \frac{2G}{c^4 R}\ddot{\mathcal{I}}_{ij}^{TT} , \tag{1}$$

G being the gravitational constant, c the speed of light, and where $\mathcal{I}_{ij}^{TT} = \mu(x_i x_j)^{TT}$ is the source mass-quadrupole moment, with the reduced mass $\mu = m_1 m_2/m$, the total mass m and the relative orbital separation components x_i. Using the equation of motion $\ddot{x}_i = -Gmx_i/r^3$, the second derivative of the mass-quadrupole moment reads

$$\ddot{\mathcal{I}}_{ij} = 2\mu(\dot{x}_i\, \dot{x}_j - \frac{GM}{r^3} x_i\, x_j) , \tag{2}$$

and one can notice that the second term in eq. (2) vanishes for $t \to \pm\infty$ since it falls off as $1/r$, while the relative velocities approach the finite value $\mathrm{v}_\infty = \sqrt{2E/\mu}$.

Therefore, at Newtonian order the memory effect for a hyperbolic encounter can be predicted through the simple formula

$$\Delta h_{ij} = 4\frac{G\mu}{c^4 R}\Delta(\dot{x}_i\, \dot{x}_j) , \tag{3}$$

where Δ denotes the difference between the initial and final state. The velocity directions at early and late times will thus obviously play a key role.

Let's suppose that the observer's line–of–sight is perpendicular to the orbital plane, say on the x_3 axis. Then only the components $\mathcal{I}_{11}$, $\mathcal{I}_{22}$ and $\mathcal{I}_{12} = \mathcal{I}_{21}$ are non-zero. For the x_1 and x_2 components at initial and final times $t \to \pm\infty$, when the polar angle is ϕ_0, we have

$$\lim_{t\to\pm\infty} \dot{x}_1 = \mathrm{v}_\infty \sin \pm\phi_0 , \quad \lim_{t\to\pm\infty} \dot{x}_2 = \mathrm{v}_\infty \cos \pm\phi_0 . \tag{4}$$

Using some trigonometric identities and rearranging, one can write the expected amplitude of the memory in the different components as

$$\Delta h_{ii} = 0 , \text{ for } i = 1, 2 \qquad \text{and} \qquad \Delta h_{ij} = 8\frac{G}{c^4}\frac{E}{R}\sin 2\phi_0 , \text{ for } i \neq j . \tag{5}$$

One can notice that according to this crude estimate, at Newtonian order and without spin, there is no memory effect in the plus polarization, while there is some difference between initial and final state in $h_\times$, depending on the initial settings of the binary.

Journal of Physics: Conference Series **610** (2015) 012048 doi:10.1088/1742-6596/610/1/012048

2. Gravitational waves from hyperbolic encounters

In this section, we will present how we computed PN accurate GW templates for both spinning and non-spinning binaries, in order to study the outcoming memory effect and compare our result with the theoretical predictions above.

2.1. Non-spinning binaries

By introducing polar coordinates (r, ϕ) and the constant inclination angle θ between the observer and the orbital plane, as well as the notation $z = Gm/r$, one can show that the resulting polarization states $h_{+,\times}$ can be written as a series of functions in terms of $r, \dot{r}, \phi, \dot{\phi}$ as $h_{+,\times} = -\frac{G\mu}{c^4 R}\left(h^{\mathrm{N}}_{+,\times} + \frac{1}{c} h^{0.5\mathrm{PN}}_{+,\times} + \frac{1}{c^2} h^{1\mathrm{PN}}_{+,\times}\right)$ where the single contributions can be found in Ref. [5]. As an illustration, the Newtonian expressions read

$$h^N_+ = 2r\dot{r}\dot{\phi}(1 + C^2_\theta)\sin 2\phi + (1 + C^2_\theta)\left(z + r^2\dot{\phi}^2 - \dot{r}^2\right)\cos 2\phi + S^2_\theta\left(\dot{r}^2 + r^2\dot{\phi}^2 - z\right) , \tag{6a}$$

$$h^N_\times = 2C_\theta\left(z + r^2\dot{\phi}^2 - \dot{r}^2\right)\sin 2\phi - 2C_\theta 2r\dot{r}\dot{\phi}\cos 2\phi , \tag{6b}$$

We now want to impose the temporal evolution of the dynamics. Considering the 1PN hyperbolic counterpart of the Kepler equation, and using 1PN accurate expressions for the orbital elements we can compute r, $\dot{r}$, ϕ and $\dot{\phi}$ in terms of the eccentricity e_t, the hyperbolic mean motion $\bar{n}$, the eccentric anomaly v and the symmetric mass ratio $\eta = \mu/m$:

$$r(v) = \frac{Gm}{c^2}\frac{1}{\bar{\xi}^{2/3}}(e_t\cosh v - 1)\left\{1 + \bar{\xi}^{2/3}\frac{2\eta - 18 - (6 - 7\eta)e_t\cosh v}{6\,(e_t\cosh v - 1)}\right\} , \tag{7a}$$

$$\dot{r}(v) = \bar{\xi}^{1/3}\frac{c\,e_t\sinh v}{e_t\cosh v - 1}\left\{1 - \bar{\xi}^{2/3}\frac{6 - 7\eta}{6}\right\} , \tag{7b}$$

$$\phi(v) = 2\arctan\left[\left(\frac{e_\phi + 1}{e_\phi - 1}\right)^{1/2}\tanh v/2\right]\left\{1 + \bar{\xi}^{2/3}\frac{3}{e_t^2 - 1}\right\} , \tag{7c}$$

$$\dot{\phi}(v) = \frac{\bar{n}\sqrt{e_t^2 - 1}}{(e_t\cosh v - 1)^2}\left\{1 - \bar{\xi}^{2/3}\frac{\left[3 - (4 - \eta)\,e_t^2 + (1 - \eta)\,e_t\cosh v\right]}{(e_t^2 - 1)\,(e_t\cosh v - 1)}\right\} , \tag{7d}$$

where $\bar{\xi}$ stands for $Gm\bar{n}/c^3$. Imposing eqs. (7) in the expressions for the polarization states (6), we obtain the gravitational wave signals from non-spinning compact binaries during hyperbolic encounters, at full 1PN order. In Fig. 1 we display some results for two different values of e_t.

A close look at Eqs. (6) allows us to extract the limit case where t (or v) goes to $\pm\infty$. For simplicity, let's consider again the case where the observer line-of-sight is perpendicular to the orbital plane, i.e. $\theta = 0$. It is not very difficult to infer that the dynamical variables $z(v) = Gm/r(v)$ and $\dot{\phi}(v)$, as well as the product $r(v)\dot{\phi}(v)$ go to zero. However, the expression for $\dot{r}(v)$ does not vanish as $t \to \pm\infty$, but rather tends to the finite value v_∞. This forces the expressions for $h_+|_Q(t)$ and $h_\times|_Q(t)$ at late times to be

$$h_+|_Q = 2\frac{G\mu}{c^4 R}\mathrm{v}^2_\infty\,\cos 2\phi_0 , \tag{8a}$$

$$h_\times|_Q = 2\frac{G\mu}{c^4 R}\mathrm{v}^2_\infty\,\sin 2\phi_0 . \tag{8b}$$

Note that the right hand side of Eq. (8a) is an even function of ϕ due to the presence of $\cos 2\phi$. However, the right hand side of Eq. (8b) is an odd function of ϕ. This implies that $\lim_{t\to+\infty} h_+|_Q(t) = \lim_{t\to-\infty} h_+|_Q(t)$, while $\lim_{t\to+\infty} h_\times|_Q(t) = -\lim_{t\to-\infty} h_\times|_Q(t)$.

Journal of Physics: Conference Series **610** (2015) 012048 doi:10.1088/1742-6596/610/1/012048

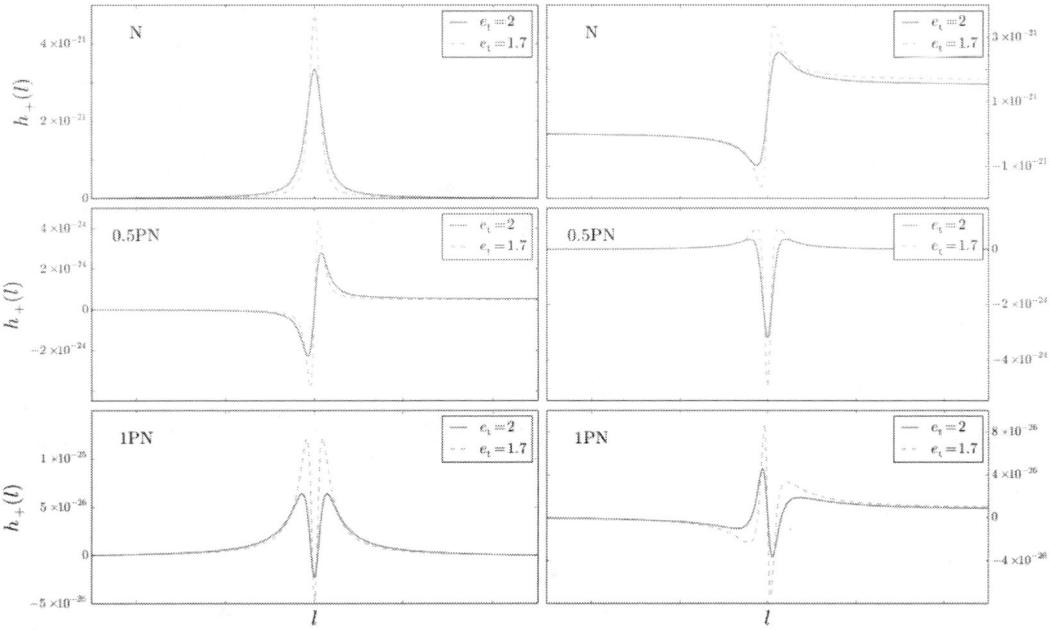

Figure 1. Polarization states at Newtonian, 0.5PN and 1PN order, as functions of the scaled time l for non-spinning compact binaries with periastron distance $r_{min} = 1.9 \times 10^9$ m and masses $m_1 = 8M_\odot$ and $m_2 = 13M_\odot$, as in [4]. The non-vanishing amplitude of the $\times$ polarization at late times, the so-called memory effect, is clearly visible at Newtonian order.

In particular, using again $\mu\, \mathrm{v}_\infty^2 = 2E$, the amplitude difference between late and early times is

$$\Delta h_+ = 0\ , \qquad \text{and} \qquad \Delta h_\times = 8\frac{G}{c^4}\frac{E}{R}\sin 2\phi_0\ , \tag{9}$$

which is essentially the same result as in Ref. [13] reported in eq. (5). This leads to the linear memory effect as visible in the $h_\times|_Q(l)$ plots in our Fig. 1.

2.2. Spinning binaries

In our recent work [5] we also included leading order spin-orbit interaction effects. This cannot be achieved in a fully analytical way, since one obtains a set of 12 coupled differential equations which have to be solved at the same time. First, one needs to introduce three Eulerian angles Φ, α and ι, since the orbital plane is precessing due to spin effects. Then we need to evolve the phasing angle Φ, the spin vectors $\boldsymbol{S}_1$ and $\boldsymbol{S}_2$ and the orbital momentum $\boldsymbol{L}$. We were able to include radiation reaction effects, a 2.5PN order correction, by adding two more differential equations for the evolution of the eccentricity e_t and the mean motion $\bar{n}$. The system contains therefore differential equations for $\dot{\Phi}$, $\dot{\boldsymbol{S}}_1$, $\dot{\boldsymbol{S}}_2$, $\dot{\boldsymbol{L}}$, $\dot{e}_t$ and $\dot{\bar{n}}$. In Sec. II of Ref. [5] we show these relations explicitly and explain their derivation in great detail.

The key result, as one can see in Fig. 2, is that spinning binaries leave a memory imprint on the metric in both the plus and the cross polarizations, due to the spin induced precession of the orbital plane and the consequent asymmetry of the initial and final state of the system. Although

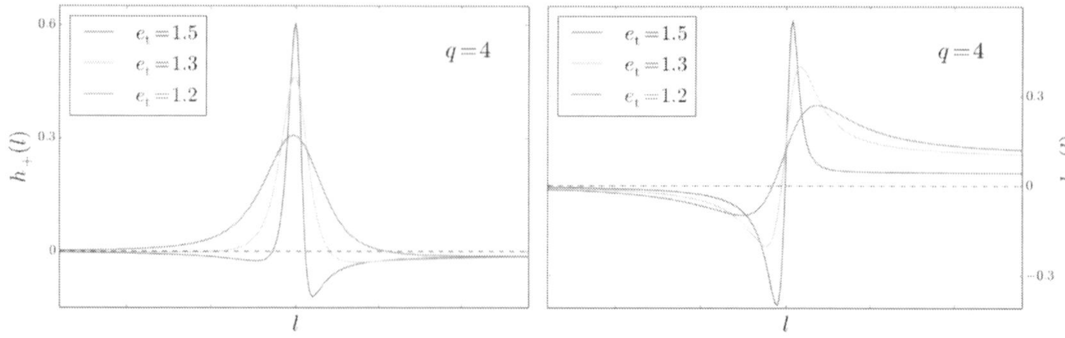

Figure 2. The gravitational wave signal of maximally spinning compact binaries on hyperbolic orbits, in units of Gm/c^2R, with mass ratio $q = 4$, impact parameter $b = 30\ Gm/c^2$ and for three different values of e_t. The orbital evolution is fully 1.5PN accurate and the influences of GW emission are taken into account. The linear memory effect causes the solid line waveform plots to depart from the dashed line after the hyperbolic passage.

small, the memory effect can be seen on the plot by comparing the initial value, reported by the dashed line, with the final value of h_+ and $h_\times$.

3. Conclusions

We described a recently developed prescription to obtain ready–to–use PN-accurate GW templates for spinning compact binaries on hyperbolic orbits, including leading order spin-orbit coupling contributions and radiation reaction effects. In the results we notice that for spinning binaries both the plus and the cross polarization states exhibit a memory effect, i.e. a non-vanishing amplitude at time $t = +\infty$, whereas for non-spinning systems only the cross polarization shows such a behaviour. Given the interesting feature of the memory in the GW signals, we also briefly reviewed the concept following the literature and providing a crude estimate of its amplitude at Newtonian order, which is proven to agree with our results.

Since the waveform is strongly influenced by spin effects, the structure of such an asymmetric burst-like signal carries many informations about the system which could be useful for parameter estimation. It will be interesting to incorporate the 2PN order non-spinning contributions to our 1.5PN-accurare orbital dynamics, as well as dominant order spin-spin interactions.

4. References

[1] M. Turner, *Astrophysical Journal*, **216**, 610-619 (1977)
[2] M. Turner, C. Will, *Astrophysical Journal*, **220**, 1107-1124 (1978)
[3] W. Junker, G. Schaefer, *Mon. Not. R. Astr. Soc.*, **254**, 146-164 (1992)
[4] J. Majár, P. Forgács, M. Vasúth, *Phys. Rev. D*, **82**, 064041 (2010)
[5] L. De Vittori, A. Gopakumar, A. Gupta, P. Jetzer, to be published in Phys. Rev. D.
[6] R.V. Wagoner, C.M. Will, *Astrophysical Journal*, **210**, 764 (1976)
[7] L. Blanchet, G. Schaefer, *Mon. Not. R. Astr. Soc.*, **239**, 845 (1989)
[8] L. De Vittori, P. Jetzer, A. Klein, *Phys. Rev. D*, **86**, 044017 (2012)
[9] C.P.L. Berry, J.R. Gair, *Phys. Rev. D*, **82**, 10751 (2010)
[10] Ya.B. Zel'dovich, A.G. Polnarev, *Astron. Zh.*, **51**, 30 (1974)
[11] V.B. Braginsky, L.P. Grishchuk, *Sov. Phys. JETP*, **62**, 427-430 (1985)
[12] V.B. Braginsky, K.S. Thorne, *Nature*, **327**, 123 (1987)
[13] M. Favata, *Phys. Rev. D*, **80**, 024002 (2009)

Journal of Physics: Conference Series **610** (2015) 012049 doi:10.1088/1742-6596/610/1/012049

Prospects for Detection of Extragalactic Stellar Black Hole Binaries in the Nearby Universe

Matthew Benacquista[1], **Jesus Hinojosa**[1], **Alberto Mata**[1], **Krzysztof Belczynski**[1,2]

[1] Center for Gravitational Wave Astronomy, University of Texas at Brownsville, One University Blvd, Brownsville, TX 78520

[2] Astronomical Observatory, University of Warsaw, Al. Ujazdowskie 4, 00-478 Warsaw, Poland

E-mail: `matthew.benacquista@utb.edu`

Abstract. Stellar mass black hole binaries have individual masses between 10-80 solar masses. These systems may emit gravitational waves at frequencies detectable at Megaparsec distances by space-based gravitational wave observatories. In a previous study, we determined the selection effects of observing these systems with detectors similar to the Laser Interferometer Space Antenna by using a generated population of binary black holes that covered a reasonable parameter space and calculating their signal-to-noise ratio. We further our study by populating the galaxies in our nearby (less than 30 Mpc) universe with binary black hole systems drawn from a distribution found in the Synthetic Universe to ultimately investigate the likely event rate of detectable binaries from galaxies in the nearby universe.

1. Introduction

Space-based gravitational wave interferometers of the LISA design will be sensitive to mHz gravitational waves. The latest design for eLISA consists of a two-arm interferometer with 1 million km arm lengths (1). The majority of stellar mass compact object binary sources for eLISA will be the Galactic population of close white dwarf binaries. With a typical chirp mass of $\sim 0.5\ M_\odot$, white dwarf binaries are unlikely to be detected from extragalactic distances since their signal strength will be too low. It is possible that a few binary black holes in the Milky Way may also be in the eLISA band (2). These systems may have much higher chirp masses (up to $\sim 60\ M_\odot$) and may therefore be detectable at Mpc distances.

In order to find a first estimate of the likelihood of observing extragalactic stellar mass binary black holes with eLISA, we have performed a population synthesis of these systems within the local universe out to 30 Mpc. We have used the Gravitational Wave Galaxy Catalogue, a sample of galaxies out to 100 Mpc, (3) to obtain the spatial distribution of host galaxies within this range. We then populated the galaxies using a binary black hole population synthesis available from the Synthetic Universe, a population synthesis of binary black holes generated by Belczynski and collaborators using StarTrack and publicly available at `SyntheticUniverse.org`. Finally, we estimate the probability of detection for each binary within the eLISA band.

2. Population Synthesis

The Gravitational Wave Galaxy Catalogue (GWGC) was put together in order to assist in electromagnetic follow-up searches for gravitational wave triggers generated by ground-based

Journal of Physics: Conference Series **610** (2015) 012049 doi:10.1088/1742-6596/610/1/012049

detectors such as LIGO and Virgo (4). These triggers were expected to arise from the inspirals of binary black holes and neutron stars. The catalogue is "a more complete, up-to-date sample created from a variety of literature sources extending out to 100 Mpc, which is as unbiased as possible to a particular kind of source." (4) The GWGC contains sky position, distance, blue magnitude, major and minor diameters, position angle, and galaxy type for 53,225 galaxies out to 100 Mpc (3). We have made a distance cut to galaxies within 30 Mpc. This results in a population of 8,947 galaxies to populate with binary black holes.

In order to populate the galaxies with binary black holes, we grouped the galaxies in the catalog into three basic galaxy types: Spiral, Elliptical, and Irregular. We populated these galaxies using a population of binary black holes from the Synthetic Universe. There are 32,496 binary black holes in this population synthesis. In order to determine the number of binaries to include in each galaxy, we assume that the blue luminosity correlates to the total binary black hole population. Based upon a population synthesis for the Milky Way which gives roughly 10^6 binary black holes (2), we have assigned 10^6 binary black holes for every Milky Way equivalent blue luminosity ($L_{\mathrm{MWEG}} = 1.7 \times 10^{10}\ L_{B,\odot}$) for each galaxy in the catalogue. Thus, for a galaxy in the catalog with a blue luminosity of L_{B}, the number of binary black holes is assigned as $N_{\mathrm{BBH}} = (L_{\mathrm{B}}/L_{\mathrm{MWEG}}) \times 10^6$.

Once we have determined the number of binary black holes to assign to each galaxy, we then select a star formation history for each galaxy based on its morphological type. We assume that spiral galaxies have had a constant star formation rate over the past 10 Gyr. Elliptical galaxies are given a burst of star formation between 9 and 10 Gyr, while irregular galaxies have been assigned a recent burst of star formation over the past 3 Gyr.

Based on these star formation rates, we then populate each galaxy by randomly selecting a binary from the pool of binary black holes taken from the synthetic universe and giving it a random age within the bounds governed by the star formation history appropriate to its morphological type. We repeat this process until we have selected the number of black holes expected by the blue luminosity. The orbital period of each binary is then propagated forward to the present day under the influence of the emission of gravitational radiation. We retain only those binaries with a final orbital period below 2000 s. These systems will be at gravitational wave frequencies above 1 mHz and therefore they will lie above the confusion-limited noise from the Galactic population of close white dwarf binaries. This resulted in 45,112 potential eLISA sources in galaxies within 30 Mpc. The populations of each galaxy type are shown in Table 1.

Table 1. Number of galaxies of each morphological type within 30 Mpc and the number of binary black holes that are potential eLISA sources within each type.

Galaxy type	Number of galaxies	Number of binaries
Elliptical	689	708
Spiral	7137	22128
Irregular	1121	22276

3. Detection Likelihood

Once we have determined the population of potential eLISA sources, we need to determine the likelihood that these binaries will be detectable by eLISA. As this is a preliminary study, we have not done multiple realizations of the population. Instead, we have run a Monte Carlo study of 10^9 binaries with an even coverage over the parameter space of black hole masses, orbital period, and distance. The black hole masses cover the range from $M \in (6, 80)$ $M_\odot$, the orbital periods cover the range $P_{\rm orb} \in (0, 2000)$ s, and the distances are less than 30 Mpc. The orbital inclinations and sky locations were randomly chosen. All of these binaries were then run through an eLISA simulator to determine the expected signal to noise ratio. We include the gravitational chirp in the evolution of this sample of binaries. Only those binaries that remain within the eLISA band for a full year are included in the signal to noise ratio calculations. Using the initial pool from the Monte Carlo study and the pool of binaries with a signal to noise ratio above the threshold, we can determine the likelihood of detection for any binary with a given chirp mass, orbital period, and distance. Projections of the likelihood are shown in Figure 1.

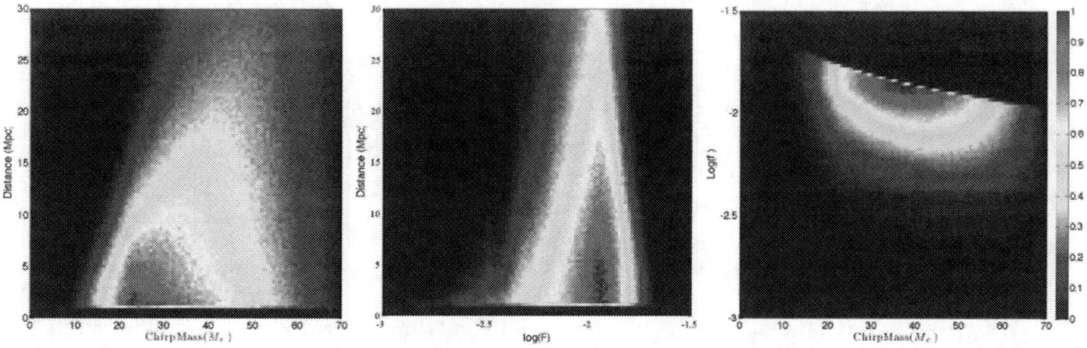

Figure 1. Likelihood of detection for binary black holes obtained from the Monte Carlo study. The likelihoods have been marginalized over frequency on the left, chirp mass in the center, and distance on the right. The abrupt cutoff in the likelihoods on the right is due to systems that evolve out of the eLISA band in one year.

Using the known locations of the populated galaxies in the catalogue along with the binary properties to the potential eLISA sources, we can then estimate the likelihood of detection for each binary in the population. We obtain a first order estimate of the number of observable extragalactic binaries through this approach, which averages over orbital inclinations and does not take into account the sky location of each host galaxy. There were no observable binaries within the elliptical galaxy population, five potentially observable binaries in the spiral galaxy population, and five in the irregular galaxy population. Out of the ten, two of these binaries had a greater than 50% probability of detection. The results are shown in Table 2

Table 2. Probability for detection from the 10 binary black holes.

Galaxy type	Detection Probability (%)				
Spiral	89.6	2.2	2.2	1.7	0.3
Irregular	51.1	17.2	9.7	2.1	2.1

4. Conclusions

We have used a very simple population synthesis technique to perform an initial investigation into the possibility of observing stellar mass binary black holes at extragalactic distances. For one realization of the population, using simple prescriptions for the population as a function of the blue luminosity and ignoring metallicity and specific star formation rates for individual galaxies, we have estimated the likelihood of observing an extragalactic binary black hole with eLISA. The results from this single realization indicate one or two binary black holes will be observed to within 30 Mpc, although there is a non-zero probability of up to 10 binary black holes being observed.

Our simulation did not include the population of black hole binaries that may arise in the globular cluster systems of the galaxies within the Gravitational Wave Galaxy Catalog. Globular cluster systems will likely come from lower metallicity populations and therefore may contain more high mass black holes which will enhance the number of observable black hole binaries. These results are promising enough to warrant further study into the population of black hole binaries within the local universe.

Acknowledgments
This work was funded by the Center of Gravitational Wave Astronomy, NSF and NASA, through grants NASA NNX09AV06A and NSF HRD0734800. JH and AM also acknowledge the support of the Arecibo Remote Command Center scholars program, funded by NSF AST0750913.

References
[1] Seoane P A, Aoudia S, Audley H, Auger G, Babak S, Baker J, Barausse E, Barke S, Bassan M, Beckmann V, Benacquista M, Bender P L, Berti E, Binétruy P, Bogenstahl J, Bonvin C, Bortoluzzi D, Brause N C, Brossard J, Buchman S, Bykov I, Camp J, Caprini C, Cavalleri A, Cerdonio M, Ciani G, Colpi M, Congedo G, Conklin J, Cornish N, Danzmann K, de Vine G, DeBra D, Dewi Freitag M, Di Fiore L, Diaz Aguilo M, Diepholz I, Dolesi R, Dotti M, Fernández Barranco G, Ferraioli L, Ferroni V, Finetti N, Fitzsimons E, Gair J, Galeazzi F, Garcia A, Gerberding O, Gesa L, Giardini D, Gibert F, Grimani C, Groot P, Guzman Cervantes F, Haiman Z, Halloin H, Heinzel G, Hewitson M, Hogan C, Holz D, Hornstrup A, Hoyland D, Hoyle C D, Hueller M, Hughes S, Jetzer P, Kalogera V, Karnesis N, Kilic M, Killow C, Klipstein W, Kochkina E, Korsakova N, Krolak A, Larson S, Lieser M, Littenberg T, Livas J, Lloro I, Mance D, Madau P, Maghami P, Mahrdt C, Marsh T, Mateos I, Mayer L, McClelland D, McKenzie K, McWilliams S, Merkowitz S, Miller C, Mitryk S, Moerschell J, Mohanty S, Monsky A, Mueller G, Müller V, Nelemans G, Nicolodi D, Nissanke S, Nofrarias M, Numata K, Ohme F, Otto M, Perreur-Lloyd M, Petiteau A, Phinney E S, Plagnol E, Pollack S, Porter E, Prat P, Preston A, Prince T, Reiche J, Richstone D, Robertson D, Rossi E M, Rosswog S, Rubbo L, Ruiter A, Sanjuan J, Sathyaprakash B S, Schlamminger S, Schutz B, Schütze D, Sesana A, Shaddock D, Shah S, Sheard B, Sopuerta C F, Spector A, Spero R, Stanga R, Stebbins R, Stede G, Steier F, Sumner T, Sun K X, Sutton A, Tanaka T, Tanner D, Thorpe I, Tröbs M, Tinto M, Tu H B, Vallisneri M, Vetrugno D, Vitale S, Volonteri M, Wand V, Wang Y, Wanner G, Ward H, Ware B, Wass P, Weber W J, Yu Y, Yunes N and Zweifel P 2013 *ArXiv e-prints* (*Preprint* 1305.5720)

[2] Belczynski K, Benacquista M and Bulik T 2010 *Astrophys. J.* **725** 816–823 (*Preprint* 0811.1602)

[3] White D J, Daw E J and Dhillon V S 2012 *VizieR Online Data Catalog* **7267** 0

[4] White D J, Daw E J and Dhillon V S 2011 *Classical and Quantum Gravity* **28** 085016 (*Preprint* 1103.0695)